Nortel Networks Troubleshooting and Optimization

RAGHO **MAHALINGAM**

Osborne/**McGraw-Hill**

Berkeley New York St. Louis San Francisco
Auckland Bogotá Hamburg London Madrid
Mexico City Milan Montreal New Delhi Panama City
Paris São Paulo Singapore Sydney
Tokyo Toronto

Osborne/**McGraw-Hill**
2600 Tenth Street
Berkeley, California 94710
U.S.A.

For information on translations or book distributors outside the U.S.A., or to arrange bulk purchase discounts for sales promotions, premiums, or fund-raisers, please contact Osborne/**McGraw-Hill** at the above address.

Nortel Networks Troubleshooting and Optimization

1234567890 AGM AGM 01987654321

ISBN 0-07-212654-X

Publisher
Brandon A. Nordin
Vice President & Associate Publisher
Scott Rogers
Acquisitions Editor
Steve Elliot
Project Editor
Janet Walden
Editorial Assistant
Alex Corona
Technical Editors
Mark Irivinti, David Callahan, Khoa Vo
Copy Editor
Judy Ziajka
Indexer
Jack Lewis
Computer Designers
Jani Beckwith, Melinda Moore Lytle
Illustrators
Robert Hansen, Michael Mueller
Cover Design
Amparo Del Rio
Series Design
Peter F. Hancik

This book was composed with Corel VENTURA™ Publisher.

Copyrights and Trademarks

"This book is a valuable reference for 2nd and 3rd tier engineers and network managers. Ragho has compiled years of experience and knowledge into a very technical, but readable resource. He de-mystifies the operation of BayRS, breaking it down into manageable components, and drawing understandable parallels to other operating systems. For a network architect and senior troubleshooter, this knowledge is integral when designing new networks and isolating elusive issues in a complex environment."

Mike Duckworth
Principal Consultant
BlueStorm Inc.

About the Author

Ragho Mahalingam is a Principal Network Engineer and a Member of the Technical Staff at Nortel Networks Global Customer Care Services, specializing in high-speed backbone architectures and redesigning traditional networks with advanced traffic engineering techniques. He has over eight years experience in supporting mission-critical data networks and troubleshooting issues with routed and switched networks. He is an active member of the IEEE Communications Society and the Network and Systems Professional Association.

This book is dedicated to my parents. —RM

AT A GLANCE

CONTENTS

FOREWORD

The Internet is changing the way the world communicates, enabling people to exchange information more effectively, work more productively and share ideas more easily. The Internet is changing the way companies compete and grow their business. New net-based business models are humbling companies that were once unconquerable. Speed rather than size has become critical and changed the competitive landscape for many companies. The opportunities to expand and grow businesses or people's capability are limited only by human imagination.

As the Internet transforms the world, the infrastructure to make it work continues to expand and grow as well. The heart of the Internet infrastructure is networking devices such as routers and switches. A massive amount of data travels the Internet today, connecting businesses, consumers, and organizations from all walks of life. The dependable access of the Internet is largely reliant on the sustained availability and reliability of the networking equipment. Nortel Networks has established itself as a leader in providing a comprehensive portfolio of networking products for the Internet—from connecting individual users to building large enterprise and service-provider networks. The commitment of Nortel Networks is to build the high-performance Internet, to make it reliable and dependable for all users.

This is the tenet behind Nortel Networks' campaign slogan: What do you want the Internet to be? Building the high performance Internet and improving the way people work, learn and play is a challenging ambition and promise. Accomplishing this task means more than simply engineering and producing the networking products—the Nortel Networks promise is to complete this portfolio with the best in class Service. As the Author calls the Internet, this 'Machine' needs to be available 24x7x365, reliable, and provide dependable service. As Networking Engineers, Managers, and Consultants, you are responsible for the vital task of providing ubiquitous Internet service and help maintain its availability at all times because companies must be open for business 24 hours a day, 365 days a year to take full advantage of the Internet. So as the world has been transformed by the Internet, so has networking service and the service professional whose job has grown in scope and complexity. The different types of networking devices, protocols, customer configurations, vendor equipment dependencies, etc., have created a challenge for service professionals to resolve complex networking problems. Plus, adding the dimension of time—speed to resolution—has created even more of a challenge.

This book is an excellent reference and service manual for administering, troubleshooting and optimizing networks built with Nortel Networks BayRS routers. The Author has complemented traditional configuration expertise with practical knowledge for making networks function better and faster—making troubleshooting network problems easier and quickly resolving impairments with the least impact. Achieving the promise of the high performance Internet and complementing it with a high availability such as 99.999% is a remarkable challenge. So, let's get to work and make the Internet work, enabling customers to answer: What do you want the Internet to be?

Matt Kochan
Vice President, Global Customer Care Services
Nortel Networks

ACKNOWLEDGMENTS

This book is a product of ultimate teamwork. Numerous people have contributed to the success of this book in various stages, from conception to publication. First, I must thank Steven Elliot for putting forth the faith and effort in this project, without whose first step this book would have never come to fruition; and to Danny Shurett who put me in touch with Steven. I would like to thank (in alphabetical order) Fitzroy Brown, David Callahan, Tien Do, Mike Duckworth, Mark Irivinti, Bridget O'Rourke, Brian Revaz, and Khoa Vo for their invaluable reading, editing, and comprehensive input with this edition. My special thanks go to Kevin Duffy for his expertise with drawings and for providing many hardware figures used in this book; to John Manaras, whose prompt and effective approval allowed this book to progress without any legal wrangling; and a particular mention of appreciation to James Knapp for providing his insight and valuable advice for publishing the book.

My appreciation also goes to the Nortel Networks Global Customer Care Services, whose support has allowed me to capitalize on laboratory resources and network simulations to produce a practical book. In particular, I would like to thank Matt Kochan for graciously authoring the foreword; Edward Boulter and Philip Kubat for their support; and my unfailing gratitude to Anthony Giovino, who has been an excellent manager, a constant source of encouragement, and a trusting friend. This acknowledgement would not be complete without my appreciation for the numerous engineers, consultants, and customers with whom I have worked during the past several years, and who have made network design and troubleshooting an excellent source of challenge and learning.

The folks at Osborne/McGraw-Hill deserve a thumbs-up for their contribution as well; special thanks to Franny Kelly, who has been the go-to person for all my arcane questions and for providing a new light into the realm of publishing; to David Zielonka, Peter Hancik, Jennifer Perillo, Lisa Bandini, Jean Butterfield, Alexander Corona, Hector De Jean, Jody McKenzie, Beth Young, and William Chan for their roles in putting the book pieces together; to Janet Walden and Madhu Prasher for their scrupulous project management and pleasant approach; to Judy Ziajka for her meticulous copyedits; and to many others who have been involved with the editing and marketing of this book. My appreciation goes to World Gym of Nashua, for providing an environment to keep a sound body after long workdays. Plenty of thanks to Ammu and Dylan for simply being the best niece and nephew, for being there and being someone to look forward to seeing every week, and for getting me hooked on Mario Kart 64.

Last, but the most important, my thanks and appreciation to Kierstin who has been wonderfully supportive of this project; for providing the encouragement and excitement on the weary days to keep going, for putting up with my incredulous schedule and sacrificing countless hours, and for being there when I needed her most; for her dependable advice, fine dining, great laughs, friendship, and love.

INTRODUCTION

There is always a saying or two about the best thing since sliced bread. Many IT and Networking professionals will say that the Internet fits this bill. The Internet has ushered in a new era of communications revolution. Networking is undoubtedly one of the most important advances made in recent decades.

With the advancements in computer networking, the complexity of networking has grown from simple taps into the coaxial ethernet to multilayered protocols and a multitude of abstractions. The demand for networking has given rise to multiprotocol networks, which in turn has givben rise to a new type of professional: the Network Engineer. Networking-savvy individuals, whether they may be addressed by the title of Network Engineer, Architect, Consultant, or some variant thereof, have been in more increasing demand from the labor market in recent years. The technical demands of complex, interwoven networks are immense; designing these vast networks is an intricate task, as is the task of maintaining them.

Networks constantly evolve and change to meet the ever-growing demands of consumer and business traffic. With these come the growing pains of network troubles and uncharacteristic 'paranormal' phenomena, which require an even more complex knowledge of and expertise in the inner-workings of various networking layers. Abstractions and interfaces at various layers have made the technology simpler, yet the mission of network analysis and troubleshooting requires an aptitude beyond that which is generally applied with network design. Networking itself is a technology and a science—network troubleshooting is a scientific art. There are no established methodologies for network troubleshooting, but the underlying networking principles are well known. Different methodologies have been used in the past, one of which relies on working one's way through the layers of the OSI model.

About Network Troubleshooting

In the past several years, working with numerous network engineers, architects, consultants, and customers alike, I have seen that those who are involved frequently with troubleshooting are very resourceful and practical experts in networking. They have a deep exposure to networking principles; even more useful is their experience with practical network phenomena, dynamic network changes, upgrades, and all other uncommon issues that might appear long after the network has been installed and certified for deployment. Each network has its own criteria and design plans, sometimes leading to idiosyncrasies that are unique to the installation, which can be looked at in both positive and negative ways.

The first step in managing and troubleshooting your network is to *know* your network. What are the critical components of your network? What are the redundant interfaces and circuits provisioned on a core router? What are the weak points? On which key backbone links must performance be monitored continuously? Knowing your network comes with exposure and experience. Troubleshooting and resolving network issues generally provide a better understanding of your network, yet a proactive approach in analyzing and understanding the network is useful in resolving issues quickly.

Is network troubleshooting that much of an important task? Yes. Surely, you are browsing this preface or have bought this book because you believe in some aspect of troubleshooting. A small configuration error can have a drastic impact on the network. Reacting quickly and applying troubleshooting techniques can help re-establish network connectivity with minimal impact. The goal of this book is to teach practical methods for troubleshooting problems at the physical, datalink, and network layers. Most networking technologies and protocols are common among different products—Ethernet is Ethernet, and similarly interoperable IP implementations work the same. Most similarities end there because each implementation may use different techniques to arrive at the same solution. This book orients itself toward BayRS routers, where the fundamental knowledge of how the product operates is most useful in troubleshooting them.

What's in this Book?

This book is all about understanding and troubleshooting BayRS networks. It provides the knowledge about the available tools and tried techniques that will help you become better, if not adept, at networking and network troubleshooting in particular. I embarked on this project primarily because troubleshooting has not been a key publishing aspect—books on protocols, technologies, and products are rampant. Yet there are few books on the market that tell you what to do when something goes wrong, or even more challenging, when something doesn't work the way you want it to.

While this book is all about BayRS, it doesn't contain everything there is to BayRS. With the support for innumerable protocols, technologies, and services, a comprehensive and detailed coverage would have spanned several volumes and taken years to complete. I have strived to provide a broad, yet in-depth coverage on commonly used technologies and protocols. My intent has been to provide an easily readable text without sacrificing accuracy in detail. The book is not a primer on the protocols or technologies covered—there are existing books on the market that provide detailed coverage on individual topics. Excellent references are provided at the end of each chapter for readers who are more closely involved with certain protocols. In the objective of composing readable text, I have made simple generalizations that protocol- and technology-savvy experts might occasionally find dubious. This is not an intentional attempt to provide erroneous information, but brief technology overviews have helped segue into the more complex troubleshooting aspects. Available references on individual topics should be used to obtain details on exceptions and intrinsics.

The dynamism of networking technology has forced me to use a reference software version, for which I have used BayRS 14.00. Many techniques and commands are equally applicable with earlier versions, although some new commands may be unavailable or minor differences in syntax may apply. To avoid the risk of packing the book with notes and cautions, I have provided important differences as inline notes only when necessary.

How to Read this Book

One size fits all. It is a good theme (at least when buying socks :-)), which is partly the notion I have applied to the outline of this book. The reading audience is diverse, from office managers who administer the network part-time to full-time network architects. With the intent on making this first edition a practical one, I have tended to focus on the needs of a fledgling network administrator as well as on those of the advanced networking professional. The book is broken into eight chapters:

- **Chapter 1** provides a broad overview of all BayRS routers, from the Access family to the Backbone family. The recently introduced Passport 2430 and 5430 platforms have been omitted because the book had been partway complete when they were released along with BayRS 14.10. This chapter is intended to provide an overview of the software and hardware architectures at a high level.
- **Chapter 2** covers BayRS basics. The GAME operating system is discussed to outline how BayRS defines certain concepts and how the interaction of a symmetric multi- processing system allows rapid forwarding. The essentials of the Simple Network Management Protocol (SNMP) and the Management Information Base (MIB) are dis- cussed to lead into the Technician Interface (TI) and the Bay Command Console (BCC).
- **Chapter 3** focuses on router administration at the nodal level as well as on performing upgrades. The configuration and provision of system services are described, as are administrative commands for re-initialization and scheduled boots. Software upgrades are a part of everyday life and are discussed in detail, including boot and diag PROM upgrades.
- **Chapter 4** begins the troubleshooting theme by describing the basic techniques. Basic data gathering and symptom analysis are essential. First, general boot, hardware, and operational issues are discussed. Next, we talk about runtime operational issues such as memory problems and subsystem faults. Later in the chapter we step into the first layer of the OSI model, the physical layer. Resolving media and cabling issues, and performing loopbacks are covered.
- **Chapter 5** delves into the troubleshooting of switched (dial) services and data link issues. Various dial solutions and various troubleshooting approaches are discussed first. Then, we transition into discussing WAN protocols such as PPP, Frame Relay, Bay Standard HDLC, SMDS, and ATM DXI. WAN protocols are one of the key troubleshooting interests of many readers because routers are usually deployed at the core, interconnecting numerous sites over the wide area. ATM applications for both the local area and wide area are also covered in this chapter.
- **Chapter 6** moves further up the OSI model by focusing on network layer protocols and routing protocols. The main emphasis is on the Internet Protocol (IP), followed by troubleshooting techniques for Internetwork Packet Exchange

(IPX). Routing protocols such as RIP, OSPF, and BGP are discussed. The topic of using Packet Capture (PCAP) is also discussed, which is sometimes necessary in troubleshooting both datalink and network layer issues.

- **Chapter 7** discusses an important concern for network managers and architects— security. The identification of security threats, their symptoms, and security measures such as firewalls, traffic filters, encryption (WEP, IPSec), and access policies are covered. The troubleshooting of security protocols is also discussed.
- **Chapter 8** is dedicated to performance management, tuning, and optimization. We discuss methods to baseline router, protocol, and interface statistics. Next, we talk about common methods to optimize the operating system, protocol, and interface-level performance.

The chapters are somewhat loosely organized into a sequential order. However, you can easily skip chapters to focus on something that specifically interests you. Wherever necessary, I have provided a cross-reference to sections in other chapters. Following the chapters are Appendices A through H, in which I have presented frequently referenced information:

- **Appendix A** lists the software suites, their availability on different platforms, and supported application groups. The current BayRS release policy is also reprinted with permission.
- **Appendix B** provides comprehensive coverage of all visual status indicators on all platforms. All status indicators are illustrated with images, including front panel and back panel LEDs, processor LEDs, link/net module, and interface LEDs.
- **Appendix C** summarizes modem configuration parameters for all platforms and the required console/modem cables.
- **Appendix D** provides a reference to Technician Interface commands and syntax.
- **Appendix E** details all available event log entities, their entity codes, and a description of where they are used.
- **Appendix F** summarizes all available TI and BCC show commands for obtaining statistics, management, and other configuration data during runtime operation.
- **Appendix G** details the contents of corporate suite images by listing all component applications and drivers, along with their applicable dependencies and their main function.
- **Appendix H** provides a glossary reference for computing and networking acronyms.

Criticism and Feedback

I have always been excited about writing papers, creating seminars, and giving lectures. Still, this is the largest written work that I have ever undertaken. There have been many exciting and challenging aspects of this project—the final result has been more than worthwhile and it has satisfied most of my goals. It has allowed me to discuss troubleshooting BayRS routers overall and present the discussion in a concise fashion for readers from all walks of networking. This work was based on topical interest expressed by many networking experts, and I hope it has addressed some of your interests. For those readers who might have wished to see a different type of book, I apologize, but please do send me your feedback.

I have put forth an inordinate amount of time and effort in eliminating technical errors and inconsistencies, but I am certain that there are glitches lurking in some pages. A current list of errata and updates are posted to the book's website at http://www.mahalingam.com/.

Readers who are interested in certain topics are welcome to mail me; perhaps that may form the basis of a future edition or a new book. I would like to hear from readers who find errors or topical inconsistencies, so that these may be corrected in a future edition.

Ragho Mahalingam
Nashua, New Hampshire
ragho@mahalingam.com

CHAPTER 1

Overview of BayRS Routers

...it is much easier to have some vague notion about any subject, no matter what, than to arrive at the real truth about a single question, however simple that may be.

—René Descartes,
Rules for the Direction of the Mind

Today's networks represent one of the most marvelous technological achievements in the past two decades. They transport immense amounts of data traffic that has taken business and consumer interaction to new levels. No longer is the network a simple cable connection, but it forms the backbone of a new paradigm for the way people work, learn, and play. The network has become the Machine itself, and this Machine is a ubiquitous collection of networking devices.

BayRS routers have been at the forefront of this communications revolution. As the quotation at the beginning of this chapter succinctly emphasizes, it is necessary to have a basic and essential understanding of BayRS routers. This chapter provides this important overview of BayRS routers, which are widely used for network routing in corporate networks and the Internet. We first cover a technical outline of the available hardware architectures, followed by the system and networking services of the BayRS software, and subsequently many general troubleshooting resources.

BayRS FEATURES AND HISTORY

Networks are deployed in all walks of life, in industries such as healthcare, telecommunications, banking and finance, manufacturing, retail sales, publishing, entertainment, computing, and energy, just to name a few. Although their requirements vary in different environments, the use of internetworks in these environments over the past decade has led to the delineation of several rigorous basic requirements and even more stringent demands for the future. These requirements include reliability, availability, scalability, quality of service (QoS), performance, manageability, and security. BayRS has consistently evolved to meet the growing needs of networks. The following sections discuss the capabilities and features of BayRS.

Reliability

Almost all businesses use their networks for conducting business transactions every day, and most of these businesses operate around the world and around the clock. Network reliability is key to these deployments, and glitches or large failures either directly or indirectly affect the operation and performance of the entire business. Unexpected or frequent outages on the part of key components may make networks less reliable and force businesses to conceive alternative plans. In some cases, network reliability is so vital to businesses that they cannot conduct business without the network.

Network reliability is ensured by the reliability of the constituent network devices themselves. Router hardware that runs BayRS has been designed with reliability in mind and have a high mean time between failures (MTBF). Such hardware is resilient to environmental factors, such as thermal fluctuations, fluid spills, and impact vibrations. Other factors such as power fluctuations are handled well by power supplies that protect against surges and regulate voltage.

Availability

Availability complements reliability. Certain unfortunate incidents, including flood damage, lightning strikes, and fatigue failure, can cause one or more network devices to stop functioning properly. In events such as this, the network operation should be unaffected, and users and business transactions must continue smoothly.

This need is ensured by redundancy features that are designed into the software, hardware, and also the network topology. BayRS software is able to automatically isolate failed components and reroute data traffic dynamically using new optimal paths. Software components such as protocols are automatically isolated and react to the unavailability of another component. Hardware components such as processors in the BayRS symmetric multiprocessing architecture are also able to isolate failures and react appropriately.

Corporate networks occasionally need to undergo changes, which must be made while the network is fully operational because 24x7x365 availability is required. Dynamic reconfiguration is often performed in BayRS routers, and its effects are always restricted to the component being changed—such as one port, protocol, slot, and so on. This minimizes the effect of the change to the component as the software concurrently reroutes traffic around the briefly disrupted component. The software and hardware architectures are such that there is no *single* point of failure. Redundancy in both software and hardware elements makes network availability high.

Availability is also ensured by another form of redundancy, using backup circuits and services. Backup circuits, such as those using ISDN or analog lines, can provide redundancy and augment availability for conditions that are external to the router. For example, a temporary outage on a DS1 (Digital Service 1) circuit to the main office may cause a significant disruption. However, BayRS backup circuits automatically react to the downed circuit and bring up a dial circuit to reroute around the outage.

Scalability

The needs of future networks will be much more demanding than those of current networks. However, it is today's networks that will need to evolve into the next generation. Corporations invest in networking gear and services with the expectation that they can be scaled to meet future demands. The goal of a successful business is to grow revenue profitably and efficiently. Revenue and earnings growth often result in proportional growth—and in many cases, exponential growth—in networking needs. The flexibility and capability of BayRS routers allows them both in hardware and software realms to scale effectively.

For example, consider Mahalingam Foods, Inc., which has several regional offices converging in a Backbone Node at the central headquarters office. As part of its Internet strategy, Mahalingam Foods is preparing to connect to the Internet and sell its products over the Web. The Backbone Node router can be scaled from the current 600 routes in the corporate network to a full feed of 60,000 or more routes from the Internet using BGP (Border Gateway Protocol). BayRS has a rich feature set of scalable software protocols and hardware capability to support extensive scalability. Corporate networks using 64 Kbps and T1 leased lines can easily evolve to T3 and OC3 speeds using the same BayRS platform.

Quality of Service

Today's network traffic is a convergence of voice, video, and data. This requires higher bandwidth and better quality of service than legacy networks. Traffic such as voice and certain video streams require real-time handling, which means that the network should have the least delays on a consistent basis. BayRS supports various well-known standards and features to provide a consistent quality of service (QoS). This includes protocols such as ST2, IntServ, and DiffServ. Resource Reservation Protocol (RSVP) is an IETF standard that enables multimedia applications to reserve bandwidth dynamically on an end-to-end basis to ensure a reliable quality with guaranteed throughput. Traffic shaping is another common feature; it is used to effectively alleviate congestion by shaping the flow of traffic through heavily utilized and congested lines. DiffServ (using TOS) and IntServ (using RSVP) are some of the recently developed and evolving standards that assist in ensuring QoS for traffic that requests it.

Performance

Unlike the previous concepts, performance is gauged by a composite of several measures. In the context of this book, performance is, in essence, a combined metric of throughput, reliability, availability, and service. BayRS hardware and software both have been measured to provide high performance in a wide range of implementations. The Backbone Concentrator Node delivers throughput of over 1 million packets per second using several high-density interfaces such as FDDI, T3, OC3, and Fast and Gigabit Ethernet. High-throughput processor modules such as the ATM Routing Engine (ARE) are capable of forwarding over 176,000 packets per second. The Fast Routing Engine-4 (FRE4), another high-density card, can route 400,000 packets per second on Fast Ethernet and Gigabit Ethernet ports. These processor cards perform well in a highly reliable fashion, sustaining uniform throughput and maintaining consistent levels of availability and service. BayRS software, similarly, has been designed and optimized with high-performance metrics.

Manageability

Networks involve machines containing several hundreds or thousands of routers, and managing them should not add to their complexity. Ease of management is brought into an elaborate network by the support and deployment of one or many network management systems. BayRS routers are fully manageable using the Simple Network Management Protocol (SNMP). Numerous applications, such as Network Management Systems (NMS) and Site Manager, can be used to manage BayRS. BayRS also provides the Technician Interface (TI), which can be accessed for in-band (in-traffic) management using telnet and for out-of-band management using a console interface. The Technician Interface features configuration management, monitoring, and diagnostics. NMS applications such as Optivity, HP OpenView, and Sun SunNet Manager can be used for configuration management, accounting, performance management, and fault management.

Available in BayRS releases 12.00 and above is an embedded web server that renders certain management features, particularly monitoring capabilities. Using a Hypertext Transfer Protocol (HTTP) browser such as Netscape or Internet Explorer, tasks such as monitoring router operation, viewing statistics, and basic troubleshooting can be accomplished. Also, in recent releases, the Bay Command Console (BCC) is available as a superset of the TI for comprehensive configuration management and statistics.

Security

Data traffic in recent times has been subject to attacks by malicious hackers expecting to compromise the integrity and use of the network. Because of this, security, privacy, confidentiality, and ethics (SPCE) in the transport and handling of data has become paramount. Networks must guarantee SPCE for traffic requiring it. For example, highly confidential email such as executive briefings and transfer of personnel data should be kept secure so that it is received and processed by the intended recipients only.

BayRS supports several aspects of SPCE to maintain data and network integrity. BayRS uses the WAN Encryption Protocol (WEP), which features DES-40 (Data Encryption Standard) and –56 bit encryption for all traffic. Encryption is also available in protocols such as IP Security (IPSec), which uses industry-standard algorithms such as DES (40- and 56-bit) and 3DES. Authentication of data in less stringent applications is provided by algorithms such as Message Digest-5 (MD5) and Secure Hashing Algorithm (SHA-1).

Another aspect of security relates to access through and to the router itself. Strict networks use carefully designed traffic filters that allow only authorized and legitimate traffic to pass through the router. The modern firewall is a recently developed product that uses several hundreds or even thousands of filters to validate traffic on a packet-by-packet basis. Apart from the traffic filters, BayRS also incorporates the Check Point Firewall-1 product, which provides a scalable firewall and traffic filter solution.

BayRS SPCE also allows authentication of users connecting to the router, using a local user list or using an authentication database such as RADIUS. SecurID is another authentication mechanism available to validate user logins. Configuration changes are recorded in the event log and can also be audited on Site Manager platforms. Overall, BayRS routers provide a comprehensive security blanket to prevent and deter violations of policy and access.

Software Suites

BayRS is released in several software suites, each of which contains a standard set of features. These features can be tailored within each suite as appropriate to the router's environment. Software suites on different platforms are designed to be appropriate for their use, and they vary accordingly. Table 1-1 lists the software suites and their availability. Components of the software suites are detailed in Appendix A, which also describes the software release policy.

Protocols

To support a wide variety of network infrastructures and requirements, BayRS supports a wide range of protocols, of which the TCP/IP suite is most commonly used in corporate networks. BayRS supports the following feature sets and protocols:

- **Network Layer Protocols**
  Internet Protocol (IP)
  Open Systems Interconnect (OSI)
  Novell Internetwork Packet eXchange (IPX)
  DECnet Phase IV and V
  Banyan VINES
  Xerox XNS
  Appletalk Phase 1 and 2
- **IBM SNA Integration**
  Source Route Bridging
  LAN Network Management Agent
  Data Link Switching (DLSw)
  SDLC to LLC2 Conversion
  BSC Pass-Through
  Transparent Synchronous Pass-Through
  Advanced Peer-to-Peer Networking (APPN)

	AN	ARN	ASN	BN	System 5000
LAN			√	√	√
WAN			√	√	√
System			√	√	√
IP Access	√	√			
Remote Office	√	√			
Corporate	√	√	√	√	√
ATM				√	
Virtual Network Router (VNR)				√	√

Table 1-1. Software Suites and Platform Availability

- **Bridging**
  Transparent (Ethernet and FDDI)
  Translation (Ethernet, Token Ring, and FDDI)
  Native-Mode LAN
- **Wide Area Networking**
  HDLC Encapsulation (Bay/Wellfleet Standard)
  Point-to-Point Protocol (PPP)
  Frame Relay
  Switched Multi-megabit Data Service (SMDS)
  X.25
  ATM DXI
  ISDN BRI and PRI
- **Dial-Up Services**
  Dial-on-Demand
  Dial Backup
  Bandwidth-on-Demand
  Hot Standby
- **Traffic Management**
  Data Compression (PPP, X.25, and Frame Relay)
  Traffic Prioritization
  Uniform Traffic Filters
  Multiline Circuits
- **Node Management**
  EZ Install/EZ Update
  Dynamic Loader
  Remote Monitoring

Appendix A provides comprehensive discussion of the protocols supported in each of these categories.

BayRS History

BayRS obtains its name from Bay Networks, formed by the unification of WellFleet Communications and SynOptics. In 1998, Bay Networks was acquired by Northern Telecom, and subsequently called Nortel Networks.

The original Nortel Networks routers, the 5 series, ran protocol suites on top of a third-party kernel called VRTX. The 5 series software mostly ran on Versa Module Europa–based (VME-based) routers. In the next, evolving generation, Nortel Networks developed both the software and hardware from the ground up to meet the needs of today's aggressive and demanding networks. Internally called the Harpoon Project, this represents the Gate Access Management Entity (GAME) operating system and protocol suites in the software realm. Newer, faster, and more reliable hardware was also developed and categorized into Access and Backbone router families. VME-based routers are still in existence today, and the GAME operating system has been designed to fully support this legacy hardware. BayRS has a modular, layered software architecture based on the GAME operating system.

Usually, BayRS itself is treated as a complete operating system whose architecture is succinctly illustrated in Figure 1-1. The underlying GAME OS manages all system resources, such as the CPU, memory, backbone, and interprocess and interprocessor communications in a symmetrical multiprocessing hardware architecture. Modular device drivers operate numerous media interface modules and interconnect with the Data Path (DP) service to route and bridge multiprotocol traffic. Higher-layer services such as routing protocols, routing table management, and transport and session protocols function in a distributed form on multiple processors. Finally, distributed system management is provided by systemwide SNMP and MIB services, along with distributed table management services for routing and service table maintenance.

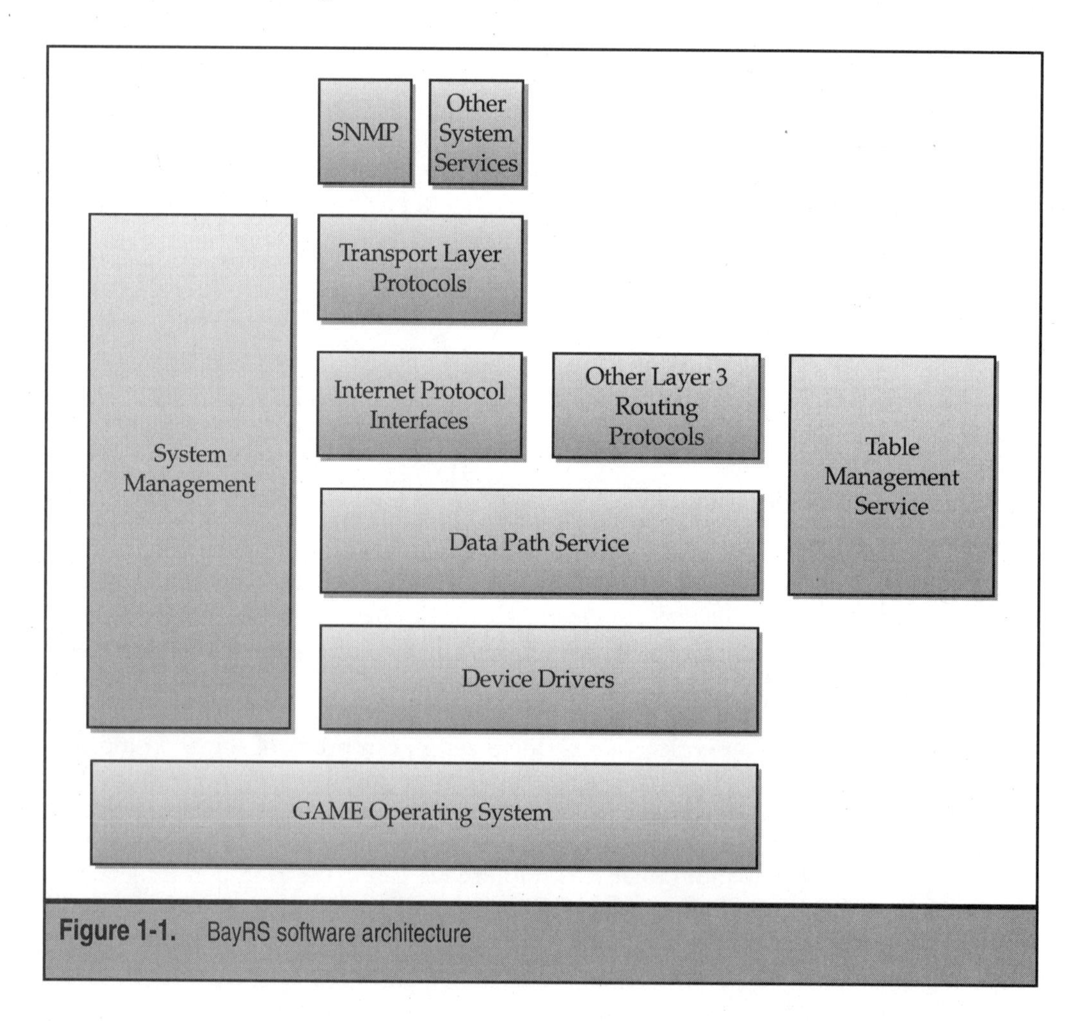

Figure 1-1. BayRS software architecture

VME ROUTERS

VME routers are aptly named because they use the Versa Module Europa bus. Based on the Eurocard standard, which was the most popular in the industry, they support many network layer protocols. These routers consist of the VME bus present in the chassis midplane, which mates with cards called the processor modules and link modules. The processor modules operate symmetrically, performing route calculation and forwarding functions, while the link modules provide media controllers and PHY/PMD services.

The processor modules are called the Advanced Communications Engine (ACE). The ACE contains dual Motorola 68020 microprocessors, one operating at 25 MHz for protocol processing and another operating at 12.5 MHz for DMA and messaging. Two versions were manufactured, one with 5 MB and another with 2 MB of memory. Several link modules were made supporting many media, including synchronous, Ethernet, Token Ring, FDDI, and T1. One link module mates with one processor module, and a link interface without an ACE processor is called a stand-alone unit and provides spare functionality.

The VME-based routers are no longer in production, although they are still operational today. They are mentioned here for completeness and are not discussed in detail.

ACCESS FAMILY

The Access family provides a series of routers that are typically used in small offices, branch sites, and smaller regional offices. The Access Node (AN) and Access Node Hub (ANH) are the simplest in this family and typically come in fixed configurations with an option to add an expansion module. The Advanced Remote Node (ARN) provides a higher level of implementation by providing a base unit with different types, two adapter slots, and one parallel expansion module. Bridging the gap between the Access and Backbone families is the Access Stack Node (ASN). It is a stackable unit based on the Fast Routing Engine (FRE) platform, which forms the basis of the Backbone architecture. ASNs provide four net module slots to support a variety of interface types. Multiple ASNs stacked together can operate as a single unit, providing higher throughput to regional offices and larger local offices. All Access family routers run the BayRS operating system and support the respective protocol suites.

Access Node and Access Node Hub

The AN and ANH share a common motherboard design based on the Motorola MC 68360 (Quad Integrated Communications Controller, or QUICC) microprocessor, although they are physically different. The AN contains at the minimum an Ethernet or Token Ring interface, or both, along with two serial interfaces. In the ANH, the Ethernet interface is replaced by an 8-port or 12-port hub. An expansion slot is available, where an additional Ethernet, Token Ring, serial, CSU/DSU, fractional T1 (or E1), and ISDN BRI interface can be installed, except that Ethernet base units cannot be expanded to install a Token Ring interface. Finally, an Ethernet data collection module (DCM) can be installed in most configurations to gather remote monitoring (RMON) statistics. All ANs and ANHs provide a console port for out-of-band management of the router.

The Access family routers, including the AN and ANH, have a unique panel of LEDs to indicate the runtime operational status of the device and components. LEDs and their status indicators for all platforms are described in Appendix B.

The AN has been reengineered to contain a real-time clock (RTC) and to support a removable PCMCIA flash card, whose function was previously provided by an internal SIMM flash installed directly on the motherboard. This enables easier upgrading of flash capacity and file system management. The PCMCIA flash eliminates complexity because the files can be copied to the new flash without the need to open the unit or interrupt network operation. To differentiate the new from the original AN routers, the new AN is called the AN2, and the original AN is called the AN1. AN2s are easily distinguishable by the presence of a PCMCIA slot in the front. Figure 1-2 shows front and rear images of AN, ANH-8, and ANH-12 routers.

Advanced Remote Node

The ARN, or Advanced Remote Node, router is a scalable and expandable router suited for contemporary access applications and for tomorrow's high-growth networks. It features a Motorola MC68360 (QUICC) and a MC68040 microprocessor, and three types of base units are available: Ethernet, Fast Ethernet, and Token Ring. An expansion module is available, which provides a choice of Ethernet and serial, Token Ring and serial, or simply seven serial interfaces. Two optional WAN adapter modules can also be installed; supported modules include serial, V.34 analog modem, 56/64K DSU/CSU, ISDN BRI (U and S/T), and fractional T1 (or E1). The modular design of the unit allows for future upgrades to new adapter and expansion modules featuring newer access technologies. As with the AN, both Ethernet base units and Ethernet expansion modules can be installed with a DCM for gathering RMON statistics. Figure 1-3 shows an example of an Ethernet base unit containing adapter modules and an expansion module.

The ARN provides two out-of-band accesses for management: a built-in V.34 console modem or a console/modem port that is factory installed. The console/modem port provides two connectors, one marked "console" and another marked "modem," of which only one can be used at any time.

Access Stack Node

The most expandable and scalable router in the Access family is the ASN. It provides large levels of growth for large offices and regional branch locations. The distinguishing feature of the ASN is the ability to stack multiple units for operation and management as a single router. Powered by a Motorola MC68040 microprocessor, the ASN motherboard is identical to the Fast Routing Engine-2 (FRE-2) architecture, which is covered in the "Backbone Family" section later in this chapter. Figure 1-4 shows an ASN router.

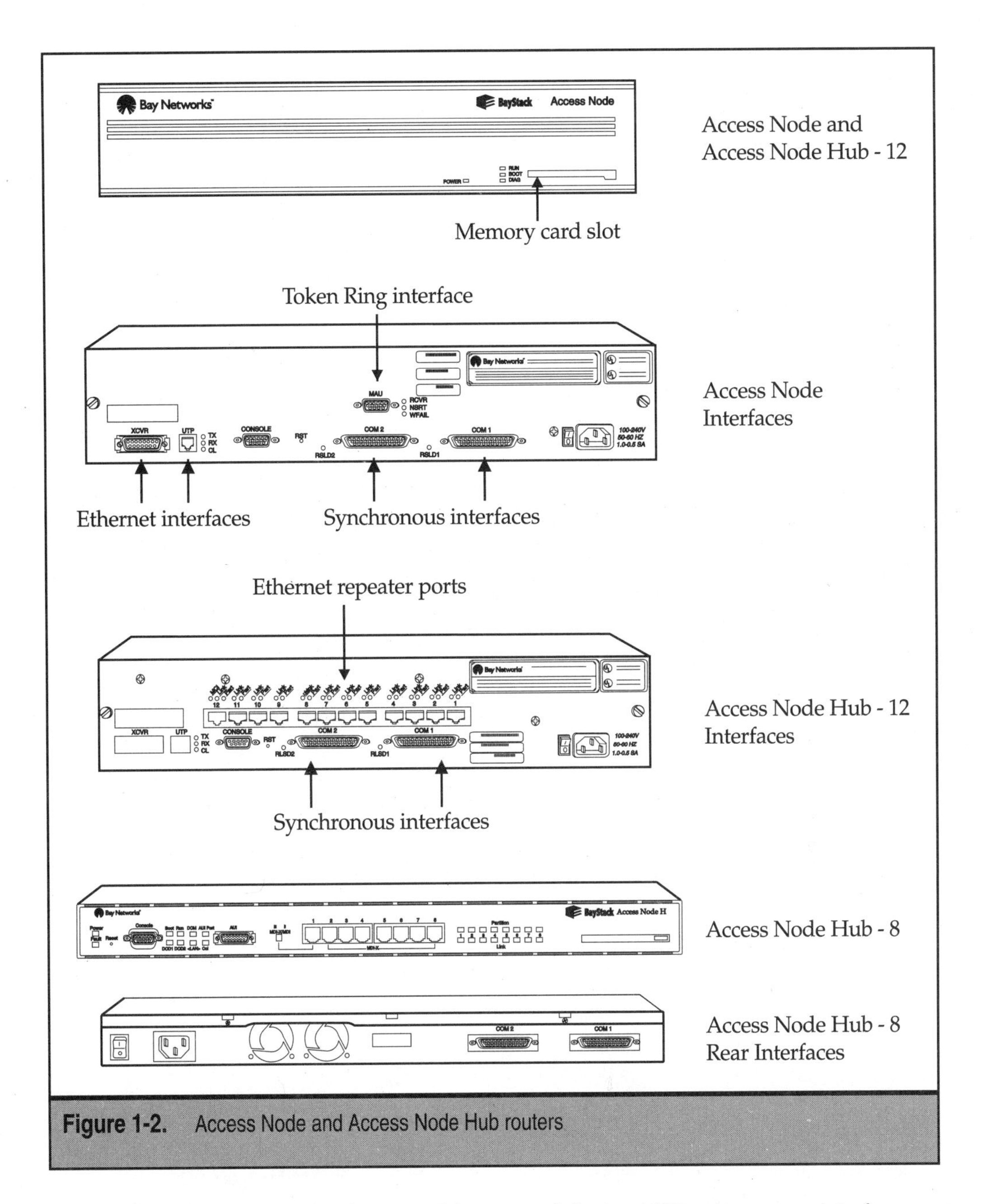

Figure 1-2. Access Node and Access Node Hub routers

Unique to the ASN is the concept of the *net module*. An ASN unit can consist of a combination of four net modules, which can support Ethernet, Fast Ethernet, Token Ring, FDDI, MCT1/MCE1, ISDN BRI/PRI, and HSSI interfaces. A special net module, called

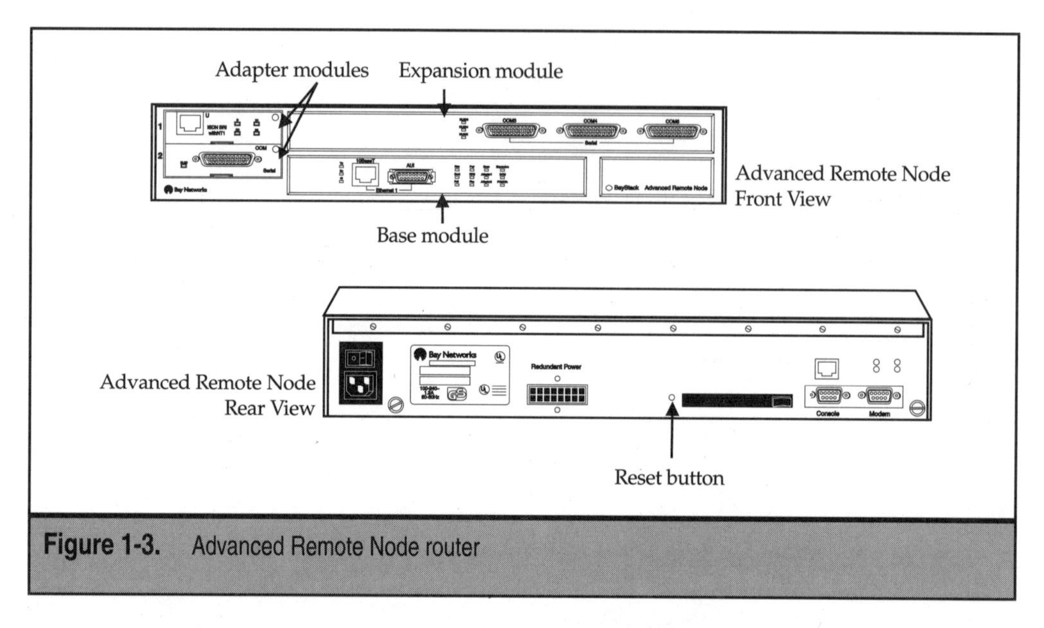

Figure 1-3. Advanced Remote Node router

the Stack Packet Exchange (SPEX), can be installed in net module slot 4 to add a unit to a stack. The SPEX module provides features similar to the Parallel Packet Express (PPX)

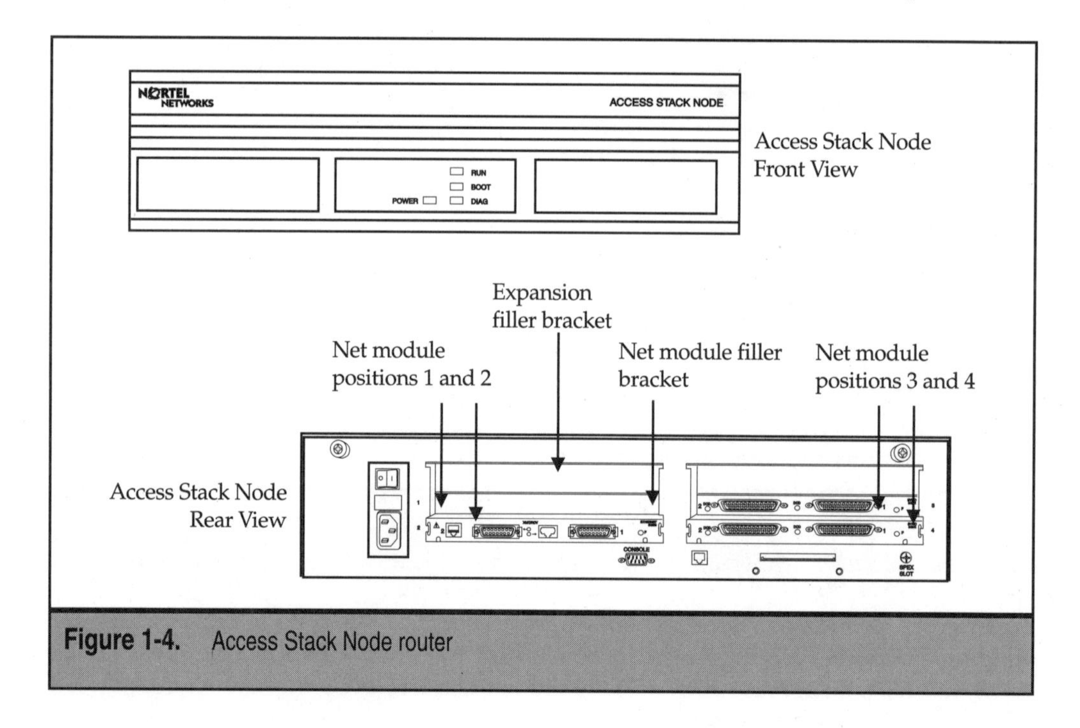

Figure 1-4. Access Stack Node router

discussed later, by providing a 160 Mbps interconnect channel between the units. It provides the functionality of a backplane bus without requiring rigid internal backplane attachments. If a unit is to be stacked, then a SPEX or SPEX-HS module must be installed in net module slot 4. It is recommended that units in a stack be powered on within 30 seconds of each other, starting with the lowest unit in the stack. Units are individually numbered using a slot selection switch in the rear that gives each a unique slot ID.

The SPEX-HS is a next-generation SPEX that provides hot swap (HS) capability for units in the stack and 256 Mbps throughput. Stacks containing original SPEX modules interconnected via SPEX cables must be powered down before a new unit is added. With the SPEX-HS net module, a new unit can be added to the stack dynamically during runtime without adversely affecting router operation. The new unit being added must be powered on only after insertion into the stack.

Figure 1-5 shows the differences between SPEX and SPEX-HS modules and connectors. A SPEX net module has two ports, labeled IN and OUT. In the unit at the bottom of the stack, the IN port is terminated, and similarly in the unit at the top, the OUT port is terminated, using the 7140 terminator. The SPEX cable (7141) is connected between the OUT port of a unit and the IN port of the unit immediately above it. In the case of the SPEX-HS net module, two ports labeled A and C are present. When only one interconnect bus is used, a single SPEX-HS cable (7166) connects all the ports labeled A. To enable dual bus configuration, another SPEX-HS cable is used to interconnect ports labeled C.

The ASN can employ an optional Fast Packet Cache (FPC), which is a bank of high-speed static RAM. This is dual-ported memory that is used to accelerate frequently accessed portions of global packet buffers. Most commonly, frame headers frequently reread by different layers and protocols are cached for faster forwarding. This optional cache is a standard component on the Backbone processor modules.

The ASN2 is an enhanced design of the original ASN (the ASN1). The redesign enables the motherboard to support a redundant power supply unit (RPSU). The Fast Packet Cache has also been redesigned for the ASN2. Although the functionality is the same, the FPC for the ASN2 cannot be installed in an ASN1, and vice versa.

BACKBONE FAMILY

The Backbone family consists of the high-capacity chassis-based routers typically deployed in central offices and enterprise core locations. Backbone nodes operate in a symmetric multiprocessing environment, providing high levels of forwarding and route calculation capacity. To achieve this, multiple processor cards called fast routing engines are deployed in individual slots. These routing engines control and operate their own link modules for LAN and WAN interfaces. In addition, they also are responsible for interoperating with other slots to provide a seamless, high-performance, single-router environment. The combination of a routing engine and a link module is called an Intelligent Link Interface (ILI).

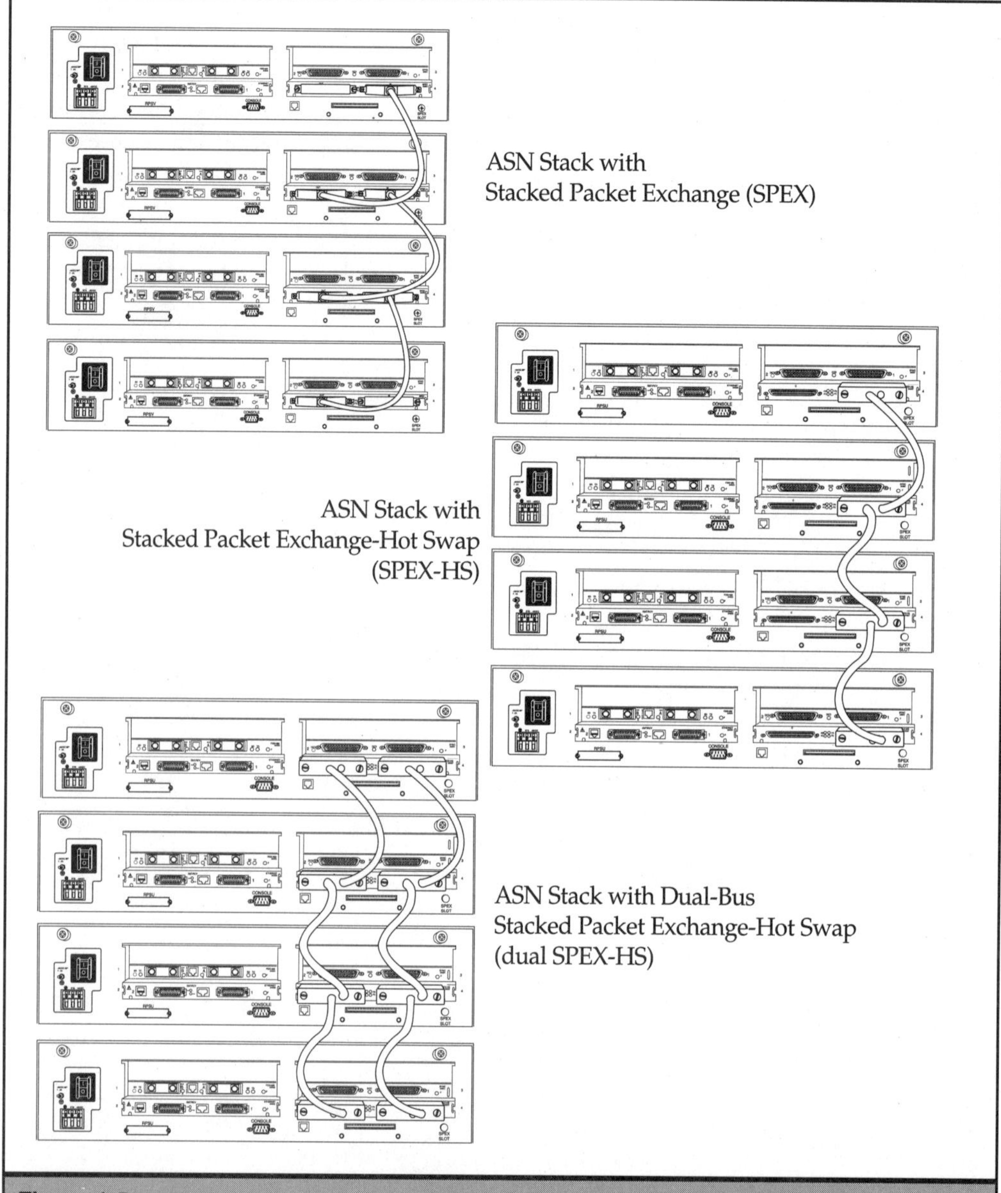

Figure 1-5. Stacking with Access Stack Node routers using SPEX, SPEX-HS, and dual SPEX-HS

Routing engines are installed in the front and link modules are inserted in the back, and both are separated by the Parallel Packet Express (PPX) midplane (see Figure 1-6). All routing engines interconnect at the high-speed PPX midplane, which is used for inter-slot communication and forwarding.

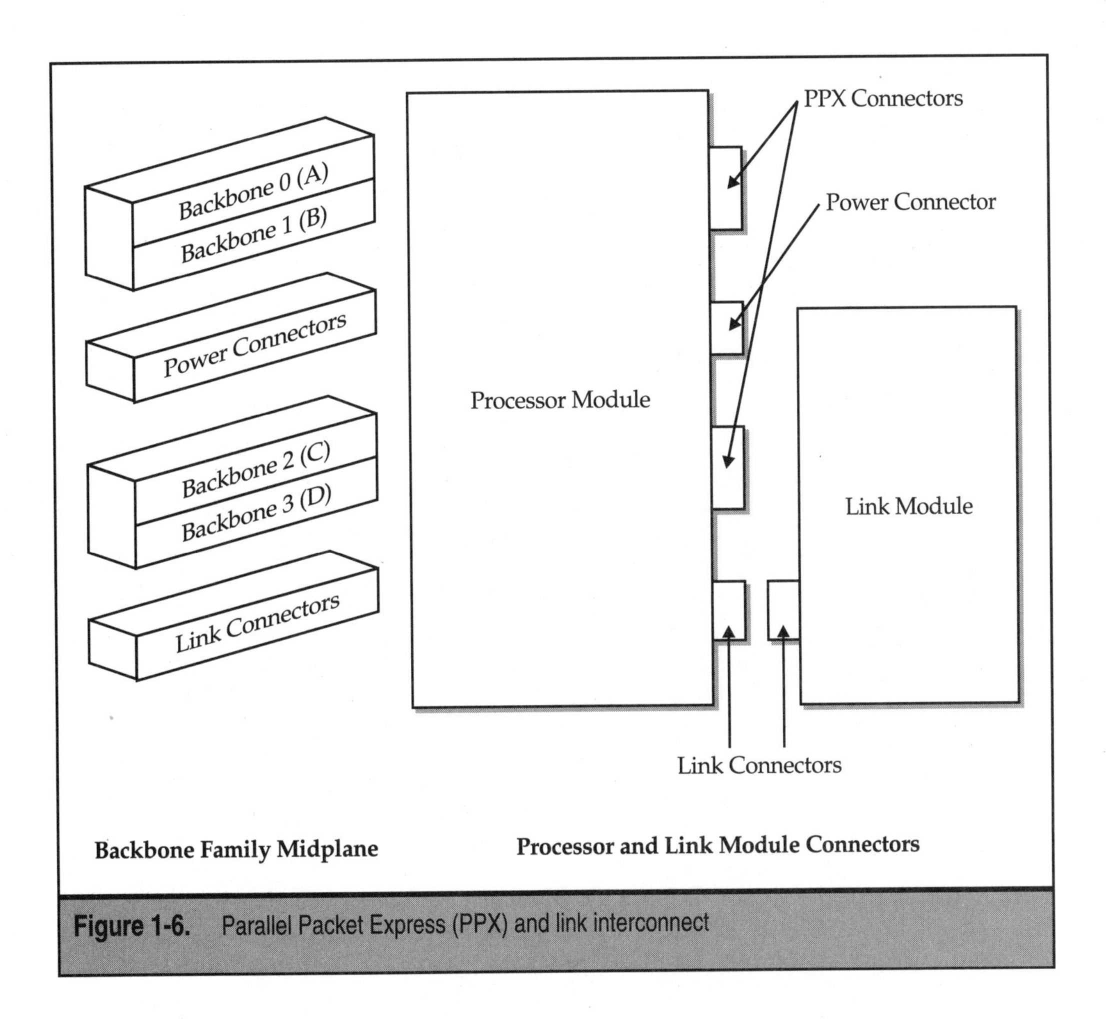

Figure 1-6. Parallel Packet Express (PPX) and link interconnect

When generally referring to a router in the Backbone router family, the name Backbone Node (BN) is usually used. Figures 1-7 and 1-8 show the BLN1, BLN2, and BCN members of the Backbone family.

As with the Access family, Backbone routers contain LED indicators on the base unit, processor, and link modules. LED status indicators for the Backbone family are described in Appendix B.

Parallel Packet Express

The PPX is a proprietary high-speed bus that was specially designed for the Backbone architecture. It provides tremendous advantages over the previous VME-based routers be-

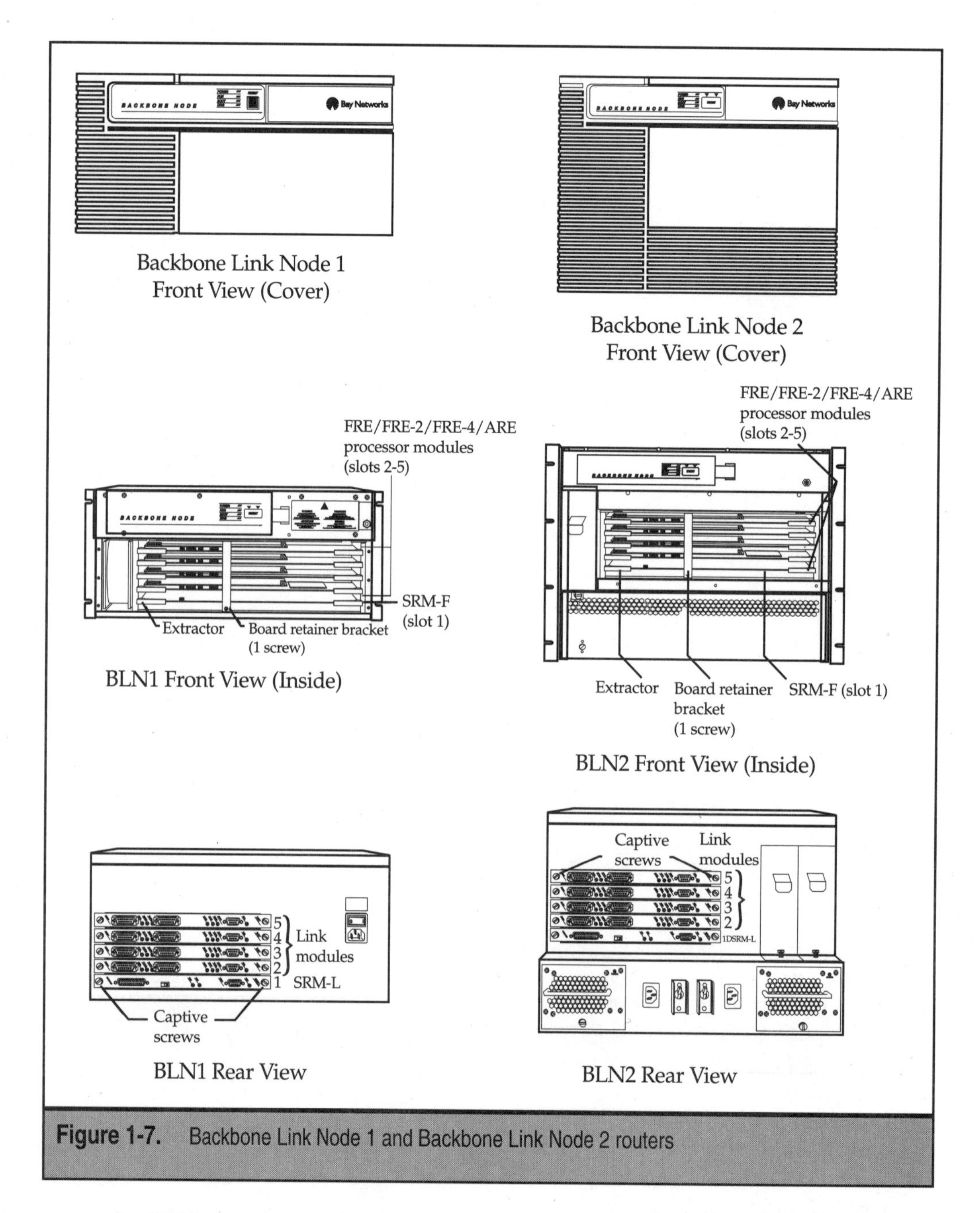

Figure 1-7. Backbone Link Node 1 and Backbone Link Node 2 routers

cause the PPX was designed with reliability and high forwarding rates in mind. It consists of four rails, commonly designated as A, B, C, and D (or 0, 1, 2, and 3). Each rail operates independent of the other and provides redundancy for both data and control

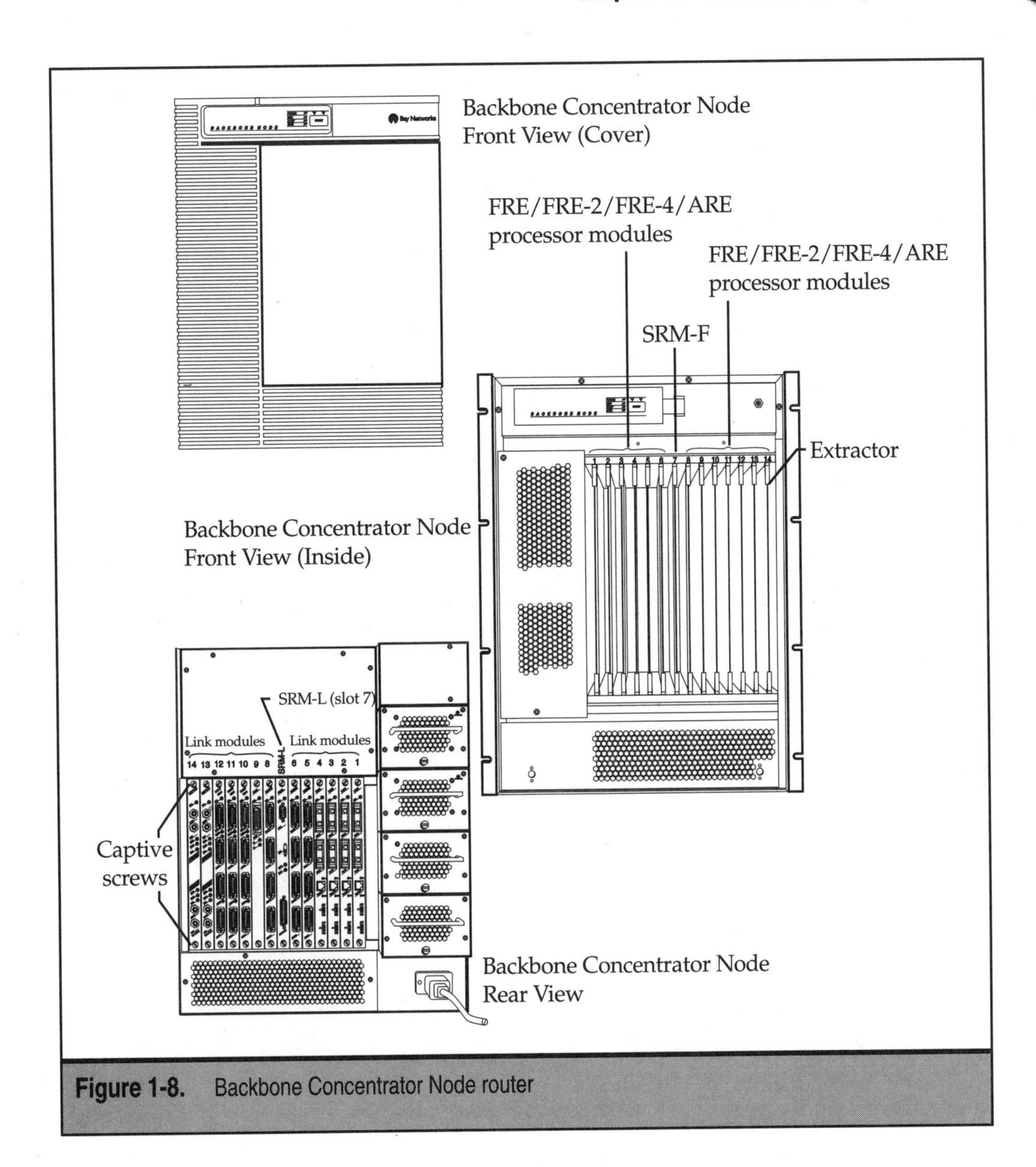

Figure 1-8. Backbone Concentrator Node router

traffic. When all rails are operational, the PPX load shares traffic across all lines. Since the PPX rails represent a bus, only one transmitter is active on any rail at a given time. However, all receivers simultaneously receive traffic on all rails. All active processor modules act as peers on the PPX. Each rail is capable of 256 Mbps throughput, which supplies an aggregate PPX throughput of 1 Gbps over the midplane.

System Resource Module

All buses require an arbitration mechanism. On the PPX, arbitration is performed under the auspices of special modules called the System Resource Modules (SRMs).

System Resource Module–Link (SRM-L)

The SRM-L is usually installed in all Backbone routers and provides arbitration over rails A and B, for a total throughput of 512 Mbps. This module is always installed because it also provides the console port attachment for out-of-band management access via the Technician Interface (TI). Since the SRM-L controls only two rails, only half of the PPX throughput is utilized when only the SRM-L is used.

System Resource Module–FRE (SRM-F)

Complementing the SRM-L, the SRM-F provides all the necessary arbitration functions on rails C and D, for an additional 512 Mbps throughput. Using both the SRM-L and SRM-F, the full PPX capacity of 1 Gbps is available on all four rails. Backbone Nodes fully populated with high-speed link modules usually require the use of both SRMs for maximum bandwidth for inter-slot forwarding and communication.

Link Module

As mentioned previously, link modules occupy the rear of the Backbone Node and mate with their respective routing engines. Numerous link modules provide different types of physical media attachments to the LAN and WAN. Apart from the physical interface itself, the link module provides line driver and link controller functions and high-speed DMA transfer to and from the routing engine. All link modules need a controlling processor module to fulfill their functions; they are nonfunctional without a processor and can be used as spares in this configuration.

Routing Engines

Synonymous with the term *processor card*, the routing engine is the heart of the routing and forwarding function. Multiple routing engines can operate symmetrically in a highly efficient, distributed software architecture that enables the forwarding, filtering, and management functions to be balanced across all processor modules in the router. The routing engine, when combined with a link module, forms the ILI.

Fast Routing Engine

The first generation of routing engines was called the Fast Routing Engine (FRE). The FRE contains a 33 MHz Motorola MC68040 microprocessor that features 4K instruction and data caches for high-performance iterative forwarding functions. Each routing engine contains a PCMCIA slot for installing a flash card, which usually contains the runtime image files, configuration files, scripts, and log files.

Each FRE module can operate independently in the absence of other routing engines. The FRE has midplane attachments to the PPX, managed by the Parallel Packet Express Interface (PPXI). The PPXI on each slot has a transmitter and receiver that is interfaced to all four rails. When transmitting, a FRE is able to transmit on only one rail at any time. The transmit rail is chosen using a random selection algorithm that ensures fair load sharing across the rails. During reception, the PPXI receives traffic simultaneously from all rails. Transmit and receive functions of the PPXI are separated so that all other rails are able to receive while one is transmitting from the same slot.

The FRE also has another attachment called the Link Interface, which connects directly to the link module. Through this high-speed interconnect, the link and the processor modules share the global memory, containing packet buffers. The link module has a DMA controller, which provides direct memory access to the high-speed global packet buffers for efficient forwarding.

Fast Routing Engine 2

The original FREs, the FRE-1s, typically came with a standard configuration of 8 MB or 16 MB of dynamic RAM. A memory upgrade required the replacement of an onboard memory daughterboard. The Fast Routing Engine 2 processors are built to enhance this design to support simple field upgrades and faster processors.

FRE2-040 The FRE2-040 (the 040 designation refers to the MC68040 microprocessor) was a redesign of the FRE-1, supporting variable and easily upgradable memory configurations with improved performance on the PPX bus. The memory on these cards can be upgraded to up to 32 MB using industry-standard single inline memory module (SIMM) chips. In addition, the FRE2 operates on less power (10A, 5V) compared to the FRE-1 (13A, 5V).

FRE2-060 Subsequent to the widespread availability of Motorola's MC68060 microprocessor, the FRE2 was improved to accommodate the higher-speed, higher-performance microprocessor. A 60-MHz MC68060 microprocessor is the core of the FRE2-060, where the 060 distinguishes it from the 040. This processor is optimized to operate high-density link modules such as FDDI, HSSI, and Octal Sync. A majority of the currently deployed routing engines are composed of FRE2-040s and FRE2-060s. It is common to refer to all these processor modules simply as FRE2s.

FRE2-060E Compression of data improves network efficiency and makes better use of network bandwidth on WAN links. A hardware co-processor module is available that can be installed as a daughtercard on the FRE2 for WAN data compression. Functions of this hardware co-processor module have been integrated into the FRE2-060 as an accelerated compression co-processor, resulting in a high-speed next-generation routing engine.

ATM Routing Engine

The ATM Routing Engine (ARE) is the processor module designed specifically for high-speed forwarding with ATM link modules. It contains dual PowerPC 604 microprocessors for parallel processing and dedicated local memory ranging from 8 MB to 64 MB. Unlike the FRE, global memory buffers are not used, and a new virtual buffer memory (VBM) is used to manage variable-sized buffers for efficient packet forwarding. The concepts of local memory, global memory, and VBM are discussed in "Memory," later in the chapter.

Fast Routing Engine 4

The successor to the FRE2 is the FRE4. (A dual-processor engine called the FRE3, similar to the ARE in design, was conceived but never produced.) The FRE4 contains a single Motorola PowerPC 750 processor operating at 266 MHz. With the advent of Gigabit Ethernet, a processor capable of higher forwarding rates was needed. The FRE4 is capable of higher forwarding rates than its predecessor and can be deployed only with the newer high-speed, four-port 10/100M or Gigabit Ethernet link modules. The PowerPC 750 provides greater computational ability than the MC 68060, so that processor-intensive protocols such as OSPF and BGP run faster and more efficiently on this platform.

The FRE1, FRE2, ARE, and FRE4 can all interoperate in the same Backbone Node chassis. Figure 1-9 shows the processor boards and their important components.

Backbone Link Node

All the components—the PPX, SRM, routing engine, and link module—are combined to form the Backbone Node routers. The Backbone Link Node (BLN) is a tabletop or rack-mountable chassis, housing up to four Intelligent Link Interfaces and an additional slot for installing the SRMs. Slot 1 is reserved for the SRMs, and slots 2 through 5 contain the ILIs. The chassis contains one surge-protected 620W power supply.

Backbone Link Node 2

The BLN2 is similar to the BLN and provides the same number of slots for ILIs. The chassis is redesigned to accommodate a redundant power supply, which is hot-swappable. The redundant power supply can provide full power to keep the unit operational if the primary supply fails. The cooling system consists of three fan trays that can be hot-swapped without affecting the unit operation. The BLN2 features the hot-swappable redundant power supply and the cooling fan trays in addition to the features provided by the BLN.

The original BLN is sometimes referred to as the BLN1 to distinguish it from the BLN2. Since both the BLN1 and BLN2 provide almost the same type of features, they are often collectively referred to as BLN in the context of network operation. When referring to a unit in the context of power supply or cooling issues, it is essential to identify a unit as either a BLN1 or BLN2.

Backbone Concentrator Node

The largest of the Backbone Nodes, the BCN is another rack-mountable chassis consisting of 14 slots—13 for ILIs and 1 for SRMs. The System Resource Modules occupy slot 7,

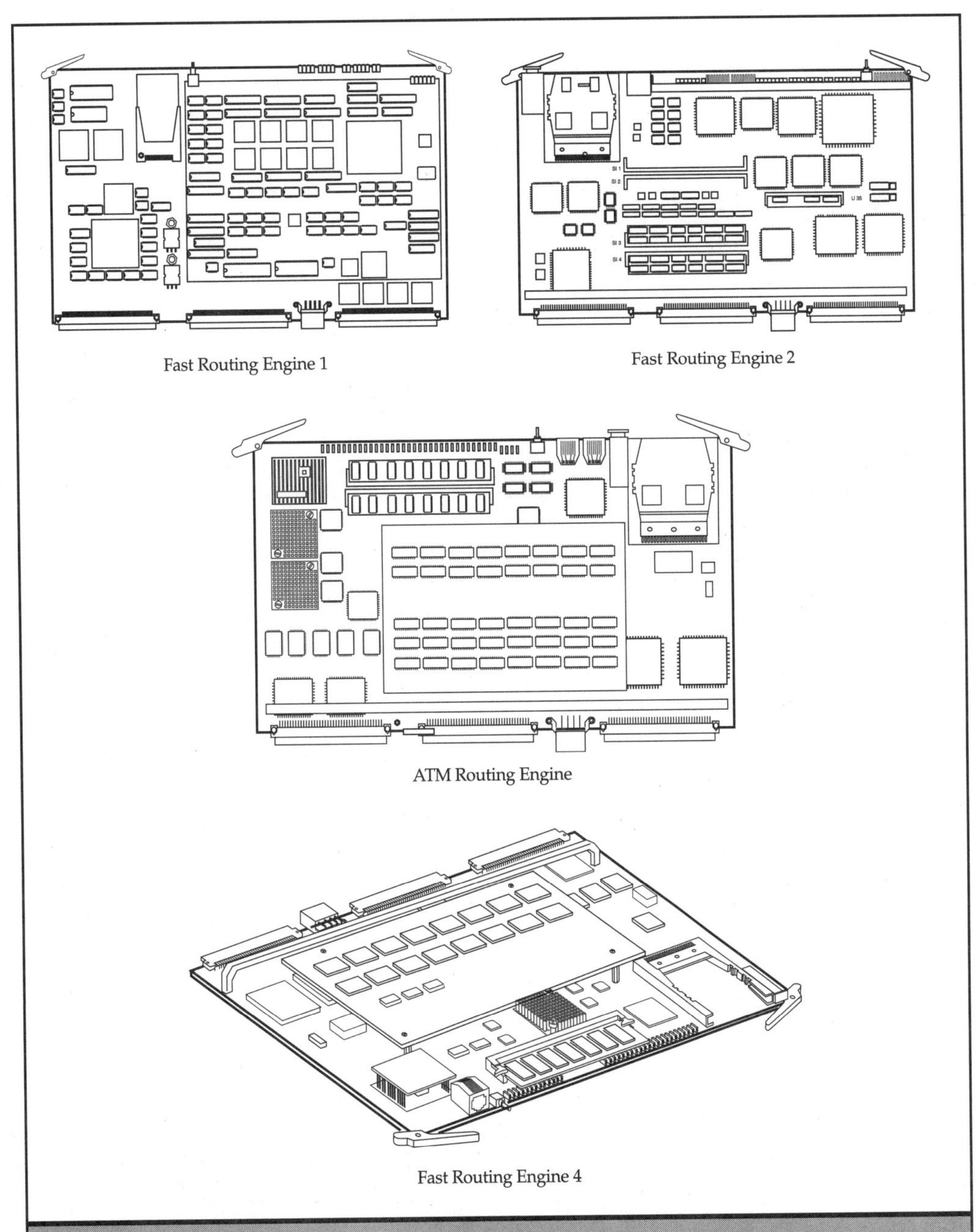

Figure 1-9. BN processor modules FRE1, FRE2, ARE, and FRE4

which is equidistant from the farthest slots on the left and right. The 13 ILIs that can be installed provide numerous interconnections and can support up to 52 LAN and 104 WAN interfaces. At least one 620W power supply is installed in the unit, and additional power supplies, up to a total of four, can be installed. A fully populated BCN requires three 620W units, and the fourth can be used for load sharing and N+1 redundancy. The BCN and BLN2 have identical power supplies and are interchangeable.

The BCN also has a fan tray at the top of the chassis for cooling and air circulation. The fan tray is serviceable, but not hot-swappable due to the size and location. There are four fans in the fan tray assembly, and even if one fan fails, the other fans continue operation and provide cooling to the unit.

SYSTEM 5000 FAMILY

BayRS routers have been adapted for use in System 5000 and 5005 chassis. System 5000 and 5005 chassis traditionally provided high-density hub and switch solutions, and the addition of BayRS routers in these chassis provide routing capability to switched and shared media. The System 5000 routers are of three types: Ethernet, Token Ring, and ATM. They are named based on the type of backplane segments they attach to.

5000 routers contain variable DRAM, from 8 MB to 32 MB, and also a flash slot for a PCMCIA flash.

Ethernet (5380) Router

All 5000 routers are designed around the PPX backplane and in essence represent an advanced FRE-2 processor module with ASN-like features. The 5380 Ethernet router has the capability to provision two net modules similar to the ASN and can route between other media interfaces, including synchronous ports. The presence of the optional PPX backplane in the chassis allows multiple 5000 routing modules to interconnect and operate as a single router. If the PPX backplane is absent, each router operates as a stand-alone unit, routing between the attached segments and interfaces.

The System 5000 has 12 Ethernet segments that span the entire chassis. The 5380 has four backplane ports that can be attached to any 4 individual segments for routing. On the front, the two net modules can provide additional interfaces for LAN and WAN interconnectivity.

Token Ring (5580) Router

The 5580 Token Ring router is similar to the 5380, except that it provides four backplane Token Ring ports that can be attached to any four backplane Token Rings. The Token Ring backplane on the 5000 can contain five or nine rings, where the fifth ring spans the entire chassis and the rest span either the left or right half.

ATM (5782) Router

The 5782 ATM router is commonly referred to as the Virtual Network Router (VNR) or Centillion Multiprotocol Engine (CME). It requires the presence of an ATM backplane in the chassis. In essence, it is an OC3 connection to the ATM fabric and can provide virtual routing across emulated LANs, classical IP routing, and multiprotocol transport.

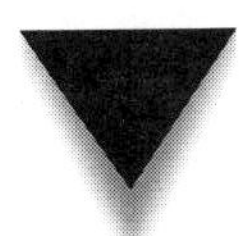

CAUTION: The 5782/VNR is installed in the 5000BH chassis. The 5780 ATM router is installed in the older 5000AH chassis, which cannot be installed in slot 14. The BH chassis contains two ATM backplanes, one spanning slots 2 to 7 and another spanning slots 8 through 13. The 5782 can be installed in any of these slots containing the ATM backplane and will provide routing functions on the attached fabric.

The 578*x* ATM router does not support any net modules and attaches directly to the ATM and the PPX backplanes.

MEMORY

All FRE and ARE processors contain a variable amount of dynamic RAM (DRAM). Older processor cards may contain as little as 8 MB of DRAM, and newer processors may contain as much as 128 MB of DRAM. On all processor cards except the ARE, this memory is divided into local memory and global memory. On the ARE, all the DRAM memory is dedicated to local memory.

Local Memory

In the default configuration, 75 percent of the memory is allocated to local memory. For example, a processor card containing 16 MB of DRAM will have 12 MB of local memory (75 percent of 16 MB). This local memory is so called because it is accessible by the local processor only. Local memory contains the runtime image and protocol applications and the internal data structures for the calculation, maintenance, and updating of routing entries. The GAME operating system is loaded into local memory at startup, and all loadable modules are loaded during dynamic load. The use of local memory can be compared to the normal use of DRAM memory on UNIX and Windows workstations.

Global Memory

Memory left unallocated to the local memory pool is called global memory. In the default configuration, the remaining 25 percent of total memory is occupied by global memory. In the example of the 16 MB processor card, there will be 4 MB of global memory (25 percent of 16 MB). Global memory is shared among the processor, link modules, and processors in other slots, which is why it is called global. This global memory is divided into packet buffers for high-speed handling and forwarding of data traffic. All inbound packets are copied by the link module directly into global memory. Likewise, the processor module delivers outbound packets to the link module by way of global memory.

The total DRAM memory can be configured into local and global memory. The process of reconfiguring memory allocation is called *carving* and is discussed in detail in Chapter 3.

Virtual Buffer Memory

On the ATM routing engine, all the DRAM is simply used for local memory. A new static RAM (SRAM), called virtual buffer memory (VBM), is available on the ARE. It varies from 1 MB to 6 MB and provides the same functionality of global packet buffers with increased efficiency. The VBM is divided into pages, enabling the allocation of a variable number of pages to a single packet buffer. This is particularly useful for the simultaneous processing of frames and cells in the ARE.

FILE SYSTEM

All BayRS routers use a flash-based, nonvolatile file system. The flash EEPROM is the nonvolatile storage for system files such as the runtime image, configuration files, and other files such as scripts and event logs. A removable PCMCIA flash (see Figure 1-10) can be installed on all routers except the AN1, which uses a SIMM flash. The flash can be formatted for use with the proprietary file system, and many of the conventional file system commands are available for managing it. Files can be transferred to and from the flash using TFTP or FTP while the router is operational. The PCMCIA flash file system can be externally managed from a Windows laptop supporting the PCMCIA standard, using the BNFS utility.

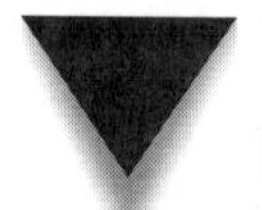

TIP: BNFS, the Bay Networks File System utility, is an unsupported, third-party application distributed by Router Management Labs (RML). See the book web site (http://www.mahalingam.com/) for details on downloading the latest version of BNFS.

Flashes can be obtained in various storage capacities; the routers support 1 MB through 32 MB. The file system used is simple and easy to manage. Unlike random-access file systems such as floppy disks, flashes are programmable devices and require a different management system. When a new file is added, it occupies the area immediately following the current last file on flash. The new file is linked to the previous one in the file system chain. If an existing file B is removed, it is simply unlinked from the file system list; and file A is subsequently relinked to file C in the chain. This is an important task that is required in file system maintenance: cleanup. The space occupied by file B must be reclaimed from the flash, using a procedure called *compacting*. Flash management and troubleshooting is discussed in Chapter 3.

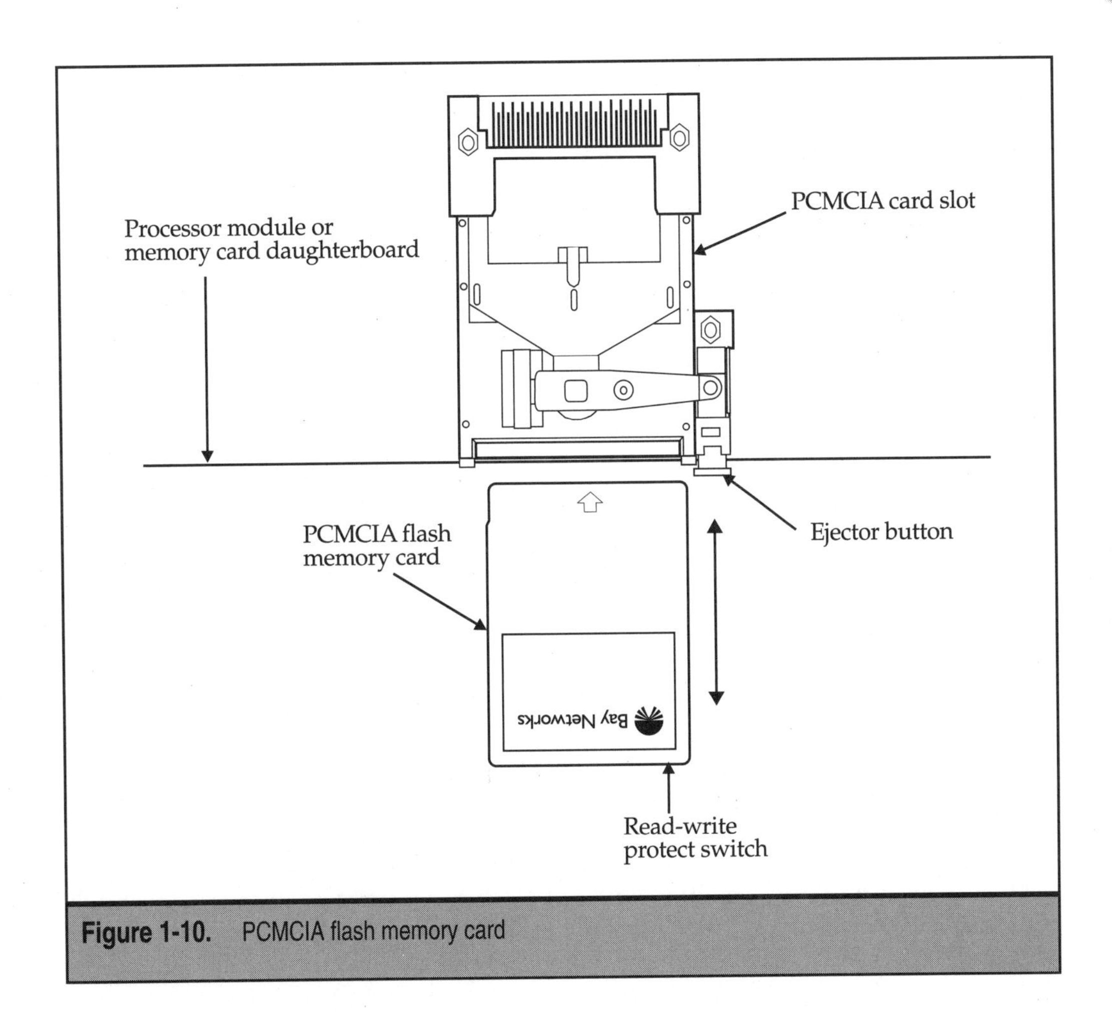

Figure 1-10. PCMCIA flash memory card

Multiprocessor routers such as the ASN and the BN can contain multiple PCMCIA flashes. They may be used for redundant image and configuration storage and also to store backup images and historical configurations.

CAUTION: When using multiple flashes in a router, take care to avoid the possibility of individual slots booting off different software image versions and to avoid loading an older image or configuration file. To prevent these problems, make sure that archived files do not bear default image or configuration file names. Also, redundant image and configuration files stored in multiple flashes must be kept synchronized.

POWER-UP SEQUENCE

When a router is first powered up, the process is termed a cold start. As the router starts to come out of reset, it executes a basic set of tests to ensure system integrity and to identify and verify all components. Sometimes called the power-on self test (POST), this procedure is commonly called the diagnostics (or diags for short). All processor motherboards have a programmable read-only memory (PROM) device that stores a diagnostics image for execution. This is a memory device that can be programmed with data, usually executable code that can be read subsequently several times. These devices are not directly writable, which means that their contents cannot be arbitrarily changed; instead, instructions must be programmed into them. Typically, these devices are programmed once and read several thousand times without alteration. Programming these PROMs require appropriate hardware and software.

Diagnostic PROM

The diagnostic PROM, commonly called the diag PROM, is programmed with code that is specifically designed to run diagnostics on the unit. On the Access products, the motherboard contains the diag PROM, and on the Backbone products, each processor module has an individual diag PROM. On the BN, each slot is responsible for executing diagnostics on its own processor module, link module, and PPX. The diagnostics software contains a command-line interface (CLI) called the Diagnostics Monitor, where specific commands can be issued.

The diagnostics code is usually responsible for identifying all components, such as the unit type, interface count, interface types, memory size, peripherals, daughterboards, component revisions, and serial numbers, to name only a few. It is also responsible for testing the proper operation of all onboard components that were identified. Failure during these tests indicates a hardware problem, and diagnostic messages are printed to the console indicating the condition. A hardware replacement may be necessary to rectify a diagnostics failure. On the Access products such as the ASN, diag failure of one component (such as a specific net module) will not affect the diagnostics process, which will continue. The faulty component is marked defective and will not be operational during runtime. In the case of the BN, failure of a component such as a link module will halt diagnostics and transition the slot to a partial boot state, where the operating system will bypass the configuration for the slot until the defective component is replaced.

During the test execution, diagnostics print detailed messages on the status of each test. These messages can be clearly seen on all Access routers, such as the AN/ANH, ARN, and ASN. On the Backbone routers, several diagnostics sessions run simultaneously on each slot, and no individual slot has access to the console. In this unique case, it is normally not possible to observe the test status on all slots simultaneously through the console port.

The Harpoon Diagnostics Console Monitor (HDCM) port (Figure 1-11) can be used to observe diagnostics messages for an individual slot. This port can be accessed by authorized Nortel Networks service personnel. In the special case in which FREs run diagnostics firmware v4.12 and later, these diagnostics messages from a remote slot can be viewed using a TI session.

Success of all diagnostics indicates proper operation of the hardware. The diag PROM will invoke the code stored in the boot PROM and will supply the boot PROM with de-

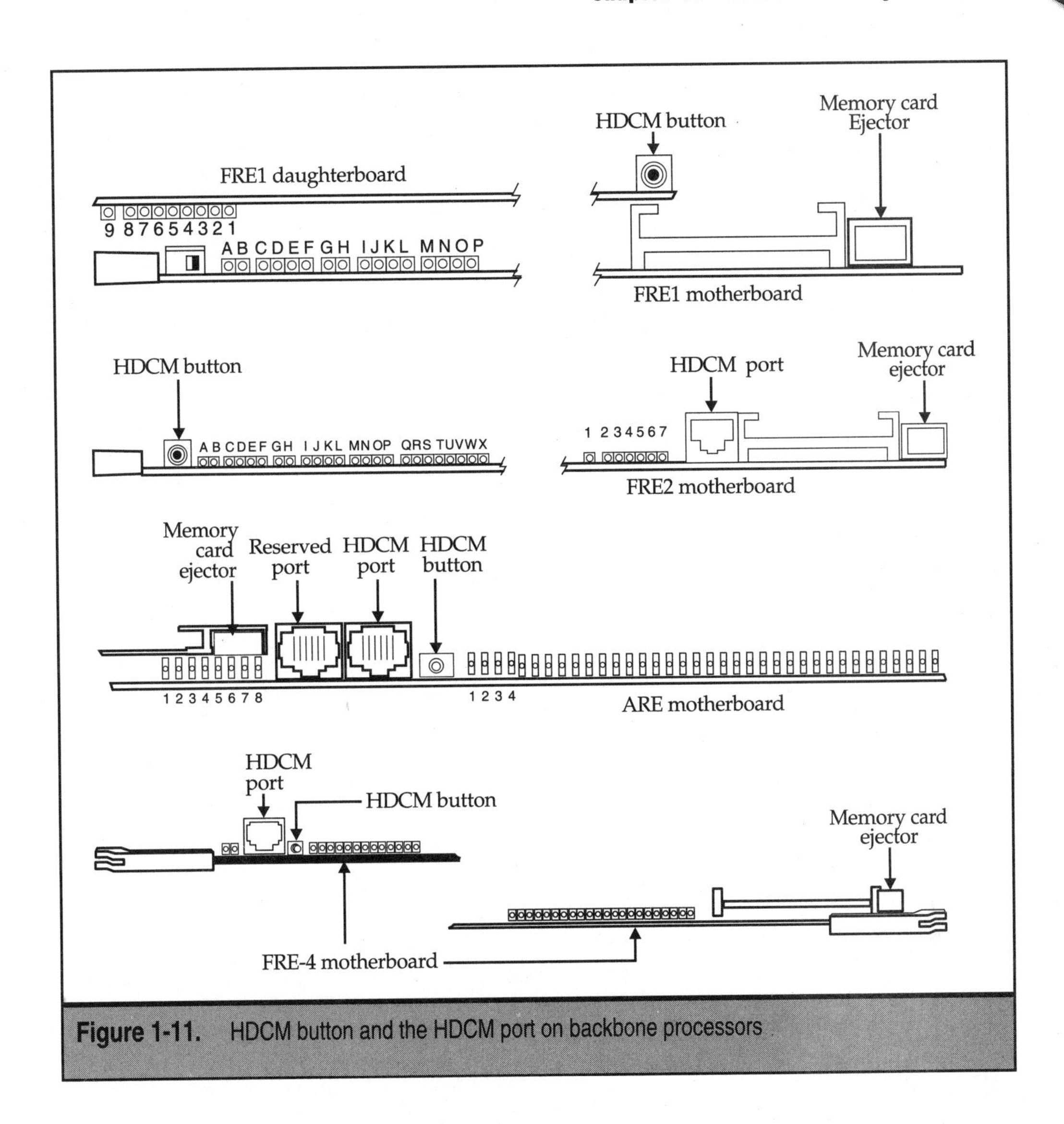

Figure 1-11. HDCM button and the HDCM port on backbone processors

tails of the unit and its components. The diag PROM will not be used subsequently during the runtime operation of the router; however, it will be used again during the next power reset.

Bootstrap PROM

The bootstrap PROM, containing the bootstrap image, is frequently called the boot PROM and is an identical PROM device as the diag PROM. Its code and function are very different, however. The boot PROM is a miniature version of the GAME operating system, the core component of BayRS. This miniature version of the GAME OS is commissioned with the task of bootstrapping the unit to a well-known state. The boot PROM contains the bare essentials to achieve this state. It has a functional mini-GAME, a TCP/IP stack, and drivers, among other components. For example, all Access routers can perform an automatic network bootup process called EZ Install. EZ Install automatically acquires an IP address using BootP, if necessary, and will employ TFTP on the runtime software from a TFTP server. This EZ Install process requires the use of GAME, an IP/UDP stack, BootP and TFTP protocols, media drivers (Ethernet, Token Ring, and Serial), and BayStandard and Frame Relay protocols, among several features. All of these minimal features are present in the boot PROM code.

The boot PROM relies on data previously supplied by the diag PROM. For example, once the diag PROM has identified the interfaces and their types, the boot PROM will be able to attempt to apply EZ Install on all the identified interfaces. If a local boot is being performed, the boot PROM has the necessary code to perform file system operations on the flash. It will read the configuration and the image into memory, decompress and run checksum on the image, and relinquish operation to the entry point in the kernel. The full-fledged GAME kernel will proceed to configure the appropriate protocols and circuits and enter the runtime state.

Boot Process

Access routers support several remote boot options, namely EZ Install, Netboot, Directed Netboot, and also a local boot. Backbone routers are usually required to promptly load the operating system and become operational with minimal service disruption, and thus they support local booting only. The following sections describe all of the available boot process choices.

EZ Install

The goal of EZ Install is to deploy access routers at remote sites with minimal initial configuration. When the router is installed right out of the box, it will load the software image from the flash and attempt to automatically acquire an IP address from another BayRS router using a proprietary BootP packet. The EZ Install process will attempt to use several protocols such as BayStandard, Frame Relay Annex D, Frame Relay Rev1 LMI, and Frame Relay Annex A across synchronous lines to obtain the IP address. If Ethernet interfaces are available, then the IP address request is also made across them. The configuration file is then downloaded from the BootP server, and the router starts to perform routing and bridging operations as required by the configuration.

Netboot

In the Netboot process, an IP address is configured on the router so that it can perform BootP operations across the configured interface. Using the TFTP parameters specified by the BootP server, the router will download the configuration or the software image, or both, from the TFTP server. Once these files are downloaded, the router will restart using the newly downloaded software and apply the configuration file for routing and bridging functions.

Directed Netboot

In the Directed Netboot process, the BootP step is eliminated by configuring the IP address of an interface and also supplying the following TFTP parameters: server IP address, configuration file location, and/or software image file location. As the boot PROM executes, the boot configuration is read, and the router initializes the interface and downloads the specified files directly from the TFTP server. Again, the router will restart using the downloaded files and initialize to the runtime configuration.

Local Boot

The local boot process is the quickest boot routine. It involves the simple loading of the software and configuration file directly from the local flash. Since the router does not have to seek these files externally, they are easily loaded, and the control is quickly transferred to the runtime operating system. In this scenario, the image and configuration files must be actively maintained and kept synchronized locally.

For the AN, ANH, ARN, and single-unit ASN routers, the local boot process is straightforward (refer to Figure 1-12). Only one flash card is present, and the router loads the software image and the configuration file from this default volume. In the case of a multi-slot router such as a stacked ASN or BN, the local boot process follows another procedure. Once the diagnostics are complete, each processor module broadcasts a sequence of boot requests on the backplane. In the BN, these requests are through the PPXI, and in the case of an ASN stack, they are sent using the SPEX.

If the backplane boot requests are unanswered after five tries, a processor module will attempt to look for a local flash to load the image. At least one processor module will locate a local boot image on flash and will load the software into memory. This processor module now becomes the *image server* for all other modules in the chassis or stack. Modules that are unable to find a local boot medium will return to the backplane boot requests and will now find the image server. The image is directly loaded across the backplane from the image server.

NOTE: The diagnostics process may take different times to complete, depending on the type of processor module, amount of memory, presence of daughtercard, and type of link module, because each module uses different tests. Hence, each slot will come out of diagnostics at a different time. Backbone processors with both SRMs installed will run tests on all four rails, which will take longer than when just one SRM is installed. If two slots containing flashes come out of diagnostics within a short period of each other, they may potentially boot off their respective flashes.

Figure 1-13 shows the sequence of a local boot operation on stacked ASN and BN routers. Following the image load, the processor module broadcasts a sequence of configure requests over the backplane. If the configure requests are unanswered after five tries, the processor will attempt to load the configuration file from the local flash. If there is no

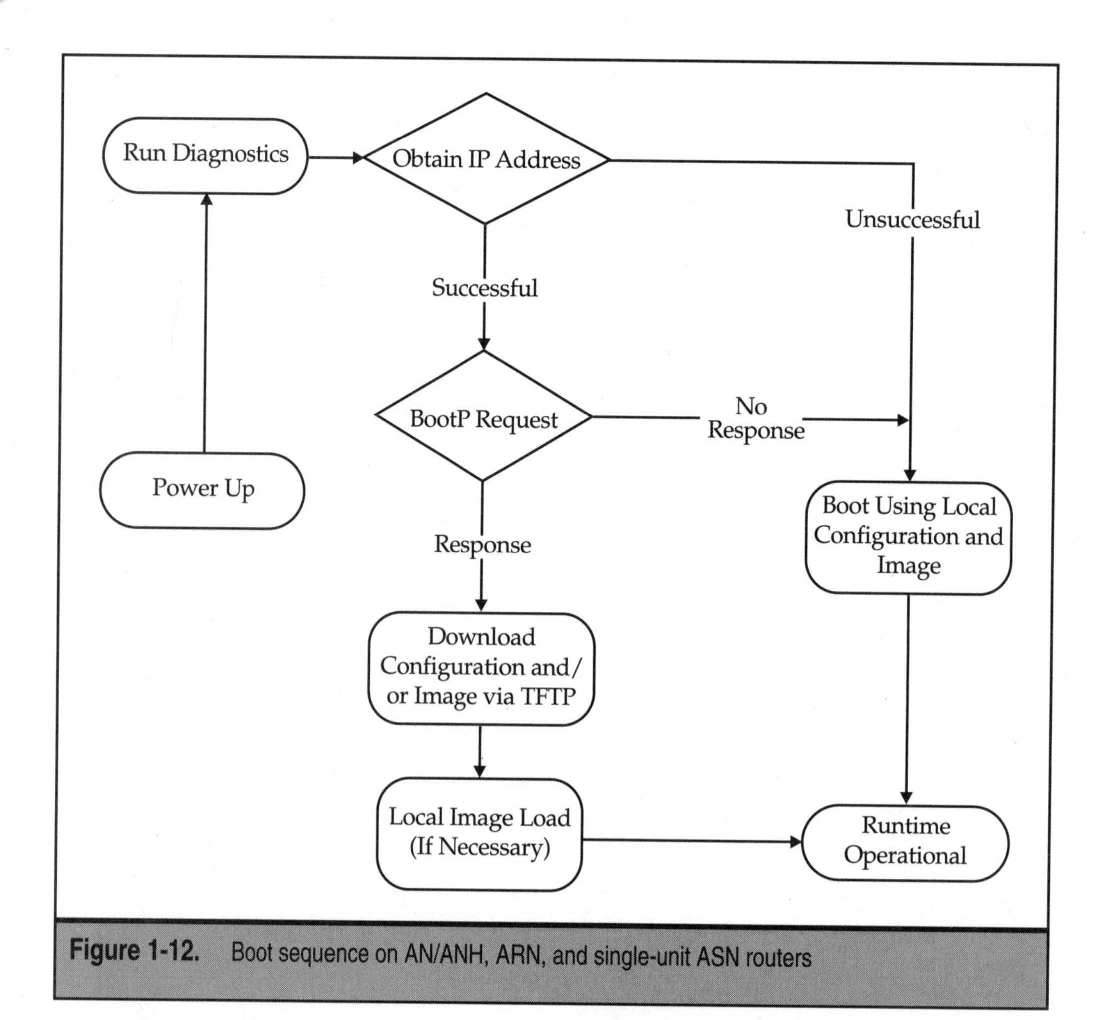

Figure 1-12. Boot sequence on AN/ANH, ARN, and single-unit ASN routers

configuration file on the local flash, the processor returns to the backplane to look for a configuration file. When a processor is able to load the configuration file from its local flash, then it becomes the *configuration server* for all other modules in the chassis or stack.

CAUTION: If two slots boot with different image versions from different flashes, then the operation of the router may be adversely affected. Also, this presence of multiple image servers will enable remote slots to boot using the first response from one of the image servers. While the ability to install and boot from multiple flashes allows redundancy, care must be taken to keep all bootable flashes synchronized with each other.

If there is no configuration file on any of the flash cards, then after an extended time-out period, the ASN will initialize with a default configuration. The default configuration does not have any routing and bridging interface entries. BN processor modules will not

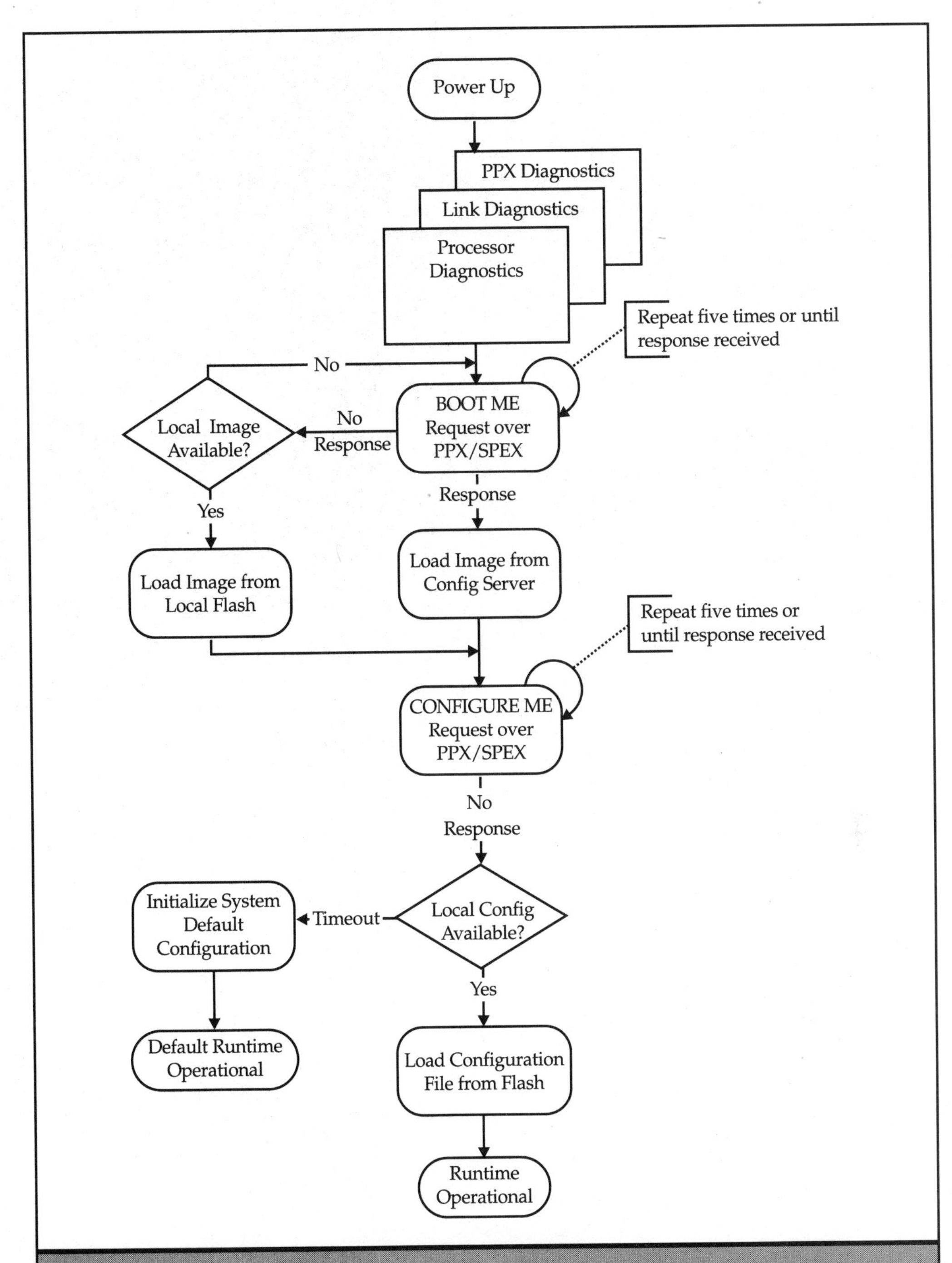

Figure 1-13. Boot sequence on stacked ASN and BN routers

initialize until a flash card with a configuration file is inserted and the router or processor module is reset.

GAME Initialization

As discussed previously, the software image can be loaded in one of several ways. Except in the case of a backbone load, the software image obtained is in a compressed archive form. The image is first decompressed and a checksum is calculated. The image contains one system kernel and several applications. Only one kernel is present in each archive, and this is the GAME operating system image. This kernel is unique to each platform, as shown in Table 1-2. The GAME kernel file is akin to the vmunix and genunix kernel files in UNIX.

The bootstrap process selectively loads the kernel image file into memory first. Subsequently, the kernel becomes operational and initializes all required system functions. This includes systemwide functions such as the Management Information Base (MIB) subsystem, Table subsystem (for the maintenance of different table types), Loader subsystem (for the dynamic loading of applications), and Technician Interface subsystem. The GAME kernel will also initialize global memory buffers and other system resources for proper use.

GAME and BayRS concepts as applied to configuration, performance, and fault management are discussed in the next chapter.

TROUBLESHOOTING AND SERVICE RESOURCES

Well, you've bought this book to get better at BayRS administration, troubleshooting, and optimization, and it is likely that you already know many of the details we covered in the

Platform	Kernel Image	Default Archive Image	Default Config File	Secondary Config File
AN, ANH	krnl_an.exe	an.exe	config	ti.cfg
ARN	krnl_arn.exe	arn.exe	config	ti_arn.cfg
ASN	krnl_asn.exe	asn.exe	config	ti_asn.cfg
FRE1 and FRE2	krnl_bn.exe	bn.exe	config	ti.cfg
ARE	krnl_bn.ppc	bn.exe	config	ti.cfg
FRE4	krnl_fre4bn.ppc	bn.exe	config	ti.cfg
5x80	krnl_s5000.exe	s5000.exe	config	ti_s5000.cfg
578x	krnl_s5000.ppc	s5000.exe	config	ti_s5000.cfg

Table 1-2. Platform-Specific Kernel, Archive, Default Image, and Config Files

foregoing sections. In the remaining chapters, this book will take you far from the daily rigors of network administration, both with theoretical knowledge and demonstration using practical situations. However, you will occasionally need to use other resources to help complete your task or project. This section lists some common resources that you will find useful.

Global Customer Care Services

Nortel Networks provides direct services and support for BayRS software and hardware. Current customers with a valid service and support contract with Nortel Networks can open cases (trouble tickets) for issues that cannot be resolved easily. Contact Global Customer Care Services, Enterprise Solutions, at (800) 2LANWAN or support@baynetworks.com to obtain technical assistance 24 hours a day, 7 days a week, 365 days a year. Outside the U.S. and Canada, you can call (33) 92-968-968 (Valbonne, France), (612) 9927-8800 (Sydney, Australia), or (81) 3-5402-0180 (Tokyo, Japan) for the nearest service center.

Online Library

The Online Library is a collection of documentation published by Nortel Networks. Whenever a major or functional BayRS version is released, corresponding document sets are available in Portable Document Format (PDF) on the Web. This library is also issued in CD form, providing a recent snapshot of the document set as applicable to the newly released version. This CD is shipped to customers who have an active software maintenance and update agreement with Nortel Networks.

This online documentation is also available free at the Global Customer Care Services, Enterprise Solutions, web site (http://support.baynetworks.com/).

Hard-Copy Documentation

When you are stuck inside a cramped closet or tangled in a maze of cables, hard-copy documentation will come in handy. Although Nortel Networks distributes documentation primarily in soft form, hard-copy manuals can also be ordered for reference and daily use. The documentation not only describes the configuration and management of protocols, but also provides overviews of network protocols for engineers and administrators being newly exposed to a particular protocol. This is a very good place to start for readers interested in learning or implementing a new network protocol.

Book Web Site

This is the web era, and we are dealing with changes in web time. This book sports a web site, where updates, errata, scripts, and other tools are posted. You can also go to the web site to send feedback directly. Links to the foregoing resources are also available at the web site. The book web site is located at http://www.mahalingam.com/.

SUMMARY

In this chapter, we completed a brief overview of BayRS software and hardware architectures. We discussed BayRS history and software features and the various hardware architectures that BayRS runs on. We also discussed the bootstrap and initialization process for all BayRS routers and the use of permanent storage. Succeeding chapters draw upon all of these concepts as applied to BayRS as a foundation for administration and troubleshooting. Because BayRS features and functionality evolve over time, you should look for changes by consulting the Online Library or the book web site whenever possible.

REFERENCES

1. *Installing and Maintaining ASN Routers and BNX Platforms,* Publication 109351-F Rev A.
2. *Installing and Maintaining BN Routers,* Publication 109352-E Rev A.
3. *Using the Model 5380 Ethernet and the Model 5580 Token Ring Routers,* Publication 112768-A Rev B.
4. *Using the Model 5782 ATM Virtual Network Router,* Publication 119557-A Rev 00.
5. *Installing and Operating BayStack AN and ANH Routers,* Publication 114113-B Rev 00.
6. *Installing and Operating BayStack ARN Routers,* Publication 114200-D Rev 00.
7. *Installing SRM-F and FRE/FRE-2 Processor Modules,* Publication 115621-B Rev 00.
8. *Installing ATM Link Modules and Routing Engines in BN Platforms,* Publication 114951-A Rev A.
9. *Installing FRE-4-PPC Processor Modules in BN Platforms,* Publication 302155-A Rev 00.
10. Product Reference Guide, May 1999.
11. J. Knapp, *Nortel Networks: The Complete Reference*, Osborne/McGraw-Hill, November 1999.

CHAPTER 2

BayRS Basics and Tools of the Trade

Man is a tool-using animal... Without tools he is nothing, with tools he is all.
—Thomas Carlyle, *Sartor Resartus*

The availability and knowledge of networking and troubleshooting tools is a key requirement for effective network administration. BayRS supports many administrative techniques and provides various troubleshooting tools for easy administration and troubleshooting. The first step in productive troubleshooting is to know the available tools and techniques at your disposal—like any new or unfamiliar tool, you must understand the functionality and the proper considerations and techniques for its use. In this chapter, we present the fundamental concepts of BayRS and the tools available at the system level for administration and troubleshooting.

GAME and BayRS

The Gate Access Management Entity (GAME) is an embedded operating system responsible for systemwide control and management of internal functions and resources. This operating system is common across all BayRS routers and therefore makes routing and forwarding services possible for all supported network layer protocols on all platforms.

The operation and functions of GAME are usually transparent to the network engineer or administrator, who typically is concerned with the normal operation of network layer protocols. This is similar to the situation of the Windows or UNIX user, who is mostly absorbed in using application tasks and is inattentive to the underlying operating system. However, there are occasions where you may find it helpful or necessary to have a minimal understanding of the inner workings of the OS. With this intent, this section provides a primer on the fundamentals of GAME.

Features

GAME provides many of the common features found in today's operating systems. Some of the dominant features are noted here:

- Multiprocessing, which enables the router to operate thousands of processes simultaneously
- Symmetrical processing, which allows multiple CPUs to execute concurrent tasks on the same slot
- Messaging, which enables fast inter-slot communication between processes
- Rapid encapsulation and decapsulation for forwarding packets
- High-performance multi-slot forwarding

- High availability, which provides independent hardware and software fault management
- Dynamic reconfiguration, which allows isolated addition, removal, and reconfiguration of a component without affecting others

In a comparison with a robust UNIX operating system, GAME has many similarities and some differences. All processes, called gates in GAME, have a hierarchical structure that maintains ancestry, as in UNIX. These processes do not run in their own separate memory space or user mode, but instead run in a common space and operate in the same supervisory mode as all other gates. Each gate (process) runs in the global supervisory mode, which provides unrestricted access. All gates are scheduled on a first-come, first-serve basis, which is different from the process priorities enforced in UNIX. Gates usually run to completion, until they voluntarily give up the CPU; UNIX limits process execution to time slices, based on priority classes. Except in the case of CPU or hardware exceptions, there is no context switching in GAME until a gate voluntarily relinquishes the CPU. The UNIX kernel, however, retains CPU control at all times with the exception of when certain interrupts occur.

Gates

The individual granular processing component in GAME is called a *gate*. The gate can be compared to a less-stateful process or a complex thread in UNIX operating systems. As mentioned before, the ancestry of gates provides a hierarchical structure. A newly spawned gate is considered to be the child of its parent gate, and it may in turn spawn other gates. Each gate is uniquely identified by a gate identifier (GID) and by a gate handle (GH), which uniquely identify it across all slots.

```
#    92: 01/10/00 18:01:24.078  DEBUG    SLOT  2  LOADER              Code:  2
Loader starting service gate 0x0000c @ 0x30044760 (env=0x00000000, flags=0xc1)
```

Certain gates are termed well-known gates because they have predefined gate identifiers. GAME subsystems and protocol masters usually have well-known GIDs. Others, such as interface gates, usually have gids dynamically assigned by GAME.

Soloist Gates

Certain gates need to be instantiated only once to perform certain tasks on a boxwide basis. In this context, it is important to recognize the symmetrical processing of gate instances on multiple slots. An example of symmetrical processing is that of the IP Routing Table Manager (RTM), which usually has its own instantiation on each slot. It receives and calculates the best routes submitted by individual routing protocols such as RIP, OSPF, and BGP.

Unlike the IP RTM, the OSPF protocol runs as a *soloist* to perform route calculations and update functions. Since the Dijkstra SPF is a CPU-intensive algorithm, the OSPF soloist is designed so that a single slot is responsible for performing this function. Other functions of OSPF, such as forming adjacencies and exchanging updates, are still performed by local slots. Some other common soloist processes include the following:

- Circuit gates
- Technician Interface
- BGP in ISP mode

A soloist gate runs after an initial election process in which all eligible slots participate. Each slot is made eligible by configuring the slot's bit in a bitmask value, called the *slot mask*. Theoretically, each eligible slot has an equal probability of being elected to run a soloist. Practically, slots that are the first to boot or quick to respond (because their load is low) typically tend to run soloists that they are eligible for.

```
#     1: 03/13/00 20:56:04.566   DEBUG     SLOT  3   GAME                Code:  23
SOLO (0x00057): election OPENING 38000000/38000000 (38000000) vote=00000000

#     2: 03/13/00 20:56:08.914   DEBUG     SLOT  3   GAME                Code:  84
SOLO (0x00057): election WON 0x10000000 (repl=90000009/38000000)
SOLO (0x00057): election CLOSING 18000000/18000000 (18000000) vote=00000000

#     3: 03/13/00 20:56:08.960   DEBUG     SLOT  3   GAME                Code:  84
SOLO (0x00057): election CLOSED 0x10000000 (repl=90000009/18000000)
```

If the soloist gate goes down either on its own accord or because the elected slot went down, another slot will be elected and continue to perform the same tasks.

Slot Mask

The slot mask, which will be frequently referred to in this book, is a 32-bit value where the most significant 14 bits each represent a unique slot number. The most significant bit (that is, the leftmost bit) of the slot mask represents the slot number 1, the next represents slot number 2, and so on. If a particular slot bit contains the value 1, then the slot's value in the slot mask is considered enabled. Observe that only the leading 14 bits are used, and the trailing 18 bits are currently unused. The slot mask is illustrated in Figure 2-1.

Slot masks are used in the configuration and loading of protocols and drivers. In Figure 2-1, observe that the slot mask is obtained by summing individual slot masks representing each slot. Readers who find it cumbersome to work with binary numbers may find the slot mask calculator provided on the web site useful for encoding and decoding the slot mask. Using a specified slot mask, the dynamic loader can load an application onto specific slots. We will revisit the use of slot masks in detail in later chapters.

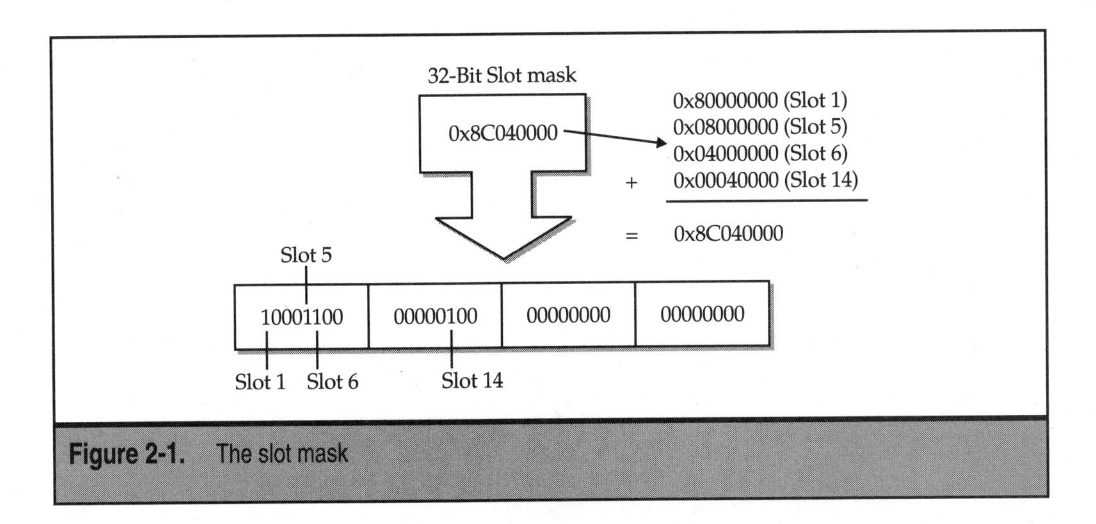

Figure 2-1. The slot mask

Mappings

Gates often may need to interact and react to state changes in other gates. Particularly, a new gate creation or the death of an existing gate may need appropriate reaction from another gate. Mapping is the method by which gates keep track of the state of other gates. In UNIX, wait3( ) calls are used by processes to wait for child processes to terminate, which is similar to the way mappings operate. But mappings can be made by any gate to any gate. UNIX processes that wait are usually suspended, while mapping (and mapped) gates can continue execution of their tasks once registered. Mappings are mostly used to deal with reconfigurations and failures that occur at the hardware or software level.

Global Memory Buffers

Global memory is divided into packet buffers of a specific size. The default packet buffer size on most platforms is 4,800 bytes, which allows for the handling of the largest frames received on LAN and WAN interfaces. Table 2-1 contains a listing of default buffer sizes by platform. The AN/ANH, ARN, and ARE can be reconfigured to have smaller buffer sizes, if they do not have Token Ring interfaces or ELANs. For example, the buffer size of 1,824 bytes can be configured on AN/ANH and ARN routers; on the ARE, the minimum buffer size is 2,048 bytes. Using a smaller buffer size in permissible configurations allows the global memory to be divided into a larger number of packet buffers. The availability of packet buffers also increases proportionally with the total amount of global memory available. Increasing the buffer size is also possible on high-density interface cards; for example, AREs can be configured to support buffers up to 10K in size.

Platform	Buffer Size	Configurable
AN/ANH (Ethernet)	1,824	Yes
ARN (Ethernet)	1,824	Yes
AN/ANH (Token Ring)	4,800	No
ARN (Token Ring)	4,800	No
ASN	4,800	No
ARE	10,240	Yes
FRE-1/2	4,800	No
FRE-4	5,376	No

Table 2-1. Buffer Sizes on Platforms

Both the link module and the processor module have direct access to global packet buffers. On the Backbone Nodes and the ASN with FPC, the high-speed SRAM is used to cache frequently accessed portions of packet buffers, thereby drastically improving forwarding performance.

Local Memory

Local memory, otherwise commonly known as private memory, is accessible for use only by the local CPU. It is used by the runtime operating system and loadable modules for code, stacks and heaps, internal databases, routing and forwarding tables, among other things. While global memory is exclusively used for the handling of packet buffers, local memory is used for everything else. This memory is highly variable, varying from 1 MB (on the original ACE processor for VME) to 128 MB (on the FRE-4).

Processor cards with higher amounts of local memory can run more protocols and handle larger routing and service tables. This is because the addition of protocol services requires the loading of new modules and processing of appropriate tables required by them. For example, adding AppleTalk to the configuration will require the router to load at.exe (AppleTalk application) and maintain tables as required by RTMP and ZIP.

Local memory requirements are usually well known at the time of design and deployment. This memory is frequently allocated and freed dynamically by applications as they require. It is similar to dynamic memory management in UNIX environments, where memory is malloc( )ed and free( )d and where ownership of memory is always well known. This dynamic memory management and knowledge of network growth requirements allows a network administrator to carefully select a minimum level of local memory required. Such granular memory management permits certain slots to have smaller amounts of memory, while some protocol-intensive slots can be installed with larger amounts of memory.

Inter-gate Communication

Inter-gate (inter-process) communication (IGC) is an essential component in most of today's operating systems. Apart from the obvious need for communication, it enables efficient sequencing and scheduling of tasks in a real-time environment. Some of the most common ways processes communicate are via signals, semaphores, mutexes, pipes, and shared memory. In GAME, IGC achieves three main tasks: namely, buffer delivery, signal delivery, and mapping. From the user context, buffer delivery is available for both local and remote inter-gate communication. For example, an IP protocol gate may transmit a buffer locally to another gate (such as an IP interface) and also transmit some buffers to the IP protocol gate executing on another slot. Buffer delivery is accomplished in several ways and is normally considered to be one of two types: reliable or unreliable. Critical buffers that need to be delivered reliably in a timely fashion are delivered using the reliable mechanism. The differences applicable in the user context are discussed in Chapter 4.

It is necessary however, to mention the reliable delivery mechanism used with reliable forwarding and remote procedure calls (RPCs). Again drawing a parallel with UNIX, the GAME RPC is used in a client-server context by gates to request the service of other gates in remote slots. GAME RPC, however, is not used for communication with gates on remote routers—they remain localized within the symmetrical multiprocessing system that represents a single router. These calls are delivered using a reliable mechanism on a systemwide basis.

```
#     30: 01/10/00 17:58:21.558  DEBUG     SLOT  2  GAME                 Code:  84
FWD [ 2-0018] (a0000008->88000009): down=08000000, no_ACK=00000000, NAK=00000000

#     31: 01/10/00 17:58:21.636  DEBUG     SLOT  2  GAME                 Code:  84
RPC [ 2-001b] (a000400f->a8000009): down=08000000, no_ACK=00000000, NAK=00000000
```

Signal delivery is a communication primitive that is used locally within each slot. Signals are used for a variety of purposes, particularly timer expirations, software signals, hardware interrupts, and data transfer. A majority of networking protocols operate on the basis of timers. For example, each OSPF interface is required to send a periodic hello update (for the period of the hello interval) to all SPF routers. After the expiration of the hello interval time, a timeout signal is delivered to the appropriate gate for the generation and multicasting of the hello update. Other protocols may perform bulk internal data transfers within gates responsible for the same protocol. Protocol- and driver-specific signals are used throughout for local management. Hardware interrupts are always intercepted by the GAME kernel and are translated into software signals for appropriate processing. And as many UNIX-savvy users may already know, signal delivery is also used to kill gates that are no longer required. The kill signal in GAME is akin to the SIGTERM in most UNIX platforms, where the disposition of the signal can be predefined so that all gates are able to terminate gracefully.

Last but not least is the use of mappings for IGC. Mappings can be generated and activated on local gates and across remote gates. As mentioned earlier, mappings are used to keep track of gate state changes. Certain gates often need to react to a change in state of related gates. For example, the Virtual Router Redundancy Protocol (VRRP) will dynamically need to react to state changes in the slotwide IP protocol gate and also the associated IP interface gate. Mappings are used by the VRRP interface gate to appropriately react to the mapped gates going down or coming up.

Dynamic Loader

The loader is a dynamic mechanism used to load applications and drivers based on configuration parameters. Loadable applications are independent software components that behave like a kernel module in UNIX. Software applications such as routing protocols, chipset drivers, and link module drivers are independently linked. One of the main benefits of dynamic loading is the ability to reconfigure a particular application dynamically without affecting other components and the efficient use of memory. Dynamic loading also aids in fault isolation and recovery for any application.

```
#    86: 01/10/00 17:58:30.355  DEBUG    SLOT  2  LOADER             Code:  43
Image ip.exe loaded successfully from 2:bn.exe

#    88: 01/10/00 17:58:30.378  DEBUG    SLOT  2  LOADER             Code:  15
Loader starting application ip.exe, address 0x30a7e190, gate id = 0x00048
```

Applications are independently loaded on each slot by its loader, based on the configuration. Local memory use by dynamically loaded applications is optimized, and smaller memory segments are needed to load the modules. Dynamic loading and unloading of applications using configuration records is discussed in Chapters 5 and 6.

Scheduler

BayRS features a simple yet robust scheduler for scheduling gates. In a traditional operating system used for conventional computing, processes are usually scheduled based on their class and priority. This is important particularly because user space processes are most common in these environments. In BayRS, the scheduler is designed for efficiency in forwarding path gates—gates for the most part share the same priority and are scheduled on a first in, first out (FIFO) basis. The scheduler's operations are mostly isolated internally and thus may not be of interest to many users. Its basics are mentioned here for the purpose of discussing watchdog faults later.

BayRS Concepts

In designing or troubleshooting BayRS routers, several basic concepts must be understood. Although many readers may be familiar with them or may know them from other products, they are presented here for comprehensive coverage.

Lines

A BayRS line is a physical or logical component that is capable of independently carrying data and whose data is uniquely observable at the physical layer.

Examples of physical lines are Ethernet and serial lines. Examples of logical lines are bearer channels on ISDN lines or an individual timeslot on T1 lines. In the case of both physical and logical lines, data received and transmitted is processed by unique line drivers. Data processed by these independent line drivers can be observed at or below the line driver level, details of which are discussed in "Packet Capture and Analysis" in Chapter 6.

```
#   180: 01/10/00 17:58:44.499  DEBUG    SLOT  2  DP                  Code:  40
DP gate rcv'd LINE message for cct 2.
dp_line_msg: CCT: 2 Slot: 2 Port: 1. DRIVER MAC: a200a0f3

#   181: 01/10/00 17:58:44.503  DEBUG    SLOT  2  DP                  Code:  23
Creating Circuit 2 soloist with line GH 0x404e
```

Line Numbers

Each line is uniquely identified by a line number. A line number may have up to 10 digits and has the format described in Table 2-2.

Field	Length	Description
Reserved	1	Reserved nibble always set to zero.
Channel	2	Channel index of the line on links with multiple lines, such as an MCT1.
Type	2	Physical media type of the line; valid values are: 1 Ethernet 2 Synchronous 3 T1 4 E1 5 Token Ring 6 FDDI 7 HSSI 9 MCT1 12 Asynchronous 13 ISDN D-channel 14 ATM 15 Bisynchronous 16 V.34 modem
Slot	2	Slot number where this line is present.

Table 2-2. Fields in a Line Number

Field	Length	Description
Module	1	Module number on which the line is present. On ARNs and ASNs, the net module number takes a value from 1 to 4. On all other platforms, this field is always set to 1.
Connector	2	Connector number on which the line is present.

Table 2-2. Fields in a Line Number *(continued)*

```
#   161: 01/10/00 17:58:39.492  DEBUG    SLOT  2  PCAP                Code:  67
interface became local - line 102101
intf gate spawned - line 102101

#   162: 01/10/00 17:58:39.496  DEBUG    SLOT  2  PCAP                Code:  65
wait_state - line 102101
```

In most cases where physical line information is needed, it is obtained from the slot, module, and connector numbers. Once the line number is formatted, leading zeros are usually dropped. For example, line number 0000101102 (which refers to the physical Ethernet line in slot 1, port 2) is usually written as line 101102. The line number is the handle used in performing packet captures (which is discussed in Chapter 6).

Circuits

A circuit component is responsible for providing essential data path services for protocol data delivered to and received from a line. Data received from lines is decapsulated, classified, and delivered to the appropriate network layer protocol. Data received from network layer protocols is appropriately encapsulated and delivered to the line. Each circuit is identified by its descriptive circuit name.

```
#   188: 01/10/00 17:58:44.570  INFO     SLOT  2  DP                  Code:   3
Circuit 2 up.
```

In uniline circuits, circuits are usually named so that the physical medium, slot, module, and connector numbers are easily discerned. For example, the circuit name E12 refers to a circuit associated with an Ethernet line on slot 1, port 2.

Occasionally, the term *interface* is used synonymously with the term *circuit*. This is discussed in more detail in the "Interfaces" section, later in this chapter.

Multiline Circuits

Each circuit may contain one or more lines. A circuit containing multiple lines is called a multiline circuit. (There are two types of PPP circuits: multiline and multilink. Both are multiple line circuits, and their differences are discussed later.) In a multiline circuit, frames delivered by network layer protocols are transmitted to a selected line. Multiline circuits are usually named based on the service and destination, because it may not be possible to include all the associated line names. Logical views of uniline and multiline circuits are shown in Figure 2-2.

The selection of the transmission line is made using a configured line chooser algorithm. Two choices are available: random and address based. From the available operational lines, the random-based chooser picks a line at random. The address-based chooser computes a hash so that data traveling between a source and destination pair always travels over the same line. The random chooser makes fair scheduling use of all available lines but may cause packets to arrive out of sequence, which must be handled appropriately by a higher-layer protocol. Address-based choosers ensure that packets arrive in sequence but may not make fair use of all available lines; certain traffic patterns could cause some lines to be overutilized while others are underutilized.

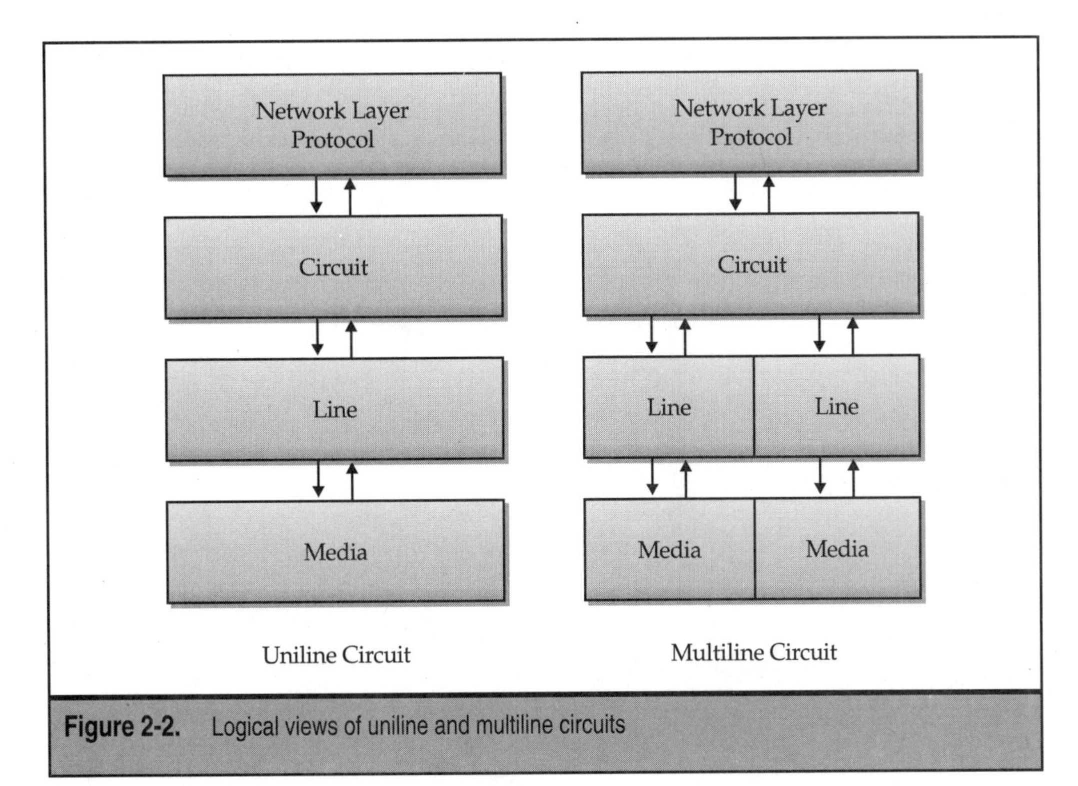

Figure 2-2. Logical views of uniline and multiline circuits

Subcircuits

Subcircuits operate similar to circuits, but they do not interface directly with higher-layer protocols. Instead, they communicate with a parent circuit, as shown in Figure 2-3. When subcircuits are present, the parent circuit usually communicates with the subcircuit first before delivering data to the line. Subcircuits are most commonly seen on frame relay circuits; each direct-mode VC is separately serviced by a subcircuit. A VC by itself does not represent a logical line because traffic on the VC is not exclusively observable on the physical medium without decoding data. The abstraction is seen only at the subcircuit level. Subcircuits are discussed in more detail when we discuss troubleshooting frame relay circuits in Chapter 4.

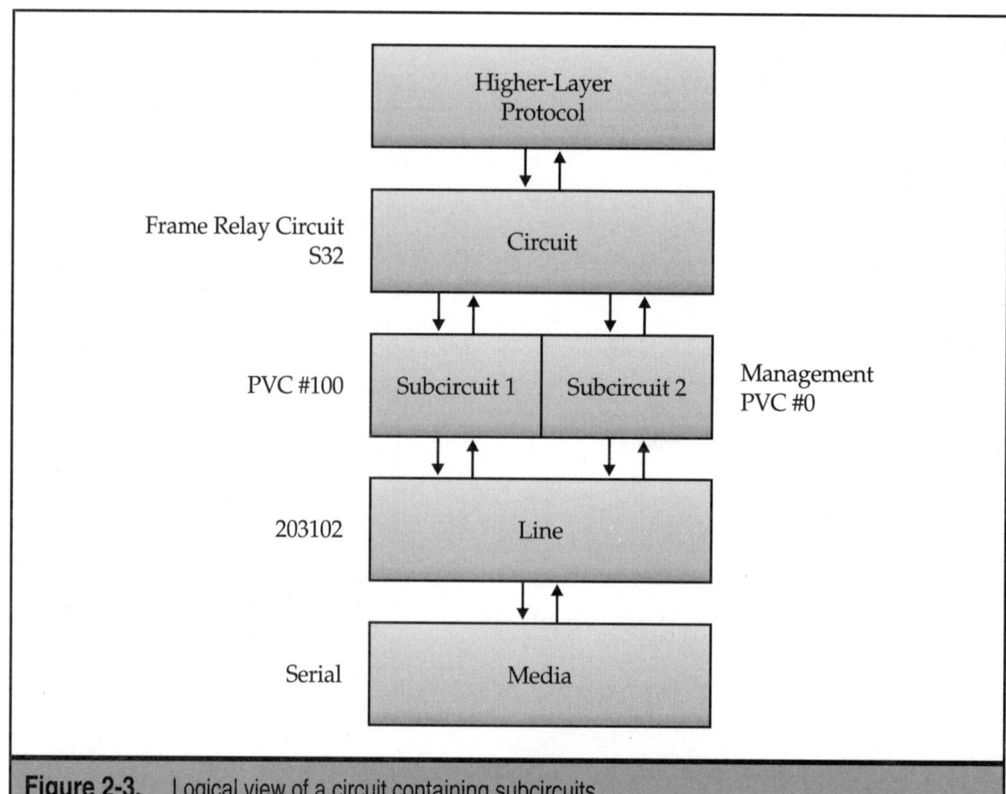

Figure 2-3. Logical view of a circuit containing subcircuits

Interfaces

Higher-layer protocol services are usually referred to as interfaces. An interface running the IP protocol is called an IP interface, and an interface running the IPX protocol is called an IPX interface. When identifying an interface, the associated protocol and network address are usually referenced. An example is IPX interface 0x00AA007E. Multiple protocol interfaces can be configured on a circuit, as shown in Figure 2-4. This is the basis for the term *multiprotocol router*.

```
#   190: 01/10/00 17:58:44.574   INFO     SLOT  4  IP                Code:   2
Interface 10.3.92.1 up on circuit 2
```

As mentioned earlier, the terms *interface* and *circuit* often are used interchangeably. The term *circuit* is treated as a synonym for *interface* as long as the specific network layer protocol is mentioned. Conversely, the term *interface* is equivalent to the term *circuit* as long as the specific circuit name is mentioned. For example, IP circuit 10.3.92.1 refers to the same element as IP interface 10.3.92.1, and Ethernet interface E41 refers to the same element as Ethernet circuit E41.

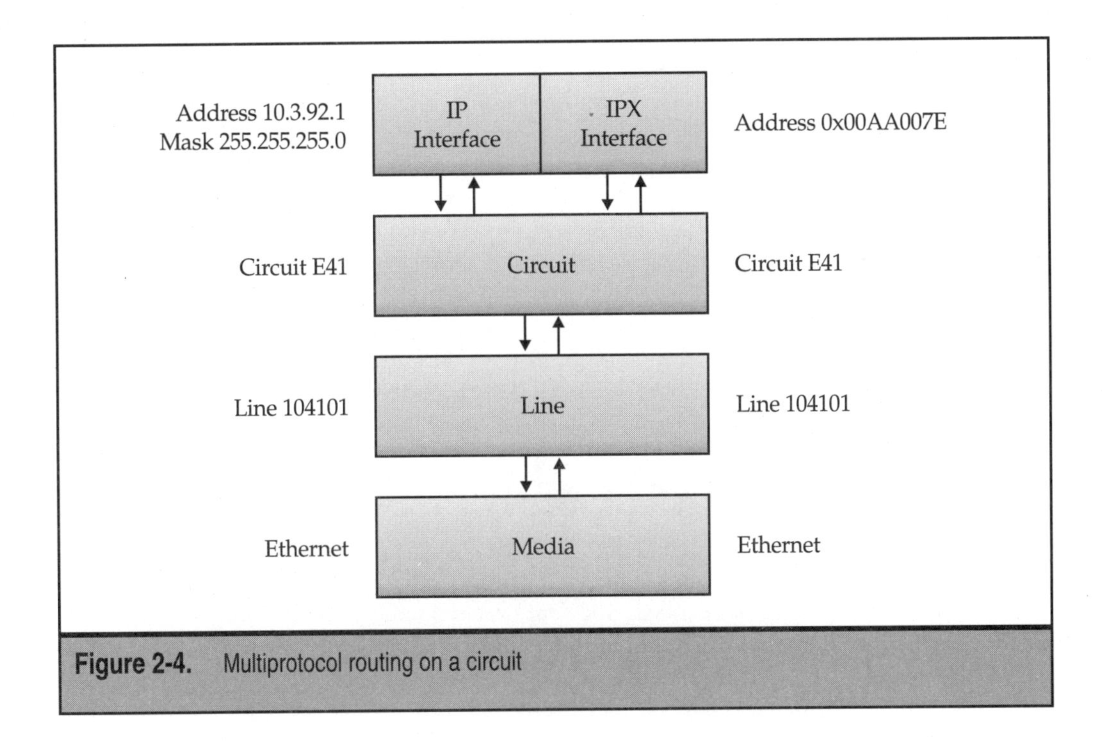

Figure 2-4. Multiprotocol routing on a circuit

In this book, unless otherwise noted or where interpretation is unambiguous, a circuit refers to a component providing data path services, and an interface refers to a component providing network layer protocol services.

Traffic Prioritization

Data traffic for applications and protocols has varying requirements, some requiring stringent delivery times and others flexible in their delays. Time-sensitive and mission-critical traffic often needs to be given priority over other traffic, such as throughput-sensitive traffic. For example, IBM SNA traffic is very time sensitive and is frequently given priority over bulk transfer traffic such as FTP. BayRS features *protocol prioritization* (PP), which allows prioritization of traffic based on protocol or other user-defined criteria.

Data is first classified and assigned to one of three priority queues: the high queue, normal queue, or low queue. Data that cannot be classified according to the specified filter criteria is automatically placed in the normal queue. Using criteria defined in *priority filters*, packets can be queued into high and low queues as desired. The classified traffic is then dequeued based on either a bandwidth allocation algorithm or a strict dequeing algorithm.

Traffic Filters

Chapter 1 discussed the use of traffic filters for the purpose of establishing an effective security strategy and developing an efficient method of managing bandwidth. Uniform traffic filters can be easily implemented for both bridged and routed traffic on inbound and outbound interfaces. As with traffic prioritization, data is classified based on one or several filter criteria. A traffic filter may be configured to take one of the following actions on a matching packet—drop or accept—with the option of logging that action. Filters are quite versatile and can be configured and managed either using predefined criteria (such as well-established protocol fields) or user-defined criteria (for very specific parameters). They are applied on a per-interface basis, and packets matching several filters simultaneously can be acted upon by the filter with the highest precedence.

SIMPLE NETWORK MANAGEMENT PROTOCOL

The Simple Network Management Protocol (SNMP) is a comprehensive protocol for the management of multi-vendor devices under a common IP framework. BayRS routers can be configured and monitored using applications that use SNMP. Site Manager and Optivity are examples of SNMP-capable network management applications.

Architecturally, the operation of SNMP requires two components. An SNMP *manager* is a device that is capable of configuring, monitoring, and receiving traps from network devices operating as SNMP agents. In most cases, SNMP managers are management applications running on a workstation. Site Manager, Optivity, HP OpenView, Sun

SunNet Manager, and IBM Tivoli TMS10 are examples of SNMP managers used with BayRS routers.

The other component of the SNMP architecture is the SNMP *agent*. The agent is responsible for accepting and processing requests from the manager and also may asynchronously send trap messages to the manager. In most cases, the interaction between the manager and agent can be described as a client-server relationship, where the manager fulfills the client role and the agent fulfills the server role.

In the seven-layer OSI model, SNMP resides on the session layer (layer 5). Beneath the session layer, UDP is used as the transport mechanism. SNMP transactions are maintained using a database called the Management Information Base (MIB). Using the Structure of Management Information (SMI) and Abstract Syntax Notation One (ASN.1) presentation formats, the MIB defines and represents all the manageable components and variables on any managed device. Both the MIB and SNMP have been universally accepted and are available on almost all networking devices that are capable of supporting them.

SNMP Packet Types

An SNMP packet sent by a manager typically contains three important fields: the version, community string, and SNMP message (data). BayRS supports v1 (RFC 1157) and v2C (RFCs 1902-1907, RFC 1901) of SNMP. The community string is used as a basic authentication mechanism by the manager and the agent. The SNMP message, called the protocol data unit (PDU) can be of five types: GetRequest, GetNextRequest, GetResponse, SetRequest, and Trap. The get PDUs are used to fetch information from the agent, and the set PDU is used to modify an attribute value. The trap PDU, as mentioned previously, is sent unsolicited to the manager by the agent to indicate certain critical conditions.

Regular SNMP traffic uses a predefined UDP port for communication, port 161. Trap messages are sent on another predefined port, port 162. A simple view of the SNMP frame format is shown in Figure 2-5.

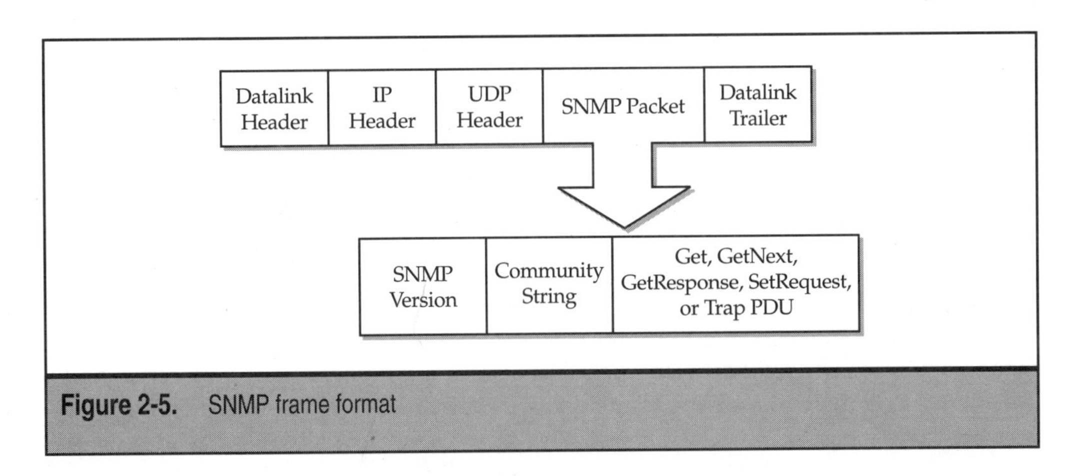

Figure 2-5. SNMP frame format

Traps

Traps are unsolicited messages that are asynchronously generated by the agent and sent to SNMP managers. They indicate a critical condition or a change in state that has significantly affected the operational behavior. For example, if a device is reset accidentally by someone's tripping over the power cord in the closet, then network connectivity is interrupted until the unit comes back to full operation. Network managers and operation centers observe unscheduled changes in network operation and react to them. In this case, the device that momentarily lost power will automatically send a cold start trap message to the network management system, which generates alarms using visual displays, pagers, and so on. Six generic standards-based traps are defined in RFC 1157: namely, coldStart, warmStart, linkUp, linkDown, authenticationFailure, and egpNeighborLoss. BayRS generates these generic traps and also can also be set up to generate enterprise-specific traps. Enterprise-specific traps can be generated for a broad range of events.

MANAGEMENT INFORMATION BASE

The Management Information Base (MIB) is a simple, hierarchical database. Ideally, a MIB is a tree of objects maintained in a hierarchy. Each object is a unique node in the tree and may contain other child objects or attributes. In display notation, an object or an attribute is uniquely identified by an object identifier (OID). The OID is constructed by simply concatenating the hierarchical object numbers in the path from the root to the object, each object number separated by a period (.).

An object in the hierarchy can either be referred to by a named identifier (for example, iso.org.dod.internet.mgmt.interfaces) or by the enumerated object identifier (for example, 1.3.6.1.2.2). When an object is referred to by name, it can be referred to by the full name identifier, or parent.child form, or simply by its unique name. Each object and attribute name is assigned a unique name in the hierarchy. For example, the identifiers iso.org.dod.internet.private.enterprises.wellfleet, 1.3.61.4.1.18, enterprises.wellfleet, and wellfleet all represent the same object. A high-level view of the private wellfleet (Nortel Networks) MIB objects is illustrated in Figure 2-6.

The MIB tree can generally be divided into two main categories:

- Standards-based objects
- Enterprise/private objects

Standards-based MIB objects are so called because they are defined by IETF Standards and RFCs. Many standards-based MIBs are broadly supported by numerous network devices from many vendors, thereby providing a comprehensive management capability using a single network management platform. Enterprise MIB objects are also called private MIBs because they are defined under the iso.org.dod.internet.private branch, beneath which each vendor implements and maintains a tree to support its networking products. BayRS supports most standards-based MIBs and also has an extensive private MIB used for configuration and management.

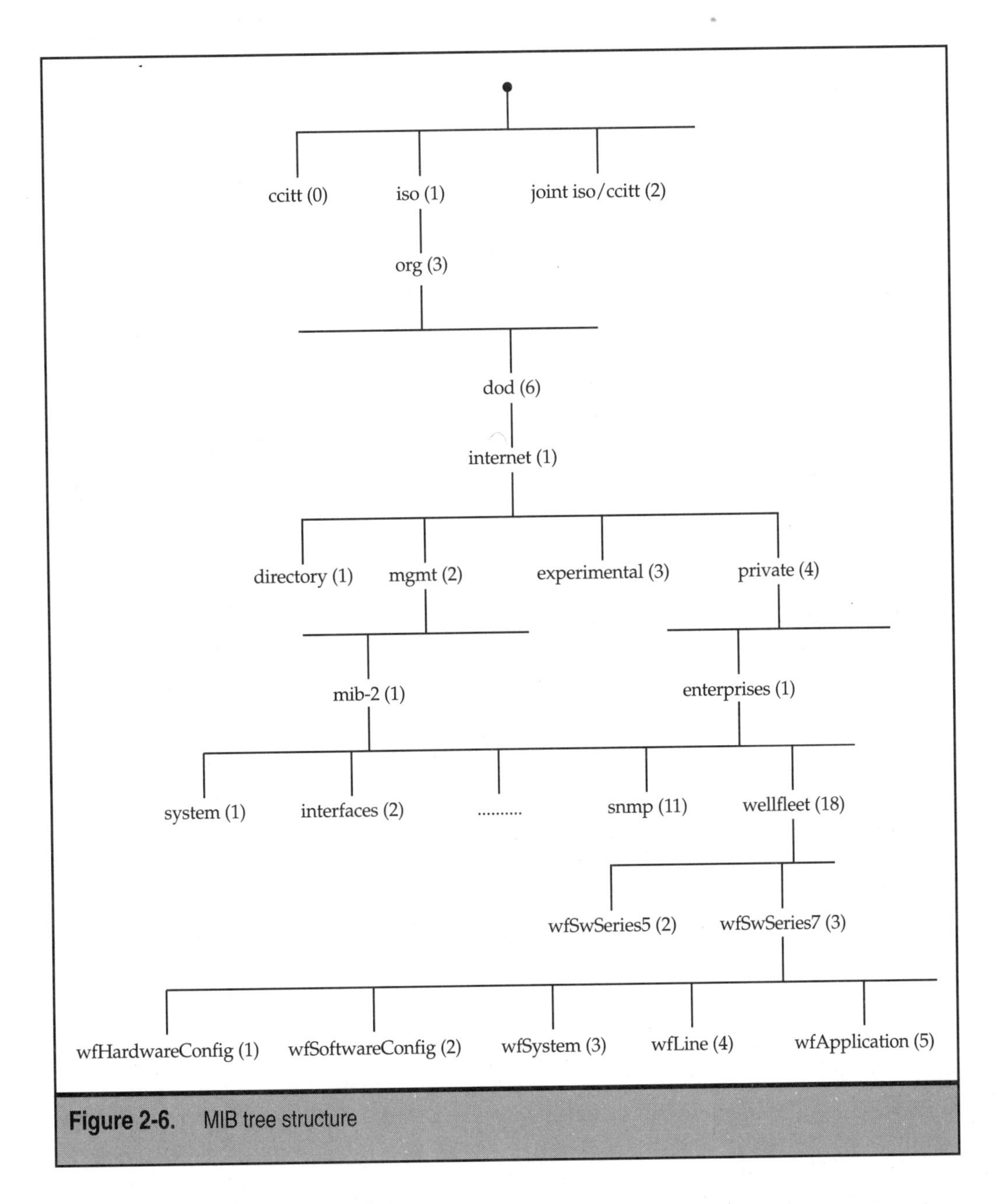

Figure 2-6. MIB tree structure

Abstractly, an *object* contains attributes or other objects, in which case it is called an *object group*. For example, `ifEntry` (interface entry) is termed an object, and `ifAdminStatus` is said to be an *attribute* of that object. Here, it is important to visualize the concept of an *instance*. An instance is an operable instantiation of an abstract object.

The `ifEntry` object just mentioned is an abstract component because it exists only conceptually. When the `ifEntry` object is instantiated, it creates an instance—example instances of the interface object are Ethernet `ifEntry` and serial `ifEntry`. As discussed before, each object contains attributes. The `ifEntry` object, for example, contains attributes such as `ifMtu` (maximum transmission unit) and `ifSpeed` (interface speed). Each instance embodies the same set of attributes and has its own values. The concepts of object, attribute, and instance are illustrated in Figure 2-7.

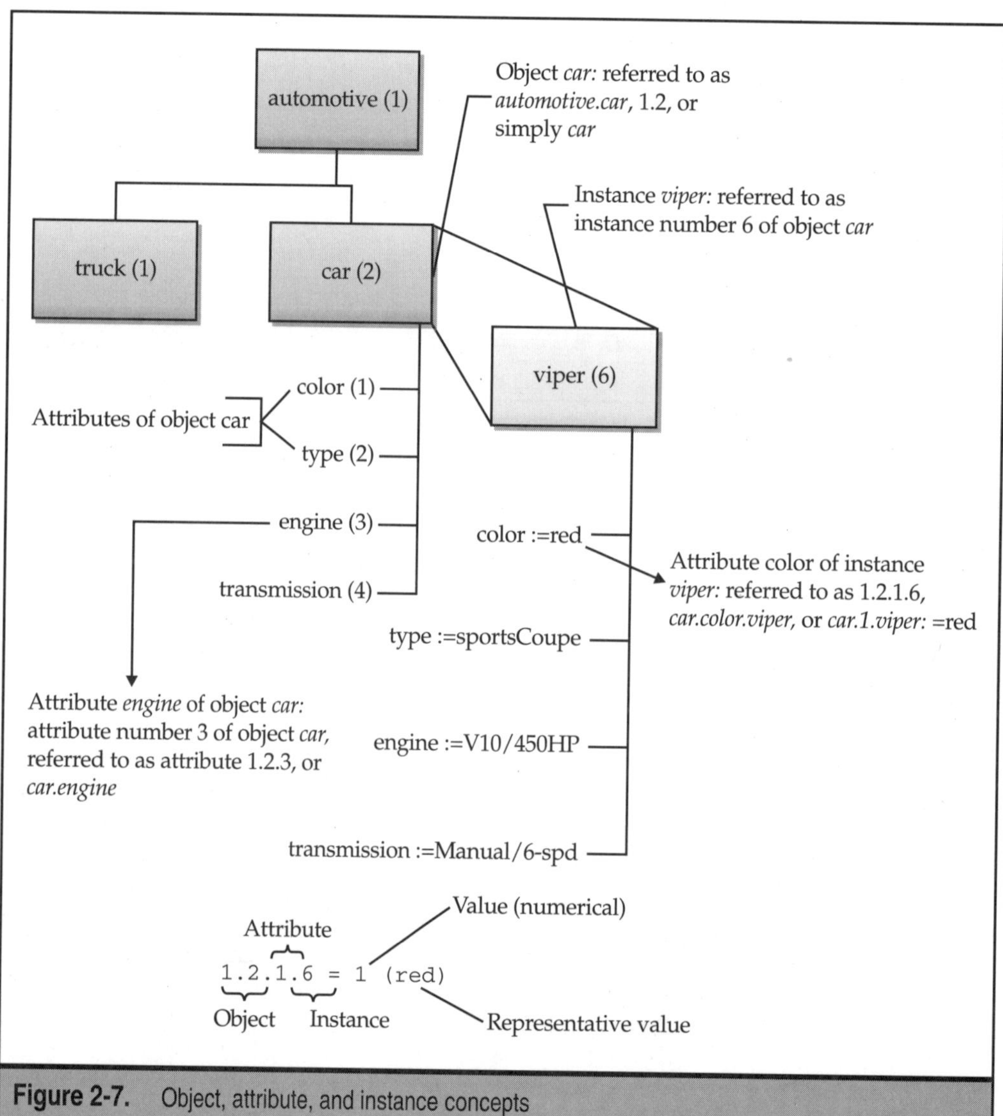

Figure 2-7. Object, attribute, and instance concepts

A particular attribute of an object instance is referred to using the form *object.attribute.instance*. In the car object example, the color attribute of instance viper is referred to as car.color.viper. Each instance of an object is referred to by its instance identifier, which is numeric. For example, the viper instance is identified by the instance ID 6. The engine attribute (3) of instance ID 6 (viper) is identified by the OID 1.2.3.6. The attribute values may be of different data types. One of the most common data types is the enumerated integer, where each data value is represented by a numerical value—for example, the value red for color is represented by 1, blue by 2, black by 3 and so on.

The enterprise branch supported by Nortel Networks routers begins with the OID prefix 1.3.6.1.4.1.18 (iso.org.dod.internet.private.enterprises.wellfleet). In particular, BayRS routers support the object branch under `wfSwSeries7` (wellfleet.3). All BayRS versions above 7.00 are called Series 7 software because they represent the new generation of routing software and hardware platforms. The previous-generation software MIB, called the Series 5 (`wfSwSeries5`, wellfleet.2) MIB, is unused in concurrent versions. The top-level object groups under the `wfSwSeries7` object are as follows:

- `wfHardwareConfig` (`wfSwSeries7.1`) consists of objects relating to hardware configuration.
- `wfSoftwareConfig` (`wfSwSeries7.2`) consists of objects relating to software protocols and drivers.
- `wfSystem` (`wfSwSeries7.3`) consists of objects relating to the system record, console, GAME kernel, statistics, circuit names, file systems, syslog facility, and other systemwide objects.
- `wfLine` (`wfSwSeries7.4`) consists of objects relating to line drivers such as serial, Ethernet, and all data-link layer lines.
- `wfApplication` (`wfSwSeries7.5`) consists of objects relating to protocol configuration parameters.

Private MIBs used by BayRS routers are often used by network administrators as reference, by NMS applications for management, and by MIB browsers for object and attribute detail. These MIBs are documented in MIB files that are provided along with the Site Manager suite. They are installed in \wf\mibs on Windows and /usr/wf/mibs on UNIX. If the default install directory is not used, then the mibs directory in the install directory will contain the MIB files. All files are written in ASN.1 notation following SMI in ASCII format and can be viewed using an ordinary text editor.

NOTE: The latest MIB files and a web-based MIB browser tailored for use specifically with BayRS MIBs are provided on the book web site. The browser is based on Scotty, a TCL-based network management software, which must also be installed on the web server. Current MIB files can also be downloaded from ftp://ftp-support.baynetworks.com/MIBs/.

MIB Access

Site Manager features the Quick Get application, which can be used to view objects and traverse the MIB tree on a BayRS router. Featuring a simple MIB browser, the application will fetch object-, instance-, and attribute-specific information such as the access type, data type, and syntax. The utility is useful for quick browsing of the MIB, but it does not allow you to make changes to the MIB. A full-fledged MIB browser in a Network Management System (NMS) can be used to make simple modifications to MIB attributes.

The standards-based and wellfleet-specific MIB objects can be accessed and manipulated using the Technician Interface. Commands for working with MIB objects are discussed in the next section.

THE TECHNICIAN INTERFACE

The Technician Interface, commonly called the TI, is the management interface on a BayRS router. It is accessible via a direct (or dial) connection to the console port and via telnet. In-band telnet connection is available when the router is configured and operating correctly. Direct-dial connection is available at all times, even when the router has no network configuration. When a router arrives from the factory, it does not have any configuration; the TI console is used to provide basic configuration that will allow an SNMP session to be established using Site Manager. Although you may already be aware of the TI and its basic use, it is an essential foundation for this book and so is covered in this chapter.

Apart from its use in the initial installation task, the Technician Interface is used to maintain and diagnose router operations. Many network operation centers regularly use the TI to monitor circuits and interfaces and to disable and enable lines and circuits. With an in-depth understanding of BayRS operation and the MIB structure, the TI can be a more valuable tool in troubleshooting and resolving operational issues. This book focuses on achieving this task with many protocols.

At first look, the Technician Interface can be compared to a command-line interface (CLI). In addition to CLI-like commands, the TI provides a comprehensive scripting language that can be used to extend the functionality. The BayRS MIB is the foundation for all configuration parameters and is covered in the next section. These MIB objects used for configuration are accessible at the most granular level using the TI. Hence, it is possible to selectively debug, troubleshoot, and optimize various hardware and software components.

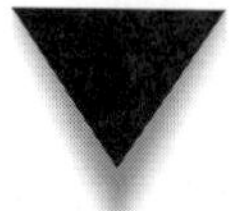

CAUTION: The Technician Interface provides minimal type checking of input to ensure that data is of the right data type. It does not provide the consistency checking or verification of values, as performed by Site Manager. The flexibility and control that the TI provides also allows a user to set faulty values that may adversely affect router operation. Router administrators should be aware of the hazards involved in committing unverified changes.

Accessing the Technician Interface

Console and dial (out-of-band) connections are established using the console port. For the Backbone Node family, a 25-pin console connector is available on the SRM-L, and for the Access family, a 9-pin connector is available on the base unit. A VT-100–compatible terminal is attached to the console port using one of the RS-232 cables listed in Table C-1 (see Appendix C). The VT-100 terminal parameters are set to 9600 baud, 8 data bits, no parity, and 1 stop bit. Flow control is either unused or set to xon/xoff.

Once the terminal is connected to the router, press ENTER and a login prompt appears. Administrators can log in for monitoring purposes as User, which allows read-only access, or as Manager, which allows privileged and unrestricted access. All Technician Interface interactions are case sensitive, including passwords.

```
Login: User
or
Login: Manager
Password:
```

After you enter the username, the router requires you to enter a password if the system is configured with one. Routers arriving from the factory do not have any password configured. If the router is configured for SecurID authentication, it will request you to enter a secure passcode when accessing it via telnet. Once the password (and passcode, if applicable) is entered, the Technician Interface displays a default welcome notice or a router-specific custom notice and a command prompt.

```
System  : Orlando-AN2 running 11.03/9, 16MB DRAM
Contact : NOC, x5000 or Entab, x5150
Location: Network Closet, 2nd Floor, Glades Ave
Serial No: AEX087239

Welcome to the Technician Interface

OR-an2(1)#
```

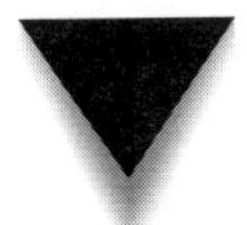

TIP: The default welcome notice can be replaced with a router-specific custom notice. An ASCII format text file titled ti_msg.txt can be installed in the system flash; this will then be displayed instead of the default welcome notice. A custom notice is particularly useful for promptly identifying a router as soon as you are logged in.

Login Scripts

User and Manager logins can be set up with automatic login scripts that are run when the user logs in. Manager and User both have their own autoscripts, which can be resourcefully written to suit the network and router needs. For example, instead of using the custom notice file, the autologin script can dynamically obtain the system name, description, and contact information directly from the system record. In this fashion, networkwide autologin scripts can be deployed without having to customize them for each router.

The default autologin script for the Manager login is called automgr.bat, and that for the User login is called autouser.bat. The script names can be changed, and they are searched on all available file system volumes, unless the file is referenced from a specific volume.

Remote Access

Routers critical to network operation should always be accessible remotely. In situations where network operation is considerably impacted, the TI may be accessible only through a dialup session. Hence, it is recommended that critical routers be set up with a dialup modem or connected to a terminal server. When connecting a modem to the router using a modem cable, you should enable the console's *modem enable* port attribute. Configuration parameters for modems are listed in Table C-2 in Appendix C. It is possible to innovatively set up remote access to the console interface, for instance, using an X.25 PAD connected to an X.25 network.

Accessing the TI via telnet requires the router to be configured as a telnet server. Using a telnet client on a workstation, the router will allow remote access to the TI through the telnet server. Several TI sessions can be established via telnet, depending on available network and local resources.

TI Soloist

Recall that the Technician Interface operates as a soloist process, as discussed in the section on soloist gates. In most cases, the slot that comes out of diagnostics and boots first will be the TI soloist. If the slot running the TI soloist is reset, then another soloist is automatically elected to assume the tasks. Existing sessions that were serviced by the previous soloist are terminated.

The TI soloist process controls the console port used for out-of-band management. The telnet TI process, on the other hand, will run on the slot that owns the interface IP address being telnetted to.

TECHNICIAN INTERFACE FUNDAMENTALS

Once you are logged into the Technician Interface as User or Manager, a default prompt appears similar to this:

```
[x:y]$
```

The slot running the TI soloist process is indicated by the value *x*, and the port number it is connected to is indicated by *y*. On single-slot routers such as the AN, the slot number is always 1. On multi-slot routers, the slot number will vary depending on where the TI soloist is executing currently. The port number in most cases is 1 except in routers such as a stacked ASN or ARN, which may have multiple out-of-band console ports.

The TI prompt is customizable; in fact, you may have already customized your router prompts. One of the reasons to do so is to clearly identify each router when multiple TI sessions are established to several routers. This assists in monitoring and troubleshooting processes. Customization of TI prompts is discussed in the next chapter.

TI sessions are terminated by executing the **logout** command at the prompt. Commands are always executed in the foreground. If a TI session is terminated abruptly, the command may or may not run to completion. A command that is currently running can be terminated by issuing the break sequence, which can be accomplished by pressing CONTROL-C or CONTROL-BREAK (at some terminals). Once the command is terminated, the TI returns to the command prompt.

Help and Navigation

The **help** command is the most frequently used command in the TI. To obtain help on a specific command, the following syntax is used:

```
[2:1]$ help [-all|<command>]
```

A complete list of commands and their syntax is provided in Appendix D, which can be obtained by entering the command **help help**.

Certain commands, if invoked incorrectly, will automatically discard the arguments and display the help text. The most recent command can be reissued by using the **repeat** command (!) which can be followed by an optional number to indicate the repeat count. Recent software versions support the use of the up arrow (CONTROL-P) and down arrow (CONTROL-N) to navigate the history list. See the following example.

```
[1:1]$ get wfSyncEntry.37.1.1
wfSyncEntry.WfSyncRxFrames.1.1 = 4568232
[1:1]$ !
wfSyncEntry.WfSyncRxFrames.1.1 = 4568236
[1:1]$ ! 3
wfSyncEntry.WfSyncRxFrames.1.1 = 4568242
wfSyncEntry.WfSyncRxFrames.1.1 = 4568248
wfSyncEntry.WfSyncRxFrames.1.1 = 4568251
[1:1]$
```

Each TI session maintains a history window of previously issued commands. Commands that were syntactically invalid are not maintained in the history. The default history window is set to 20 commands, which can be increased up to 40, for each TI telnet and TI console. For the TI console, use

```
[1:1]$ set wfSerialPortEntry.wfSerialPortHistoryDepth.instance 40; commit
```

and for TI telnet sessions, use

```
[1:1]$ set wfTelnet.wfTelnetHistoryDepth.instance 40; commit
```

To view the command history buffer, issue the **history** command without any arguments. Each command in the history list is preceded by a history event number. The event number can be used as an argument to the **history** command to rerun it. Commands can be abbreviated as long as the shortened form is unique in the command list.

```
MX2-BCN> get wfSyncEntry.37.2.2
wfSyncEntry.wfSyncRxFrames.2.2 = 9723989
MX2-BCN> get wfSyncEntry.39.2.2
wfSyncEntry.wfSyncTxFrames.2.2 = 10243632
MX2-BCN> history
    1   get wfSyncEntry.37.2.2
    2   get wfSyncEntry.39.2.2
MX2-BCN> his 1
get wfSyncEntry.37.2.2
wfSyncEntry.wfSyncRxFrames.2.2 = 9724524
```

Display and Command-Line Editing

The default screen display is 80 columns wide and 24 lines long. Depending on the terminal emulation parameters, the screen display rows and columns may be adjusted. At times, data may scroll past the screen's display lines. By default, paging of output is enabled for the TI every 24 lines. This can be changed, or the pagination can be turned off, using the **more** command:

```
{2}BCN2-Con$ more on 20
  More Mode: ON
  Lines per screen: 20
{2}BCN2-Con$ more off
  More Mode: OFF
```

MIB Commands

Enterprise-specific and standards-based MIB objects can be manipulated using the following set of TI commands:

- **list**
- **get**

- set
- commit

Object and attribute names are case sensitive and must be entered correctly when retrieving or setting attribute values. The TI will generate an "invalid object or attribute" warning if the object or attribute name is spelled incorrectly.

List

The **list** command can be used for the following tasks: listing all accessible objects, listing all accessible objects with a specific prefix, listing attributes for a specific object, and listing configured instances of a specific object. It is most often used as a first step to identify attributes, objects, and instances. All accessible objects at the terminal nodes can be referenced just by the object name, without needing to refer to the entire OID either in named or dotted-decimal notation. The **list** command can be abbreviated using the letter l.

```
$ help list
Usage:
list      [[<instances> [<obj_name>]]]
```

Using the **list** command without an argument lists all accessible MIB objects.

```
$ list
wfCSMACDEntry = 1.3.6.1.4.1.18.3.4.1.1
wfCSMACDAutoNegEntry = 1.3.6.1.4.1.18.3.4.16.1.1
wfFddiEntry = 1.3.6.1.4.1.18.3.4.4.1
wfFddiSmtEntry = 1.3.6.1.4.1.18.3.4.15.1.2.1
.........
```

With a partial prefix and an asterisk suffix as shown here, the **list** command will display all MIB objects starting with the specified prefix. This example lists all MIB objects starting with dot1d, the 802.1d Spanning Tree MIB.

```
$ list dot1d*
dot1dBasePortEntry = 1.3.6.1.2.1.17.1.4.1
dot1dStp = 1.3.6.1.2.1.17.2
dot1dStpPortEntry = 1.3.6.1.2.1.17.2.15.1
dot1dTp = 1.3.6.1.2.1.17.4
dot1dTpFdbEntry = 1.3.6.1.2.1.17.4.3.1
dot1dTpPortEntry = 1.3.6.1.2.1.17.4.4.1
dot1dStatic = 1.3.6.1.2.1.17.5
dot1dStaticEntry = 1.3.6.1.2.1.17.5.1.1
```

When a complete object is specified, the **list** command displays all attributes accessible for the specific object. In the following list, the attributes for the translation bridging global object (wfBrXb) are displayed.

```
$ l wfBrXb
wfBrXbBaseDelete = 1
wfBrXbBaseDisable = 2
wfBrXbBaseEthernetLanId = 3
wfBrXbBaseMaxRifEntries = 4
wfBrXbBaseCurrentRifEntries = 5
wfBrXbBaseAgeTime = 6
wfBrXbBaseBcastAddressConversionDisable = 7
wfBrXbBaseDefaultMode = 8
wfBrXbBaseBcastMode = 9
wfBrXbBaseSaps = 10
wfBrXbFddiBridge = 11
```

The last syntax uses the –i (instances) option, to list all configured instances for a specific object. The following example lists all instances of the IP interface (wfIpIntfCfgEntry) configuration object.

```
$ l -i wfIpIntfCfgEntry
inst_ids   = 192.168.43.1.1
             204.98.183.95.2
             204.98.99.12.3
```

Get

The **get** command is used for the following tasks: to display the value of an attribute for one or all object instances and to display the values of all attributes for a specific object instance. The **get** command can be abbreviated as **g**. The syntax for the **get** command is shown here.

```
$ help get
Usage:
get        {<obj_name>|<obj_id>}.{<attr_name>|<attr_id>|*}[.{<inst_id>|*}]
```

In the first syntax below, we use **get** with a wildcard attribute to obtain all attribute values for the global instance of the BGP version 4 (wfBgp4) object.

```
$ g wfBgp4.*.0
wfBgp4.wfBgp4Delete.0 = 1
wfBgp4.wfBgp4Disable.0 = 1
wfBgp4.wfBgp4State.0 = 1
```

Here, we use the wildcard in the instance position for a specific attribute: the physical address of all ARP entries stored in the wfIpNetToMediaEntry object.

```
$ g wfIpNetToMediaEntry.2.*
wfIpNetToMediaEntry.wfIpNetToMediaPhysAddress.1.57.82.98.33 =
        x00 x00 xA2 x09 xF5 x89
wfIpNetToMediaEntry.wfIpNetToMediaPhysAddress.1.57.82.98.40 =
        x00 x00 xA2 xCB xC8 x81
```

As shown in the preceding examples, the wildcard character (*) can be used in place of the attribute (to get all attributes) or in place of the instance (to get all instances). Instances may contain a partial prefix or suffix along with the wildcard to narrow the get query. The **get** command will disallow the use of the wildcard for both the attribute and instance fields. This is because objects containing numerous attributes and several instances will quickly produce a tremendous output listing that may consume significant memory.

```
$ g ip.ipDefaultTTL.0              (Symbolic object and attribute, numeric instance)
$ g 1.3.6.1.2.1.4.2.0              (Numeric object, attribute, and instance)
$ g ip.2.0                         (Symbolic object, numeric attribute and instance)
$ g 1.3.6.1.2.1.4.ipDefaultTTL.0 (Symbolic attribute, numeric object and instance)
ip.ipDefaultTTL.0 = 30             (All four commands return the same value.)
```

The attribute being fetched can be referenced in any legal MIB fashion (either numerical or symbolic), as shown in preceding examples, although it is most common and simpler to use the object name instead of the OID. If a certain object has only one instance, as is the case with most base records for global objects, then the instance specification in the **get** command is optional.

Set

The **set** command is used for one specific task: modifying the value of a single attribute of a specific instance. As with the **get** command, the attribute is specified in *object.attribute.instance* format, but the wildcard character must not be used. The value being specified must have the same data type as the attribute. Table 2-3 lists the common input data types, their formats, and if applicable, their ranges.

Data Type	Format and Range
Integer	Unsigned integer, theoretical range 1 to 4294967295, but range may be limited. Consult the attribute syntax in the MIB.
Octet String	Hexadecimal number, starting with 0x. Commonly used by MAC addresses. Consult the attribute syntax for length.
IP Address	Dotted-decimal IP address in the form *a.b.c.d*.
Display String	Alphanumeric string enclosed in quotation marks.

Table 2-3. Common Input Data Types and Their Formats

The syntax and examples of the **set** command are shown here.

```
$ help set
Usage:
set       {<obj_name>|<obj_id>}.{<attr_name>|<attr_id>}.<inst_id> <value>

$ set wfTftp.wfTftpDefaultVolume.0 1
$ set 1.3.6.1.4.1.18.3.5.3.6.2.0 1
$ set wfTftp.2.0 1
```

It was mentioned previously that care should be taken when modifying an attribute's value using the TI. Since the TI does not perform any validity checking, the user must verify the validity of a value before committing the change. To aid in the verification of the value being set, consult the MIB for valid values and ranges for that specific attribute.

Commit and Save

MIB attributes that are set in-band using SNMP are immediately committed to memory and activated. However, MIB attributes that are set using the TI command must be activated using the **commit** command. This is done because multiple **set** commands are often issued in sequence, and issuing a single **commit** command will activate the changes all at the same time.

```
[6:1]# commit
```

Changes committed to memory are active only while the router is operational. If the router is reset, the changes are lost unless they are saved permanently in a configuration file. The following commands perform two tasks: saving the currently used configuration (in memory) in a temporary staging configuration file, and backing up a known, good working configuration to the flash, from which the router was initially booted.

```
[6:1]# save config 6:staging.cfg          (Temporary staging config file)
[6:1]# copy 6:config 6:bkup101299.cfg     (Dated backup of config file)
```

Configuration changes usually should not be saved permanently in the default configuration file unless the changes were minimal and their effects are well known. The two commands in the preceding example will retain the existing configuration file, backup a working copy, and save a copy of the configuration in memory. If the router resets because of a configuration error or some other event, it will automatically reinitialize with the known working configuration. During this time, if the router reboots because of an external event such as a power failure, it will automatically revert to the known working configuration file (config). This procedure is crucial for core routers and unattended remote office routers. Once the committed changes have been verified to work over the period of a few days, then the staging configuration file can be copied to the default configuration file.

```
[6:1]# copy 6:staging.cfg 6:config
```

The File System

All BayRS routers use a PCMCIA flash for file storage, except the AN1, which features a flash SIMM. Both flashes are formatted using the same proprietary file system, which has been developed to suit programmable flash devices. The file system can be managed using the Files Manager utility in Site Manager or using the Technician Interface.

File System Commands

File system access is similar to that for the DOS file system. All flashes are removable and are numbered based on the slot number where they are currently installed (for example, 1:, 2:, 5:, and so on); this is similar to DOS, which uses an alphabetical volume sequence (such as a:, b:, c:, and so on). Backbone Node processor modules can each be installed with a flash card, and system controller cards (in VME routers) can contain up to two flash cards. Earlier VME routers such as the AFN and LN feature a floppy-based DOS file system, whose file system commands are the same as those of DOS. Commands used in file system manipulation are listed in Table 2-4.

Command	UNIX Equivalent	DOS Equivalent	Task
cd	cd	cd	Change current working directory to a new volume
directory	ls	dir	List contents of a volume or specific files
copy	cp	copy	Copy one or more files within the same volume or across volumes
delete	rm	delete	Permanently remove a file from the file system
compact	n/a	n/a	Compact a file system to reclaim unused space
dinfo	df	chkdsk	View complete file system status on one or more volumes
format	format	format	Format a single volume using the proprietary file system

Table 2-4. General TI File System Commands

Command	UNIX Equivalent	DOS Equivalent	Task
partition	mkfs	fdisk	Divide a single volume into two separate file system partitions
record	script	n/a	Record TI input and output to a file
mount	mount	n/a	Mount a file system from a newly inserted flash
unmount	umount	n/a	Unmount a file system so that the media can be removed
type	cat	type	Output contents of a file, in ASCII and/or HEX format
xmodem	sb/sz	n/a	Transfer files to and from the file system using an out-of-band serial connection

Table 2-4. General TI File System Commands *(continued)*

Except for the DOS file system on the VME routers, the file system is not hierarchical and does not contain directories. The volume being accessed is the root directory and contains files and no subdirectories.

```
$ cd 2:
New Present Working Directory: 2:

$ dir
 Volume in drive 2: is
 Directory of 2:

File Name              Size     Date        Day      Time
--------------------------------------------------------
laneipvc.bat           8125  05/24/1999  Mon.    09:58:14
atm_monitor.al         6572  06/21/1999  Mon.    14:54:08

16777216 bytes - Total size
16762519 bytes - Available free space
```

```
8969193 bytes - Contiguous free space

$ dinfo
VOL     STATE        TOTAL SIZE     FREE SPACE     CONTIG FREE SPACE
---------------------------------------------------------------------
 1:     FORMATTED     8388608        6162126           6162058
 2:     FORMATTED    16777216        16762519          8969193
 3:     CORRUPTED

$ format 3:
The format will erase the ENTIRE contents of file system 3:.
Do you wish to continue? [y|n]: y
Formatting file system on volume 3: ...
Formatting completed

$ compact 2:
Compacting file system on volume 2: ...
This may take several minutes...Please wait...      100% Complete
Compaction completed

$ partition create 1:
```

When creating a partition, the current file system on the existing volume cannot exceed one-half the total media size. The file system on the existing volume will remain intact. Once the partition has been created, the new volume is referenced as <volume>b:, and the existing volume can be referenced as either <volume>a: or <volume>:. When a partition is deleted, all files on <volume>b: will be lost since that volume is formatted. The **partition** command is supported only on the Access routers and System 5000-based ATM routers.

After the **format**, **compact**, and **partition** commands, the file system appears as shown here:

```
$ dinfo
VOL     STATE        TOTAL SIZE     FREE SPACE     CONTIG FREE SPACE
---------------------------------------------------------------------
 1a:    FORMATTED     4194304        1967822           1967754
 1b:    FORMATTED     4194304        4194304           4194304
 2:     FORMATTED    16777216        16762519          16762519
 3:     FORMATTED     8388608        8388608           8388608

$ record open 1b:ti_audit.log
$ copy 2:atm_monitor.al 3:
$ delete 2:*.al
```

A simple **rename** or **move** command is not available; these tasks must be accomplished by using a combination of the **copy** and **delete** commands, as shown in the preceding listing.

```
$ mount 5:
Mounting new volume...
Device label:
Directory: 5:
```

The **mount** command is used to explicitly engage the router to use a newly inserted flash file system. After the **mount** command is issued as shown above, volume 5 becomes the active volume. If a file system command is executed before the volume is mounted, the volume is mounted automatically.

```
$ unmount 3:
$ record close
$ type 1b:ti_audit.log
... ...
```

File Naming

Files on the flash can have a maximum of 15 characters, including a period, which is optional. They can be named using alphanumeric characters and the underscore (_) and period (.) characters. Spaces in file names are not allowed. Files names must begin with an alphabetical character, and the file extension is optional, although it is conventional to use one of the extensions listed here:

- **exe** Router images are usually named with the .exe extension, such as bn.exe or asn.exe.
- **out** Images for VME-based routers usually have the .out extension.
- **cfg** Binary configuration files can be identified easily using the extension .cfg. The default configuration file is called config.
- **log** Event log files can be saved to the flash using the .log extension.
- **bat** Script files installed on the flash can be automatically executed using the prefix if they contain a .bat extension.
- **mnu** Display menu for script files are usually named with the .mnu extension.
- **al** Aliases are permanently saved in files with the .al extension.

Wildcards

Certain commands, such as **copy**, **delete**, **dir**, and **type**, will accept file name arguments containing the ? and * wildcards. The ? wildcard represents any single character in that position, and the * wildcard represents any number of characters.

```
[1:1]# dir 2:*.bat
Volume in drive 2: is
 Directory of 2:

File Name                Size      Date        Day        Time
---------------------------------------------------------------
laneipvc.bat             8125   05/24/1999   Mon.     09:58:14

[1:1]# dir 3:*.??
Volume in drive 3: is
 Directory of 3:

File Name                Size      Date        Day        Time
---------------------------------------------------------------
atm_monitor.al           6572   06/21/1999   Mon.     14:54:08
```

Transferring Files

Operating system flashes shipped from Nortel Networks already contain the system image, PROM images, configuration file, and other files. During the course of router operation, it is usually necessary to transfer files to and from the flash. Most commonly, configuration and log files are transferred from the flash file systems.

Trivial File Transfer Protocol Files can be transferred to flash file systems using two methods. The first and simplest method is the use of the Trivial File Transfer Protocol (TFTP). In the default configuration, BayRS runs both a TFTP client and server so that files can be transferred using both TFTP get and put routines. From the Technician Interface, the TFTP client has the following syntax and can be used to get and put files from a TFTP server on the network.

```
# help tftp
tftp      {get|put} <name|address> <vol>:<file_spec> [<vol>:<file_spec>]
# tftp put 57.87.192.20 1:config ORan2bkp.cfg
```

Back up the current configuration to the TFTP server as ORan2bkp.cfg.

```
# tftp get 57.87.192.20 ORan2new.cfg 1:staging.cfg
```

Get a new configuration file and save it as 1:staging.cfg to perform staging on the router. Similar commands can be issued from a TFTP client on a workstation to get files from the router or put files onto it. Notice that the local file is prefixed with the volume number, which if absent, will imply the currently active volume. When you connect from a workstation to the BayRS router acting as a TFTP server, the TFTP default volume will be accessed unless another volume is explicitly specified.

Security File transfer using TFTP does not require authentication on the part of the TFTP client. This raises a security concern on routers behaving as TFTP servers because a malicious client can transfer files and damage the integrity of the network. Networks with high security requirements may restrict TFTP access using access filters or disable it in favor of some other mechanism such as FTP.

File Transfer Protocol An alternative to TFTP is FTP, the File Transfer Protocol. FTP uses TCP (Transmission Control Protocol) as the transport layer protocol, which provides reliable transport using acknowledgments. Most important, users must first be authenticated before performing file transfers. In contrast, TFTP does not require authentication and uses UDP (User Data Protocol), which is a connectionless transport mechanism. BayRS routers can behave only as FTP servers, and an external FTP client (such as one running on a workstation) is usually used to perform file transfers.

Bay Networks File System BNFS, the Bay Networks File System utility, can be used to manage flash file systems using Windows laptops. Available from the Router Management Labs web site at http://www.nortelnetworks.com/rml, the BNFS utility provides commands to perform the same set of file system operations as can be performed using the Technician Interface. Flashes can be formatted and files can be transferred to and from the volume with ease. BNFS provides an offline, alternative method for transferring files to the flash. This utility is unsupported because of the myriad types of laptops and PCMCIA drivers available today.

Scripts and Aliases

Aliases are custom-defined commands that are used to define a complete list of other commands in a sequence. They take the place of long or multiple command sets and provide flexibility in using the Technician Interface. Aliases are created using the **alias** command and removed using the **unalias** command. Current aliases in memory can be stored in a file and reloaded later into another TI session using the **save** and **source** commands. Aliases provide a way to enhance the TI and customize its behavior.

We mentioned the login scripts earlier, which function like the .login (or .profile) files in UNIX and the autoexec.bat file in Windows. The login scripts are a subset of a broader TI script system. Script files fall into two categories: embedded scripts and batch scripts. Embedded scripts are applications that are developed and built into the software image. Batch scripts are customizable and are used for a variety of purposes, including monitoring statistics, enabling and disabling circuits, and viewing configuration and state information for protocols and services. The **show**, **enable**, **disable**, and **monitor** command scripts are widely used with protocols, drivers, and services.

Script files can be installed in flash cards, and BayRS will automatically locate and run scripts from flashes that have been included in the search path. Network administrators and managers can write scripts to perform specific tasks, particularly to aid in customized monitoring and troubleshooting. Writing scripts for these purposes is discussed in the troubleshooting chapters (4–6).

Reinitializing Slots

Slots can be reinitialized from the TI in four ways, and the behavior of each method is subtly different. The following four commands can be used to reinitialize slots: **boot**, **diags**, **reset**, and **restart**.

boot

The **boot** command warm-starts the router and reloads the image and the configuration file. The system log is preserved in this process. As shown here, if the **boot** command is invoked without any arguments, then the default image and configuration files are used. The default files can be overridden using an optional image file and/or configuration file name. Using the hyphen (-) character for the file name indicates the default files, as listed in Table 1-2.

```
# help boot
Usage:
boot      [<vol>:<image_name>|- <vol>:<config_name>|-]

# boot 2:staging_bn.exe 3:staging.cfg     (Explicitly specified image and config files)
# boot - 2:staging.cfg                    (Default image file, staging config file)
# boot 4:staging_bn.exe -                 (Staging image file, default config file)
```

The **boot** command will verify that the configuration and image files are accessible before warm-starting the router. It has a systemwide impact and reboots all slots in a router. Router reboots can also be scheduled for a future time.

diags

The **diags** command cold-starts a router or the specified slots. Cold-starting the router invokes the diagnostics, and the subsequent bootstrap operation will load the default image and configuration files. It is important to ensure that a boot image server and config server are available; alternately, a flash file system must be accessible containing the default image and configuration files.

```
# help diags
Usage:
diags   [options]   [<slot ID>]
# diags                          (Can be omitted on single-slot routers only)
# diags 5-7                      (Execute diags on slots 5, 6, and 7)
# diags 8,12                     (Execute diags on slots 8 and 12)
```

A complete set of diagnostics, including the CPU, memory, backbone, and link tests (as applicable) are performed. If the slot ID argument is the keyword all, then diagnostics are run on all slots. On the ARN platform, the **diags** command can be invoked with the –on or –off option to toggle execution of the command subsequent to a cold start.

On Backbone Node routers, the **diags** command (with diagnostics monitor version 4.12 or above) can be used to selectively run diagnostics on the link module or processor module; the options –lm (link module) and –pm (processor module) are followed by the slot number. These options will be used later in the troubleshooting chapters (4–6).

reset

The **reset** command causes the selected slot or a range of slots to be warm-started. Unlike the **boot** command, which affects the entire unit, the **reset** command affects only the specified slots. The log is preserved during the reset. Resetting slots will automatically load the image and configuration using the backplane. On single-slot routers, reset will load the image and configuration from the flash; this behavior is identical to when you issue the **boot** command without any arguments.

restart

The **restart** command clears all operational tables on the specified slots and reinitializes the currently running copy of the image. This is by far the fastest method for reinitializing a slot because the image is not reloaded from the backbone or the flash. As with all other warm starts, the event log is preserved during the restart.

Table 2-5 summarizes the behavior of the four reinitialization commands.

Bay Command Console

The Bay Command Console (BCC) is the most recent addition to BayRS. BCC provides next-generation features to the Technician Interface. It has a hierarchical, context-sensitive user interface based on a Tool Command Language (TCL) shell. It features a comprehensive, easy-to-use command set for configuring protocols and services. Also, TCL scripts can be written to augment configuration management, performance management, and fault management tasks.

The primary focus and function of the Technician Interface has traditionally been performance and fault management, and Site Manager has typically been used for configuration management. BCC, on the other hand, can be used for all comprehensive management

Action	boot	diags	reset	restart
Warm start	Yes	No	Yes	Yes
Reinitialize log	No	Yes	No	No
Reload image	Yes	Yes	Yes	No
Reload configuration	Yes	Yes	Yes	No

Table 2-5. Behavior of Reinitialization Commands

purposes. Although BCC is a recent and evolving management application, it covers a major portion of all manageable protocols and services. Since it represents the future of console-based management, a majority of the troubleshooting tasks in the remaining chapters will be demonstrated using both BCC and TI commands.

BCC is launched by typing the **bcc** command at the TI prompt.

```
Router1> bcc
Welcome to Bay Command Console!
     * To enter configuration mode, type config
     * To list all system commands, type ?
     * To exit the BCC, type exit
bcc>
bcc> config
box#
```

BCC Modes

BCC has two modes: command mode and configuration mode. The command mode in BCC limits a user to read-only operations, such as monitoring statistics and viewing configuration details. Modification of the configuration requires you to sign on in the configuration mode, which can be entered by typing **config** at the command mode prompt. Prompts for the command and configuration modes are unique; the read-write capability is indicated by the trailing # sign in the prompt.

The command mode is accessible by both User and Manager logins, while the config mode can be accessed only by the Manager login.

BCC Concepts

Unlike many command-line interfaces, which use a linear concept, BCC is hierarchical. It exemplifies the underlying tree structure used in the configuration of BayRS routers. The configuration tree structure can be compared to the MIB tree in organization. The top-level component in the BCC hierarchy is the *box*, which represents the entire device. Under the box object, subcomponents such as global protocols and interface configurations are represented in tree form (see Figure 2-8).

Object An object is a data structure representing a configurable physical or logical entity such as a physical interface or a protocol on an interface. Every configurable object belongs to a specific class that defines its characteristics.

Class A class is a template for a configurable object (such as Ethernet or the protocol IP). When you add a new object to the configuration of a device, the BCC creates a copy (an instance) of the appropriate template. Each instance is an object uniquely identifiable within the total device configuration.

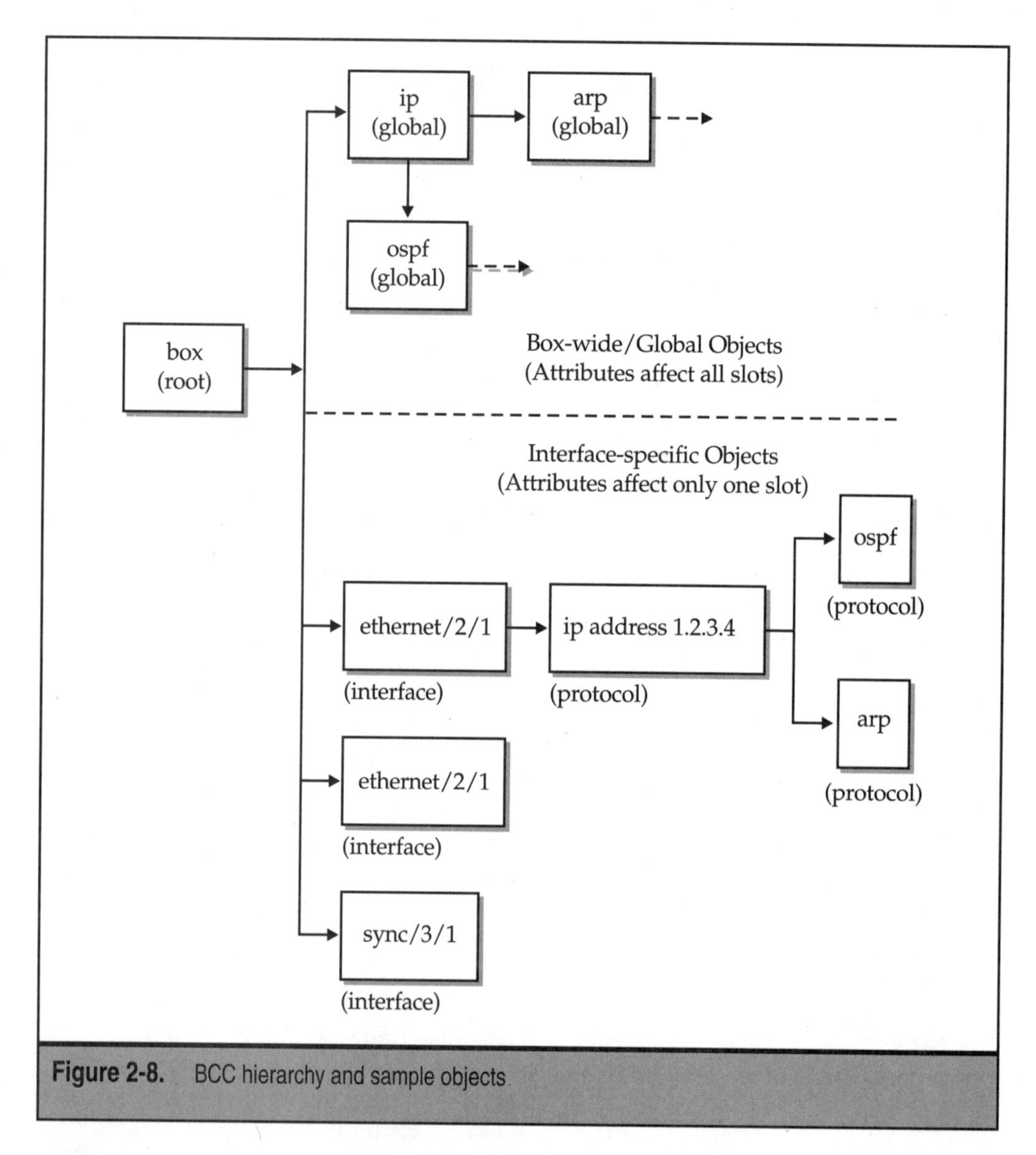

Figure 2-8. BCC hierarchy and sample objects.

Parameters Parameters are attributes or properties of a configurable object. Parameters may be required, derived, or optional.

- **Required parameters** The minimum set of parameters for which the BCC requires you to supply values. For example, the required parameters of a physical port are slot and connector.
- **Derived parameters** Parameters for which the BCC supplies a value. For example, a derived parameter of OSPF (global) is router-id. In this case, the

BCC derives a value for router-id from the address of the first IP interface configured on the device.

- **Optional parameters** The set of parameters for which you can optionally specify customized values, replacing any default values set by the system. For example, an optional parameter of an Ethernet interface is bofl-retries. This parameter normally has a system default value of 5 (5 retries), but you can change this to another numeric value.

Instance Identifier The instance identifier uniquely identifies a single instance of an object configured on a Bay Networks device. The BCC instance ID consists typically of the name of an object combined with the values you specify for its required parameters. For example, the BCC instance ID for an Ethernet interface consists of ethernet/<*slot*>/<*connector*>, as in ethernet/2/1.

Context The context is the working location or level within the BCC configuration tree. Just as a UNIX file system has a current working directory within which you can add, modify, or delete files, the BCC configuration tree has a current working context within which you can add, modify, disable, enable, or delete objects. The BCC indicates the context by displaying a context-sensitive prompt in configuration mode. The prompt identifies the configured object at your current working location or level within the configuration hierarchy.

For more information on BCC concepts and BCC for management tasks, refer to the latest version of "Using the Bay Command Console," linked on the book web site (http://www.mahalingam.com/).

TI Commands in BCC

Certain commands are marked for use from the Technician Interface only. These commands are not available in the same form in BCC, although the same functionality is available in other BCC commands. Occasionally, you may need to temporarily invoke TI commands from BCC, which you can do using the **tic** (TI command) call.

For example, the TI command

```
box# tic get wfIpBase.2.0
wfIpBase.wfIpBaseEnable.0 = 1
```

is equivalent to the BCC command

```
box# mget wfIpBase.2.0
wfIpBase.wfIpBaseEnable.0 enabled
```

BCC MIB Commands

The TI MIB commands **list**, **get**, and **set** are used frequently throughout this book. BCC also provides equivalent commands: **mlist**, **mget**, and **mset**. The **commit** command can be used in both the TI and BCC, with the same behavior.

THE EVENT LOG

The BayRS event logging facility is a troubleshooting tool that provides a historical record of events and a concurrent view of the events taking place. It is also a debugging tool that often helps avoid expensive and time-consuming lab work, because the events provide a complete history. While some may find this view a bit odd, it is exactly this type of detail that eliminates complexity and expedites a solution.

Event Log Fundamentals

The event log is facilitated by both the GAME operating system and the router hardware platform. It provides a comprehensive summary of events, including the entity type, severity, time of occurrence, and descriptive text. Users who are familiar with the UNIX syslog facility will find that the event log facility provides similar features. The underlying mechanism, however, is different; integration of the BayRS event log and syslog is discussed later in this chapter.

The maintenance of any event history requires considerable memory storage space. BayRS routers do not have unlimited storage space and use a finite portion of volatile memory to store events. This finite memory area will store a scrolling window of events as they are written to it. Multi-slot routers have dedicated event memory on each slot that saves events pertaining to the local slot.

Table 2-6 summarizes the amount of event memory available in each processor type.

Processor Module	Log Memory
AN1 and AN2	32K (16K if DRAM size is 2 MB)
ARN	32K
ASN1 and ASN2	64K
FRE1	64K
FRE2	64K
ARE	64K
FRE4	64K

Table 2-6. Event Log Memory Available on Various Platforms

Operation of the Log

On multi-slot routers such as the ASN and BN, each slot has individual memory dedicated to log messages. Events occurring in a slot are logged in that slot only, thus allowing each slot to have a unique view of its own events. These entries are maintained in the log even during system reboots, application restarts, and software crashes. However, the log is lost after hardware reinitialization, which can be brought about by administrative diagnostics or a cold start.

Most events written to the log have a predefined code representing the event. Events that have a code have a specific number of arguments. Instead of writing full-text entries, the system logger will simply enter the event code and the argument values into the log. This allows a larger number of events to be stored in a static-size log. When the log is viewed using the **log** command in the Technician Interface or using Events Manager, the event code is used to format the display text based on the stored arguments. When the log becomes full, the earliest entry is deleted, and the new entry is added. This maintains a sliding history of events occurring in each slot.

Each event entry in the system log contains several components, which are listed in Table 2-7. The anatomy of an event log entry is shown in Figure 2-9.

Entry Item	Description
Severity Type	Impact of this event entry (Fault, Warning, Informational, Trace, or Debug)
Entity Responsible	The software component responsible for generating the event
Event Code	An event number, which along with the entity uniquely identifies the event
Message	Contains arguments for display with predefined event codes, or a full-text message for messages logged dynamically
Time Stamp	The time at which the event was logged; the precision depends on the platform
Sequence Number	A number assigned to the log entry to order events in the display

Table 2-7. The Event Log Entry Format

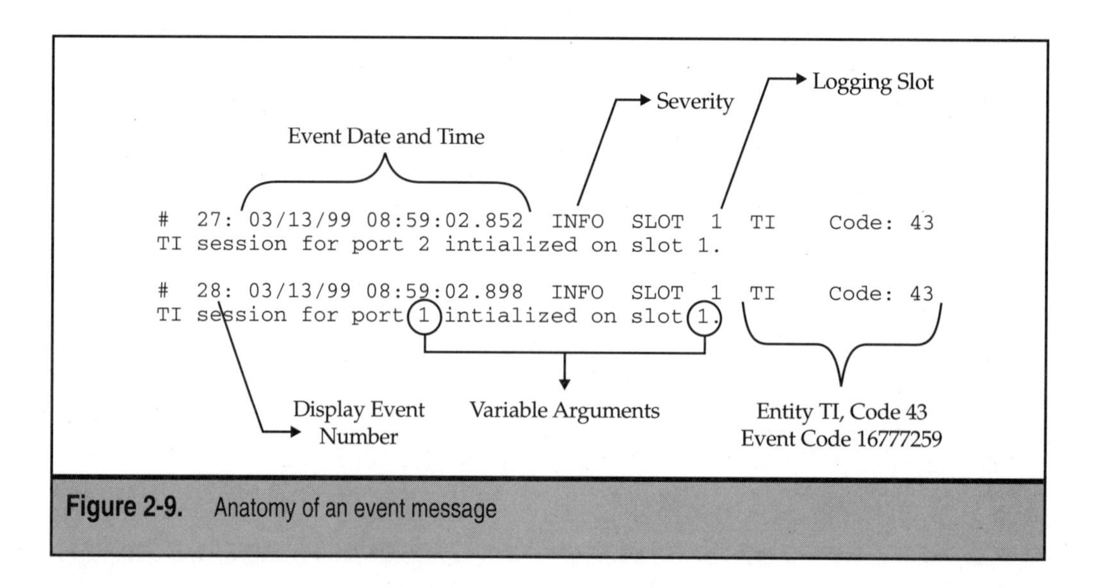

Figure 2-9. Anatomy of an event message

Severity Levels

The severity of an event can be of five possible values: Fault, Warning, Info, Trace, and Debug.

Fault messages indicate a critical system or subsystem error that has a major impact and affects system or network operation. Hardware failures and unexpected software errors are examples of errors that generate fault messages.

A warning message indicates an error with a minor impact that is recoverable and may not disrupt network operation; however, the problem may turn into a higher-impact issue if left unaddressed. Events such as diagnostic failures, invalid configuration files, and duplicate network addresses are flagged with warning messages.

Informational messages indicate normal operational events on the router. Trace messages indicate events that occur as a result of network activity. For example, initialization of the IPX protocol is logged as an information message, and the establishment of an OSPF adjacency is logged as a trace message. Normal monitoring of the router requires the viewing of fault, warning, and informational messages. Trace messages, however, are useful when diagnosing network abnormalities, specifically with operation of protocols over the network.

Debug messages indicate nonstandard events that do not fall into one of the other severity types, but aid in the debugging and troubleshooting of problems. Some debug messages contain readable text information, while others contain hexadecimal dumps that need to be interpreted. These messages are not analyzed in the regular course of router operation, but they may be useful in providing insight into a problem when other events messages provide limited leads. Users are encouraged to read and scrutinize fault, warning, info, and trace messages before looking into debug messages because these other messages usually indicate the symptoms of the problem.

Table 2-8 summarizes the severity levels.

Abbreviation	Level	Interpretation
F	Fault	Critical system error or subsystem error; major impact
W	Warning	Caution; unexpected subsystem behavior; minor impact
I	Information	Routine and normal operation; no impact
T	Trace	Tracking of protocol behavior, notably packets; diagnostics use
D	Debug	Granular messages showing specific actions; troubleshooting use

Table 2-8. Severity Levels and Interpretation

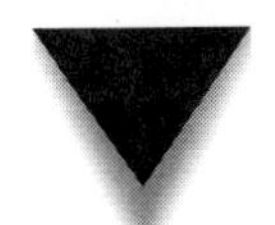

TIP: Debug messages are generally used under the guidance of Nortel Networks Customer Service. However, many debug messages are easily interpreted and incorporated into daily troubleshooting use and are discussed throughout the book.

Entity

An entity is an independent software component, such as a protocol, driver, or service. When an entity logs a message using the system logging facility, its associated entity code is automatically stored with the event entry. For example, ATM, FR (frame relay), and X25 (X.25) are individual entities that are uniquely associated with their events. A full listing of all entities is available in Appendix E, Table E-1.

Event Code

Each entity has a range of valid event codes. Each combination of entity and event code is unique and represents a specific event message. This combination is used to generate enterprise-specific traps that are transmitted by a BayRS router. The router generates the trap as a 32-bit integer, of which the trailing two bytes are composed of the entity code (higher byte) and the event code (lower byte).

Message

Most event codes, except for some debug messages, have a predefined text message with some variable values. As shown in Figure 2-9, variable values are stored only in the message area. Some dynamically generated debug messages are stored as full text in the log; in these instances, there is only one string variable containing the full event text. When the event messages are viewed using the **log** command or Events Manager, the variable values are automatically substituted in the corresponding event message template. The template messages are stored in string files, which have a .str extension and are included in the platform image archive (such as bn.exe). The fully formatted message is in readable form and can be used to comprehend router operation.

Time Stamp

The system logging facility stores the time when the event entry was logged by the entity responsible. The time stored is exact to the precision supported by the underlying hardware platform—events can be distinguished in fractions of a second. On all platforms, the time is accurate down to the millisecond, and on platforms such as the FRE, it is more precise. For the purpose of using the event log for monitoring, diagnosis, and troubleshooting, viewing the event times down to the milliseconds is sufficient. When you view event messages, the time stamp shows the milliseconds.

Sequence Number

The sequence number is tagged to each event by the system logging facility. Since the log is a systemwide service, multiple entities can simultaneously call the system logger to record an event. The system logger sequences events in the order they are received, and the TI **log display** command and the Events Manager use this sequence to order events. The sequence number is different from the display event number.

Viewing the Log

As mentioned earlier, the router log can be viewed in two ways. The Events Manager utility in the Site Manager suite provides a graphical user interface into viewing and custom filtering events from a router. It also allows a previously saved binary or text log file to be viewed offline. You can also use the **log** command from the Technician Interface; this is the most common method used.

```
[2:1]$ log
```

To view the log, type **log** at the TI prompt followed by an optional set of filter arguments. You can view a previously saved binary log by providing it as an argument to the **log** command. Events can be filtered based on the following fields:

- Entity name
- Severity
- Slot number
- Event date
- Event time
- Event code

The syntax of the **log** command is as follows:

```
log       [<vol>:<logfile>] [-d<date>] [-t<time>] [-e"<entity>"]
         [-f<severity>] [-s<slot ID>] [-p[<rate]] [-c<code #>] [-w]
```

- **-eENTITY** ENTITY can be any valid system entity that generates event logs. All entity names are represented using capitalized words. A full listing of entities is provided in Appendix E. This option can be specified multiple times as required. When this option is specified, only events specific to the entities are shown. The default is to show messages relating to all entities. Example: `-eIPX` will show all IPX related events.
- **-fSEVERITY** SEVERITY can be any combination of the characters in the list fwtid. The severity values represent (f)ault, (w)arning, (t)race, (i)nfo, and (d)ebug, respectively. The default severity events include all but debug messages. Example: `-ffw` will show only fault and warning messages.
- **-sSLOT** SLOT is specified as a number, ranging from 1 through 14 depending on the platform. This option can be specified multiple times as necessary. For example, `-s5 -s6` will show all events from slots 5 and 6. If the slot filter is unused, events from all slots are shown by default.
- **-cCODE** CODE is specified like the slot specification and can be specified multiple times. It is normally used in conjunction with the ENTITY option, to filter specific event codes for a single entity. Example: `-eBGP -c5 -c10` will display BGP events with codes 5 and 10.
- **-dDATE** DATE is specified in *mm/dd/yy* format. It displays events since that date only. If left unspecified, events in the entire log window are shown.
- **-tTIME** TIME is specified in *HH:MM* or *HH:MM:SS* format to show events only since that time. This option is usually used in combination with the DATE option. If the DATE option is omitted when using the TIME option, today's date is assumed.
- **-pPOLLRATE** POLLRATE polls the event log for new messages every POLLRATE seconds. All new messages are periodically polled and retrieved for display on the screen.

The option –w will result in a wide output, where the entry header is immediately followed by the event message in the display. This view is optimal for terminals with wide displays and for text-processing scripts. If the first argument to the **log** command is a file name as shown in the syntax, then events from the file are read and displayed instead of the current event log in memory.

Log Write Filters

The log memory in each slot is of finite size. Occasionally, you may want to filter out certain events before they are written to the log to provide more memory for other events, and filtering is necessary in several troubleshooting situations. Many troubleshooting situations are analyzed by setting the extended debug level of certain protocols and services. Enabling these debug levels allows an entity to log more detailed events on its operation. However, this extended debugging of protocols with numerous interfaces may generate a substantial number of event messages. You can provide more event memory for these extended debug messages by filtering out other messages.

Similarly, frequent network changes or certain talkative protocols may generate a lot of event messages that a network administrator may not find necessary. Again, events can be filtered out to provide a comprehensible log with events of interest.

CAUTION: Write filters should not be enabled for extended periods or on a permanent basis. Doing so may eliminate valuable log events from other entities, which may be needed for a subsequent troubleshooting task. Debug messages in particular should not be permanently filtered from the log, even though they may not be used on a regular basis.

Configuring Write Filters

The **log** command is used to configure write filters. The syntax is slightly different than that used to view the log.

```
log         [-x|-i] [-e<entity>] [-f<severity>] [-s<slot ID>]
```

The –i option configures the system logging facility to accept just events that match the filter criteria. The –x option excludes events that match the specified criteria. The –s option applies the filter only to the specified slot.

The –i option without any arguments includes no events, and the –x option without any arguments excludes none of the events. For example, to include only frame relay events (and exclude everything else), you can use the following commands:

```
[2:1]: log -i                              (Include no events)
[2:1]: log -i -eFR                         (Include FR entity events)
```

Displaying Write Filters

Write filters currently configured on each slot can be displayed using the following syntax:

```
log       -z [-s<slot ID>]
```

If the slot filter argument is not specified, then the filter configurations for all slots are displayed.

Preserving the Log

The log is a sliding window of events. Earlier events are removed and replaced with more recent events when the log memory space becomes full. A router that has been operational for several weeks or months will typically have a full log memory and will constantly replace old events with new ones. In some cases, the log contents need to be viewed offline or stored for a specific reason, typically for troubleshooting and analysis. In these cases, the log memory contents can be saved in a file in binary form. Log contents should be saved soon after an event or set of events occur; if the saving of the log is delayed, then the events of importance may be aged out.

Using the Events Manager, the current log file is first fetched (File | Get Current Log File). The log can then be saved on the local disk (File | Save Output To Disk). Alternately, the **save log** command can be issued from the TI to save the log to a flash in binary format.

```
[3]MX5> save log 3:mx5_120498.log
```

Saving the log in binary format improves efficiency, because the events are stored in native format. The **save log** command can be used with the same filter arguments as the **log** command, to save selected events matching specified criteria. Log filters are discussed in detail in the previous section.

If the log is to be saved soon after a reset or a systemwide fault on a slot, then you must wait until the slot is fully operational again before saving. Attempting to save the log prematurely may not allow the **save** command to retrieve the log events from the slot that was reset. You should monitor the status of the slot in the following manner, if visible LEDs cannot be observed:

```
[3]MX2> list -i wfHwEntry
inst_ids = 1
           3  ← slot 2 missing, disconnected
           4
[4]MX2> list -i wfHwEntry
inst_ids = 1
           2  ← slot 2 reconnected
           3
           4
```

Automatically Saving the Log

Events in the log memory can periodically be saved to flash when the log is full. This can be achieved by using the autosave feature, which automatically preserves the log entries to flash. Automatic saving is most useful in cases where the log has to be monitored for an extended period of time, and where the use of write filters is unsuccessful.

You enable autosave by setting two attributes, which are parameters of the `wfSerialPortEntry` object. The first is the `wfSerialPortAutoSaveNumFiles` attribute, which indicates the number of times the log will be written to the flash. The maximum value for the attribute is 99, and the autosaving to flash will cease if the file system becomes full. The following command sets the `wfSerialPortAutoSaveNumFiles` attribute to 50:

```
bcc->box
mbf02# mset wfSerialPortEntry.33.1 50
```

The log files are written in the form autoX.log, where X represents the sequence number of the autosave file. You set the file system used to save the log files by configuring the `wfSerialPortAutoSaveVolume` attribute, which is set to volume 1 in the following example.

```
bcc->box
mbf02# mset wfSerialPortEntry.34.1 "1"
```

On the ASN and 5000 routers, the instance ID 1 is replaced by the slot number of the router whose log is to be autosaved. On all other routers, there is only one instance of the `wfSerialPortEntry` object with instance ID 1.

Clearing the Log

Events that have been reviewed in the log and are no longer necessary can be cleared by issuing the **clearlog** command in both the TI and BCC.

```
[4:1]$ clearlog 2
[4:1]$ clearlog 5-8
[4:1]$ clearlog 10,12
```

It is prudent behavior to save the log to a file before clearing it, should you need to review it later. The **clearlog** command is often issued prior to making a configuration change, as with an MIB **set** command. Clearing the log makes analyzing the log later easier because the log will contain only events after the MIB **set** command.

SUMMARY

This chapter described many important BayRS basic concepts. GAME is the proprietary operating system that schedules gates (processes) and provides common system functions. Lines, circuits, and interfaces form the basic building blocks for multiprotocol routing and bridging services. The MIB subsystem is essential for storing and retrieving configuration details as well as statistical values maintained dynamically for use by SNMP applications. BayRS reports events of varying severity using the system logging facility, from system-affecting faults to minor trace and debug events. The TI is used for out-of-band administration and troubleshooting, as is BCC, which also provides configuration and monitoring capabilities.

REFERENCES

1. *Configuring WAN Line Services*, Publication 308648-14.00 Rev 00.
2. *Using Technician Interface Software*, Publication 308657-14.00 Rev 00.
3. Event Messages Database, http://support.baynetworks.com/library/tpubs/events
4. M.A. Miller, P.E, *Managing Internetworks with SNMP*, 3rd Edition, IDG Books, November 1999.
5. D. Perkins and E. McGinnis, *Understanding SNMP MIBs*, Prentice-Hall, December 1996.
6. *Troubleshooting Routers*, Publication 308656-14.00 Rev 00.
7. *Configuring and Managing Routers with Site Manager*, Publication 309081-14.00 Rev 00.
8. *Using Technician Interface Scripts*, Publication 308658-14.00 Rev 00.
9. *Reference for BCC IP show Commands*, Publication 308603-14.00 Rev 00.

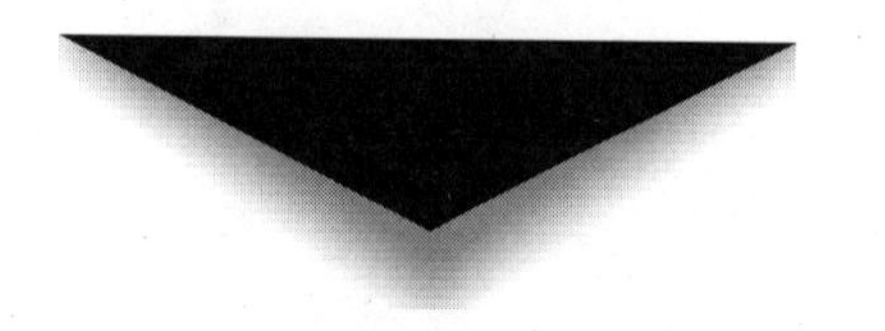

CHAPTER 3

Router Administration and Software Upgrades

You can only form the minds of reasoning animals upon Facts: nothing else will ever be of any service to them. Stick to Facts, sir!

—Charles Dickens, *Hard Times*

Network administrators are responsible for maintaining a fully functional network at all times. At the nodal level, this maintenance requirement translates into the task of administering individual routers. Some routers will require simple and infrequent management, while others that undergo changes may require closer administration. In this chapter, we will cover many important administration tasks, including software upgrades. The most common tasks, such as file system management, use of the system event log, and MIB access through the TI/BCC, were discussed in Chapter 2.

BAY COMMAND CONSOLE

Veterans of the Technician Interface may already be using and experienced with the Bay Command Console (BCC). Key fundamentals of the BCC were discussed in Chapter 2, which users unfamiliar with BCC should read to get a quick start. BCC is very simple and intuitive to use, and it has a context-sensitive interface that users will promptly become accustomed to and master without any help. For this reason, use of BCC is not discussed in great detail here.

Help and Navigation in BCC

Chapter 2 discussed the BCC context and hierarchical structure. Commands in the history list can be recalled using the Up Arrow (CONTROL+P) and the Down Arrow (CONTROL+N) keys. Contexts and commands are automatically completed if a partial string is entered, as long as the string is unique, so the command or context can be identified unambiguously. Table 3-1 lists the syntax and functions of frequently used commands, with examples.

NOTE: The **stack** context is specific and unique to the ASN router. Throughout the remainder of this book, the **box** context is used to represent the stack context as well as the box context. Readers should interpret references to box to apply equally to the stack context. The distinction between box and stack is enumerated whenever necessary.

The configuration mode provides the privilege to edit, create, or delete configuration objects and to set or change attribute values. In the BCC paradigm, when the exact command sequence or the associated object is not known clearly, a typical cyclic behavior is followed to determine the context and attribute of interest and valid values. Figure 3-1 illustrates this cycle.

Syntax	Function	Example

?
help object
	Context-sensitive help. If the object is in a different tree structure, then the **help** command can be used.	`box->snmp`
		`fp3-x[3]# ?`
		`box->snmp`
		`fp3-x[4]# help ip rip`

attribute-name ?	Show the current, legal, and default attribute values or value range for the attribute.	`box->http`
		`mx2(101)# authentication ?`
		`Current value: basic`
		`Legal values: basic,digest`
		`Default value: basic`

info	Display all attribute values for the current object.	`box->http`
		`mx2(102)# info`
		`  port 80`
		`  authentication basic`
		`  domain-name {}`
		`  help-base-url {}`
		`  state enabled`

lso	List (ls) all child objects (o) for the current object.	`box->snmp`
		`mx2(36)# lso`
		`community/pr1vate`
		`community/n3+adm1n`
		`community/1nf0+3ch`

show config [-r] [-all]	Show the configuration of the current object and all child objects if –r(ecursive) is used; –all displays the complete configuration.	`box->ip`
		`mx2(65)# show config -r`
		`...`

pwc	Print the working context.	`box->ip->rip`
		`mx2(54)# pwc`
		`box; ip; rip`

back [*x*]	Go back up *x* levels in the context tree; defaults to going back up one level.	`box->snmp->community/n3+admin->`
		`manager/n3+admin/10.47.10.45>`
		`mx2(25)# back`
		`box->snmp->community/n3+admin`
		`mx2(26)# back 2`
		`box->`

Table 3-1. Frequently Used Help and Navigational Syntax in BCC

Syntax	Function	Example

box
stack
	Go directly to the root context; stack is used on ASN routers, and box is used on all others.	`box->ip->bgp`
		`mx2(28)# box`
		`box`
		`mx2(29)#`

disable [*object*]
enable [*object*]
delete [*object*]
	Disables, enables, or deletes the specified child object and its children; applies to the current working object if no object is specified.	`box->ip->rip`
		`mx2(30)# delete`
		`box->ip`
		`mx2(31)# box; disable ip`
		`box`
		`mx2(32)#`

Table 3-1. Frequently Used Help and Navigational Syntax in BCC (continued)

The file `autoexec.bcc`, if available, will automatically be sourced as a TCL script when BCC is invoked using the **bcc** command. BCC commands in this book are executed using a customized parser that is loaded using `autoexec.bcc` (check the book's web site to obtain a recent version).

MULTIUSER ACCESS

Chapter 2 mentioned that there are two default user logins: **Manager** and **User**. The multiuser access feature enables the creation of several user groups and assigns usernames, passwords, and levels of privilege to them. The Manager and User passwords are stored in NVRAM, and the multiuser access configuration detail is stored in the MIB configuration. Users can be authenticated either using the encrypted password stored in the local MIB or using RADIUS.

Privilege Levels

Each user is assigned a privilege level: User, Manager, or Operator. The Operator level is intermediate between the User and Manager levels and provides limited read-write access to the configuration and access to most system-level commands. A user with low-level access attempting to execute a higher-level command will see an error message indicating that the privilege level is insufficient.

Auditing

Commands entered by users are logged in the system event log for auditing. These events can be trapped and sent to an NMS and saved in a permanent record file. Events in the

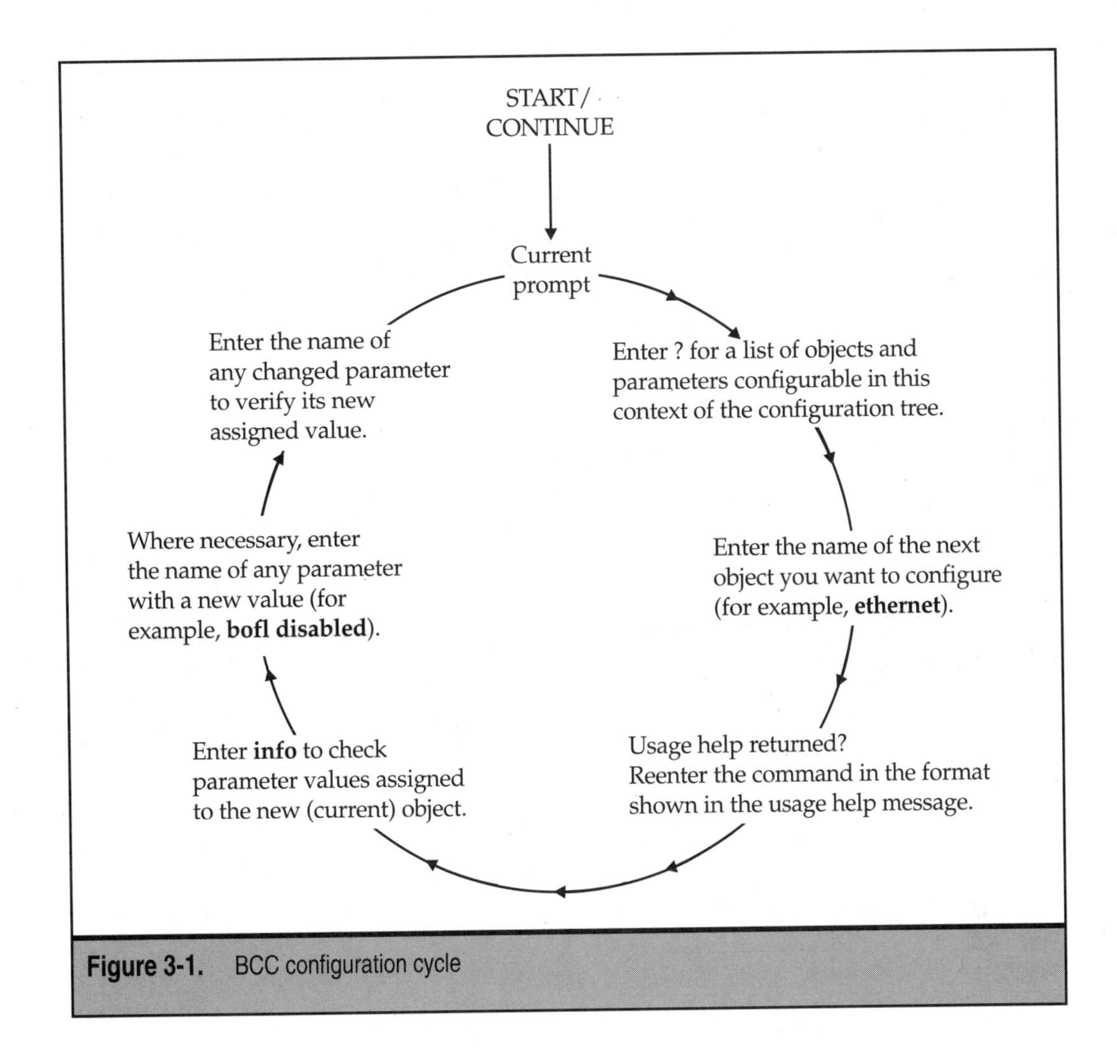
START/
CONTINUE
Current
prompt
Enter ? for a list of objects and
parameters configurable in this
context of the configuration tree.
Enter the name of the next
object you want to configure
(for example, **ethernet**).
Usage help returned?
Reenter the command in the format
shown in the usage help message.
Enter **info** to check
parameter values assigned
to the new (current) object.
Where necessary, enter
the name of any parameter
with a new value (for
example, **bofl disabled**).
Enter the name of
any changed parameter
to verify its new
assigned value.

Figure 3-1. BCC configuration cycle

router's log can age out over time and can also be cleared by a user with Manager privileges; hence, logging important events to an SNMP manager is essential in environments with high security needs.

Configuring Multiuser Access

The first step in configuring multiuser access is to create the `access` object. This object is not created by default during initialization; only the User and Manager logins exist initially. Logging in as Manager, the `access` object must be created initially by the administrator. It can be later accessed by any user with Manager-level privileges. Minimum and maximum field lengths can be altered for multiuser access, and the `minimum-password-length` value should be changed in secure routers. By default, the router allows a user to have a null password. If you want to set up RADIUS authentication, then you must enable the `radius-server-accounts` attribute. Creation of the access object is illustrated here with the default attribute values.

```
box
mx2(12)# access

box->access
mx2(13)# info
  maximum-login-length 16
  minimum-login-length 4
  maximum-group-length 16
  minimum-group-length 4
  maximum-password-length 16
  minimum-password-length 0
  configuration-accounts enabled
  radius-server-accounts disabled

box->access
mx2(14)# minimum-password-length 6
```

Auditing can be enabled on a system-wide basis for different privilege levels. By default, auditing is disabled. When an audit configuration object is created, by default auditing is performed for all privilege levels (User, Manager, and Operator). The following commands create the audit object and change the `audit-level` attribute to include only the Manager and Operator levels.

```
box->access
mx2(16)# audit

box->access->audit
mx2(17)# audit-level "manager operator"

box->access->audit
mx2(18)# info
  audit-level {manager operator}
  state enabled
```

Creating Groups and Users

Before creating a new username, you must create an access group, which encompasses a set of users with a certain privilege level. An individual user can belong to multiple groups, in which case commands will be executed based on the privileges granted by the group with the highest privilege level for those commands. In the following example, a group named **local-admin** with a group-id attribute of 1 is created, and username **ragho** is created and assigned to the group. By default, usernames initially have a null password. The administrator can assign a default password, or the user can be advised to change the password upon the first login.

```
box->access
mx2(27)# groups

box->access->groups
mx2(28)# group name local-admin group-id 1

box->access->groups->group/local-admin
mx2(29)# info
  group-id 1
  name local-admin
  privilege-level user
  audit-level {manager user operator}
  state enabled

box->access->groups->group/local-admin
mx2(30)# privilege-level manager

box->access->groups->group/local-admin
mx2(40)# back 2; users

box->access->users
mx2(41)# user ragho

box->access->users->user/ragho
mx2(42)# true-name "Ragho Mahalingam"

box->access->users->user/ragho
mx2(43)# info
  login-id ragho
  encrypted-password p0WakbRn/3
  group-ids 1
  audit-level {manager user operator}
  true-name {Ragho Mahalingam}
  state enabled
```

The users, groups, and audit multiuser access configuration objects are illustrated in Figure 3-2.

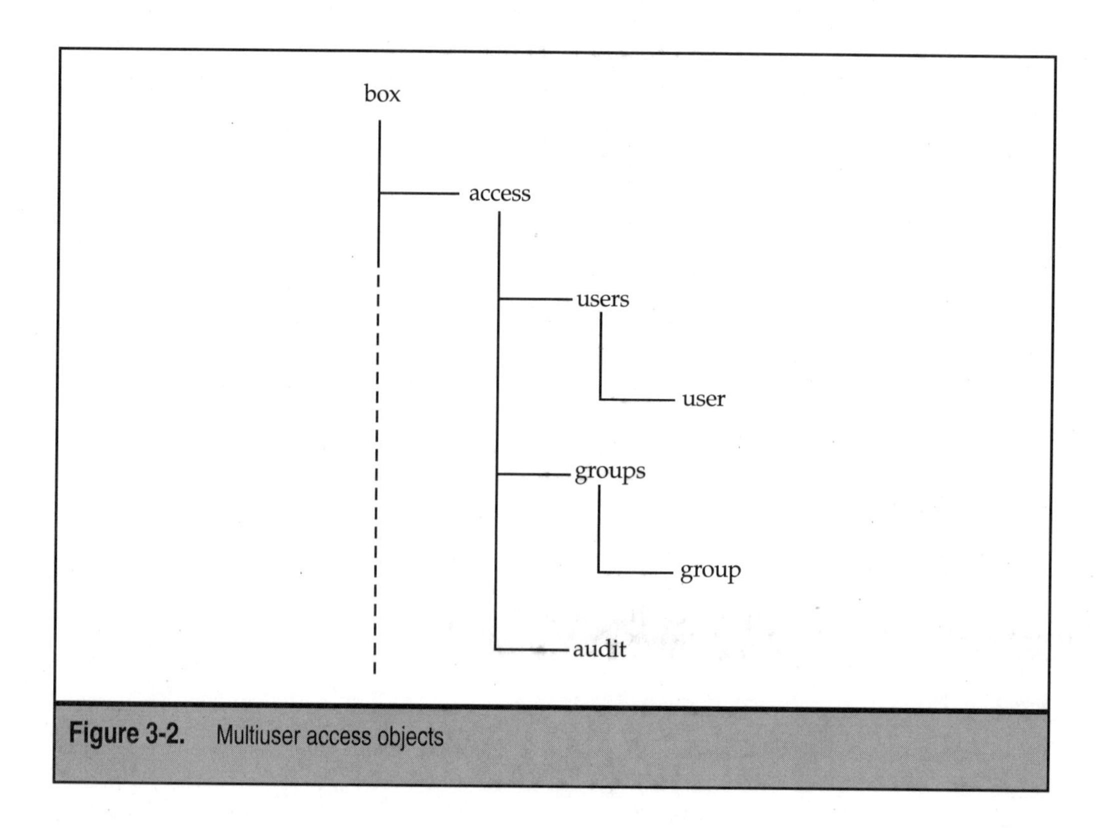

Figure 3-2. Multiuser access objects

Changing the Password

Passwords can be changed using the **password** command. The User and Manager login passwords are stored in battery-backed nonvolatile RAM (NVRAM), and nondefault logins created for multiuser access are stored in the MIB in encrypted form.

The User and Manager passwords are stored even when the router does not have power or if the configuration file on flash is modified. On multi-slot routers, each slot stores the Manager and User passwords on its own NVRAM. When the password is changed, GAME automatically propagates the password from the soloist slot to all remote slots. Removing a processor module from a slot allows the module still to retain the configured password; if the same processor is installed in another router and becomes the TI soloist, then it will accept only the password stored in its NVRAM. Hence, it is important to keep passwords synchronized on all processor modules when they are newly added or removed.

An administrator logged in as User can change the password only for the User login; however, an administrator with the Manager login can change the password for both standard users (User and Manager). Passwords for multi-level users can be changed by the individual user or by a user with Manager privileges. All passwords are stored in the

MIB in encrypted form and cannot be seen in unencrypted form by any user, including those with Manager privileges. For reasons of security, a Manager-level user can change another user's password only if the previous password is known; otherwise, the user's account must be deleted and re-created to set a new password. In the following example, the password for the user "ragho" is changed by the Manager login.

```
box->access->users->user/ragho
mx2(45)# password ragho
Changing password for ragho
Old password:
New Password:
Retype new password:
ragho password changed.
```

CONSOLE AND TELNET TI

The TI soloist, as discussed in Chapter 2, controls out-of-band management access through the console. It is a secondary process compared to the forwarding functions of the router, and it requires significant memory. Certain slots that are dedicated to forwarding only and that should not run the TI soloist should be removed from the `console-slot-mask` attribute in the box object, as shown in the following example. Refer to the section "Soloist Gates" in Chapter 2 for details on the function of soloist processes.

```
box
mx2(8)# console-slot-mask {2 3 5}
```

In the preceding example, slot 4 is eliminated from the TI soloist slot mask. Notice that following the TCL syntax, the curly braces { and } are used to include a list of values in BCC. The `console-slot-mask` attribute affects only the TI soloist for the console port. To disallow the telnet TI from running on a specific slot, the slot must be disabled in the `wfProtocols.wfTelnetLoad` attribute. The `wfProtocols` MIB object is global and thus contains only one instance with instance ID 0. In a manner similar to that in the `console-slot-mask` example, slots 2, 3, and 5 can be allowed to accept telnet sessions as follows:

```
box
mx2(9)# mset wfProtocols.wfTelnetLoad.0 0x68000000; commit
```

See the "Slot Mask" section in Chapter 2 for details on the use of the slot mask. In the preceding example, when a user attempts to telnet to an IP address in slot 4, the session will be refused because it has been excluded. Recognize that the telnet service must also be enabled globally for the other slots to accept telnet sessions.

Chapter 2 discussed the need to administer routers remotely. Routers installed in remote offices may be connected to a modem for remote dial-in access. In setups where a modem is attached to the console port, the modem attribute for the console object must be enabled. The following two equivalent commands can be used; one is a BCC attribute and the other is a direct MIB attribute setting.

```
box->console/1
mx2(55)# modem enabled

box->console/1
mx2(56)# mset wfSerialPortEntry.12.1 1
```

SYSTEM INFORMATION

Every router in the network has a system name. The system name unambiguously identifies the router within the network. Accompanying the system name is the system location and the administrative contact. This information is usually set by the administrator during installation but may require verification and change if location or organizational changes occur in the network. All three attributes are part of the system record (**wfSys**), which can be accessed under the box object.

```
box
unknown(100)# system-name mx2

box
mx2(101)# location "Network Room P120-10, 138 Amherst Ave"

box
mx2(102)# contact {Network Ops, x9824 or Ammu, x9872}

box
mx2(103)# info
   ... ...
   contact {Network Ops, x9824 or Ammu, x9872}
   system-name mx2
   location {Network Room P120-10, 138 Amherst Ave}
   mib-counters enabled
   console-slot-mask {2 3 5}
```

SIMPLE NETWORK MANAGEMENT PROTOCOL

Chapter 2 described the basic concepts of SNMP. In SNMP, a manager attempting to access the agent must provide basic authentication information in the form of a *community string*. The community string is essentially a simple string provided to a manager and also configured on the agent so that it can perform basic verification of the identity of the sender. BayRS verifies the identity of the manager using the IP address and grants a privilege level based on the community string provided.

Community Strings

Within the SNMP configuration, each community string is assigned a level (read-only or read-write), and several manager IP addresses are tagged to it. Managers with these specific IP addresses are allowed the privilege level of the community string. A manager with the correct community string and an incorrect IP address is denied access; so is a manager with the correct source IP address but an incorrect community string. In the BayRS SNMP configuration, community strings are created soon after SNMP is created globally. A hierarchy of SNMP objects is shown in Figure 3-3. In the following example, the global `snmp` object is created with default attribute values. A community label n3+adm1n with read-write privileges is also created under the `snmp` context.

```
box
mx2(2)# snmp

box->snmp
mx2(3)# community label n3+adm1n

box->snmp->community/n3+adm1n
mx2(4)# access read-write

box->snmp->community/n3+adm1n
mx2(5)# info
  label n3+adm1n
  access read-write
  scope-type {}
```

The default access level for a community string is read-only. Changing the `access` capability of a community string to read-write allows an SNMP manager to make configuration changes for the device.

Managers

Managers are individually added to the list of devices allowed to communicate with the router using SNMP. Using the IP address 0.0.0.0 as the manager address will allow any SNMP manager to connect to the router using the community string. In the following example, any device using the n3+adm1n community string is allowed to manage the router.

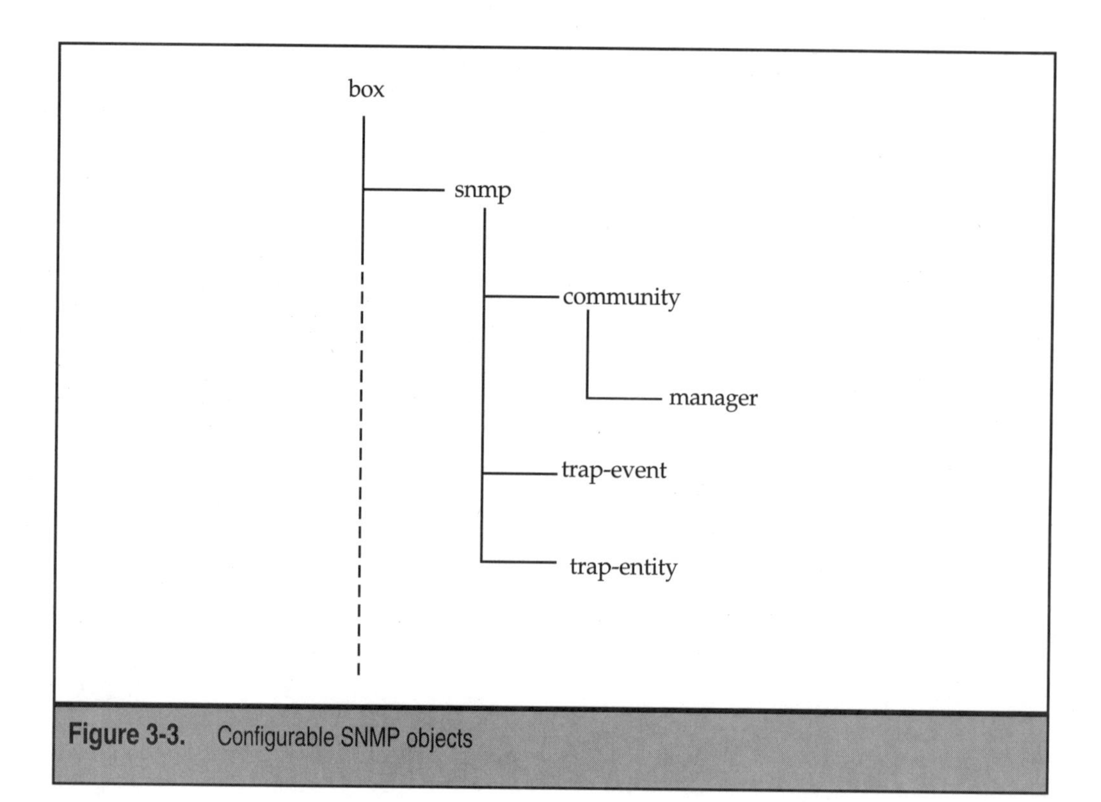

Figure 3-3. Configurable SNMP objects

```
box->snmp->community/n3+adm1n
mx2(40)# manager 0.0.0.0

box->snmp->community/n3+adm1n->manager/n3+adm1n/0.0.0.0
mx2(41)# info
  address 0.0.0.0
  trap-port 162
  traps generic
```

If the all-encompassing 0.0.0.0 address mask is not used, then a specific manager address must be used. Multiple managers can be configured for each community string, and several community strings can be created for each router. If you want to allow or deny SNMP access for a range of addresses, you can do so in two steps: first by allowing the manager 0.0.0.0 access and then by selectively accepting or denying SNMP sessions through the use of access policies (discussed later under "Access Policies").

SNMP managers can also be limited to viewing only certain MIB tree structures. The **snmpserver** views allow you to configure several agent views that can be used to provide unrestricted privileges, but only for a specific set of MIB objects. The **snmpserver** command is covered in Chapter 7, when we discuss SNMP security and the troubleshooting of SNMP issues.

Trapping Events

Log events entered into the system log can be monitored and, if they match certain criteria, generated as enterprise-specific traps. Each manager has a `traps` attribute, which defaults to generic traps only; both enterprise-specific and generic traps can be enabled by setting traps to all.

```
box->snmp->community/n3+adm1n->manager/n3+adm1n/0.0.0.0
mx2(41)# traps all
```

Events can be trapped in two ways. The first is based on the entity and the event severity level; for example, fault and warning messages for a specific set of entities or all entities can be trapped by this method. This method is commonly referred to as *trapping by entity*. Entity names are documented in Appendix E. Alternately, events can be specifically trapped based on the event code and the entity it is logged for. Called *trapping by event*, this method traps only a single event as opposed to trapping by entity, which traps a range of events matching the severity levels. Of the two examples here, the first demonstrates trapping by entity—any fault event for all entities on slot 4. The special entity name `all` is used to represent all entities.

```
box->snmp
mx2(79)# trap-entity entity all slot 4

box->snmp->trap-entity/all/4
mx2(80)# fault-log on

box->snmp->trap-entity/all/4
mx2(81)# info
  entity all
  slot 2
  fault-log on
  warning-log off
  information-log off
  trace-log off
  debug-log off
  state enabled
```

The next example illustrates trapping by event. Event code 3 of entity IP is logged whenever an IP interface fails. This specific event is trapped as follows, using the `trap-event` object:

```
box->snmp
mx2(90)# trap-event entity ip event 3

box->snmp->trap-event/ip/3
mx2(91)# info
  entity ip
  event 3
  state enabled
```

It is generally recommended that fault events be trapped to monitor system integrity, because a router in steady-state operation does not log fault messages except when a critical error occurs.

SNMP-related events can be viewed under the `SNMP` entity, and the configuration and statistics are displayed using the **show snmp** command. Security considerations with SNMP are discussed in Chapter 7.

COMMON SYSTEM SERVICES

Many management tasks on the router require some essential services, and other services are desirable as well. Services such as TFTP or FTP, Telnet, and SNMP are essential for performing file transfers, remote administration, and network management. Other protocol services such as HTTP, NTP, DNS, and syslog are beneficial in maintaining a healthy network.

System services such as SNMP, TFTP, DNS, NTP, and syslog use the User Datagram Protocol (UDP); and other protocols such as Telnet, FTP, and HTTP use the Transmission Control Protocol (TCP). Forwarding protocols such as DLSw also require the use of TCP on the router. UDP is always enabled on the router to support SNMP and TFTP services, whereas TCP is an enhanced service that must be enabled to support the optional system and forwarding services. Figure 3-4 illustrates the use of UDP/TCP by various protocols and the assigned port numbers. Note that some protocols are designed to use either TCP or UDP, in which case the transport protocol used by BayRS is shown in the figure.

The hierarchy and location of top-level system service objects is illustrated in Figure 3-5. Child contexts are discussed separately in each service section.

Transmission Control Protocol

A reliable transport layer protocol, TCP is enabled by creating `tcp`, a child object of `ip`. Creating a service such as Telnet, which requires the use of TCP, will automatically instantiate TCP on the router. Attributes specific to TCP can be modified under the `tcp`

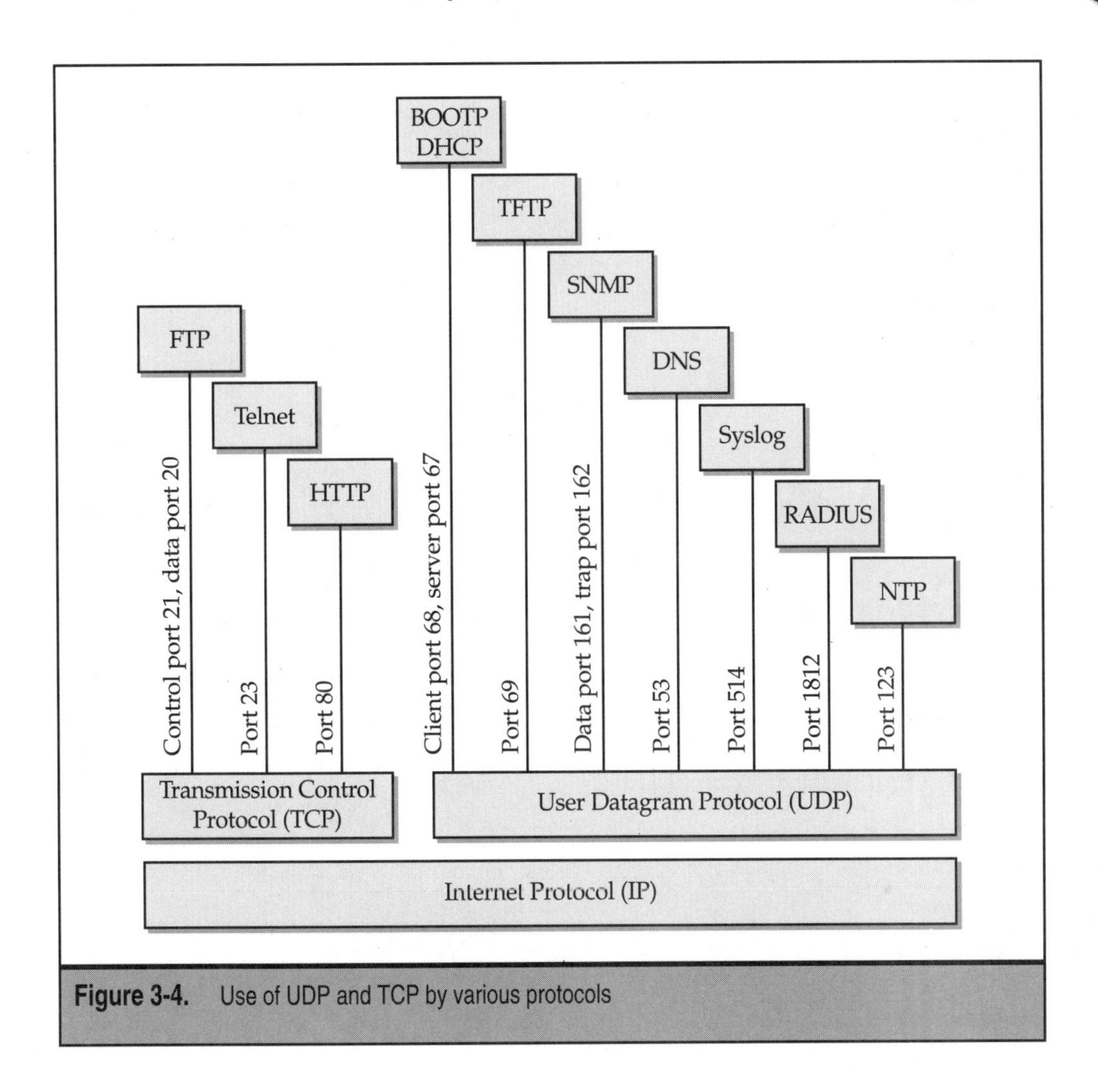

Figure 3-4. Use of UDP and TCP by various protocols

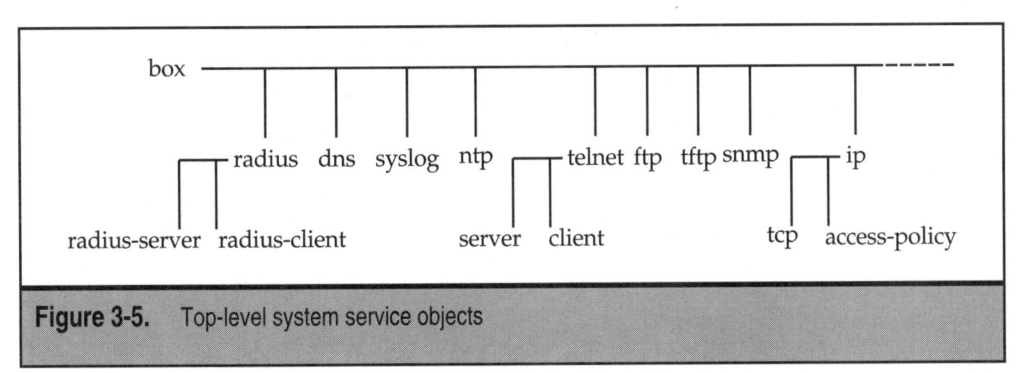

Figure 3-5. Top-level system service objects

object; these usually affect several protocols. The creation of the global `tcp` object with a listing of configurable attributes is shown here:

```
box
mx2(1)# ip; tcp

box->ip->tcp
mx2(2)# info
  min-rto 250
  max-rto 240000
  max-win 4096
  state enabled
```

Trivial File Transfer Protocol

TFTP is a simple file transfer protocol used to move files to and from flash file systems. (File transfer mechanisms were discussed in Chapter 2, in the section "Transferring Files.") For the router to use either the TFTP client or server, the TFTP protocol must be globally configured and enabled. The `tftp` object resides under the global box object, and simply creating the object automatically instantiates and enables it. The `default-volume` attribute is set to the current volume and can be changed as necessary. The following listing illustrates the creation of the `tftp` object globally, and shows the default attribute values.

```
box
mx2(6)# tftp

box->tftp
mx2(7)# info
  default-volume 1
  retry-timeout 5
  close-timeout 25
  retry-count 5
  state enabled
```

Security issues with TFTP were discussed in Chapter 2. Access to TFTP can be tightly controlled by implementing IP access policies, covered later in this chapter.

Both the **tftp** configuration object and the **tftp** file transfer command are accessed using the `tftp` syntax, albeit with different arguments. Configuration contexts are given higher priority than commands, so running a **tftp** file transfer in configuration mode will cause an error because the command will be evaluated using the BCC configuration syntax. The **tftp** command can be run in the TI context by prefixing it with **tic** (see "TI Commands in BCC" in Chapter 2).

File Transfer Protocol

FTP provides a more reliable and authenticated mechanism than TFTP for transferring files. The FTP protocol is more complex than TFTP and thus consumes more local memory and processing power. However, certain scenarios may require the use of FTP, while others may need the use of TFTP. As with TFTP, the ftp server object is created under the global box object. Once created, an FTP client can connect to the FTP server on the router and perform file transfers. FTP on the router operates in server mode only. The creation of the `ftp` object in BCC and a listing of default attribute values is shown here.

```
box
mx2(9)# ftp

box->ftp
mx2(10)# info
  default-volume 2
  login-retries 3
  idle-timeout 900
  max-sessions 3
  tcp-window-size 60000
  state enabled
```

The `default-volume` attribute specifies the root file system that is used by default when a user logs in. An FTP user can switch to a different file system using the **cd** command.

Telnet

The router's Technician Interface can be accessed remotely using the Telnet protocol. Using a telnet client on a workstation, a telnet session is established to any one of the router's operational IP interfaces. On the BayRS router, both the telnet server and the telnet client are available. The telnet server accepts sessions from clients and allows access to the TI. The telnet client allows the TI user to remotely telnet to another telnet server from the local router. Both the `client` and `server` are child contexts of the `telnet` service, which resides under the box object. In the following example, the two child objects (`client` and `server`) are created under the telnet context in BCC; default attribute values are listed using the **info** command.

```
box
mx2(0)# telnet

box->telnet
mx2(1)# client
```

```
box->telnet->client
mx2(2)# info
  debug-log-flag off
  remote-port 23
  prompt {}
  state enabled

box->telnet->client
mx2(3)# back; server

box->telnet->server
mx2(4)# info
  manager-script automgr.bat
  lines 24
  more enabled
  prompt {mx2> }
  login-timeout 1
  password-timeout 1
  command-timeout 15
  login-retries 3
  auto-user-script {}
  force-logout disabled
  history 20
  state enabled

box->telnet->server
mx2(5)# back; lso
client  server
```

As with the tftp client, the telnet client is usually not run in BCC configuration mode and will be interpreted to mean the telnet object instead. If you need to run the telnet client in configuration mode, it can be run in the TI context. The customized parser described earlier will help differentiate these unique contexts and commands based on the arguments and will reevaluate the command using **tic** if necessary. (See the book's web site for more information. The parser is a user-provided, unsupported third-party tool.)

Hypertext Transfer Protocol

Abbreviated as HTTP and most commonly known as the protocol enabling the WWW, Hypertext Transfer Protocol service on the BayRS router provides universal access to simple administration tasks. Using a web browser such as Netscape Navigator or Microsoft Internet Explorer, router administrators can connect to any configured IP interface on the router to perform file system management, view the event log, monitor interface and protocol statistics, observe faults and warnings, and toggle interface or line states. Creation of the global http object is shown here, along with a list of default attribute values.

```
box
mx2(6)# http

box->http
mx2(7)# info
  port 80
  authentication basic
  domain-name {}
  help-base-url {}
  state enabled
```

HTTP runs on top of TCP using the assigned port number 80. The default port number can be changed to improve web security. Also for security reasons, the HTTP server will disallow users with Manager privileges and a null password. Manager-level users must have a nondefault password to gain authenticated access to the web interface. All HTTP-related events are logged under the HTTP entity. The configuration and statistics can be viewed using the **show http** command.

Domain Name Service

DNS, or Domain Name Service, is commonly used in most enterprise networks to map IP addresses to meaningful names. Using UDP as the transport protocol, it uses the predefined port 53 to resolve names to IP addresses and vice versa. In a client-server relationship, a DNS client such as a router will contact a DNS server to perform hostname and address resolutions.

DNS Client

Routers ordinarily have several IP interfaces, and each address can be uniquely mapped to a unique name. In DNS, each host is part of a name service domain encompassed by the whole network or a portion of the network. One or more DNS servers operate together and provide authoritative name-to-IP-address resolution and vice versa. For example, the router named **mx2** has three circuits—E11, E12, and H31—and a virtual IP interface. It is installed at Mahalingam InfoTech, the IT arm of the Mahalingam group of companies, represented by the DNS domain infotech.mahalingam.com. Each circuit has its own IP interface and is mapped in the server configuration as shown in Table 3-2.

BayRS routers can be configured to act as DNS clients, which can connect to servers and resolve hostnames to addresses. The two most common commands using DNS name resolution are **telnet** and **ping**. The DNS client is easily set up on the router by first creating the DNS service and then configuring the domain name of the system and at least one name server to connect to for name resolution. An example DNS configuration is shown here using the dns object and a single name server.

Circuit	Address	Name	Fully Qualified Domain Name (FQDN)
mx2-virtual	192.168.3.1	mx2	mx2.infotech.mahalingam.com
E11	192.168.4.1	e11-mx2	e11-mx2.infotech.mahalingam.com
E12	192.168.5.1	e12-mx2	e12-mx2.infotech.mahalingam.com
H31	192.168.6.1	h31-mx2	h31-mx2.infotech.mahalingam.com

Table 3-2. Example of DNS Hostname Mappings

```
box
mx2(106)# dns

box->dns
mx2(107)# domain-name infotech.mahalingam.com

box->dns
mx2(109)# info
  ...
  domain-name infotech.mahalingam.com
  recursion enabled
  ignore-truncation enabled
  authoritative-only disabled
  use-default-domain enabled
  hosts-file {}
  state enabled

box->dns
mx2(110)# name-server number 1 address 192.168.9.51

box->dns->name-server/1
mx2(111)# info
  number 1
  address 192.168.9.51
  port 53
```

DNS servers cache information from servers responsible for other domains, and they provide clients with this detail in a nonauthoritative fashion. If you want the DNS client to accept only authoritative responses, then enable the `authoritative-only` attribute. Similar to the `/etc/hosts` file in UNIX, a local file containing IP addresses and names can be specified in `hosts-file`, which will be used if a DNS lookup fails. The default domain name suffix is appended to a name service query to obtain a fully qualified name if the query fails with just the name itself.

DNS Proxy

BayRS routers can also be configured to act as proxy servers for DNS. They can relay cached name resolutions and can also send proxy requests to a list of DNS servers. This function is especially useful in networks where network address translation is used to translate private IP addresses. It also helps limit DNS traffic traveling over WAN links by reusing cached name and address resolutions. Both DNS client and proxy event messages can be viewed under the `DNS` entity in the event log.

Syslog

Syslog, traditionally found in most UNIX systems, is a software service that receives and logs messages from other local services or remote applications. Messages can be logged locally to a file, displayed at a terminal, or sent to a printer or another remote syslog service. Almost all concurrent UNIX operating systems have a syslog daemon (server), which can receive and store event logs for extended periods of time. For example, router events for several weeks or many months can be stored in a log file on disk for future auditing and analysis. The interaction of syslog on the router and the syslog daemon on the workstation is shown in Figure 3-6.

Configuring syslog is a simple task. The `syslog` object is first created under box, and a set of syslog hosts is configured to receive the messages. By default, the log is polled every five seconds, and events matching the syslog filter criteria are sent to the specified syslog hosts. The use of smaller poll periods and broad filter criteria will increase syslog load and network traffic; hence, care should be taken when changing the poll period and configuring the filters. Syslog-related messages can be viewed in the event log using the entity name `SYSLOG`.

The following example creates a syslog host 57.82.152.139, enables time sequencing of events, and creates an event filter named faults to send fault events only. At least one filter must be created for a syslog host to receive events.

```
box
mx2(13)# syslog

box->syslog
mx2(14)# info
  maximum-hosts 5
  log-poll-timer 5
  state enabled

box->syslog
mx2(15)# log-host 57.82.152.139
```

```
box->syslog->log-host/57.82.152.139
mx2(16)# time-sequence enabled

box->syslog->log-host/57.82.152.139
mx2(17)# info
  address 57.82.152.139
  udp-port 514
  log-facility local7
  time-sequence disabled
  state enabled

box->syslog->log-host/57.82.152.139
mx2(27)# filter name "faults" entity all

box->syslog->log-host/57.82.152.139->filter/57.82.152.139/faults
mx2(28)# severity-mask fault

box->syslog->log-host/57.82.152.139->filter/57.82.152.139/faults
mx2(28)# info
  name faults
  entity all
  fault-map critical
  warning-map warning
  info-map info
  trace-map debug
  debug-map debug
  severity-mask fault
  event-lower-bound 0
  event-upper-bound 255
  slot-lower-bound 0
  slot-upper-bound 0
  state enabled
```

As with other event filters, a syslog event filter can separate events based on the entity, slot, severity, and range of event codes. Syslog events are logged under the `SYSLOG` entity.

DHCP Server

BayRS has been integrated with another Nortel Networks product, NetID, which provides expansive and cohesive management of IP addresses dynamically using DHCP and DNS. The NetID server manager dynamically assigns IP addresses to end stations requesting address assignments or other dynamic attributes and can interact with several BayRS servers to manage dynamic IP addresses. Clients sending DHCP discover requests are recognized by the router, and the request is either automatically processed or

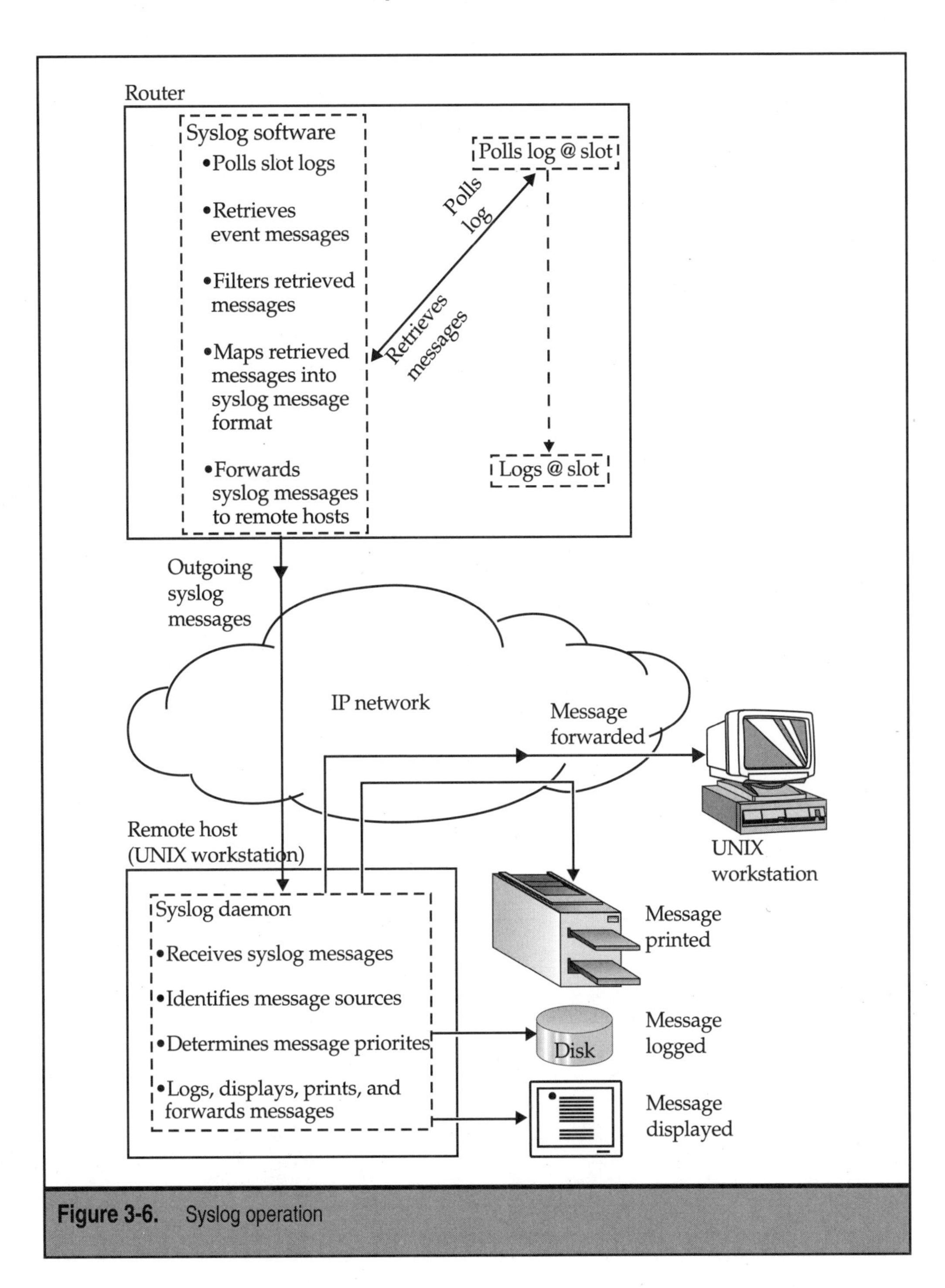

Figure 3-6. Syslog operation

proxied to the NetID server manager. The proxying of a DHCP discover request is shown in Figure 3-7, where the router waits for a successful lease commit acknowledgment from the server manager before assigning an IP address. This mode of operation is called the *safe mode*.

The DHCP server can also automatically assign an IP address first and then request the lease commitment from the server manager. This can be achieved by disabling safe mode operation. Currently, the DHCP server can be configured only using Configuration Manager and will be configurable using BCC in the future. Using Configuration Manager, first create the DHCP server object by choosing Protocols | Global Protocols | DHCP | Create DHCP. DHCP server runs using TCP, so you must enable TCP before creating the DHCP server. You then enable the server at the interface level by setting the **DHCP Server Enable** parameter in the BootP interface window: choose Protocols | IP | BootP | Relay Agent Interface Table. (This parameter needs to be set because it is disabled by default.) Configure the NetID server manager address and other server attributes in the DHCP global service window: choose Protocols | Global Protocols | DHCP | Global.

The event log entity for the DHCP server is `DHCP`, and DHCP-related parameters and statistics can be viewed using the **show dhcp** command.

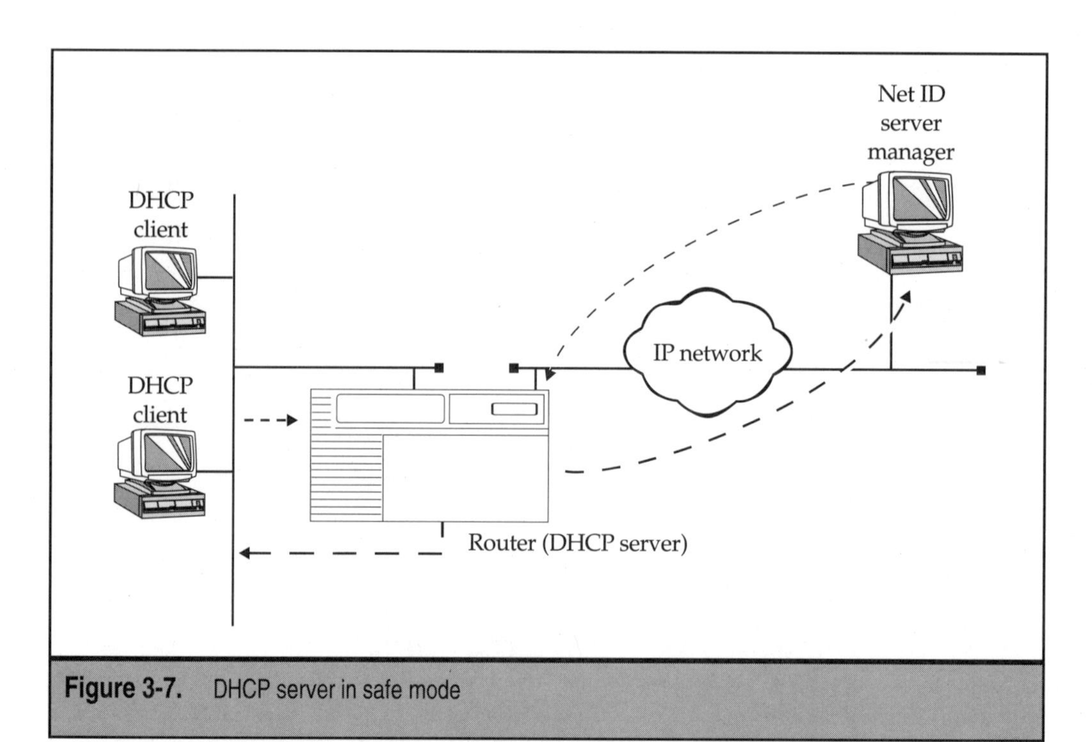

Figure 3-7. DHCP server in safe mode

Network Time Protocol

BayRS supports the use of the Network Time Protocol (NTP) to synchronize the router's time with that of a high-precision stratum clock in the network. Using NTP version 3, BayRS time clients can maintain accurate clock detail with an NTP server and provide networkwide time precision on the order of milliseconds or microseconds. Time synchronization on all internetwork routers is very useful for sequencing network events. On BayRS routers, synchronizing time using NTP is useful for automatic rekeying of ciphers using the WAN Encryption Protocol (WEP) and also helps ensure that scheduled boots occur at the correct time. A simple NTP relationship with a peer 105.95.128.50 is configured in the following example.

```
box
mx2(138)# ntp

box->ntp
mx2(139)# peer address 105.95.128.50

box->ntp->peer/105.95.128.50
mx2(140)# peer-preference enabled

box->ntp->peer/105.95.128.50
mx2(141)# info
  address 105.95.128.50
  peer-preference enabled
  mode server
  src-ip-address 0.0.0.0
```

The global NTP object called `ntp` also resides under the box context. BayRS supports NTP in client mode only, and at least one server peer must be configured. Peers are polled in unicast mode periodically to synchronize the time. When multiple peers are configured, the `peer-preference` attribute is enabled to accept time from a preferred peer; all peers are polled periodically, however. By default, the source IP address of the NTP client's polls will be that of the outgoing interface. If the NTP server is configured to accept polls from specific IP addresses only, then the `src-ip-address` attribute is set accordingly.

You can use the `NTP` log entity to view log events and the **show ntp** command to display the configuration and statistics.

Remote Authentication Dial-in User Service

Remote Authentication Dial-in User Service (RADIUS) provides a mechanism to centralize the authentication and accounting detail for services requiring it. Access to computers and applications customarily requires authentication, and centralized management of these functions provides greater security and permits better management.

BayRS routers act as RADIUS clients and connect to a RADIUS server to authenticate a user dialing in on analog (POTS) or digital (ISDN) lines. Along with dial-in users, multiuser access users can also be authenticated using RADIUS. At least one slot is configured with a RADIUS client, and at least one RADIUS server's IP address is configured for authentication and accounting. Multiple slots can be configured with clients, and several servers can also be configured—one server is marked primary, and if this server is unavailable, the client can authenticate users with one of several alternate servers. Here, the configuration of the global `radius` object with a client on slot 3 is illustrated.

```
box
mx2(16)# radius

box->radius
mx2(17)# radius-client slot 3 address 80.23.10.1

box->radius->radius-client/3
mx2(18)# authentication enabled; accounting enabled; info
  slot 3
  address 80.23.10.1
  debug-message-level no-debug
  authentication enabled
  accounting enabled
  accounting-direction all
```

The `radius` object resides under the global box context. At least one `radius-client` instance and one `radius-server` instance is created for authentication and accounting (see Figure 3-8). A client by default is not set to perform authentication or accounting; each of these can be enabled individually. In the preceding example, a client with IP address 80.23.10.1 is set up on slot 3 with both authentication and accounting. The client's IP address must be a previously configured IP interface on the router. Following the client creation on a slot, the server configuration is set to point to a central RADIUS server. Here, a primary authentication and accounting server with IP address 192.168.123.115 is set up with default values.

```
box->radius->radius-client/3
mx2(22)# back; radius-server address 192.168.123.115

box->radius->radius-server/192.168.123.115
mx2(23)# authentication-server-type primary

box->radius->radius-server/192.168.123.115
```

```
mx2(24)# accounting-server-type primary

box->radius->radius-server/192.168.123.115
mx2(25)# primary-server-secret x87KlpQcM

box->radius->radius-server/192.168.123.115
mx2(26)# info
  address 192.168.123.115
  authentication-server-type primary
  accounting-server-type primary
  server-mode both
  authentication-udp-port 1645
  accounting-udp-port 1646
  primary-server-secret x87KlpQcM
  response-timeout 3
  retry-count 2
  reset-timer 10
  automatic-reset disabled
  state enabled
```

Many RADIUS servers use port 1645 and 1646 for authentication and accounting based on the earliest implementations. Certain stricter and flexible RADIUS server implementations adhere to using port 1812, which can be set up using the `authentication-udp-port` and `accounting-udp-port` attributes.

Events are logged under the `RADIUS` entity; configuration and statistics can be viewed using the **show radius** command.

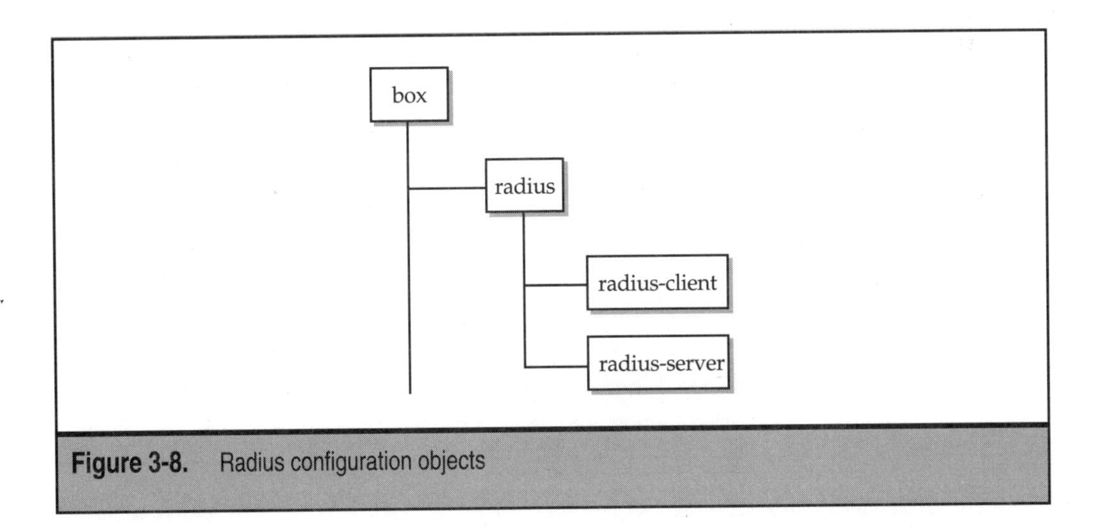

Figure 3-8. Radius configuration objects

Access Policies

Two distinct techniques can be applied to control access to system services on a BayRS router. The first approach is to apply traffic filters to selectively accept protocol service requests on all inbound interfaces. While this method is comprehensive, one or more filters must be configured on all circuits and must be applied on all inbound packets. In certain implementations, the traffic filter approach may be a bit inefficient; for these cases, the second method—access policies—is more aptly utilized.

All system services provided by a BayRS router are based on the IP protocol, and access policies hence are configured as subcontexts under the global ip object. Access policies can be applied to Telnet, FTP, TFTP, SNMP, NTP, and HTTP services. In the following example, a low precedence access policy named limit-http is created to universally deny access from any device to the HTTP server on the router.

```
box->ip
mx2(53)# access-policy polname limit-http

box->ip->access-policy/limit-http
mx2(54)# service http

box->ip->access-policy/limit-http
mx2(55)# network address 0.0.0.0 mask 0.0.0.0

box->ip->access-policy/limit-http->network/limit-http/0.0.0.0/0.0.0.0
mx2(56)# back; info

box->ip->access-policy/limit-http
mx2(57)# info
  action deny
  log on
  service http
  precedence 0
  polname limit-http
  state enabled
```

Following this, another policy, called allow-http, with a higher precedence, is created to enable HTTP sessions from the 192.168.83.0/24 subnet only. A service request matching two or more access policies is acted upon by the policy having the highest precedence. In this example, the allow-http policy is configured with the higher precedence.

```
box->ip
mx2(66)# access-policy polname allow-http

box->ip->access-policy/allow-http
mx2(67)# service http; action allow
```

```
box->ip->access-policy/allow-http
mx2(68)# network address 192.168.83.0 mask 255.255.255.0

box->ip->access-policy/allow-http
   ->network/allow-http/192.168.83.0/255.255.255.0
mx2(69)# back; precedence 1

box->ip->access-policy/allow-http
mx2(74)# info
  action allow
  log on
  service http
  precedence 1
  polname allow-http
  state enabled
```

Access policies with the opposite effect (that is, that deny access from specific subnetworks and accept access from all others) can also be created and applied.

Virtual IP Interface

Chapter 2 discussed circuits and interfaces. A virtual IP interface is one that is not associated with any individual circuit. Occasionally called the **circuitless IP interface**, the virtual IP interface can dynamically run on any slot because it does not require a circuit or an underlying physical or logical line to run. BayRS routers currently support the use of one boxwide virtual IP interface that runs as a soloist on a single slot. The virtual IP address facilitates universal access to the router as long as a viable route is available; this means that as long as at least one circuit is up, the router's virtual IP interface is available.

The virtual IP interface can be used for management tasks such as SNMP, FTP, and Telnet operations without regard to the administrative or operational status of specific circuits on the router. It operates in forwarding mode and can run OSPF and BGP routing protocols. This configuration is desirable in networks where many OSPF neighbors or BGP peers can communicate with the virtual IP interface to manage routing information. A virtual IP interface 192.168.85.1/30 is configured below, with the operational slot mask set to include slots 2, 3, and 4.

```
box
mx2(0)# virtual

box->virtual
mx2(1)# circuit-name "mx2-virtual"

box->virtual
```

```
mx2(2)# ip 192.168.85.1/30
box->virtual->ip/192.168.85.1/255.255.255.252
mx2(3)# slot-mask {2 3 4}

box->virtual->ip/192.168.85.1/255.255.255.252
mx2(4)# info
  slot-mask {2 3 4}
  address 192.168.85.1
  mask 255.255.255.252
  cost 1
  broadcast 0.0.0.0
  configured-mac-address {}
  mtu-discovery disabled
  mask-reply disabled
  all-subnet-broadcast disabled
  address-resolution arp
  proxy disabled
  host-cache-aging cache-off
  udp-checksum enabled
  end-station-support disabled
  redirects enabled
  cache-size 128
  mtu 0
  state enabled
```

The circuitless IP interface is used for purposes including BGP peering, OSPF adjacency, and DLSw peering and as an associated address for unnumbered IP interfaces. Management functions are also performed using the virtual IP. If multiple applications use the virtual IP, then the slot mask must be set to include processor slots with sufficient memory.

Scheduled Boot Service

Router boots can be scheduled using the RUI, or the Remote User Interface. The RUI subsystem (`wfTiRui`) provides remote access to certain TI commands, using SNMP. For example, file system operations such as file deletions and flash compaction can also be done using the RUI. Although the RUI's file system functions have been replaced by other direct operations, it is still used to schedule router boots.

Router boots are usually scheduled for a later time when an immediate boot is not possible or necessary and the timing of the boot must occur within a specific window. Since scheduled boots are configured for a specific time, it is important that the router's date and time be correct.

To create a scheduled boot configuration record, first enable RUI by choosing Platform | Scheduled Boot | Create Boot Param in the Configuration Manager window. Then choose Platform | Scheduled Boot | Boot Parameters to access the RUI Boot Parameters screen. Click Add to add a new scheduled boot record and supply the date, time, and boot image and configuration files. When specifying the boot image and configuration files, make sure that the file path includes the volume number (for example, 2:bn.exe).

Scheduled boot entries can be deleted using the same window; all scheduled boots can be disabled from the global RUI Boot Parameters window. Event messages related to scheduled boots are logged under the `RUIBOOT` entity, and general events for the RUI can be seen under the `TI_RUI` entity.

Thresholds

BayRS features a statistical subsystem that can calculate, monitor and generate alerts when certain attributes rise, fall, or reach specific values. The section "The Event Log" in Chapter 2 described how protocol and interface changes are automatically logged as informational and trace events. This behavior can be expanded and customized using statistical thresholds. For example, a network administrator may want to be notified when the number of IP routing table entries fluctuates by more than 50 routes. Similarly, network operation centers may need to be informed when traffic on important WAN lines exceeds key levels. The threshold functionality in BayRS can be used to generate log events, which can be trapped and sent to an NMS station for alarm generation, trouble ticketing, and paging.

Thresholds are currently configured using Configuration Manager: choose Protocols | Global Protocols | Thresholds. Attributes are polled every 60 seconds by default; this value can be changed in the global parameters. For a specific attribute and instance, you can configure a low-, medium-, and high-threshold value, and for each threshold, you can assign a message severity (Info, Warning, or Debug). Thresholds can be triggered either by an absolute value or by the rate of change of the value of an attribute. For example, a threshold trigger can be activated if the IPX service table exceeds 1,000 entries (absolute value), or if it fluctuates by ± 60 entries every minute (rate of change).

You add a threshold by clicking the Add button in the Threshold Interface Lists window (choose Protocols | Global Protocols | Thresholds | Thresholds). A new window appears providing a navigator for the available MIB objects. Following the MIB tree, first select the desired attribute and then fill in an instance number, which can be obtained using the TI (see "MIB Commands" in Chapter 2) or using an NMS. Once an attribute and instance are specified, the threshold values, event severity levels, actions, and calculation units are specified. A descriptive threshold label can be added to easily identify an event log entry, which is logged under the `STA` entity.

In the example shown in Figure 3-9, the number of IPX service entries (`wfIpxAdvSysServCount`, global instance identifier 1) is monitored as a statistical threshold. Using a global polling period of 60 seconds, a **warning** event is set to be generated if the service entries fluctuate by 3 per second (180 entries per minute). Debug

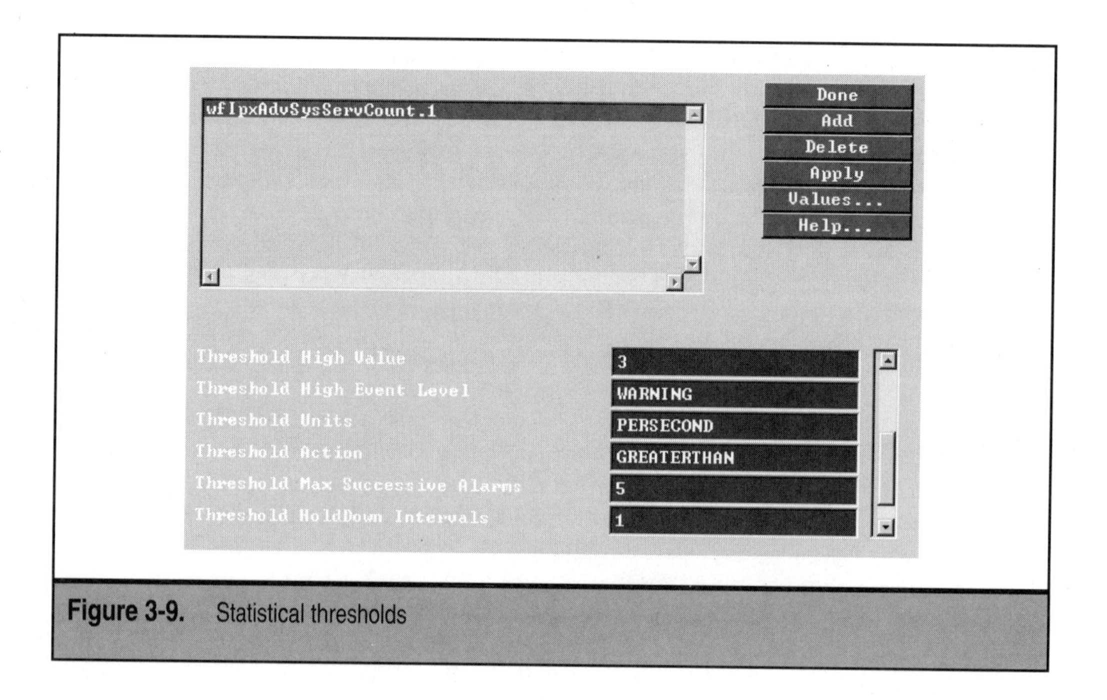

Figure 3-9. Statistical thresholds

thresholds can be triggered to generate debug messages for troubleshooting purposes, and will be discussed in the following chapters.

Using the show and ping Commands

Two commands are frequently used in the monitoring and verification of router and network operation: **show** and **ping**. The **show** command accepts several subcommands and many optional arguments. It is available using both the TI and BCC. BCC natively includes most show commands, and the TI has a few embedded commands. The TI **show** command requires batch script files supplied with the BayRS software, which must be placed in a flash file system. The show commands are used to display routing and service tables, configuration parameters, and interface and protocol statistics. Customized TI/BCC scripts can be written to obtain, format, and display any information accessible via proprietary and standards-based MIBs.

Both TI and BCC show commands produce similar output. The TI show scripts can be enhanced using custom scripts, and BCC can be enhanced using TCL procedures. Since BCC is a recent and evolving application, TI show commands are more common and widely available. Occasionally a user may come across a TI show command that has no equivalent in BCC; in such a case, custom TCL scripts can be written to perform the same function. Alternately, TI show commands can be called from BCC using the **tic** command with the following syntax:

```
bcc# tic show subcommand arguments ...
```

Append a question mark (?) to show commands and subcommands to see a list of legal options and values. Appendix F provides a list of the show subcommands currently available with the TI and BCC.

Using Reset, Restart, Diags, and Boot

Chapter 2 discussed reinitialization of slots using the **reset**, **restart**, **diags**, and **boot** commands. Each command is used in specific situations, and this section discusses when to use each.

The **boot** command is perhaps the most commonly used because users automatically associate it with reinitialization. It is most appropriate to use **boot** when the entire box must be reloaded and resynchronized with a newer image or configuration. Configuration changes that have a systemwide impact, such as those involving a network layer protocol, are usually scheduled to occur during a downtime window. Occasionally network operators reboot routers using the **boot** command when they see a networkwide issue; this should be avoided, however, unless a systemwide inconsistency occurs in a router. The **boot** command should also be used when loading an image or configuration file with a nondefault name on the router.

A more granular access to the **boot** command's functionality is available with the **reset** command. The **reset** command on single-slot routers (AN, ANH, ARN, and single-unit ASN) behaves identically to the **boot** command with the default boot arguments. On multi-slot routers, the **reset** command will reboot only a specific set of slots so that they reinitialize with the configuration and image. You must use **reset** in cases where network availability is essential for other slots. Configuration changes such as memory recarving or changes to driver buffer queues may require a reset; other configuration changes such as circuit and interface provisioning will dynamically take effect. Note that a resetting slot will automatically load the current image and configuration from the server slots. The reset process does not load a newer image file or configuration file from the flash , except in the case of a full system reset.

Network administrators observing inconsistencies in a slot's operation can either restart or reset the slot. The **restart** command reinitializes the slot using the current image, which in turn reinitializes the protocols, circuits, interfaces, and routing and service tables. The image is not reloaded, which allows a restarting slot to quickly become operational again with minimal service disruption. Restart is usually advised when minimal or no configuration changes are in effect. As with the **reset** command, image and configuration files are not reloaded from the flash.

Diagnostics must be run on a slot if the hardware configuration is changed or an existing hardware module is swapped. Modules that support the hot-swap feature will bypass diagnostics during insertion so that they initialize and become operational quickly. The Fail LED on the link module remains lighted to indicate that diagnostics have not been run; this task must be performed as early as possible following the hot-swap. Older hardware or modules that do not support hot-swapping must be inserted after a power-off to prevent damage. Some older modules may be inserted and obtain power, but the **diags** command must be specifically issued to them for them to initialize properly. Occasionally, a hardware problem may cause unexplainable aberrant behavior on a slot, which can be confirmed by running **diags**. Running **diags** on all slots is equivalent to cold-booting the entire system, which performs diagnostic tests on all components.

Out-of-Band File Transfers

A common administration task is the transfer of files to and from the flash file system using the out-of-band management port. Direct terminal and dialup TI console sessions can use the **xmodem** command to perform file transfers. This command is most useful when the only available access method involves using the console port, or when a reliable transfer process is needed because the network function is impaired.

The **xmodem** command supports XMODEM/CRC, MODEM7 block, XMODEM-1K block, and YMODEM protocols for file transfers. The most popular and contemporary protocol used is the YMODEM protocol. YMODEM provides binary file transfers with error detection. It also supports cancellation of the file transfer, which is especially useful with large files or noisy analog lines. Multiple files can be transferred using a single YMODEM transfer without regard to file type, and files are automatically named with the original file name. You should always use the YMODEM protocol in binary mode, to help ensure that file transfers are successful even on noisy lines.

The syntax of the **xmodem** command is as follows:

```
xmodem sb|st|rb|rt[ymkclwn] filename1 filename2 ...
```

The **sb** and **st** options send the specified files in the binary and text formats respectively, and the **rb** and **rt** options receive files in binary or text form. Text-format transfers should be performed only on files guaranteed to contain only ASCII text. Configuration files and images must always be transferred in binary format.

Other options can be specified; the most common are described in Table 3-3.

Login and Welcome Notices

Two unique messages can be displayed at the Technician Interface: the login notice can be displayed above the login prompt prior to a login, and the welcome notice can be displayed after a user has logged in, but before the prompt appears. The login notice is contained in a file named `ti_notice.txt` and is used to specify authorized use, system-specific detail, and contact information. The welcome notice (similar to motd, the message of the day) is stored in a file named `ti_msg.txt` and can contain system-specific information or announcements specific to the router. If the `ti_notice.txt` and `ti_msg.txt` files are absent, the default login and welcome notices are displayed.

The login notice is displayed above the login prompt (see Figure 3-10). On most terminals, the display is 24 lines high, and it is recommended that the login notice be no more than 20 lines long. If the login notice is too long, then the text at the top may scroll by with-

Option	Description
y	Use the YMODEM protocol. It is recommended that YMODEM be used for all transfers.
w	Wait 15 seconds before initiating the startup handshake. Some terminal emulation applications require you to manually invoke the matching **receive** or **send** command on the workstation. Using this option will allow you sufficient time to invoke the matching command.
p	Print the action and events of the **xmodem** command as they occur.
l	Disable logging. By default, **xmodem** will log events under the `XMODEM` entity, which may assist in troubleshooting file transfer issues. Using this option disables logging of events.
n	Allow CAN-CAN to abort file transfer in midsession. It is recommended that this option also be used, particularly with large files. The CAN character is ^X (CONTROL+X) by default, which if typed repeatedly will abort the transfer. This is helpful for recovering from an erroneous transfer or excessive errors during a large file transfer. If you omit this option, the **xmodem** command can be terminated only during the beginning of the file transfer.

Table 3-3. Common **xmodem** Command Options

out the user being able to read the content. Both the login and welcome notices should be concise and contain important and pertinent information only.

SOFTWARE UPGRADES

Software upgrades are a necessity in today's evolving networks. New network functionality requirements such as new protocols are available only in recent software updates. The Virtual Router Redundancy Protocol (VRRP), for example, is a recent protocol developed by the VRRP working group of the IETF. This protocol has received widespread acceptance for its redundancy service and was implemented in BayRS 13.10 for general availability. BayRS routers running previous versions of the software must be first upgraded to version 13.10 or above to achieve VRRP functionality. Similarly, new networking hardware requires new software to operate it. Occasionally, certain network installations may run into software bugs, which are usually resolved by a newer software release or revision. Overall, software upgrades are essential to meet corporate IT stan-

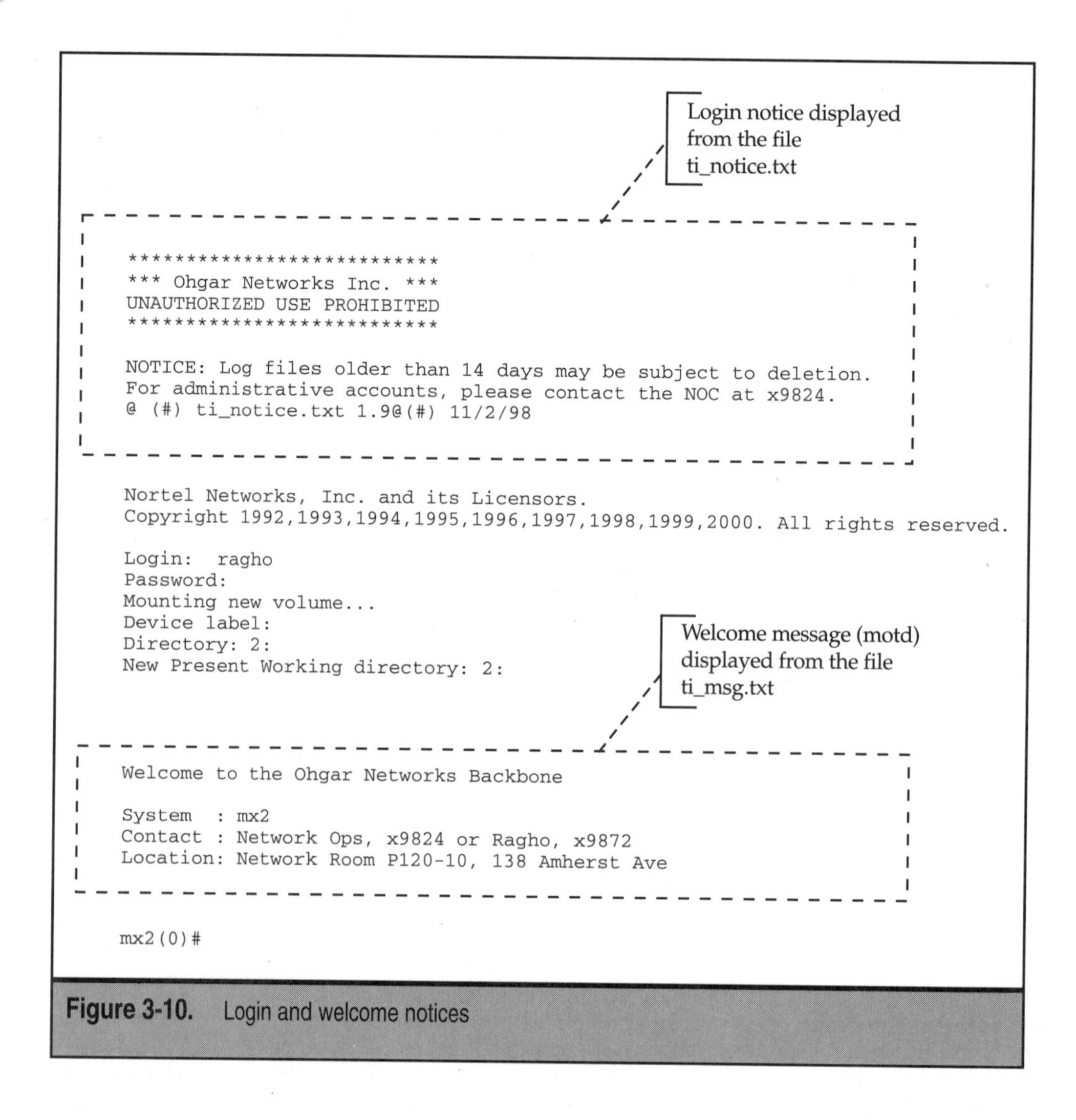
Login notice displayed from the file ti_notice.txt
****
*** Ohgar Networks Inc. ***
UNAUTHORIZED USE PROHIBITED
****
NOTICE: Log files older than 14 days may be subject to deletion.
For administrative accounts, please contact the NOC at x9824.
@ (#) ti_notice.txt 1.9@(#) 11/2/98
Nortel Networks, Inc. and its Licensors.
Copyright 1992,1993,1994,1995,1996,1997,1998,1999,2000. All rights reserved.
Login: ragho
Password:
Mounting new volume...
Device label:
Directory: 2:
New Present Working directory: 2:
Welcome message (motd) displayed from the file ti_msg.txt
Welcome to the Ohgar Networks Backbone
System : mx2
Contact : Network Ops, x9824 or Ragho, x9872
Location: Network Room P120-10, 138 Amherst Ave
mx2(0)#

Figure 3-10. Login and welcome notices

dards, meet new functionality requirements, support new network hardware and interfaces, and fix bugs.

Preparing for an Upgrade

Before you begin the upgrade process, follow the steps described here to prepare for the upgrade. Readers who are unfamiliar with the new software release policy should read it; it is included in Appendix A.

Read the Release Notes and Upgrade Documentation

The release notes for each software version contain information specific to the release. In particular, the release notes include exceptions and special considerations for installing and operating the software. Router administrators and network managers can gain insight into foreseeable issues and plan accordingly.

Check the Hardware Requirements

Another important task in the preparation process is the verification of minimum hardware requirements. If the software upgrade involves skipping one or more major versions, then it is essential that you perform this step. First verify that the flash file system meets the minimum requirements for storing the new runtime image and configuration file. You should obtain a larger flash to contain all system files if the current flash cannot hold all the files. If the current flash has enough free space but insufficient contiguous space, then the free space must be recovered using the **compact** command.

Progressively newer software versions require more dynamic RAM to run and perform routing functions. For your specific platform, check the release notes to ensure that there is no minimum memory requirement, and if there is a minimum memory requirement, take appropriate steps to obtain enough memory for the device. If you simply meet or minimally exceed the memory requirement, then consider the amount of additional memory required to load and maintain protocols that the router will be running. The more protocols and routing and service entries in the protocol tables, the more memory will be required. Do not proceed further with the upgrade if the minimum requirements are not met.

If your system files are too large to fit in the flash file system and this is your only limitation, then see "Customizing Images" later in this chapter to trim the contents of the runtime image.

Check for the Need to Upgrade Software to Support New Hardware

When you add new hardware, you may also need to perform software upgrades solely to support the new hardware. In these cases, it is vital to verify that the minimum bootstrap and diagnostics PROM versions are used. Almost always, you will need to upgrade at least the diagnostics PROM to recognize and properly initialize a new hardware component. In other cases, such as the addition of a larger flash, you will need to upgrade the boot PROM. The release notes specific to the upgrade version and the installation manual specific to the new hardware will provide information on the minimum PROM versions you need.

Create a New Router Configuration

If the software upgrade is being performed to resolve software bugs, then a new configuration is not needed. However, if you are upgrading to add new functionality, you should create a new configuration with the required functionality offline. Network downtime is usually scheduled for brief periods of time, and having a prebuilt configuration will give you more time for testing and troubleshooting.

Configurations can be built offline using Configuration Manager, a key component in the Site Manager suite. A currently existing configuration can be modified offline to support the new configuration changes. In large corporate networks, you may be able to test the configuration, along with the new software, in a lab environment. Network administrators who have this ability are encouraged to do so. This will help eliminate various verification and troubleshooting tasks in the actual device being upgraded. Also, this process will help minimize the system downtime required and bring the router more quickly to full operational status after the upgrade.

Dynamic configuration changes are usually not recommended for large-scale network reconfigurations. Minor changes can usually be made dynamically without disrupting network operation. Currently running configurations can be modified in local, remote, or cache mode so that they can be used to boot the router at a later time. Local configuration mode provides full flexibility by allowing a configuration file to be created offline, and remote mode allows the fetching of a current configuration from the flash of an operational router. Configuration Manager in cache mode obtains a dynamic copy of the MIB configuration using SNMP from the operational router and allows subsequent additions and changes offline. For more information on local, remote, and cache mode configurations, consult the Site Manager documentation for the software version you are using.

If you are not using Site Manager, textual BCC configuration commands can also be summarized in a file offline for subsequent execution as a script in BCC. BCC supports the **source** command, which takes an input file argument that is executed as if it were typed on the command line.

Create a New Topology Drawing

All networks, whether tiny or enormous, must be fully documented with a network topology drawing. Networks that are especially extensive may have their topologies documented in sections. Both minor and major changes to a network routing node or several nodes require updating to reflect a current topology. It is recommended that previous topologies be archived for future reference.

Creating Backups

Backups are essential for various reasons. Networks change dynamically and periodic backups can help restore the network to a well-known operational state following a change. Software upgrades are generally classified as network changes, and unplanned surprises can be addressed by reverting to a backup software and configuration; this will allow the network to function as before while you investigate the network issues encountered following an upgrade.

Back Up the Configuration

Regardless of the extent and scale of configuration changes, configuration files must always be archived. The most obvious reason for archiving is to enable you to restore the router to a previously known state and reestablish network operation. Also, archiving network topologies and router configurations provides a clear understanding of the evolution of the network. With proper design and archival procedures in place, a network topology can be used to reconstruct a configuration file and vice versa. Performance baselining, trending, and future scaling are also facilitated by the availability of backup topologies and configurations.

Back Up the Image and Flash Contents

As with the configuration file, all other critical file system contents should be backed up to a server. Using either FTP or TFTP (Router Files Manager), all file system contents can be copied to a workstation for future reference and, if necessary, immediate recovery. Unforeseen issues may require you to abort a software upgrade, in which case you can use the backup copy of the current file system to restore the system. A backup of the current flash may not be necessary for recovery purposes if you are using a spare flash for upgrade or if the router is a multi-slot system (see "Preparing the Flash" later in this chapter). For future review, however, the flash contents should be archived and dated.

All FTP file transfers must be performed in binary mode so that data is transferred in 8-bit clean form. Files containing binary data, such as image files, will be incorrectly transferred if ASCII mode is used. When performing the backup, compare the before and after file sizes to ensure that the files were transferred correctly and completely.

Many FTP clients vary in actual operation, although the underlying protocols and commands are the same. Several easy-to-use GUI FTP clients are available both commercially and for free. The following example shows how to back up flash contents with a traditional FTP session using a command-line interface on a UNIX workstation.

```
% ftp 198.102.32.1
Connected to 198.102.32.1.
220 WfFTP server(x14.10) ready.
Name (198.102.32.1:root): ragho
331 Password required for ragho.
Password:
230 User ragho logged in.
ftp> binary
200  Type set to I.
ftp> cd 2:
250  CWD command successful.
ftp> prompt
Interactive mode off.
ftp> mget *
200  PORT command successful.
150  Image data connection for 2:config (198.102.32.1,0)
     (3312 bytes).
226  Binary Transfer Complete.
local: config remote: config
```

```
3312 bytes received in 0.12 seconds (26 Kbytes/s)
200  PORT command successful.
150  Image data connection for 2:bn.exe (198.102.32.1,0)
    (12769744 bytes).
226  Binary Transfer Complete.
local: bn.exe remote: bn.exe
12769744 bytes received in 63 seconds (2e+02 Kbytes/s)
ftp> quit
221 Goodbye.
```

Preparing a Post-Upgrade Test Plan

A successful software upgrade not only involves a successful reboot using the new software but also entails validation of the network operation. Following the software upgrade and reboot, the router must at the very least execute routing and bridging functions exactly as it was prior to the upgrade. If new protocols were added or configuration parameters were changed, then you must ensure that these function as desired also.

To this end, you need to draft a plan to comprehensively test all protocols and services in a simple manner. Elaborate test plans consume time, and verification of a simple set of key routing or bridging tasks is usually satisfactory. For example, if your network runs both IP and IPX protocols, then you can test server connectivity to at least a few IP servers and a few IPX servers from a small number of dispersed clients.

Obtaining Software

When you purchase a particular release of the BayRS software, it is usually supplied on a CD by Nortel Networks Software Services. When you purchase a specific suite of a software version, it may be delivered on a flash. Users with a contracted login to the Nortel Networks Enterprise Solutions Support web site can download BayRS software from the Assurance Online site. (The Assurance Online web site is linked on the book web site; it can also be reached directly at http://support.baynetworks.com/svc_online/support.cgi.) Each software release contains images for each supported platform. Recall that in Chapter 1 we mentioned that software suites support a varying set of protocols. Refer to Appendix A to verify that the software suite you obtained contains the protocols required.

All software suites contain at least the following files:

- BayRS runtime image (bn.exe, asn.exe, arn.exe, an.exe, or s5000.exe).
- Default configuration file (config), which is not needed for software upgrades.
- Secondary configuration file (ti.cfg, ti_asn.cfg, ti_arn.cfg, or ti_s5000.cfg), needed for backup.
- Installation batch file (install.bat or inst_arn.bat), needed if future reinstallation is required.
- Debug aliases file (debug.al), needed for troubleshooting tasks.

- BCC help file (bcc.help or bcc_*platform*.hlp), needed for context-sensitive BCC help. (BCC is a recent feature that is not available in certain 11.*x* and 12.*x* software versions. If your software does not contain a BCC help file, you can ignore it, especially if you do not plan on using BCC.)
- One or more bootstrap PROM update files (freboot.exe, fre4boot.ppc, areboot.ppc, asnboot.exe, arnboot.exe, anboot.exe, or s5000boot.exe). (Note that the FRE4 platform was introduced in BayRS 13.20. The boot and diagnostics PROM files for the FRE4 are included in version 13.20 and subsequent versions.)
- One or more diagnostics PROM update files (frediag.exe, fre4diag.ppc, arediag.ppc, asndiag.exe, arndiag.exe, andiag.exe, or s5000diag.exe).

The Backbone family software includes PROM files for all supported processor modules for the release.

Preparing the Flash

The first step of the upgrade is to identify the PROM versions and, if necessary, upgrade them. But before performing this task, you must prepare the flash file system to receive the new system files.

The following instructions apply if you are upgrading a router with a single slot or a single flash card. This includes the AN1 router, which contains an internal SIMM flash. If the system contains multiple flashes and one flash can be prededicated to receive the new system files, then some steps in this section may be omitted.

When backing up files from the current file system, you can use TFTP, FTP, or BNFS. If BNFS is used, then you must verify that TFTP or FTP is already running on the router so that new files can be transferred to the flash. Using FTP provides a reliable method of transferring the system files, although certain routers are better upgraded using TFTP. Whenever possible, use FTP to transfer files to the flash. The administrative status of TFTP or FTP can be checked as shown here. The BCC commands on the left and the MIB get commands on the right are equivalent.

```
box# tftp; state                    mget wfTftp.1.0
state enabled                       wfTftp.WfTftpDisable.0 enabled

box# ip; tcp; state                 mget wfTcp.2.0
state enabled                       wfTcp.wfTcpDisable.0 enabled

tcp# box; ftp; state                mget wfFtp.WfFtpDisable.0
state enabled                       wfFtp.wfFtpDisable.0 enabled
```

Note that FTP runs using TCP, which must also be configured for operation. Besides verifying that FTP/TFTP is enabled, run the **loadmap** command to confirm that the protocol application is indeed loaded.

```
mx2-bln# loadmap
     File name          Load Addr     Size
---------------------------------------
--> arp.exe            0x315e15e0  0100504
--> ftp.exe            0x315d12c0  0045348    (FTP)
--> tcp.exe            0x315be7c0  0076516    (TCP, required for FTP)
--> tftp.exe           0x315dc400  0020936    (TFTP)
--> tn.exe             0x31670ed0  0044404
--> ip.exe             0x3167bc60  0454060
```

Next, you need to discard files that are no longer needed on the current flash and recover the free space. Before performing the deletion process, load the string files from the flash to memory. The string files are contained in the image archive and are used for MIB object and attribute names and other static text used by various applications.

```
box# tic string load
```

The **string load** command is a TI command and must be called using **tic** to run it in the TI context; otherwise, it will be treated as the TCL **string** command and will generate an error. At this point, all nonessential files in the flash can be deleted, and the freed space can be reclaimed. Do not delete the existing configuration file (config), as it is essential for loading the configuration parameters when the router is rebooted with the new runtime image. If a spare flash is used, then you need to copy the existing configuration file to the spare flash.

```
box# rm *.exe install.bat debug.al ti*.cfg bcc.help
Removing file: 3:bn.exe
Removing file: 3:freboot.exe
Removing file: 3:frediag.exe
Removing file: 3:areboot.ppc
Removing file: 3:arediag.ppc
Removing file: 3:ti.cfg
Removing file: 3:debug.al
Removing file: 3:bcc.help
8 File(s) deleted.
```

The **delete** command from the TI performs the same function as the **rm** command in BCC.

Compacting the Flash

Once the older files have been removed from the file system, the flash must be compacted. This makes all the free space available for use by the new system files. If the flash is a secondary flash on a multiflash system, make sure that it is empty or has sufficient space for the new files and, if necessary, compact the secondary flash.

```
box# compact 3:
Compacting file system on volume 3: ...
This may take several minutes...Please wait...      100% Complete
Compaction completed
```

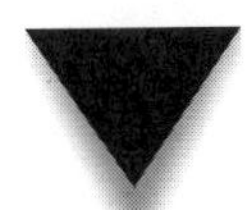

CAUTION: The compacting process takes a long time and is proportional to the storage capacity of the flash. It should not be interrupted, and the router should not be powered off during the operation. Doing so will render the flash without a usable file system, which must be reformatted before the flash can be used. Since the process involves an upgrade, the corrupt flash will be unable to initialize the router.

Performing PROM Upgrades

Many software upgrades involve a move to a subsequent maintenance release or revision. In situations such as this, it is usually not necessary to reprogram the PROMs because the bootstrap and diagnostics codes change infrequently. Even if the upgrade involves a leap from one functional release to the next, the PROM codes may not need upgrading.

NOTE: Programming the PROM device is also called burning. The two terms are used interchangeably in this chapter.

TIP: AN1 routers with motherboard revisions earlier than R14 do not have an onboard programming device. On these units, PROM upgrades must be performed by physically replacing the PROM chipsets. To identify the motherboard revision of the unit, issue the following command:

```
box# mget wfHwEntry.6.1
wfHwEntry.wfHwMotherBdRev.1 00.00.00.08
```

Identify the PROM Versions

You can check the current boot and diagnostics PROM versions using the following MIB get commands.

```
box(10)# mget wfHwEntry.19.*
wfHwEntry.wfHwBootPromSource.1 {}
wfHwEntry.wfHwBootPromSource.2 rel/8.10/freboot.exe
wfHwEntry.wfHwBootPromSource.3 rel/8.10/freboot.exe
wfHwEntry.wfHwBootPromSource.4 fix/4.21/7/freboot.exe
wfHwEntry.wfHwBootPromSource.5 rel/13.20//fre4boot.rom

box(11)# mget wfHwEntry.16.*
wfHwEntry.wfHwDiagPromSource.1 {}
wfHwEntry.wfHwDiagPromSource.2
```

```
     /harpdiag.rel/v4.12/wf.pj/harpoon.ss/image.p/frediag.exe
wfHwEntry.wfHwDiagPromSource.3
     /harpdiag.rel/v4.12/wf.pj/harpoon.ss/image.p/frediag.exe
wfHwEntry.wfHwDiagPromSource.4
     /harpdiag.rel/v4.12/wf.pj/harpoon.ss/image.p/frediag.exe
wfHwEntry.wfHwDiagPromSource.5
     rel/fre4diag_v1.14/bin/diagmake/fre4diag.ppc
```

Slot 1 on BLN routers and slot 7 on BCN routers will not contain any boot or diag PROM information because they house the SRM modules, which do not contain PROM devices. Table 3-4 lists the boot and diag PROM file names issued with BayRS for various platforms.

Validate the PROM Images

Use the **readexe** command to check the boot and diag PROM code versions supplied with the new software. Use the Router Files Manager or an FTP client on a workstation, or the TFTP client on the router, to transfer the platform-specific boot and diagnostics PROM files to flash. Then run **readexe** on the newly transferred PROM files.

```
box(23)# tftp get 132.245.155.134 freboot.exe 4:freboot.exe
Transfer with 132.245.155.134 completed successfully.
box(24)# tftp get 132.245.155.134 frediag.exe 4:frediag.exe
Transfer with 132.245.155.134 completed successfully.
```

Platform	PROM File Names
AN	anboot.exe, andiag.exe
ARN	arnboot.exe, arndiag.exe, arn_pdbrom.rom
ASN	asnboot.exe, asndiag.exe
5380/5580	s5000boot.exe, s5000diag.exe
5780/5782	s5000boot.ppc[1], s5000diag.ppc
FRE1/FRE2	freboot.exe, frediag.exe
ARE	areboot.ppc, arediag.ppc
FRE4	fre4boot.ppc[1], footnote text = , fre4diag.ppc

[1]The FRE4 platform was introduced in BayRS 13.20.

Table 3-4. **boot** and **diag** PROM File Names

In the preceding example, the frediag.exe and freboot.exe PROM images are copied to the flash from the 132.245.155.134 TFTP server. Running **readexe** on the newly transferred files will validate the integrity of the PROM images.

```
box# readexe 4:frediag.exe

Processing contents of '4:frediag.exe'...

---------------------------------
-- Module name:   frediag.exe
---------------------------------
   Validating header checksum... OK
   Validating image checksum... OK

   Program execution address space:
   --------------------------------
   Load Address: 0x00000000  Size: 244356 Bytes  Entry point:  0x00000400

   PROM storage address space:
   ---------------------------
   PROM Load address: 0x00000000

   Input file information:
   -----------------------
   Platform Key: (0101000B) BB M68000 MotherBoard (FRE FRE2 FRE2_60)
   Workspace:    /harpdiag.rel/v5.14/wf.pj/harpoon.ss/image.p
   Compression:  OFF
   Revision:     5.14
   Last Modified: Wednesday July 22 11:09:40 1998
   File type:    Executable file
   Tool name:    Oasys Linker
```

The **readexe** command must be run after the PROM files have been transferred to flash. This will verify the integrity of the files transferred, which will help you avoid inadvertently programming the PROM with an incomplete or corrupt file.

Program the PROM Devices

Once you have determined that you need to upgrade the boot and/or diag PROM, the files must be validated as discussed in the previous section. Once validated, the files can be programmed using the **prom** command. The **prom** command has the following syntax:

```
prom      [-v|-w] <vol>:<ROM file> <slot ID> [<slot ID> ...]
```

The command is first used with the **–w** (write) option to program the new PROM file. As noted in the previous section, the PROM files for each processor platform are different. Make sure that the PROM file that is about to be used is correct for the processor that will be updated. This is important on the BNs, because they can contain FRE-1/2, FRE4, and ARE motherboards, all of which need unique PROM files. In nodes with different processor types, the PROM command must be issued each time with a different file name and slot number to update all processors.

```
box# prom -w 4:frediag.exe 2

File information:
-----------------
Image:     /harpdiag.rel/v5.14/wf.pj/harpoon.ss/image.p
Created:   Wednesday July 22 11:09:40 1998

prom: This operation may take a maximum of 10 minutes to perform.
prom: slot 2 completed successfully.
```

Notice that the slot numbers can be separated using a comma, or a range can be specified using a hyphen.

```
box# prom -w 4:freboot.exe 1,2,4
box# prom -w 4:areboot.ppc 5-6
box# prom -w 4:fre4boot.ppc 3
```

CAUTION: The **prom** command, like the **compact** command, may take several minutes to complete. Keyboard interrupts are disabled during this process, and care should be taken to let the command run to completion. Failure to do so may result in an incompletely programmed PROM device, which could cause the router to fail to boot or run diagnostics.

The **prom** command must be issued twice for each slot: once to upgrade the bootstrap PROM and again to upgrade the diagnostics PROM.

TIP: Multiple processor modules of the same type can be updated using the same **prom** command, thereby saving time by performing simultaneous programming on several slots. However, interruption of the prom process due to an event such as power loss may render all of these slots inoperable at the same time. Whenever possible, it is recommended that each slot be updated and verified individually.

CAUTION: If the **prom –w** command fails, do not attempt to reboot the router to recover. Instead, transfer a new copy of the PROM update file to flash, run **readexe** to validate it, and rerun the **prom –w** command on the slot.

Verify the PROM Upgrades

Recall that in the **prom** command syntax, either the **–w** or **–v** option can be used. Once the **–w** option has been used to update the PROM, the contents of the newly programmed device can be verified using the updated file on the flash. This step reverifies the previously validated update file. Use the **prom** command again, as follows:

```
box# prom -v 4:frediag.exe 2

File information:
-----------------
Image:      /harpdiag.rel/v5.14/wf.pj/harpoon.ss/image.p
Created:    Wednesday July 22 11:09:40 1998

prom: This operation may take a maximum of 10 minutes to perform.
prom: slot 2 completed successfully.
```

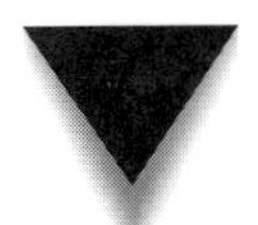

NOTE: The verification process is also sometimes used prior to burning the PROM to check whether the PROM version is already up to date. However, the verification process tells you only whether the code on the PROM device is different than the code on the flash. The **readexe** command is better suited for checking the PROM version.

If the burning of the PROM was successful, without any errors, the **prom –v** command will confirm that the programmed image and the PROM file contain the same code. Like the **–w** option, the **–v** option must be used twice: once to verify the boot PROM and again to verify the diagnostics PROM.

Remove the PROM Update Files

Once the PROMs have been burned, the PROM update files are no longer necessary. They can be removed from the flash, and the free space can be reclaimed, which you will want to do particularly if flash space is tight. The runtime image is the largest file and occupies the most space in the flash. The PROM files can be left untouched on the flash if there is sufficient space to receive the remaining system files.

```
box# rm *boot* *diag*
Removing file: 4:freboot.exe
Removing file: 4:frediag.exe
Removing file: 4:areboot.ppc
Removing file: 4:arediag.ppc
4 File(s) deleted.
```

If the PROM files are deleted, then the flash space must be recovered using the **compact** command (see "Compacting the Flash" earlier in this chapter).

Customizing Images

When the new software arrives on a medium other than a preformatted flash, occasionally the current flash file system may not be large enough to hold all the system files. The runtime archive image is the largest file and can be customized to a smaller size by removing applications and drivers that are not needed for the router. The runtime image is an archive of numerous smaller files, which are loaded individually by the dynamic loader (see "Dynamic Loader" in Chapter 2).

Certain applications and drivers are specific to one platform and are not included in other platform images. Files containing the .ppc extension are for PowerPC-based platforms (ARE and FRE4). The .ppb files are currently for use by FRE4 processors only. Using Image Builder, component files that are not needed for this implementation can be safely removed. For example, if a network runs using only IP, then other unused protocol applications such as IPX, DECnet IV, OSI, AppleTalk, and so on can be removed. Image Builder can be launched from the Site Manager screen by choosing Tools | Image Builder.

TIP: It is safe practice to keep module and chipset drivers in the image because they consume relatively small amounts of space. Router administrators are usually more likely to remember that a router does not contain a specific protocol suite than the drivers for a new link module. Many network additions or changes involve the use of a new module.

CAUTION: Unless critically necessary, you should not alter components of the kernel image within the archive. In particular, the EDL and STR files in the archive should never be deleted because they are used by many TI/BCC commands. Consult the Nortel Networks Technical Support Center prior to removing a file from the kernel image. Kernel image file names start with krnl_ and are listed in Table 1-2 .

Use Image Builder to remove files in the archive that are not needed (see Figure 3-11). This reduces the size of the final runtime image that will be transferred to the flash. Careful customization of the image will usually let you fit it into the current flash. If the image size is still too large after customization, then you will need to obtain a larger flash to hold the system files.

Steps to Customize Images

Follow these steps to customize images:

- **To load the current image file for customization** Choose File | Open. Image Builder automatically analyzes the image and lists the current components. Each component is a small ensemble of applications and/or drivers. In addition to the current components, Image Builder also searches the builder.dir directory under the installation for available components. Available components are those that are made available from previous uses of Image Builder and those that do not exist in the current image. If components were removed from the image previously, those components will be listed as available for addition.
- **To remove a component** Select a current component and click the Remove button. For example, if a router does not use the Appletalk protocol, then component 4x007 (AppleTalk Phase 2 Router Software) can be removed. When you are removing or adding a component, the current compressed image size is shown at the top portion of the screen. The compressed size of the final customized image must be smaller than the contiguous free space on the flash for the image to fit completely.

- **To add a component** Select an available component and click the Add button. This process is usually necessary if an application was inadvertently removed from the image. When a component is highlighted for addition or removal, the compressed and uncompressed image sizes are shown at the lower portion of the screen. It is important to keep the total compressed size of the full image in view when adding or removing a component.
- **To customize specific applications** Selecting the Details check box for both the available and current components lists the constituent applications of a component. For example, component 4x003 (IBM Services) contains three separate applications: APPN/CP (appn_cp.exe), APPN/LS (appn_ls.exe), and DLSw (dls.exe). If DLSw is required in the network for subarea SNA transport but not APPN, then the APPN applications can be removed from the detailed listing. Simply highlight a specific application and click Remove or Add as necessary. When you remove or add an application in a component, the component is prefixed with an asterisk (*), indicating that it was modified. Most image names easily identify the function of the application and have an associated description; consult the application description to ensure that a required application is not inadvertently removed.

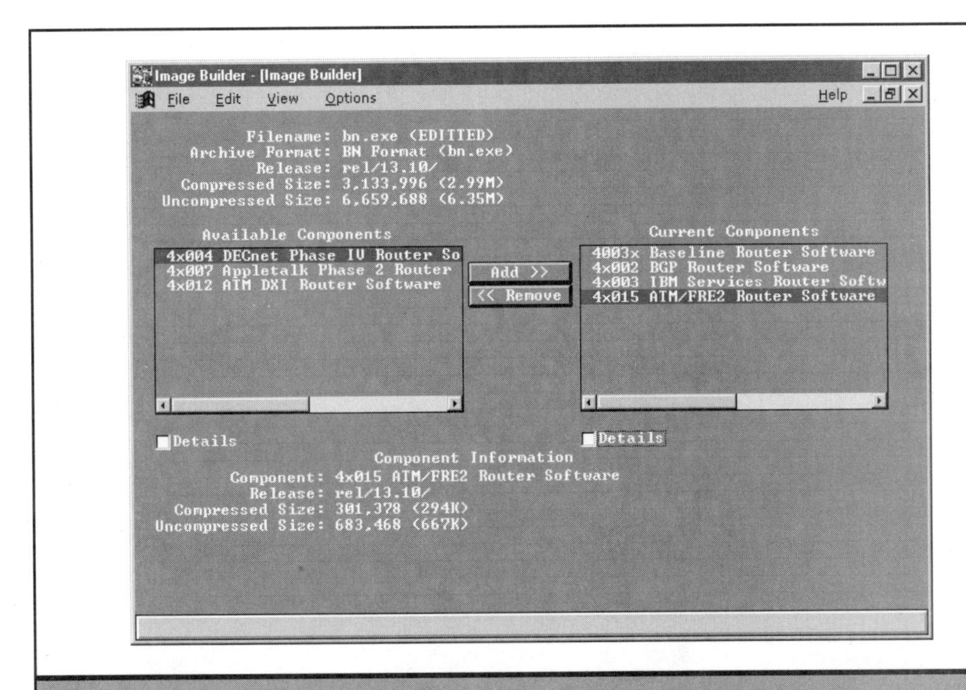

Figure 3-11. Image Builder

- **To save the image** Once the image has been customized to the specific requirements of the router, save it using File | Save (or Save As). The modified runtime image can then be copied to the flash as described in the next section.

NOTE: All applications have a platform key that indicates the platforms that the application will run on. A few platforms share common processors, but others have different processors and coprocessors and will not run another platform's image. Image Builder will disallow the addition of an incompatible platform application to the current image. For example, the BGP software from the AN platform cannot be added to a BN platform image.

NOTE: Each application has a version number associated with it. Image Builder and the dynamic loader will both enforce the requirement that an application's version must match that of the runtime kernel, which is usually the BayRS version. This check is performed to eliminate interoperability issues.

Transferring Core System Files to Flash

Following a successful PROM upgrade, the core system files can be transferred to the flash using FTP or TFTP. In simple software upgrades, the only required file is the runtime image. This is because the router can be fully operational with just two files: the runtime image and the configuration. It is, however, good practice to transfer other associated system files specific to the version being upgraded to. This includes files such as the install script, the debug aliases file, and the BCC help file.

Validating the New Runtime Image

In the same way that you validated the PROM image files, you need to validate the new runtime image. As mentioned previously, the runtime image is the largest file on the flash. FTP transfers will never cause file errors, except in the case of aborted transfers or binary transfers incorrectly performed in ASCII mode. If TFTP is used to transfer the file, the underlying connectionless UDP transport will not help recognize a lost packet that may invalidate the image. A simple error such as a lost fragment, incorrect transfer mode, or unnoticed abort will result in invalidation of the entire image. The archive image is protected by a header checksum and also an image checksum, and all component files in the archive are likewise checksum protected.

The protection mechanism is in place to prevent the accidental loading and operation of a corrupt image. Since the runtime image is key to the operation of the router, it must be inspected once it has been transferred to the flash file system. The same **readexe** command used previously is used in this task as well. However, this operation will take a long time and will produce extensive output for each individual file in the image archive. If **readexe** encounters an error in the header or image checksums, it will immediately halt operation and display a message similar to the following:

```
box# readexe bn.exe
Processing contents of '2:bn.exe'...

---------------------------------
-- Module name:   krnl_bn.exe
---------------------------------
   Validating header checksum... OK
   Validating image checksum... OK

..........
..........

---------------------------------
-- Module name:   igmp.exe
---------------------------------
   Validating header checksum... BAD
   Validating image checksum... BAD
```

TIP: Before running the **readexe** command, turn off screen paging by typing `more off`. This will prevent the **readexe** command from timing out while reading the archive image and waiting for the paging input from the keyboard. If paging is inadvertently left enabled, **readexe** may falsely report that the image is bad. To prevent this, always disable paging prior to running the **readexe** command. The output of the **readexe** command can be allowed to rapidly scroll down the screen because it will automatically stop if the command encounters an error.

The new archive image is ready to use when the last module checksum validation succeeds.

Rebooting with the New Runtime Image

If the diagnostics PROM was burned in the process of the upgrade, the reboot of the router must involve a cold start—this will run the new diagnostics image and initialize all hardware. If none of the PROMs were upgraded, or if only the boot PROM was upgraded, then the router can be rebooted using the **boot** command.

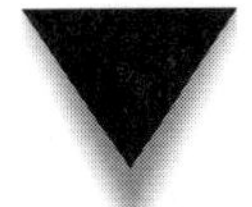

CAUTION: If a slot is warm-started instead of cold-started following a diag PROM upgrade, then the bootstrap may fail. This will hang the router if the upgrade is being carried out via telnet. The slot can be safely recovered only by physically cold-starting the slot.

If the router is cold-started, make sure beforehand that the image and the configuration have their default file names (see Table 1-2). As mentioned before, if a spare flash is being used, then it should, at minimum, contain the current configuration file and the new runtime image.

If the diag PROM was not burned, then a named boot can be issued as follows:

```
box# stamp
Image:      rel/13.00/
Created:    Sat Oct 10 20:23:06 EDT 1998

box# clearlog
clearlog: slot 1 confirmed
clearlog: slot 2 confirmed
clearlog: slot 3 confirmed

box# boot 2:bn14_00.exe 2:config
Rebooting system...
```

Notice that a **clearlog** command was issued before the **boot** command; this enables you to view a clean log containing the events following the boot. Any errors in the form of warnings can be immediately observed and diagnosed. Also, note the use of an image file name that is different than the default file name; in the example, the new image file was transferred to a spare flash using a different name. This is not required, but it will assist in differentiating the file based on the name following the boot.

Verifying the Upgraded Software Version

You can run the **stamp** command prior to the **boot** command to note the software version being upgraded from. Immediately following the boot operation, log in to the TI and run the **stamp** command again to check the version. The result should reflect the software version being upgraded to. If this is not the case, the first boot processor may have acquired the image from another flash, which may have been accidentally left in the router. See "Checking the Boot Source Image and Configuration" in Chapter 4 to verify the boot source and then repeat the reboot process after removing the unneeded flash.

```
box# stamp
Image:      rel/14.00/
Created:    Tue Dec 7 01:02:45 EST 1999
```

Testing the Upgrade

The test plan you created before performing the upgrade is now your guide to verifying the success of the upgrade. On rare occasions, upgrades may affect network connectivity, notably in cases where functional protocol changes or interoperability notes have not been adequately reviewed and addressed in anticipation of the upgrade. Perform a series of simple, quick tests such as protocol pings and file server accesses until you feel satisfied with the upgrade.

Upgrading the Configuration MIB Version

Recall that we rebooted the router using the configuration file from the old software version. The MIB objects in the configuration are based on the earlier software MIB version, which will limit the router to configuration features supported by the previous version. To enable and use features available with the new software version, you must convert the configuration file to use the MIB objects with the new version. If the configuration was created from scratch in local mode using the new Configuration Manager version, then the configuration MIB version need not be updated.

The MIB version of the existing configuration file can be checked in two ways. Using Configuration Manager, you can load the configuration file in remote mode to display the MIB version of the objects being used in the file. Alternately, you can use the **type** command to display the first 4 bytes of the configuration file in hexadecimal format, which will provide information on the MIB version. Figure 3-12 shows how to interpret the MIB version.

TIP: It is not mandatory that you upgrade the MIB version of the configuration file. However, future enhancements to the router configuration, such as the enabling of a new protocol, may be affected because the older MIB version may not support the configuration objects that the new protocol requires.

Upgrading the MIB version of the configuration file is simple. At the TI or BCC prompt, issue the **save config** command with a new file name argument:

```
box# save config 2:upgrade.cfg
box# clearlog
box# boot 2:bn14_00.exe 2:upgrade.cfg
```

Again, it is not necessary to reboot the router with the configuration file using the new MIB version. If the test plan has already been carried out successfully, then the

```
mx2(0)# type -x 2:config
File: 2:config
00000000  00 0C 00 14 00  00  00  00  AA  00 ...............
```

Functional release number in byte 4. Here, 0x14 equals 20.
Major release number in hexadecimal format, in byte 2. Here, 0x0C equals 12.

Figure 3-12. Interpreting the MIB version of the configuration file from the TI

router does not require rebooting. On the other hand, if configuration additions are planned for the router following the upgrade, you should reboot the router with the upgraded configuration.

Upgrade Checklist and Quick Reference

To begin and complete the upgrade, use the following checklist to verify that all the necessary items are available:

- Console cable and terminal (strongly advised, but not mandatory)
- Spare flash (recommended, but not mandatory)
- Software image suite (on the flash or on a workstation)
- FTP or TFTP client, or a TFTP server on the workstation (to transfer files)
- Site Manager software on the workstation (if needed)
- New hardware being installed (if any)

Table 3-5 summarizes the tasks needed to successfully complete an upgrade. Not all upgrades are identical, and some items may not be applicable in certain cases. You are least likely to encounter problems when upgrading if you start with a fully preplanned process.

	Quick Reference Task	Needed?	Done?
1	Read the release notes and upgrade documentation specific to the version being upgraded to.		
2	Obtain new hardware and check for any special hardware requirements.		
3	Create a new configuration file offline.		
4	Redraw the topology map highlighting the changes being made.		
5	Back up the current flash contents, including the image and configuration.		

Table 3-5. Upgrade Quick Reference

	Quick Reference Task	Needed?	Done?
6	Obtain the software image and customize the runtime archive. Images arriving in flashes usually do not require any customization to conserve space.		
7	Prepare a post-upgrade test plan to verify the success of the upgrade.		
8	Verify that all necessary applications (FTP/TFTP, TCP) are loaded on the router. Also load all string files into memory.		
9	Remove older system files and compact the flash to receive the new software image. Spare flashes can simply be formatted to receive the new system files.		
10	Identify the current PROM versions and transfer the new PROM images to flash. Run readexe to check file integrity and to determine whether a boot or diag PROM upgrade is necessary.		
11	Burn the new PROM images and validate them using the prom command. Remove the PROM update files and recompact the flash.		
12	Transfer the new runtime system files to the flash. Also transfer the offline configuration to the flash, if one was built. Run readexe on the archive to verify integrity.		
13	Reboot the router using the new BayRS version. If new hardware was installed, perform a cold boot to run full diagnostics. Verify that the new software version has been booted after the router is operational.		
14	Execute the post-upgrade test plan to ensure that the network is operating normally.		
15	Upgrade the configuration MIB version after the network operation has been tested. Offline configurations created using the appropriate Site Manager version need not be updated.		

Table 3-5. Upgrade Quick Reference *(continued)*

SUMMARY

The most common task of a network engineer or administrator is router administration. In the earlier portion of this chapter, we discussed many common system administration tasks: managing user accounts, enabling and configuring system services, and performing network management. Access policies help limit access to specific system services and to certain locations. We also discussed file system administration, scheduled boots, and the **show** and **ping** commands. Last, we discussed the steps for performing software upgrades; both PROM and system software image upgrades were covered.

We are now ready to discuss troubleshooting basics and procedures for resolving basic media issues, the topics of the next chapter.

REFERENCES

1. *Using the Bay Command Console*, Publication 308659-14.00 Rev 00.
2. *Configuring SNMP, BOOTP and DHCP Services*, Publication 308644-14.00 Rev 00.
3. *Managing Routers Using the HTTP Server*, Publication 308653-14.00 Rev 00.
4. *Configuring IP Utilities*, Publication 308631-14.00 Rev 00.
5. J. D. Case, M. Fedor, M. L. Schoffstall and C. Davin, *A Simple Network Management Protocol*, RFC 1157, May 1990.
6. K. Sollins, *The TFTP Protocol*, RFC 1350, July 1992.
7. M. McKenzie, *Telnet Protocol Specifications*, RFC 495, May 1973.
8. T. Berners-Lee, R. Fielding and H. Frystyk, *Hypertext Transfer Protocol*, RFC 1945, May 1996.
9. A. Frisch, *Essential System Administration: Help for Unix System Administrators*, O'Reilly & Associates, December 1996.
10. P. Albitz & C. Liu, *DNS and BIND*, O'Reilly & Associates, September 1998.
11. D. L. Mills, *Network Time Protocol (Version 3) Specification and Implementation*, RFC 1305, March 1992.
12. B. Hunt and G. Estabrook, *TCP/IP Network Administration*, O'Reilly & Associates, January 1998.

CHAPTER 4

Troubleshooting Basics and Resolving Media Issues

Ah, trouble, trouble, thee are the two different kinds . . . there's the one you give and the other you take.

—Kay Boyle, *The Crazy Hunter*

Network problems come to light in several ways. Most commonly, a network connectivity issue is isolated by a network management application that generates an alarm immediately. In some cases, problems are observed by a user unable to perform computing tasks that require the use of the network. There are myriad symptoms, with numerous causes, and usually identifying the symptom does not immediately identify the cause. Careful analysis of the symptoms is required to track down the cause of the faulty network behavior.

Troubleshooting networks is an *art* and, unfortunately, not a science. Knowledge of networking technology and protocols certainly makes the troubleshooting process easier, but it is still a challenging and exciting task. There is no one foolproof method for troubleshooting network problems; each issue can be resolved in a variety of ways, and the goal is to resolve each problem in the shortest amount of time. For this process, network administrators and operators should be skilled in the available tools, the investigative techniques, and the appropriate use of both. The skillful use of these tools and techniques is what makes network troubleshooting an art.

Before beginning to solve a problem, you should have all the details at hand. Many network details can be fetched from the network itself (such as from routers and switches), but external details aren't so readily available. Occasionally, you may be able to gather certain details only during the course of the investigation. As much as possible, you should have a complete understanding of the problem and an investigative plan before attempting to solve a network problem.

GATHERING INFORMATION

To begin the basic analysis of a network problem and determine the scope of the troubleshooting involved, you should start by answering the following questions:

- What is the exact nature of the problem?
- When did the problem first occur?
- Is this a recurring problem? If it is recurring, is it intermittent or periodic?
- If the problem occurs under certain conditions only, what are they?
- Is this a new installation or an existing network?
- What recent changes have been made to the router or the network?
- How is the problem affecting the network and the users?
- If the problem is triggered under certain conditions, can the conditions be induced for testing?
- What BayRS or Site Manager software version is being used in the network?
- What has been done, if anything, to attempt to alleviate the problem? Has it worked?

Analyzing Basic Symptoms

A comprehensive and thorough analysis of the symptoms will assist in a swift resolution to the network problem. To this end, you should determine the scope of the network problem by looking at the following issues:

- **Detailed and specific symptoms** Noting specific symptoms and gathering more details about them will help you identify the cause of the problem more quickly. For example, the symptom "cannot ping the IPX file server" is more detailed than "cannot ping." The symptom description "cannot ping the *marketing* IPX file server" is even more descriptive.
- **Recent network changes** Some network problems arise as a result of changes that were implemented recently. These effects may be immediate or delayed, in which case it is important to identify a reverse chronological list of network changes. Examples of network changes include the reconfiguration of circuits, movement of network nodes and endstations, provision of new LAN/WAN circuits, and so on.
- **Symptom observation time** Identify the time each symptom occurred or was first observed. If there are multiple symptoms, record the time each symptom occurred or was observed. Note that symptoms may not always be observed immediately after their first occurrence.
- **Extent of problem** Network problems are usually localized to a network area or protocol. Answering the following questions can provide further detail to help you identify the extent of the network problem. Does the problem occur in:
  - One protocol on a single port?
  - Several protocols on a single port?
  - One protocol on several ports in one slot?
  - Several protocols on several ports in one slot?
  - One protocol on several ports in all slots running that protocol?
  - Several protocols on several ports in all slots?
  - Several routers in the network?

OPERATIONAL PROBLEMS

An operational problem is a network problem spanning several protocols at several ports in all slots on a boxwide basis. On occasion, a boxwide operational problem may not directly affect the protocols at all ports, but the symptoms may seem to indicate that it does. For example, if a router fails to boot, one symptom may be that it does not appear to be forwarding any traffic for all the configured protocols.

This section discusses several of the most common operational problems and how to troubleshoot them.

Boot Issues

Use the console cable provided with the router to first establish a management session to observe boot messages. If possible, cold-start the router and observe the diagnostics and boot messages, particularly on the Access routers.

If the router contains a flash card, make sure that it is installed properly in the slot. Eject the card and reinsert it firmly into the slot and wait five minutes.

If the diagnostics messages indicate a component failure, then the defective component or unit may need to be replaced. Certain diagnostics failure messages, such as those indicating memory errors, may reveal problems so severe that the unit needs to be replaced immediately. On the other hand, failure of a single port in the unit may allow other ports to remain operational while a spare is being obtained. The following screen output illustrates diagnostic failure messages.

```
ASN Power-Up Diagnostics: Name: asndiag, Revision 2.24
Approximate Test Time 1 minute 30 seconds or less.
It is Thu, 12/30/1999, 21:25:02 (PM)

Testing Slot 1

Testing CPU..................................PASSED
Testing I/O Module in Position 1............PASSED
Testing I/O Module in Position 2............PASSED
Testing I/O Module in Position 3............FAILED
Testing I/O Module in Position 4............PASSED
Testing Internal Power Supply...............PASSED
Invoking Bootstrap...


ARN Power-Up Diagnostics: Revision 2.02
Approximate Test Time: 1 minute 30 seconds or less.
It is Thu, 12/30/1999, 08:41:46 (AM)
Memory Size: 32MB

Testing Base Module CPU .....................PASSED
Testing Base Module LAN .....................PASSED
Testing Adapter Module 1 ....................FAILED
Testing Expansion Module (Rev 1.06) .........PASSED
Invoking Bootstrap...
```

The flash must contain valid image and configuration files if the unit is set up to boot locally. If the default configuration file, **config**, is absent on the flash, then the router will attempt to load the **ti.cfg** file (see Table 1-2 for secondary config filenames) after 5 minutes. If both files are absent, after an extended timeout period (10 to 15 minutes) the ASN and BN routers will be initialized with a null **config** file.

The AN/ANH and ARN will fail to boot if both **config** and **ti.cfg** are absent. The **ti.cfg** file will allow the router to initialize to a minimal runtime environment that will allow access to the Technician Interface for recovery or reconfiguration.

The messages on the console may indicate that the configuration file or the image file is absent or not usable. The configuration file may be unusable if it has been incorrectly saved or corrupted during file transfer. In this case, a spare flash containing a valid configuration file can be used.

If the router image has recently been upgraded, then it is essential to verify the integrity of the image. This can be done from the Diagnostics Monitor (DM), which can be entered by pressing CONTROL-C or CONTROL-BREAK during diagnostics or bootstrap operations. From the DM, enter the **readexe** command as shown in the following example to verify the integrity of the image. If the image contains errors, then a spare flash containing a valid image file must be used to boot the system. Note that even if the image passes the file integrity tests, it must be the appropriate file for the platform. For example, an AN router cannot be booted with an ASN image.

```
AN_DIAG>> readexe an.exe
Processing contents of '5:an.exe'...

--------------------------------
-- Module name:   krnl_an.exe
--------------------------------
   Validating header checksum... OK
   Validating image checksum... OK
```

If the router is set to netboot, you should ensure that the bootable interfaces are configured and connected. Also, ensure that the BootP and TFTP servers are operational and configured correctly. If the router attempts to netboot again, there may be a network connectivity issue or server problem.

Hardware Issues

Issues relating to the network hardware can be usually identified by the diagnostics process. If a component fails diagnostic tests, the software automatically turns on the diag LED on the slot. On single-slot routers, the diag LED is on the base unit, and on multi-slot routers, the diag LEDs are on the processor and link modules. Appendix B describes the diag LEDs on all router platforms. On the Backbone routers, the diag LED on link modules is labeled FAIL.

As discussed in the previous section, "Boot Issues," it is sometimes necessary to use a console connection to isolate the failed component. For example, failed modules on an ARN and ASN are indicated in the diagnostics output.

Failed hardware can be hot-swapped if the underlying hardware fully supports this approach. The hot-swap capabilities of various hardware platforms can be obtained from the installation manuals. The current compatibility matrix for the Backbone family is available from a link at this book's web site.

Power and Fuse Issues

A common source of operational issues is power and fuse problems. Reliable power sources are paramount to efficient router operation. Power and fuse related issues, their symptoms, and resolution are discussed in this section.

Power Problems Routers in network data centers or communication closets should have a reliable power supply. In particular, the power source for Access routers should ideally be conditioned, and a surge protector must be installed on the line to protect against voltage spikes. Any router, if possible, should be connected to a constant power source such as a primary conditioned source that is backed up by a generator. Even in situations where the use of a conditioned, uninterrupted power supply is not feasible or economical, a surge protector still must be used.

Occasionally, power surges and outages may cause problems. If a router fails to boot, check the LEDs first for any activity. All routers have a power LED; if this light is on, the router's source of power is good. Certain unregulated power sources may supply a lower or fluctuating voltage than is optimal for the router's power supply, which may cause intermittent operational issues. If the unit power source has been verified and the router power LED remains off, then the base unit containing the power supply must be replaced. Adapter, expansion, and net modules in access routers may not need to be replaced in this process unless all components failed simultaneously.

On Backbone routers, the power LED is located on each processor module and also on the power supply. The BLN1 does not have a modular power supply, and the power status indicator is on the front panel. BLN2 and BCN routers have modular, hot-swappable power supplies, which have a power LED on the supply itself. If a power supply LED is off, first ensure that the power supply is installed properly and the power cable is connected securely. On routers with redundant line cords (RLC), ensure that both power lines are connected. If the power LED still remains off, then the power supply must be replaced. Modular power supplies on Backbone routers have large mean times between failure (MTBF).

Fuse Problems If all power supply LEDs are on and a few processor modules appear to have no power, then the problem may be a blown fuse on the BLN and BCN routers. Before investigating the fuse, make sure that both the processor and link modules are firmly attached to the backplane. Power to individual slots is supplied by different lines protected by individual fuses. Tables 4-1 and 4-2 list symptoms, event messages, and affected fuses.

CAUTION: Do not attempt to investigate or service the fuses when the power supplies are connected to a live power source. Backbone nodes must be serviced only by qualified personnel because of high-energy hazards. Even if you are sure that a particular fuse has blown, make sure that the chassis is completely powered off before servicing the fuses. For detailed and specific information, consult the *BLN and BCN Fuse Service Manual.*

Fuse	Volt Amperage	Function	Physical Symptoms	Event Log Symptoms
F1	25A @ +5V	Slot 1 SRM-F	All SRM-F LEDs are off.	Backbones 0 and 1 became disconnected.
F2	25A @ +5V	Slot 1 SRM-L, slot 2 link and processor modules	All processor and link LEDs on slot 2 are off, and all SRM-L LEDs are off.	Backbones 2 and 3 became disconnected. Slot 2 became disconnected.
F3	25A @ +5V	Slot 3 link and processor modules	All processor and link LEDs on slot 3 are off.	Slot 3 became disconnected.
F4	25A @ +5V	Slot 4 link and processor modules	All processor and link LEDs on slot 4 are off.	Slot 4 became disconnected.
F5	25A @ +5V	Slot 5 link and processor modules	All processor and link LEDs on slot 5 are off.	Slot 5 became disconnected.
F6	7.5A @ +12V	Media ports on slots 4 and 5	All processor and link LEDs on slots 4 and 5 are off, and the 12V 2 LED in SRM-L is off.	Backbones 0 and 1 became disconnected. Slots 4 and 5 became disconnected.
F7	7.5A @ +12V	Media ports on slots 2 and 3	All processor and link LEDs on slots 2 and 3 are off, and the 12V 1 LED in SRM-L is off.	Backbones 2 and 3 became disconnected. Slots 2 and 3 became disconnected.
F8	5A @ -12V	Technician Interface (console) port	No response at the TI console.	None.
Spare	25A @ +5V	Spare	None.	None.

Table 4-1. BLN Fuse Functions and Symptoms

Fuse	Volt Amperage	Function	Physical Symptoms	Event Log Symptoms
F1	25A @ +5V	Slot 1 link and processor modules	All processor and link LEDs on slot 1 are off.	Slot 1 became disconnected.
F2	25A @ +5V	Slot 2 link and processor modules	All processor and link LEDs on slot 2 are off.	Slot 2 became disconnected.
F3	7.5A @ +12V	Slots 1 and 3 link and processor modules	All processor and link LEDs on slots 1 and 3 are off.	Slots 1 and 3 became disconnected.
F4	25A @ +5V	Slot 3 link and processor modules	All processor and link LEDs on slot 3 are off.	Slot 3 became disconnected.
F5	25A @ +5V	Slot 4 link and processor modules	All processor and link LEDs on slot 4 are off.	Slot 4 became disconnected.
F6	7.5A @ +12V	Slots 2 and 4 link and processor modules	All processor and link LEDs on slots 2 and 4 are off.	Slots 2 and 4 became disconnected.
F7	25A @ +5V	Slot 5 link and processor modules	All processor and link LEDs on slot 5 are off.	Slot 5 became disconnected.
F8	25A @ +5V	Slot 6 link and processor modules	All processor and link LEDs on slot 6 are off.	Slot 6 became disconnected.
F9	7.5A @ +12V	Slots 5 and 8 processor modules and SRM-L	All processor and link LEDs on slots 5 and 8 are off, and all SRM-L LEDs are off.	Backbones 2 and 3 became disconnected. Slots 5 and 8 became disconnected.

Table 4-2. BCN Fuse Functions and Symptoms

Fuse	Volt Amperage	Function	Physical Symptoms	Event Log Symptoms
F10	25A @ +5V	SRM-F	All SRM-F LEDs are off.	Backbones 0 and 1 became disconnected.
F11	25A @ +5V	Slot 8 link and processor modules and SRM-L	All processor and link LEDs are off on slot 8, and all SRM-L LEDs are off.	Backbones 2 and 3 became disconnected. Slot 8 became disconnected.
F12	7.5A @ +12V	Slots 6 and 9 link and processor modules and SRM-F	All processor and link LEDs are off on slots 6 and 9, and all SRM-F LEDs are off.	Backbones 0 and 1 became disconnected. Slots 6 and 9 became disconnected.
F13	25A @ +5V	Slot 9 link and processor modules and BCN fan tray	All processor and link LEDs are off on slot 9, and all front-panel LEDs are off.	System fan module failure, one or more fans not operating properly. System temperature has risen to a cautionary range. Slot 9 became disconnected.
F14	25A @ +5V	Slot 10 link and processor modules	All processor and link LEDs are off on slot 10.	Slot 10 became disconnected.

Table 4-2. BCN Fuse Functions and Symptoms *(continued)*

Fuse	Volt Amperage	Function	Physical Symptoms	Event Log Symptoms
F15	7.5A @ +12V	Slots 10 and 12 link and processor modules	All processor and link LEDs are off on slots 10 and 12.	Slot 10 became disconnected. Slot 12 became disconnected.
F16	25A @ +5V	Slot 11 link and processor modules	All processor and link LEDs on slot 11 are off.	Slot 11 became disconnected.
F17	25A @ +5V	Slot 12 link and processor modules	All processor and link LEDs on slot 12 are off.	Slot 12 became disconnected.
F18	7.5A @ +12V	Slots 11 and 13 link and processor modules	All processor and link LEDs are off on slots 11 and 13.	Slots 11 and 13 became disconnected.
F19	25A @ +5V	Slot 13 link and processor modules	All processor and link LEDs are off on slot 13.	Slot 13 became disconnected.
F20	7.5A @ +12V	Slot 14 link and processor modules and BCN fan tray	All processor and link LEDs are off on slot 14, and all front-panel LEDs are off.	System fan module failure, one or more fans not operating properly. System temperature has risen to a cautionary range. Slot 14 became disconnected.

Table 4-2. BCN Fuse Functions and Symptoms *(continued)*

Fuse	Volt Amperage	Function	Physical Symptoms	Event Log Symptoms
F21	25A @ +5V	Slot 14 link and processor modules	All processor and link LEDs are off on slot 14.	Slot 14 became disconnected.
F22	7.5A @ +12V	Reserved for future use	None.	None.
F23	7.5A @ +12V	Reserved for future use	None.	None.
F24	25A @ +5V	Spare	None.	None.
F25	5A @ -12V	Media connectors in slots 2, 4, 6, 9, 11, and 13	None.	Media: System error, service attempting restart.

Table 4-2. BCN Fuse Functions and Symptoms *(continued)*

Certain slots are protected by more than one fuse; also, some lines supply a voltage at an amperage different from other lines as shown in the table. If the specific symptoms in the table are matched, then you can check whether the fuse has blown. Industry-standard fuses are used, as shown in Figure 4-1. The most commonly used 25A/+5V fuse has a

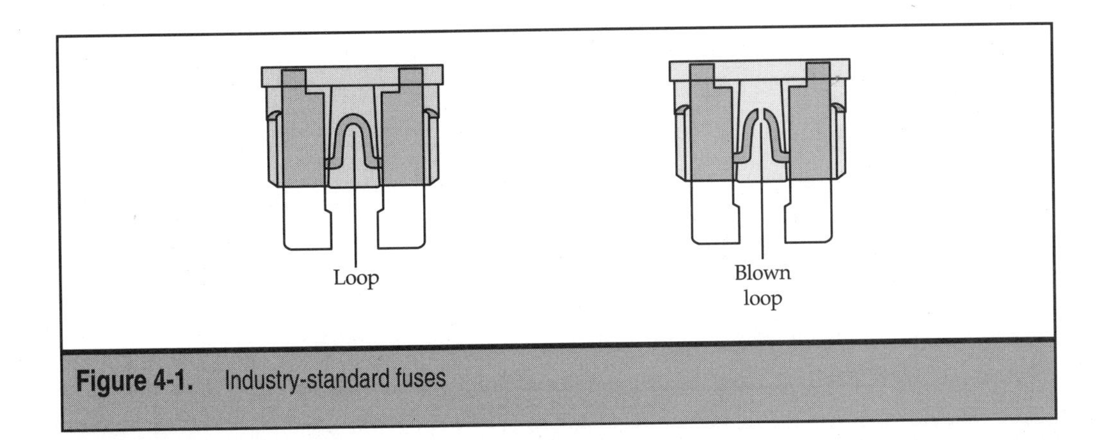

Figure 4-1. Industry-standard fuses

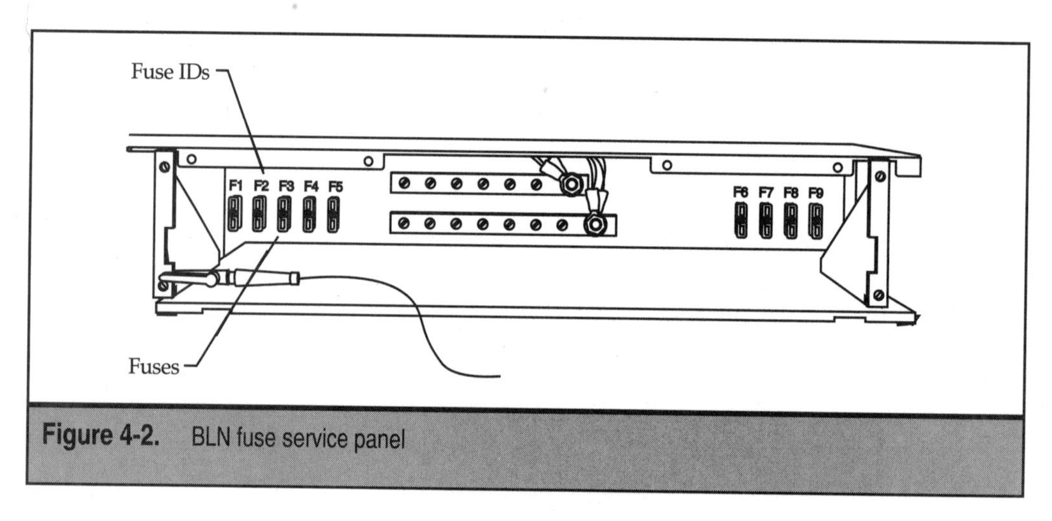

Figure 4-2. BLN fuse service panel

spare in each chassis. The fuse panels on the BLN and BCN are illustrated in Figures 4-2 and 4-3, respectively.

The BLN is protected by 8 fuses, and the BCN is protected by 23. The following events are symptomatic of a blown fuse when they accompany either a "Slot disconnected" or "Backbone became disconnected" message in the event log:

- None of the LEDs on the link module are lighted.
- Either CR2 or CR3 LED on the SRM-link module are not lighted.
- None of the LEDs on the processor module are lighted (you must remove the front panel and EMI shield to observe the processor module LEDs).

If fuse F8 is blown in the BLN (F26 in the BCN), there will not be any log messages, but the console port will not respond even with a valid connection. Once a blown fuse has been identified and replaced with an identical spare, the processor and link modules in the affected slots should become operational. If the fuse is not blown or the modules fail to come up, then try inserting the module in another slot temporarily to see if it obtains

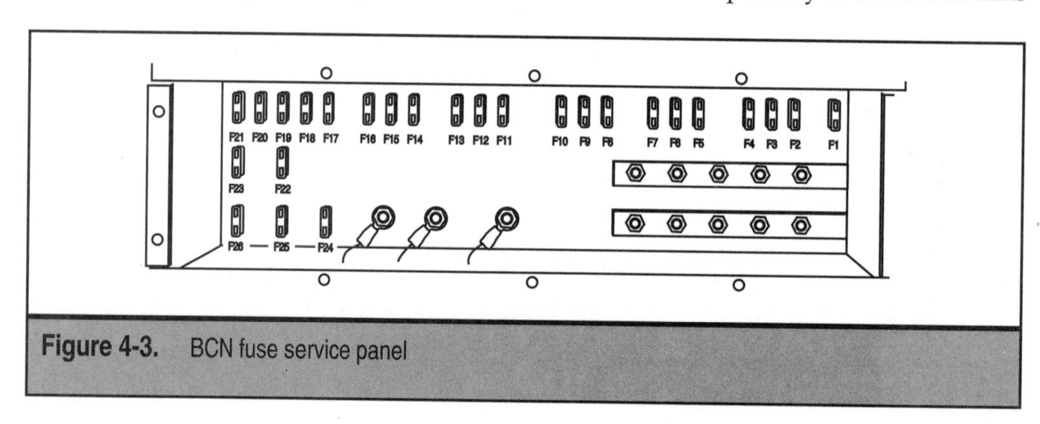

Figure 4-3. BCN fuse service panel

power. If the module obtains power, then the power connector on the previous slot is faulty; otherwise, the affected module must be replaced.

Access Stack Node Issues

The ASN is a unique architecture providing the stacking feature. Some common stack management issues are discussed in this section.

SPEX and SPEX-HS Problems As mentioned in Chapter 1, the ASN automatically bypasses net modules that fail diagnostics. Failure of CPU tests or fast packet cache (FPC) tests will stop diagnostics, and the defective motherboard or FPC must be replaced. SPEX and SPEX-HS net modules always must be installed in net module position 4; otherwise, they will not pass diagnostics. If a new unit is inserted into a stack, all units in the stack must be powered off when using SPEX net modules. When using SPEX-HS net modules, the existing stack need not be powered off, but the new unit being added must be powered on only after the SPEX-HS cables are connected to the stack. The new unit may fail diagnostic tests if it is added to the stack once it has been powered on.

SPEX-related issues are easily identified by the presence of a flashing diag LED. If the stack installation is new, make sure that each unit in the stack uses boot PROM version 8.10 or later when you are using SPEX-HS modules. In the case of a dual-bus configuration, the boot PROM version must be at least 10.00. How to check the boot PROM version is discussed in Chapter 3, in the section "Identify the PROM Versions." If the PROM versions meet minimum requirements, then the SPEX-HS connectors and terminators must be checked. If the cabling or termination is incomplete or faulty, then the SPEX-HS net modules may not recognize the bus properly and will fail diagnostics.

If the cables and terminators (SPEX-HS cables are self-terminating) are connected properly and the unit still fails diagnostics, you can bypass the unit to see that the bus is fully operational with the remaining units. On the isolated unit, remove the SPEX-HS net module and rerun diagnostics to verify that the remaining components function properly. If net modules still fail, then they must be replaced.

Slot Identifier Problems The hardware and software differentiate ASNs in a stack by the slot identifier. The slot identifier is a dial located at the bottom right on the back of each unit (see Figure 4-4). Each slot must be configured appropriately for the hardware and software configuration it supports. If a slot dial is set up incorrectly, the software will recognize that the configuration is incorrect and log several warning messages. If a slot is initialized without any configuration attributes for the installed net modules, then the following event message will indicate the error for the slot number:

```
#   112: 05/23/99 11:41:30.304  DEBUG    SLOT  2  MODULE              Code: 11
Waiting for wfModuleEntry to be created on this slot.
```

Observe that the debug event in the example is logged on slot 2, but no configuration records exist for that slot. In this example, the configuration records point to another slot, but the slot identifier was incorrectly changed to slot 2. Single-slot ASNs should always be set and configured using slot identifier 1 to allow for future growth and stacking.

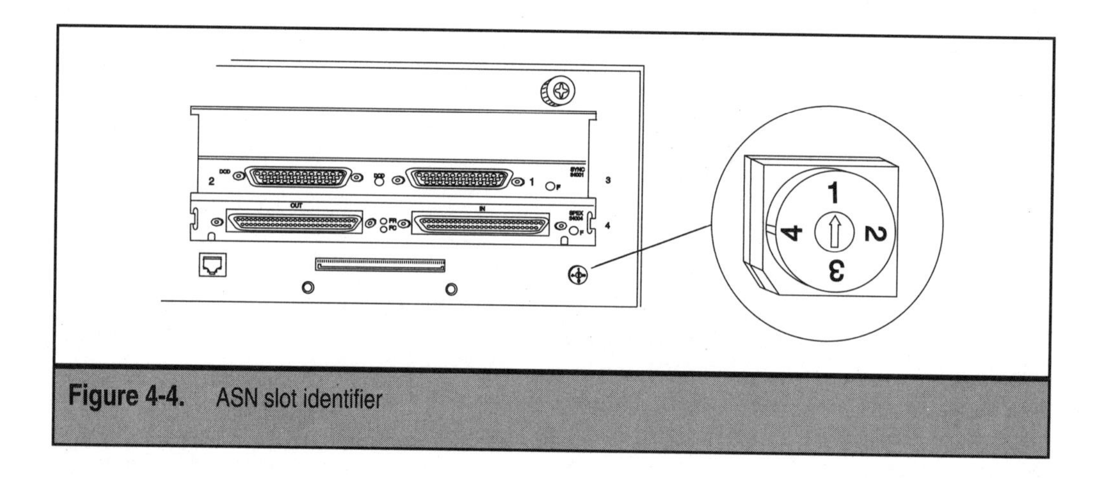

Figure 4-4. ASN slot identifier

Troubleshooting the Wrong Router

Yes, it happens! Occasionally, a brief, exasperated utterance such as "Why, I…!!" and "What the hey?" is followed by a long silence and the realization that you were troubleshooting the wrong router. Especially in large networks consisting of several hundred or several thousand routers, it is important to keep routers identified uniquely and correctly.

Before beginning to troubleshoot a software problem on a router, first ensure that you are working with the correct router by performing the following steps:

- Check the login or welcome notice displayed at startup for information specific to the router. If a router-specific notice is not available, take steps to set up each router with a unique login notice as discussed in Chapter 3.
- Check the TI prompt. Most routers have prompts that uniquely identify them, like this:

  ```
  [4]Boston2-BCN>
  ```

  If the TI prompt has not been customized to uniquely identify the router, ensure that you take steps to customize it later.

TIP: You can change the TI telnet prompt by running the following command:

```
(telnet)server# prompt "[%slot%]Boston2-BCN> "
```

You can customize the TI console prompt by running the following command:

```
console/1# prompt "[%slot%]AN-Miami> "
```

Change instance ID 1 to the appropriate port number on ASNs and ARNs.

- Check the `system-name` attribute under the `box` object if you are using BCC, or run the **show system information** command:

```
box# system-name
system-name   Boston2-BCN
```

- Check the system record if you are using the TI. Routers that are configured properly may have a system record like the one shown here:

```
[1:1]$ get wfSys.5.0
wfSys.wfSysName.0   "AN-Miami"
```

As part of the troubleshooting process, you should complete the following three steps as soon as you log in to the router. These steps will help you immediately identify several items: the system; its critical faults, if any; and the integrity of the configuration.

- Identify the system using one of the procedures just described. The simplest method is to execute the **show system info** command in the BCC, which also formats the uptime (`wfSysUpTime`) into readable format.
- Run the command **log –ff** to look for any faults that may have occurred. If the command returns without any faults, the router does not have any critical faults stored in the log, assuring you that there no critical errors on the router.
- Retrieve the `wfHwConfigFile` (26) and `wfHwActiveImageName` (28) attributes individually for all slots using the `wfHwEntry` object. Using BCC, use the **show hardware config** and **show hardware image** commands to obtain the boot sources for the configuration and image. Recall that the boot configuration and image files must be consistent and synchronized across all slots. A difference in the values, particularly in the volume number, is cause for investigation. Also check the configuration file time stamp on the flash to ensure that it has not been modified since the last boot operation.

Forgotten Passwords

Usernames and passwords are used to authorize and authenticate network engineers and administrators before providing access to the router. The security and integrity of a password is only as good as the strength of a password. For example, using a simple password such as a spouse's name or a birth date easily paves way for a security breach. To combat this, security-conscious administrators usually require a password consisting of a complex sequence of alphabetical characters and numerals.

Occasionally, users who maintain multiple passwords or infrequently use their administrative privileges may forget their passwords. Passwords on user accounts created using the multilevel access feature can be recovered only be deleting the account and re-creating the passwords.

The special system accounts, Manager and User, cannot be deleted, and their passwords are stored in NVRAM. If the User password is forgotten, then an administrator logged in as Manager can change the User password.

If the Manager password is forgotten, one of the procedures described here must be used to restore the default or recover it. The procedures cannot be performed using remote TI sessions; they must be accomplished using a directly attached out-of-band TI session. They are described here because only network administrators and engineers with physical access to the routers can restore the default password.

On the BLN and BCN routers, the password is distributed and stored on all processor modules. The processor that runs the TI soloist or the telnet TI gate will authenticate the Manager and User logins using the passwords stored locally in NVRAM. To restore the default password on these routers, follow these steps:

1. Turn off the power to the chassis and connect a terminal to the console port.
2. Remove the front panel and the shield and disengage the extractor clips on all the processor modules. Then pull each module out slightly so that it is fully disconnected from the backplane, but do not remove the processor completely from the slot.
3. Do not disconnect the system resource modules from their slots.
4. If all the slots are fully populated, then completely remove one processor module alone from its slot.
5. In an empty slot, insert a spare processor module that is known to not have a password. Processor modules shipped from the factory that have never been used and those that have never been operated in a router requiring a password will have a default (null) password.
6. Insert the system flash card from one of the other processor modules into the newly inserted processor module.
7. Reboot the router. This will boot the single processor module and initialize the TI soloist process on that slot.
8. Reinsert all the remaining processor modules and engage the thumb clips. All of these processors will perform a backbone boot using the preinitialized processor.
9. Wait for all the processor modules to come out of the boot state and enter the run state.
10. Log in as Manager on the console terminal. The router should not prompt you for a password because the TI soloist is running on the processor that does not contain a password.
11. Once you are logged in as Manager, issue the **password** command as discussed in Chapter 3, in the section "Changing the Password," and set a new password. This will automatically update all the processor modules with the new password.

Alternately, the password on the AN/ANH, ARN, ASN, and BN routers can be returned to the default state using the special password removal image that is available from Nortel Networks Customer Service. You will need a spare flash and an alternate router for accessing the flash, or the BNFS utility can be used, as discussed in Chapter 2. Use the following procedure after obtaining the special password removal utility from Nortel Networks Customer Service.

1. Turn off the power to the router and connect a terminal to the out-of-band management port.
2. Copy the password removal image to the flash using the default archive image name. Consult Table 1-2 for the default system image names.
3. Copy the existing configuration file or the default configuration file (config or ti.cfg) to the flash. The configuration file is not loaded for runtime operation, but is needed for the password removal image to initialize and perform its function.
4. Eject the current system flash and insert the new special flash into the router. On multi-slot routers, make sure that you partially eject all other flashes so that they do not interfere with the password removal image.
5. Reboot the router and check the console terminal for messages. As the kernel initializes and spawns the TI, it will automatically erase both the Manager and User passwords.
6. After password removal is complete, you will be prompted to reinsert the system flash and press the RETURN key to reboot the router. Earlier password removal images begin to alternate the front panel LEDs once the passwords are erased. Depending on the image you obtained, you may either see interactive messages on the TI console or flashing front panel LEDs. If you observe flashing front panel LEDs and do not see any text on the console, simply replace the original flash and cycle the power to the router.
7. When the router reboots for the second time and enters steady-state operation, you can log in as Manager or User; neither will have a password.
8. Issue the **password** command to set a new password.

Lack of File System Space

Flash devices are linearly programmable, which makes it difficult to perform random programming (writing) to the file system. This is why the available free space and the contiguous free space differ, as discussed in Chapter 2. If a file is deleted on the flash, the file system unlinks the file and makes it inaccessible, but the file continues to occupy the space until it is reclaimed. When a new file is stored, it is stored in the first unused space at the end of the file system. Eventually, the file system space may become exhausted because the deleted files still occupy memory. While performing file system operations, you may occasionally notice that the contiguous free space is too low to complete the current operation, as in the following example.

```
mx2(8)# cp 2:bn.exe 4:
Copying 2:bn.exe to 4:bn.exe    0% Complete
Insufficient space for new file

Output file deleted.
0 File(s) copied
```

The preceding example shows an insufficient space error while attempting to copy a file. The following example shows an aborted TFTP file transfer for the same reason.

```
mx2(9)# tftp get 192.168.18.254 2:bn.exe 4:bn.exe
Transfer with 192.168.18.254 failed with: Unknown file system error: 0.
```

An FTP session to the router may be abruptly disconnected if the destination file system has insufficient file space. The error may not be seen until all the available file space is exhausted, at which point errors similar to the following may appear.

```
ftp> put bn.exe
200  PORT command successful.
150  Image data connection for 4:bn.exe (192.168.19.254,20).
netout: Broken pipe
421 Service not available, remote server has closed connection
Lost connection
```

If you encounter errors during file operations, run **dinfo** or display a directory of the volume to check the amount of contiguous and free space available. To remedy a problem caused by low contiguous free storage, run the **compact** command with the volume number as the argument (see "File System Commands" in Chapter 2). Figure 4-5 shows the change in the file system when a file is deleted and then the **compact** command is run. After running the **compact** command, run the **dinfo** command again to verify that the contiguous free space has increased.

Memory Issues

Chapter 2 discussed the concepts of local and global memory. Global memory buffers are used to handle all traffic, and local memory is used for the operating system and routing and forwarding tables. Occasionally, as a result of network evolution or a memory configuration error, memory resources may become exhausted. Memory issues are common, though symptoms may manifest themselves in different ways. Memory problems and possible remedies are discussed here.

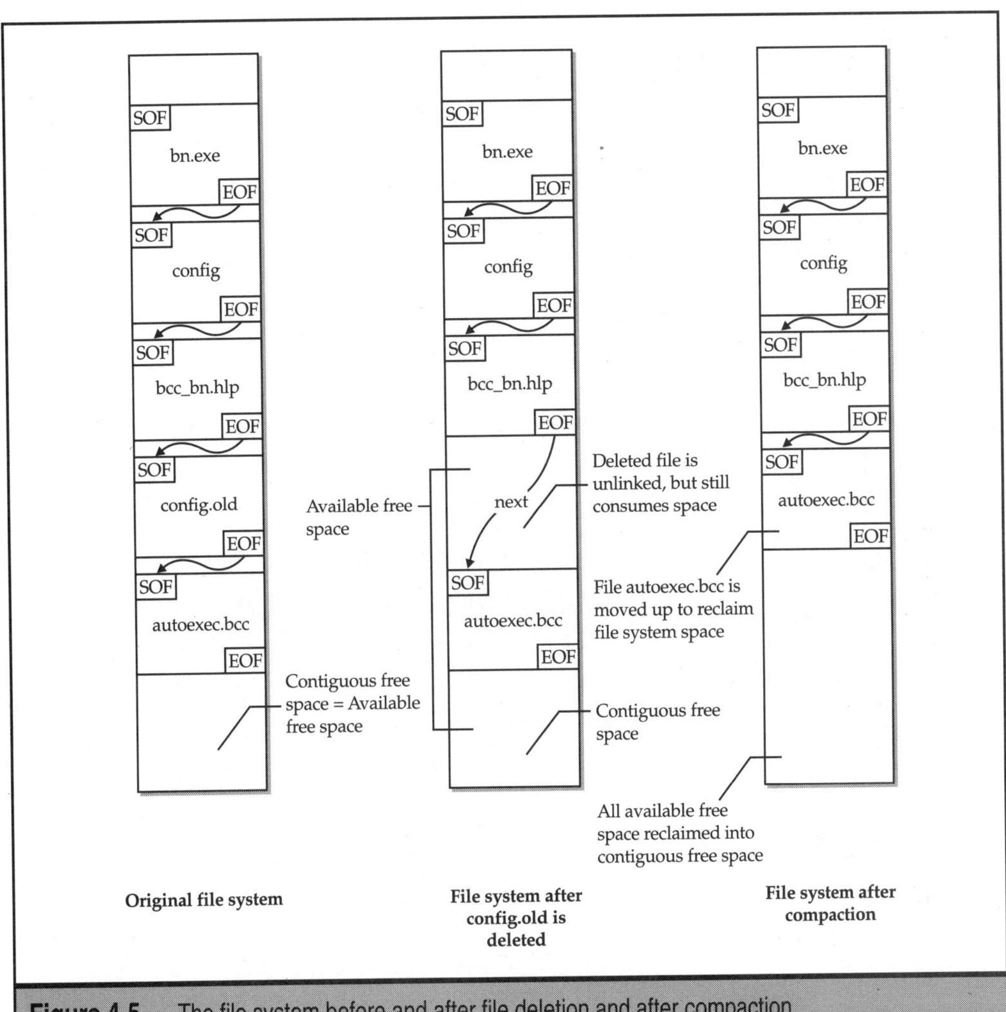

Figure 4-5. The file system before and after file deletion and after compaction

Insufficient Local Memory

The local memory in a processor module can become oversubscribed when any of the following occurs:

- Addition of one or more new protocols that maintain dynamic tables
- Growth of the network, resulting in significantly larger routing and service tables
- Configuration of the packet capture (PCAP) application, using almost all of the available local memory

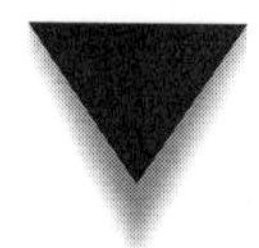

NOTE: The GAME operating system automatically enforces a restriction on the PCAP application by ensuring that at least 200K of local memory is left free after the allocation. However, subsequent memory allocations may affect system operation if the amount of free memory is small.

Recall that earlier in this book we mentioned that during the design phase, the local memory should be configured to allow for future growth and scaling. In cases where network growth is rapid and unforeseen, such as after a business merger, routing and service tables can grow quickly. In some situations, addition of a new network protocol may also result in more local memory consumption.

New protocol applications affect local memory use in two ways. The first and most noticeable change is the immediate consumption of executable code space. Subsequently, the dynamic maintenance of network tables by the new protocol will also consume memory. If insufficient memory is available to load the application in the first place, the dynamic loader logs an event:

```
#   32: 11/09/1999 09:05:24.957  WARNING  SLOT  5  LOADER               Code:  14
Dynamic Loader unable to allocate enough memory to load application
```

Other situations, such as routing protocol misconfigurations, can also result in unexpectedly large tables. The oversubscription of local memory is typically characterized by an unexpected slot reset or recurring resets.

An insufficient local memory condition can sometimes also occur during an incorrectly configured packet capture, discussed in Chapter 6. When this occurs, the event log will report an out-of-memory fault condition like the following:

```
#    4: 03/26/1999 24:28:31.665  FAULT    SLOT  6  GAME                 Code:   3
out of memory

#   51: 05/26/1999 03:29:37.709  FAULT    SLOT  3  GAME                 Code:  80
insufficient memory to continue => GAME restarts

#   17: 06/16/1999 03:44:21.656  FAULT    SLOT 14  ATM                  Code:   1
Out of Memory while allocating vc-cct table block

#   11: 07/24/1999 05:03:26.610  FAULT    SLOT  1  HELLO                Code:   2
Out of memory in g_malloc
```

Baselining local memory usage periodically is useful in determining the growth in memory use, to determine the amount of memory needed to prevent these memory problems. Slot-specific memory use can be obtained from the `wfKernelEntry` object, for which instances are created by using the slot number.

```
mx2(2)# mget wfKernelEntry.*.4
wfKernelEntry.wfKernelSlot.4 4
wfKernelEntry.wfKernelMemorySize.4 22544784
wfKernelEntry.wfKernelMemoryFree.4 8291184
... ...
```

The `wfKernelMemorySize` value is the total amount of local memory left after the platform-specific kernel has been loaded; `wfKernelMemoryFree` represents the total local memory free after all applications have been loaded and their routing, forwarding, and service tables are populated. This is the amount of memory that can be consumed by existing protocols for dynamic tables and by the configuration of additional protocols. The same memory is also used by packet capture to trace packets transiting a specific line on the slot. On slot 4 in the preceding example, about 37 percent of the local memory is free for dynamic use.

Current system memory use on each slot can also be retrieved, using the **show system memory** command:

```
show system memory                                  Mar 2, 2000 19:40:27 [GMT-5]

Memory Usage Statistics (Megabytes):
------------------------------------

Slot Total     Used      Free      %Free
---- -------- -------- -------- -------
   3  21.72 M   1.02 M  20.70 M      95%
   4  21.72 M   1.01 M  20.70 M      95%
```

Global Buffer Issues

Packet buffers are dynamically requested and allocated, and their use can vary dramatically depending on the traffic pattern. During a period of bursty inbound traffic on a half-duplex line, outbound packets are buffered until transmission is possible. This may require more than the usual amount of packet buffers for buffering. Sustained high-throughput traffic will result in higher packet buffer use. GAME performs real-time buffer allocation and reclamation. However, like local memory, packet buffers may become oversubscribed.

Peak use due to periodic or sustained needs may result in buffer starvation by certain services. For example, consider a case where a large number of outbound IP packets and many inbound IPX packets are buffered and waiting to be transmitted and processed, respectively. If another protocol such as AppleTalk requests a specific number of buffers for transmitting zone details in ZIP packets, because of the oversubscription of buffers, AppleTalk may be unable to obtain the desired number of buffers. This scenario may degrade the operation of certain protocols and severely impact others. Some sensitive protocols and operating system services are less tolerant of buffer starvation and may cause a subsystem restart:

```
#   10: 09/27/1999 13:38:29.719  FAULT    SLOT 12  NTP                Code:   2
Buffer allocation failed

#   90: 03/16/1999 07:05:48.414  DEBUG    SLOT  1  SNMP               Code:   9
Agent error, failed to get a buffer!

#  153: 09/11/1999 21:05:26.509  DEBUG    SLOT  2  TELNET             Code:  18
TN_MAPPER Manager could not allocate a buffer...Error!
```

```
#   69: 01/16/1999 09:50:34.919  DEBUG    SLOT  3  SYSL              Code:  42
Process could not allocate a buffer

#  118: 10/12/1999 10:03:55.632  FAULT    SLOT  1  S5_CHASSIS        Code:   2
Buffer allocation failed
```

While the results of buffer starvation might appear quite dramatic, the underlying problem must be quite extensive for symptoms or issues to occur. Buffer starvation will almost never occur in a well-designed memory utilization model, except if enormous, unanticipated growth occurs. GAME is responsible for handling, allocating, and reclaiming packet buffers for individual processes. Buffer starvation is usually viewed as the result of runaway processes hogging the resources; GAME may terminate processes that overutilize buffers, or an application unable to handle buffer starvation may restart in the expectation that the system condition will be normal later. Note that restarts (say, for example, of IPX) affect only the specific application; other applications will continue normally.

Hogging Buffers

As in the case of local memory, global buffer use can also be indirectly affected by the growth of routing and service tables. For example, IPX SAP packets can contain a maximum of 7 services each. Consider a situation where 2,000 global buffers are available on the local slot, and the IPX services have grown from 1,000 to 3,500 after a business merger. When the IPX SAP advertisement timer expires, a total of 500 SAP packets (that is, 3,500 SAPs divided by 7 SAPs per packet) must be sent. Assuming that there are two IPX interfaces on the slot, then a total of 1,000 (twice 500 SAPs) packets are needed to complete the operation, which is about half of the total global buffers. Any process utilizing over a third of the global buffers is monitored by GAME for buffer overutilization, and in this case, when IPX requests and uses 1,000 buffers, IPX is reported in the log to be hogging buffers.

```
#  5578: 06/13/99 12:13:17.277  DEBUG    SLOT  1  GAME              Code:  91
pending 0x0006e @ 0x3096a0ae hogs bufs: del=549, #1=  0, #2=  0, trans=  1

#  5579: 06/13/99 12:13:17.289  DEBUG    SLOT  1  GAME              Code:  91
pending 0x0006e @ 0x3096a0ae hogs bufs: del=550, #1=  0, #2=  0, trans=  1
```

This protective mechanism is in place to monitor runaway processes. In this example, the use is legitimate, and the event log points to network growth that needs attention. Persistent hogs bufs events indicate a slot or many slots requiring additional global packet buffers.

The `wfKernelBuffersTotal` and `wfKernelBuffersFree` attributes provide the slot's total buffer count and the count of the buffers currently unused. The free buffer count will vary dynamically depending on the allocation. In this example, slot 4 has about 83 percent of the total buffers free.

```
wfKernelEntry.wfKernelBuffersTotal.4 1527
wfKernelEntry.wfKernelBuffersFree.4 1274
... ...
```

The same information from the MIB can be viewed using BCC by running the **show system buffers** command. The total, used, and free buffer counts are displayed along with a calculated percentage of the number of free buffers.

```
show system buffers                            Mar 2, 2000 19:37:05 [GMT-5]

Buffer Usage Statistics:
------------------------

Slot Total Used  Free  %Free
---- ----- ----- ----- ------
   3  1527   253  1274    83 %
   4  1527   253  1274    83 %
```

Design Considerations

Certain local and global memory issues are a result of poor network and router design. For example, consider a router with two interfaces: one Ethernet (10 Mbps) and one Fast Ethernet (100 Mbps). Many higher-layer protocols can calculate round-trip times and use the bandwidth based on the network capacity. However, certain protocols unable to manage congestion may saturate the Fast Ethernet, thereby causing the router to buffer outbound packets on the 10 Mbps Ethernet line. Sustained saturation of the Fast Ethernet interface may eventually cause the buffer space to run out, which may result in dropped packets or protocol resets. In such a case, increasing the global buffers will only delay the onset of the saturation but will not eliminate it. The underlying issue is a design consideration, where the "big pipe, little pipe" condition must be eliminated to avoid the little pipe (10 Mbps) being saturated. Redesigning the network with two 100 Mbps interfaces or two 10 Mbps interfaces instead will provide better utilization of buffer resources and the bandwidth.

Memory Reconfiguration

By default, 75 percent of DRAM is allocated to local memory and 25 percent to global memory. On some platforms, this division can be reconfigured when necessary. For example, a router may require more local memory for the addition of some protocols, and you may want to allocate this from global memory, because fewer than 50 percent of the packet buffers are utilized, even during periods of peak traffic. This situation is more common than the case where highly deviant traffic patterns require the occasional use of more buffers than are available by default.

Platforms have minimum and maximum values for local and global memory, as shown in Table 4-3.

Currently, memory partitioning cannot be reconfigured on the ARN, FRE-1, ARE, and FRE-4. An attempt to reconfigure it will generate a "memory cannot be reallocated on this platform" message in the log. To reallocate memory, you need to change the

Platform	Local (Min)	Local (Max)	Global (Min)	Global (Max)	Increment
AN/ANH	1868 KB	16204 KB	180 KB	14516 KB	1 KB
ASN	4 MB	31 MB	1 MB	16 MB	1 MB
FRE-2	4 MB	48 MB	2 MB	16 MB	2 MB

Table 4-3. Local and Global Memory Configuration Limits

amount allocated to global packet buffers; the remaining memory is automatically assigned to local memory. Global memory can be changed only in granular increments supported by the platform.

```
mx2(6)# mget wfKernParamEntry.*.4
wfKernParamEntry.wfKernParamSlot.4 4
wfKernParamEntry.wfKernParamTotMem.4 32768
wfKernParamEntry.wfKernParamLocMem.4 24576
wfKernParamEntry.wfKernParamGlobMem.4 8192
```

The current memory configuration can be obtained via the `wfKernParamEntry` object. As with the `wfKernelEntry` object, the instance ID is equal to the slot number. The `wfKernParamTotMem` attribute equals the sum of `wfKernParamLocMem` and `wfKernParamGlobMem` attributes.

As mentioned earlier, the global memory can be reconfigured in specific increments, using the `wfKernCfgParamEntry` object. This can also be done via the Site Manager GUI, using the Administration menu item.

```
mx2(7)# mset wfKernCfgParamEntry.wfKernCfgParamGlobMem.4 10240; commit
mx2(8)# reset 4
```

The memory configuration is stored in two locations: in the `wfKernCfgParamEntry` object and in the NVRAM of the processor board. For the reconfiguration to take effect, the slot must be reset. If the processor module is swapped with a newer one, the memory must be reallocated, even though the `MIB` object is present. Following the reset, the runtime local and global memory configuration must be verified.

```
wfKernParamEntry.wfKernParamLocMem.4 22528
wfKernParamEntry.wfKernParamGlobMem.4 10240
```

If there are errors in the global memory configuration, the event log will display warning messages, and the configuration change will have no effect.

On certain platforms, such as the ARN and ARE, the packet buffer size can also be changed to a smaller or larger size, as discussed in Chapter 2. In the following example, the `wfKernCfgParamBufSize` attribute is changed on an ARE to support packets of up to 8,192 bytes.

```
mx2(9)# mset wfKernCfgParamEntry.wfKernCfgParamBufSize.4 8192; commit
mx2(10)# reset 4
```

Notice that increasing the buffer size reduces the total buffer count, if the global memory remains the same. Both the global memory and the buffer size should always be increased in small numbers, and the impact of the change must be analyzed prior to further reconfiguration.

Default Memory Configuration and Buffer Size Issues

In the previous section we mentioned that the memory configuration is stored not only in the MIB, but also the processor module's NVRAM. If a previously used processor is installed in another router, or if the existing memory configuration is deemed erroneous, then the memory configuration and the buffer size configuration must be reset to the defaults.

Set the default global memory size and packet buffer size as follows:

```
mx2(11)# mset wfKernCfgParamEntry.6.4 1      (wfKernCfgParamGlobMemReset)
mx2(12)# mset wfKernCfgParamEntry.8.4 1      (wfKernCfgParamBufSizeReset)
mx2(13)# commit; reset 4
```

The global memory size and the buffer size can be reset independently of each other.

Understanding Loadmap and Buffers

Local memory is used for executable images, routing tables, and forwarding tables. The loadmap is a table containing the application names, their program load addresses, and their sizes. When the dynamic loader obtains applications, it loads them in the order they are received from the archive. Hence, the order in which applications are loaded will be different with each boot operation and will be unique on each slot. The loadmap can be viewed using the **loadmap** command, followed by an optional slot number or a range of slots.

The loadmap is also stored in the log in two cases. When a fault occurs, it is saved with the current stack and the loadmap in the log. Also, when a log is saved in a file, the current loadmap is saved at the end of the log.

The following is a loadmap showing the program counter (PC) load addresses (in hexadecimal format) for each application and the application size (in decimal format) on slot 3 in a BLN.

```
fp3# loadmap 3

----------------------
Loadmap from SLOT 3:
----------------------
     File name        Load Addr    Size
------------------------------------------
--> bcc.exe           0x305c2910  1954092
```

```
--> run.exe         0x31261db0  0036448
--> arp.exe         0x3161c550  0100648
--> dns.exe         0x31634e90  0035340
--> ftp.exe         0x3163d8b0  0045348
--> tcp.exe         0x316489f0  0076516
--> tftp.exe        0x3165b4f0  0020936
--> snmp.exe        0x316626e0  0041060
--> tn.exe          0x31670780  0044404
--> ip.exe          0x3167b510  0454252
--> ilacc.exe       0x315c5230  0036156
--> qenet.exe       0x315d9270  0005288
```

Applications in the loadmap may not be listed in sequence based on the load address. The first image loaded is the kernel, which is not listed in the loadmap but occupies the lowest executable memory range. The loadmap is re-sorted based on the load address and can be used to determine the top 10 applications that consume local memory and the top 10 applications that consume buffer space.

In the preceding section, we used the `wfKernelEntry` object to look at local memory and global buffer use. A more granular look at local memory and global buffer use can be acquired using the `wfKernelBufOwnerTask` and `wfKernelMemOwnerTask` attributes. A fragment of the kernel statistics is shown here to identify the buffer and task owners and the resource use. The information is from the same slot, slot 3, on the BLN whose loadmap was just shown.

```
... ...
wfKernelEntry.wfKernelBufOwnerTask1.3 315C5DF0
wfKernelEntry.wfKernelBufOwnerTask1Bufs.3 100
wfKernelEntry.wfKernelBufOwnerTask2.3 315C5DF0
wfKernelEntry.wfKernelBufOwnerTask2Bufs.3 100
... ...
wfKernelEntry.wfKernelMemOwnerTask1.3 316B418A
wfKernelEntry.wfKernelMemOwnerTask1Size.3 12952400
wfKernelEntry.wfKernelMemOwnerTask2.3 3002C0D0
wfKernelEntry.wfKernelMemOwnerTask2Size.3 2861680
... ...
```

The loadmap is illustrated in Figure 4-6, and the top two buffer owners and local memory users are also illustrated. Notice that the applications are ordered from the lowest load address to the highest, with the kernel representing the lowest. Some applications have been omitted for clarity in the figure. If an application's load address is lower than a task address and is the closest (to the task address), then the task is said to reside in the application (see the figure). The function address `0x315C5DF0` resides in the *ilacc* driver and represents the top two buffer owners—in this case, the two sets of 100 buffers are used for the transmit and receive queues on one of the configured Ethernet circuits.

Similarly, the top two memory owner tasks are the *ip* application and the *kernel* application, respectively. Again, notice that the kernel's address `0x3002C0D0` is lower than all other application addresses. The MemOwnerTask values denote the dynamic memory allocated to the specific application; in this case, the static image footprint of the *ip* application is 454,252 bytes (from the loadmap), but it has dynamically allocated 12,952,400 bytes to maintain a full Internet routing table.

Fault Messages

Faults are errors that severely affect the state and operation of GAME or a subsystem, such as a protocol. Unlike most warnings, faults are not easily recoverable and require the subsystem or operating system to be restarted. This restart provides an error-free initialization, which is usually required to recover the subsystem or the entire slot. The tasks for interpreting, diagnosing, and remedying faults are discussed in this section.

In general, faults can be broken down into two types based on the scope and range of the impact: slotwide errors, which affect all applications on a slot, and localized errors, which affect only a specific protocol subsystem. In all these cases, diagnostics do not run, and hence the log is preserved for troubleshooting. Subsystems and slots recover immediately within a few seconds, depending on the scope of the re-initialization that may be required.

Bus Errors

A bus error usually occurs when an address pointer is invalid or inaccessible. In simple terms, one part of the software is handed an invalid reference to data that is required to

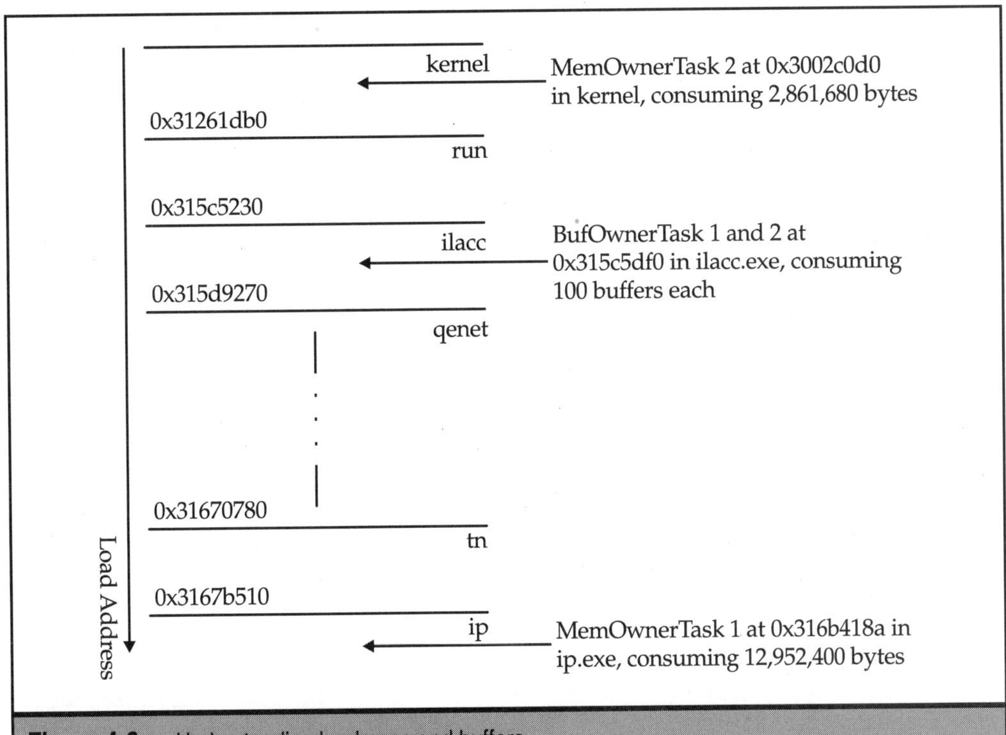

Figure 4-6. Understanding loadmaps and buffers

complete the task at hand. Commonly, this problem is the result of a wild or null pointer in a data structure. The following is an example of a bus error:

```
#   257: 04/23/98 14:41:50.324  FAULT    SLOT  1  GAME                  Code: 164
Error: exception vector 2 - Bus error
```

Bus errors may occur as a result of invalid or corrupt configuration files and sometimes as a result of a software bug. Configuration files that are corrupt may lack mandatory attributes or have values out of range. BayRS typically assumes that the configuration attributes are valid and delegates the creation and verification of configuration files to Site Manager. Data type checking is also done by BCC. As a result of a corrupt configuration file lacking a mandatory attribute, the software may access uninitialized data without checking the validity of the data or access a null location that does not contain any data.

Bus errors are rare but occasionally may follow a software upgrade. BayRS automatically loads and converts the MIB objects and attributes from the older configuration file. On occasion, upgrading from a substantially older version (for example, 7.60) to a current version (for example, 13.01) may result in incomplete conversion because of significant MIB differences between the versions. If a bus error occurs after an upgrade, you should re-create the configuration using the newer version of Site Manager or BCC. If the fault persists, then the condition may indicate a rare software bug that needs to be diagnosed and resolved by Nortel Networks Customer Service.

Watchdog Faults

A watchdog is a hardware timer that is initialized every time a gate is scheduled to run. A gate is equivalent to an individual process that executes along with other gates (processes) in a multitasking environment such as GAME. (Gates were discussed in detail in Chapter 2.) The watchdog timer watches the currently executing gate; every gate that is scheduled to run has to finish handling its buffers or signals before the timer expires. This protective mechanism is common across most run-to-completion multitasking operating systems, and GAME is no exception. It helps guarantee that other gates have a fair chance of running and completing their assigned tasks. If a gate continues to run beyond the timer, GAME assumes that it is stuck in an infinite loop and so not relinquishing the CPU. Thus, when the timer expires, GAME automatically traps the event and restarts the entire slot. This is a slotwide fault and affects all protocols and temporarily stops forwarding on the slot. A report of a typical watchdog fault is shown here.

```
#   613: 10/24/97 17:06:23.769  FAULT    SLOT  1  GAME                  Code: 123
Watchdog expired, service terminating

#   614: 10/24/97 17:06:24.542  FAULT    SLOT  1  GAME                  Code:  59
system restart

#   615: 10/24/97 17:06:24.542  DEBUG    SLOT  1  GAME                  Code: 166
System restart in tmo_wdog.c at line 70
```

If an unexpected watchdog fault occurs following provisioning of new services or circuits, there may be a scaling problem. BayRS is a real-time operating system, where the real-time feature is characterized not only by the reliable delivery of the data, but also the *timely* delivery of data. The real-time functionality of an operating system can be affected when the application stretches beyond its capabilities. For example, a high-speed HSSI line running frame relay service is theoretically capable of supporting over 2,500 PVCs with a committed information rate (CIR) of 16K. However, in practice, the management and servicing of over 2,500 PVCs will consume an extravagant amount of memory and CPU processing power. Scenarios such as this typically signify a scaling limitation inherent in the design, which can be resolved by reevaluating the new configuration changes. Normally, distribution of protocol services across two or more slots capitalizes on multiprocessing capabilities and allows greater scalability.

Tag Violations

A tag violation may occur if a device accesses a memory location incorrectly. This problem occurs only in architectures supporting memory tagging, a process that individually protects executable images, read-only memory locations, and read-write memory locations. This includes the ASN, all FRE modules, and the ARE. For example, an application's image needs only to be read after it has been loaded, and any modification to the code would be erroneous. A device attempting to do so would be performing an invalid operation, which would be detected by the tag hardware and intercepted by GAME. The entire system on the slot is reset if a tag violation occurs, which may imply a possible hardware or software anomaly.

```
#    72: 01/20/98 00:57:23.684  FAULT    SLOT  2  GAME              Code: 134
Tag violation 2 (0=RFR/1=BB/2=CPU/3=LNK) accessed 2 (0=NA/1=RW/2=RO/3=RE)
```

Troubleshooting tag violations can be quite difficult and is usually done with the assistance of Nortel Networks Customer Service.

Parity Violations

Parity hardware is similar in function to tag hardware. Parity checking is a simple process that verifies the integrity of data in memory, similar in concept to data parity checking in asynchronous lines. Parity violations usually occur because of bad memory or a bad processor board. Here is an example of a parity violation recorded in the log:

```
#    72: 11/20/97 10:57:43.614  FAULT    SLOT  1  GAME              Code: 217
Parity violation 2 (0=LNK/1=BB/2=CPU/3=INVLD) detected
```

Parity violations are easily resolved by replacing the motherboard (processor module).

ARE-Specific Faults

The unique characteristics of the ATM Routing Engine (ARE) were mentioned in Chapter 1. The ARE sports a dual PowerPC processor for parallel processing and contains the

VBM for packet buffers. These unique characteristics in the ARE architecture have special considerations; specific faults that are not seen in other platforms occur in the ARE due to these differences.

VBM Errors VBM, or virtual buffer memory, is unique hardware presently only on ARE processor modules. It provides special buffer management and variable buffer size, differing from the fixed-size buffers used in FRE modules. As is typical with most memory management techniques, the VBM is divided into pages, and each page consists of 256 bytes. Depending on the size of a buffer, it may contain one or more pages. Each buffer is terminated by a page called the guard page, which provides segmentation and prevents buffer overlaps. An incorrectly accessed buffer may cause reading or writing past the last page, into the guard page. Other protected areas of the VBM will also generate errors and can cause slot reset. A typical VBM error message is shown here.

```
#  3131: 01/02/98 03:41:42.182  FAULT    SLOT  1  GAME                 Code: 240
CPU Read of Mid page.   Not valid.   VBM error detected (0x0068)

#  3132: 01/02/98 03:41:42.198  FAULT    SLOT  1  GAME                 Code: 240
Error: exception vector 5 - VBM error
```

MCP Machine Check Errors The MCP machine check is another fault that is unique to the ARE. It is similar to a bus error on the Motorola 68K processor on the FRE-1/FRE-2, the only difference being that the error occurs on a PowerPC processor. The discussion of bus errors earlier is equally applicable to the MCP machine check, and the troubleshooting process is the same. An example of a machine check error message is shown here.

```
#    21: 11/04/97 07:11:12.412  FAULT    SLOT  5  GAME                 Code: 164
Error: exception vector 2 - Machine check (MCP)
```

Orphaned Buffers

A unique error that is classified as a fault is the orphaned buffer. Recall that in Chapter 2, we mentioned that a buffer is always owned either by GAME or by a specific gate to which it has been assigned. Buffers can become orphaned in two ways. A gate may complete execution without freeing the buffers owned, thus orphaning them. Also, a gate may receive a buffer and then orphan it because that gate is not supposed to receive buffers. This condition is uncommon because it indicates a software interaction problem.

```
#   19: 08/21/1999 17:27:13.149  WARNING  SLOT 12  GAME                 Code: 111
Scheduler exited due to orphaned buffer
```

Orphaned buffers are rare, and if you encounter an orphaned buffer fault in the log, you can address it by first resetting the entire system because an external or internal factor may have induced software instability. In most instances, orphaned buffers are not fatal to the system and will not affect connectivity unless too many buffers are orphaned, causing buffer starvation. Note that an orphaned buffer will not cause a subsystem or slot reset. A persistent orphaned buffer problem may indicate a software bug.

Scheduled Panics

Scheduled panics are restarts triggered by the code in the event of noticeable, irrecoverable errors. We discussed the most common scheduled panics earlier, which occur as a result of insufficient local memory or insufficient buffers. Operator-configured log crash points can also generate scheduled panics (performed under the direction of Nortel Networks Customer Service). Except for operator-configured panics, scheduled panics are usually a result of memory management issues or scaling limitations. Manager-configured log crash points are used to obtain crash dumps when detailed troubleshooting information is required. A crash dump is a copy of the memory image and is similar to a core dump in UNIX operating systems.

Other Faults

A few other faults can occur on the router, including hog buffers, reliable gate forwarding failures, and gate deaths. None of these constitutes a fatal error but may indicate the onset of a scaling problem (as in the case of hog buffers) or a software problem. There is no specific method to troubleshoot each of these issues individually. In general, an undiagnosable fault in the log must be reported to Nortel Networks Customer Service for further analysis.

PHYSICAL MEDIA ISSUES

As discussed in the prelude to this chapter, there are several troubleshooting methodologies. One of the most common approaches is to investigate the problem from the bottom up. This means first investigating the lowest layer of the OSI model, which is the physical layer, and then analyzing the data link and network layers in sequence. This section describes some popular techniques for troubleshooting physical media issues.

Diagnostic Issues

Symptoms of physical layer issues include periodically bouncing lines or lines that are operationally down. Many PHY (physical) or PMD (physical-medium dependent) issues also appear as frame transmission or receive errors. During diagnostics, link modules and media interfaces are run through a battery of tests to ensure their integrity. Diagnostic tests can be bypassed during a warm start, which may fail to identify a defective interface.

```
# 116: 06/19/1999 11:59:44.613  DEBUG    SLOT  8  MODULE              Code:  61
I/O module 8 has not been verified with diagnostics
```

If multiple interfaces on a link module are down, first check to see that the module is inserted properly. Entire link modules or media interfaces that fail diagnostics are reported in the log with warning messages. Media interfaces can still be initialized by the driver, and protocols can become operational on interfaces that have bypassed diagnostics, but a warning message is logged to indicate a potential problem.

```
# 185: 08/16/1999 16:23:11.279  WARNING  SLOT 11  MODULE              Code:  60
I/O module 11 has failed diagnostics
```

Interface diagnostic failure messages are unique and specific to the media type and are reported in the corresponding event log entity.

```
#  133: 05/14/1999 17:58:11.966  DEBUG    SLOT  9  ASYNC              Code:  19
Connector COM1 diagnostic FAILED with failure status 3.

#   17: 06/26/1999 16:47:25.623  FAULT    SLOT  3  ATMINTF            Code:   1
Port 1: diagnostic failed with status 1.

#   82: 07/11/1999 04:33:11.003  DEBUG    SLOT  1  BISYNC             Code:  27
Connector COM2 diagnostic FAILED.

#  167: 12/09/1999 22:31:11.103  WARNING  SLOT  5  CSMACD             Code:   5
Connector XCVR4 diagnostic failed.

#  136: 08/26/1999 08:12:33.402  DEBUG    SLOT  8  DS1E1              Code:  47
Connector COM2 diagnostic FAILED with failure status = 1.

#  143: 04/15/1999 24:34:42.748  DEBUG    SLOT  1  DSUCSU             Code:   9
DSU/CSU on COM2 diagnostic FAILED.

#   75: 04/06/1999 18:23:54.216  WARNING  SLOT 10  FDDI               Code:   3
Node 1 diagnostic failed.

#  177: 08/11/1999 05:41:35.410  WARNING  SLOT 13  HSSI               Code:   5
Connector HSSI1 diagnostic failed.

#   50: 05/23/1999 14:14:01.727  DEBUG    SLOT  1  MODEM              Code:  24
Modem on COM3 diagnostic FAILED.

#   61: 02/28/1999 23:51:12.084  DEBUG    SLOT  4  SYNC               Code:  26
Connector COM2 diagnostic FAILED.

#   99: 10/17/1999 15:44:29.915  WARNING  SLOT  6  TOKEN_RING         Code:   4
Connector MAU1 diagnostic failed.
```

Notice that several entities report similar messages, but with different severity levels. The severity level reported by the line driver depends on how the driver continues with the initialization. Some drivers, such as the line driver for SONET/SDH interfaces (logged under `ATMINTF`), will not continue initialization when a diagnostic failure is encountered; others, such as synchronous drivers, will continue the initialization process until they encounter a fatal error. An interface that has failed diagnostics may initialize and be able to provide circuit services; those that have not run diagnostics will also report an event message stating that diagnostics have not been run. For example, a Token Ring interface might report the following event:

```
#   93: 03/10/1999 11:29:20.788  WARNING  SLOT 14  TOKEN_RING         Code:   6
Connector MAU2 not verified with diagnostic.
```

If a link module or interface has failed diagnostics, then rerun diagnostics on it to verify that it is indeed defective. In some cases, a loosely connected cable may result in diagnostics test failure. The **–lm** option of the **diags** command can be useful in this regard. A faulty interface or link module must be replaced.

Media State

Media-related issues usually are noticed as a result of a problem with a higher layer (for example, the network layer). For example, you may observe that an Ethernet circuit is down only after observing that the associated IP interface is inoperative. Regardless of whether the problem is being analyzed from the top down or from the bottom up, the media state of an interface should be first checked.

Table 4-4 lists the MIB objects for all available LAN and WAN media interfaces, along with their state attributes and values.

Individual DS0s on a MCT1/MCE1 interface can be monitored using `wfLogicalLineEntry.wfLogicalLineState`, which reports values similar to other media interfaces as listed in Table 4-5. (Note that state attributes common among media are not relisted in this table. For example, the MCT1/MCE1 and serial state attributes also apply to the fractional T1/E1 media, which use the same `wfDs1E1PortEntry` and `wfSyncEntry` objects.)

Using the TI **get** or BCC **show *media* summary** (for example, **show ethernet summary**) command, you can retrieve the state of a particular media instance. A line can be in the down state if it is administratively disabled, if there is no physical connection, or if the connection or cabling is incomplete or incorrect. For example, a serial line will stay down if an incorrect cable is used.

Many media states are self-explanatory. The following paragraphs discuss the interpretation and appropriate action for various media states.

- **Initializing** A media line that is coming up for the first time will be set to the Initializing state by its line driver. If the line driver is unloaded and reloaded at runtime, the Initializing state also will be seen. This state is transient and should not persist for more than a few seconds, until the driver fully initializes and declares a finite state. If a line stays in the Initializing state for an extended period, there may be a loose cable connection.
- **Not Present** In Chapter 2, we discussed lines and line drivers. A line is said to be *configured* when it is instantiated in the MIB configuration. It is said to be *present* when the line driver is providing services on it. When a line is configured, the line driver is automatically loaded by Site Manager and BCC. On BNs, both a link module driver and the media chipset driver must be loaded. If one of the necessary drivers is unavailable, the line will remain in the Not Present state. This condition can occur if the driver image was accidentally removed during customization of the runtime image archive.

```
#  173: 01/01/2000 05:37:05.959  DEBUG    SLOT  2  LOADER              Code:  17
Could not locate application lance.exe in 2:bn.exe... not loaded
```

The Not Present state can also occur if a driver is unloaded but overlooked. No events are logged in this condition; link module drivers (wfLinkModules) and chipset drivers (wfDrivers) usually must not be unloaded without proper cause. Drivers are loaded using standard slot mask values, as described in Chapter 2, in the "Slot Mask" section. Site Manager and BCC automatically load the appropriate chipset and link module drivers when

Media	MIB Object	State Attribute (Number)
Ethernet (CSMA/CD)	wfCSMACDEntry	wfCSMACDState (3)
Token Ring	wfTokenRingEntry	wfTokenRingState (3)
FDDI	wfFddiEntry	wfFDDIState (3)
Serial	wfSyncEntry	wfSyncState (3)
Bisync	wfBisyncEntry	wfBisyncState (21)
T1[1]	wfT1Entry	wfT1State (3)
E1	wfE1Entry	wfE1State (3)
HSSI	wfHssiEntry	wfHssiState (3)
MCT1[2]	wfDs1E1PortEntry	wfDs1E1PortState (3)
MCE1	wfDs1E1PortEntry	wfDs1E1PortState (3)
DS0 on MCT1/MCE1	wfLogicalLineEntry	wfLogicalLineState (3)
ATM SONET/SDH	wfAtmInterfaceConfEntry	wfAtmInterfaceOperStatus (4)
ATM DS3/E3	wfAtmInterfaceEntry	wfAtmInterfaceState (3)
Fractional[3] T1 and Fractional E1	wfSyncEntry and wfDs1E1PortEntry	wfSyncState (3) and wfDs1E1PortState (3)
V.34 Modem	wfModemEntry	wfModemLineState (6)
56/64K DSU/CSU	wfDsuCsuIfEntry	wfDsuCsuOpState (10)
ISDN BRI S/T/U	wfIsdnBriInterfaceEnt ry	wfIsdnBriState (3)
ISDN PRI	See MCT1/MCE1	

[1] T1 and E1 media interfaces refer to digital services that terminate in 520x, 522x, and 5250 link modules. Digital carriers that terminate in a CSU/DSU and are connected to the router's DTE using a serial port must be treated as serial connections.

[2] ISDN PRI services are also provided by the DS1E1 driver used by the MCT1/MCE1 modules.

[3] This object applies to the fractional T1/E1 module for the AN/ANH and ARN routers. The fractional line is handled by the serial service, and the T1/E1 services are performed by the DS1E1 driver.

Table 4-4. Media States and Attributes

Value	Ethernet	Token Ring	FDDI	Serial	Bisync	T1/E1	HSSI	MCT1/E1 FT1/E1	SONET/SDH	V.34 Modem	56/64K DSU/CSU	ISDN BRI
1	Up	Up	Up	Up	Up	Up	Up	Up	Up	Unknown	Normal (Up)	Up
2	Down	Down	Down	Down	Down		LMI Wait	Yellow Alarm	STS/LOP	On Hook	Local Analog Loop	Down
3	Init	Init	Init	Init	Init	Init	BOFL Wait	AIS Alarm		Off Hook	Local Digital Loop	Init
4	Not Present	Not Present	Not Present	Wait	DSR Wait	Not Present	CA Wait	Red Alarm	STS/AIS	Connected	Remote Digital Loop	
5				Not Present[1]			Init			Busied Out	Telco Loopback	
6				DSR Wait			Not Present			Reset	Remote Digital Loop with Pattern	
7				Hold Down			DTE Loopback				Local Analog Loop with Pattern	
8				Remote Loop			DCE Loopback		STS/RDI		Generating Pattern 2047	
9							DCE Loopback Wait					
10							DCE Loopback Timeout	Loopback				UBC Waiting
11								BERT				
15								Init				
16									Unequipped			
19				Line Loop BOFL Test[2]			Line Loop BOFL Test					
20				Not Present	Not Present		Not Present	Not Present				Not Present
32									Signal Label Mismatch			

[1] Note that some attributes have two "not present" states. In recent software versions, the lower value is obsolete and the higher (new) value is used. Both values indicate the same cause (that is, the absence of the driver) on the different versions.

[2] The line loop BOFL test condition does not occur in BayRS, but only on BayStream software. The BayStream software provides internetwork switching services and interworking, and uses many of the same MIBs that BayRS uses. The line loop BOFL test condition is mentioned here for completeness.

Table 4-5. Media State Attribute Values

configuring a circuit. If you suspect a driver-related issue, delete the interface (or the slot in the case of the link module) and re-create the configuration for the interface (or slot).

```
box# mset wfDrivers.wfLANCELoad.0 0x40000000; commit
box# log -fd -eLOADER

#   102: 01/10/2000 17:58:32.039  DEBUG    SLOT  2  LOADER              Code:  43
Image lance.exe loaded successfully from 2:bn.exe

#   103: 01/10/2000 17:58:32.046  DEBUG    SLOT  2  LOADER              Code:  15
Loader starting application lance.exe, address 0x30a1d4a0, gate id = 0x00040
```

- **DSR Wait** This condition indicates that the router's DTE (Data Terminal Equipment) interface is waiting for the DSR (Data Set Ready) signal from the DCE (Data Communications Equipment). DCEs include modems and CSU/DSUs, which by default assert the DSR signal when they initialize. The DTE (router) will not come up until it successfully recognizes the DCE. If the interface stays in the DSR Wait state, then try resetting the modem or CSU/DSU attached to it. Verify that the cable used is correct (see Table 4-6 later in this chapter) and is attached securely and firmly.
- **Hold Down** The line driver may mark a port as down for a variety of reasons, including carrier loss, loss of clocking, and loss of DCE. Lines that have intermittent noise can cause framing and signal errors, which may cause the line driver to shut down the port. Once the signal and framing return to normal, the port is activated again. On high-speed lines with small recurring errors, lines may go down and come back up several times a second. This causes circuits to initialize only to be taken down again, which may generate a considerable load on the processor.

  To ensure that an individual port does not unfairly thrash the processor, a one-second hold-down timer is instituted on serial lines. When a line goes down, it is held down for a period of one second before the driver attempts to reinitialize it again. This condition is called the Hold Down state and is transient. The serial port will enter the Initializing state after this and may remain in the Initializing state if the line fails to come up. The `wfSyncLastState` attribute provides the last serial interface state, which can indicate the state progression in conjunction with the current state.
- **Remote Loop** The Remote Loop and Hold Down conditions are unique to serial lines. Before discussing the Remote Loop state, let us discuss BOFL packets and their use.

  On serial lines, a special breath-of-life (BOFL) packet is periodically transmitted to verify successful transmission. A BOFL packet (see Figure 4-7) contains the MAC address of the transmitting router in both the source and destination address fields, with the Type field set to 8102 (Nortel Networks). No special

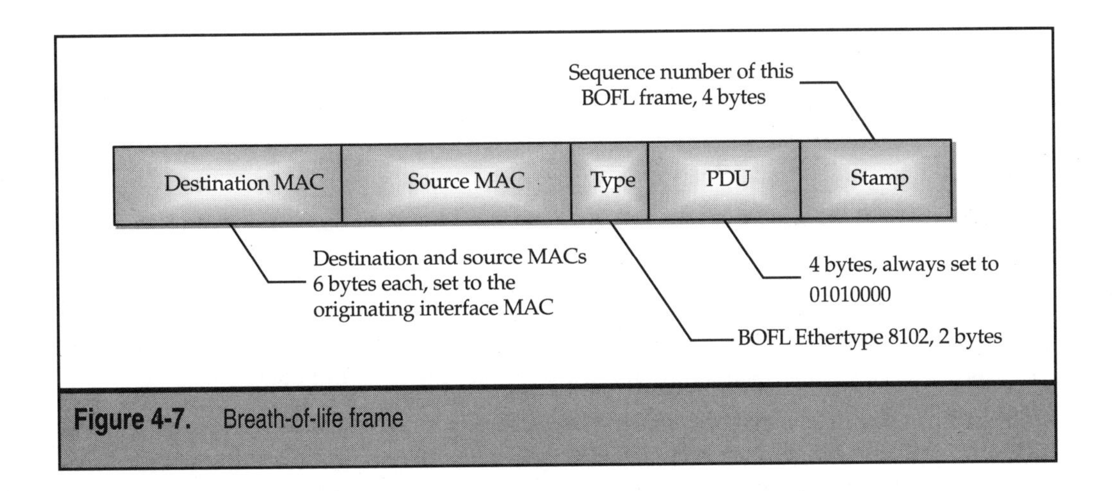

Figure 4-7. Breath-of-life frame

data is included, and these packets are used to verify link integrity. For example, the Nortel Networks proprietary HDLC protocol (usually called Bay Standard or Wellfleet Standard) regularly transmits BOFL frames and expects to receive them from the remote end—if three out of four BOFL packets are missed from the remote end, the line is considered unusable and is brought down. Lines with excessive transmission errors can cause data corruption or loss, leading to the drop of a few BOFL packets.

BOFL packets are otherwise not commonly used with other data link protocols on serial lines, except with PPP. However, they can be used to detect a loop condition or to verify that an operator-induced loop was successful. Normally, the `wfSyncEntry.wfSyncRemoteLpbkDetection` attribute is disabled; when this attribute is enabled, it allows the line driver to recognize its own BOFL packets. When a successful loopback is established, the remote device starts retransmitting all frames it receives. Note that the term *remote device* is used loosely here; it represents any device downstream to the DTE. Once loopback detection is turned on, the line driver sets the interface state to Remote Loop. This indicates that the media interface, cable, and line (up to and including the remote loop) are not defective.

Loopback tests are discussed in more detail in the next section. Remote loopback detection must be turned off when it is not in use. It should be turned on prior to loopback testing to see if the network problem being observed is due to a downstream loop that you are unaware of.

- **LMI Wait** This state is seen only on HSSI lines. WAN protocols such as frame relay, ATM/DXI, and SMDS require the use of a Link Management Interface (LMI) that communicates directly with the line driver. The operation of the HSSI driver requires that it communicate with the LMI gate. If the LMI gate is not yet operational during initialization, the HSSI line will enter the LMI Wait state.

If a line stays in the LMI Wait state, check to make sure that the higher-layer protocol (for example, frame relay) is loaded on the slot and configured on the interface. If there are no configuration or loading issues with the protocol above, then disable and reenable the HSSI interface to reinitialize the driver, and also unload and reload the protocol on the slot.

- **BOFL Wait** Following the LMI Wait state, successful dialog between the line driver and LMI will allow the driver to proceed. As for serial lines, BOFL may optionally be used with HSSI lines. If BOFL is enabled, the driver transmits a BOFL packet and waits for a BOFL packet from the remote end to indicate the presence of a link. This state is called the BOFL Wait state and is skipped if BOFL packets are disabled on the interface. It is unique to HSSI lines.

  When a HSSI line remains in this state, disable BOFL packets on the line to see if it will come up. Then perform a loopback (discussed in the next section) to see if the interface and line can transmit and receive frames without any errors.
- **CA Wait** The CA (Communications Adapter) Wait condition is identical to the DSR Wait condition that occurs on serial interfaces. This condition will be reported on HSSI interfaces and can be solved by addressing cable or CSU/DSU issues.
- **DTE Loopback** A DTE loopback is a local digital loop created without the involvement of the transmission line. Again, this state is seen only on HSSI lines that have explicitly requested a DTE loopback (loop A) or that have been inadvertently put into this mode. A successful local digital loop will result in this condition, which verifies the integrity of the HSSI port, the cable, and the premise-side components of the CSU/DSU.
- **DCE Loopback** In a DCE loopback, data frames travel beyond the local digital loop and onto the telco transmission line. Either a local line loop (loop B) or a remote line loop (loop C) will result in a DCE Loopback condition. If a local line loop is set up, then this condition verifies the proper functioning of the DTE, cable, and local CSU/DSU. On the other hand, if a remote line loop is employed, this state indicates that the transmission line and the line circuitry of the remote CSU/DSU are also good.
- **DCE Loopback Wait** A DTE loopback is easily and immediately established following a request. However, establishment of DCE loops may take longer. Following a local or remote line loop request, a HSSI line enters the DCE Loopback Wait state to confirm that the DCE sees and returns the loopback request. The router's DTE interface remains in this state for 10 seconds to see if the loopback is established.
- **DCE Loopback Timeout** After the 10-second period that a HSSI interface stays in the DCE Loopback Wait state, it declares that the loopback request has timed out. Even remote line loops, which usually take the longest to establish, do not take longer than 10 seconds to be established. If this timeout occurs

following a local line loop, reset or replace the local CSU/DSU. If it occurs following a remote line loop, investigate the transmission line and the remote CSU/DSU.

- **Red Alarm** Red alarms indicate a significant issue on the digital service link, where either a LOS (loss of signal) or OOF (out of framing) condition has occurred. LOS indicates that the port is no longer able to see valid bit transitions to determine bit values. An OOF occurs when framing bits in all frames received have errors, rendering them unusable. Both LOS and OOF signify a condition where no valid data is being received.
- **Yellow Alarm** A yellow alarm is also called a RAI, or remote alarm indication. A RAI occurs when the locally transmitted signal is not being received by the remote end. A local yellow alarm follows a remote red alarm, when the remote device transmits the RAI to identify the problem. This condition usually indicates that the transmission path from the local side to the remote side is defective.
- **AIS Alarm** Any equipment along the transmission path that detects a loss of signal (LOS) will send an AIS (Alarm Indication Signal) to the downstream device. The downstream device (an MCT1 port, for example) will assert the AIS, indicating that an upstream device is reporting a LOS condition. When this condition occurs, investigate all devices and connections upstream to the device you're working with. AIS alarms are also called blue alarms.
- **Loopback** On MCT1/MCE1 modules and Fractional-T1/E1 interfaces, a loopback at any point in the transmission will be detected by the driver, which will set the interface state to Loopback. The exact nature of the loopback can be determined from `wfDs1E1PortEntry.wfDs1E1PortLoopbackState`, which can help determine if the loop code originated at the local side or remotely, and the type of the loop.
- **BERT** The Bit Error Rate Test (BERT) is a transmission quality measurement commonly used with telecommunication circuits. High-density digital transmission services (such as T1) offer the facilities to perform BERT to determine circuit quality. While voice traffic traveling on the same telecommunication circuits is usually tolerant of small errors, digital data transmission services are adversely affected, even with error rates as low as $1x10^{-4}$ (one error in 10,000 bits). Media interfaces supporting T1/E1 interfaces, such as the MCT1 module, can initiate and run BER tests, which are discussed later, in "Loopbacks and Testing." When a port is in BER testing, the state is reported as BERT.
- **STS/LOP** A loss of pointer (LOP) error is reported on synchronous transport signal (STS) lines (SONET/SDH) when several consecutive invalid pointers or new data flags are received. These conditions usually occur if data integrity is compromised along a section or path, or if a remote device is defective. If a

SONET/SDH interface is marked as STS/LOP, then check all devices downstream and the transmission link.

- **STS/AIS** On an STS circuit, the AIS is sent as a series of 1's by the upstream device and is similar in function to the blue alarm on DS1 circuits. If an interface is in the STS/LOP state, receipt of three consecutive STS/AIS alarms will clear the LOP state. This is because the local device confirms that the defect condition lies upstream. Troubleshooting STS/AIS conditions is identical to troubleshooting AIS on DS1 circuits.
- **STS/RDI** The remote defect indication (RDI) state on STS circuits is similar to the yellow alarm on DS1 circuits. A loss-of-frame, loss-of-signal, or AIS defect alarm sent by the remote path terminating equipment is recognized, and the state of the interface is reported as STS/RDI. This indicates that the transmission path from the local to the remote path termination is defective.
- **Generating Pattern-2047** The 56/64K DSU/CSU can generate pattern-2047, a BERT pattern used to test telco circuits. The interface must manually be put into pattern-2047 mode during loopback testing, which is discussed later in this chapter.

Cabling Issues

Another step in investigating physical layer issues is verification of the cable. Cables are unlikely to fail or get damaged frequently, but these incidents sometimes occur in harsh environments. Environmental factors such as high temperatures, humidity, and pollutants can sometimes damage inadequately designed custom cables.

If a problem occurs after installing a new cable, first ensure that the cable being used is appropriate for the desired application. Consult Table 4-6 for details on using the appropriate cable for the connector and interface type you need. Check the cable by disconnecting both ends and visually inspecting it. If the cable appears to have bent or broken pins, then use another one. If it appears acceptable, then recable both ends and make sure it is connected securely and firmly.

A common mistake with cables is using the wrong interface connector. Certain modules such as the Dual Sync Dual Ethernet have four identical 15-pin connectors: two used for serial communication and two used for baseband Ethernet. Connecting a downstream cable to the wrong interface may damage the port or the board. Connector positions on link modules are unique and may be numbered differently; for example, interface 1 on one link module may be on the left, and it may be on the right on another type of link module. If the interface is down, check the connector number on the module and recable accordingly.

Some common applications such as Ethernet connections (straight-through and crossover) can be performed using industry-standard cables. The same is true for other applications such as RJ48 to RJ48 connections for T1/DS1 circuits and ISDN PRI interfaces. In these cases, make sure that you are using the correct cable type (for example, MCT1 interfaces use a standard 8-pin straight-through RJ48 cable).

Media Interface and Cable Type	Cable Number	Comments
Ethernet		
AUI to AUI	7115	15 pin, 15 ft.
Token Ring		
9-pin to MAU	7125	9 pin, 15 ft.; STP cable
9-pin to MAU	7126	9 pin, 15 ft.; ferrite bead, STP cable
9-pin to RJ11	7128	9 pin, 1 ft.
Fiber Distributed Data Interface		
FSD to FSD	7135	32.81 ft.; fixed-shroud duplex connector
FSD to ST	7136	32.81 ft.; FSD to straight-tip bayonet
SC to ST, SMF	7164	9 ft.
SC to ST, MMF	7169	9 ft.
SC to SC, SMF	7165	9 ft.
SC to SC, MMF	7163	9 ft.
ATM		
SC to ST, MMF	7169	9 ft.
HSSI		
DCE to DTE	7830	50 pin, 10 ft.
DTE to DCE	7831	50 pin, 50 ft.
T1 and E1		
9-pin to RJ45	AA0018001	9 pin, 10 ft.; MCE1[1], 120 Ohm
9-pin to unterminated[2]	AA0018002	9 pin, 10 ft.; MCE1, 120 Ohm
15-pin to 15-pin, with 9-pin timing source	AA0018006	15 pin, 0.5 ft.; 9-pin external T1 timing source, for use with T1/DS1 with external clock[3]
RJ45 to RJ45	AA0018014	8 pin, 10 ft.; MCE1, 120 Ohm

Table 4-6. Media Cables for Various Applications

Media Interface and Cable Type	Cable Number	Comments
RJ45 to unterminated	AA0018015	8 pin, 10 ft.; MCE1, 120 Ohm
15-pin to 15-pin, crossover	AA0018021	15 pin, 15 ft.; MCT1[4]
15-pin to 15-pin, straight-through	AA0018022	15 pin, 15 ft.; MCT1
9-pin clock interface to 15-pin	7150	9 pin, 15 ft.; MCT1 external clock
15-pin loopback	7151	15 pin; MCT1 loopback
RJ48 loopback	7152	RJ48, MCT1 loopback
15-pin to RJ48 crossover	7153	15 pin, 15 ft.; MCT1 crossover
RJ48 to RJ48 loopback	7154	RJ48, 15 ft.; MCT1 loopback cable
15-pin to 15-pin loopback	7155	15 pin, 15 ft.; MCT1 loopback cable
RJ48 to RJ48 straight-through	7156	RJ48, 15 ft.; MCT1 straight-through
15-pin to 15-pin straight-through	7157	15 pin, 15 ft.; MCT1 straight-through
9-pin to 9-pin crossover	7160	9 pin, 10 ft.; MCE1 crossover
9-pin to 9-pin straight-through	7161	9 pin, 10 ft.; MCE1 straight-through
75 Ohm coax	7162	Twist-lock coaxial, 10 ft.; MCE1
9-pin to 15-pin straight-through	7167	9 pin, 10 ft.; MCE1 straight-through
9-pin to 15-pin crossover	7168	9 pin, 10 ft.; MCE1 crossover
15-pin to 15-pin Framer plug (CSU DTE)	7401	15 pin, 1 ft.; DSX-1/DSX-2, T1 framer plug to CSU DTE plug
15-pin to 15-pin framer plug (CPE)	7415	15 pin, 15 ft.; DSX-1/DSX-2, T1 framer plug to CPE plug
15-pin CSU receptacle to RJ45	7650	15 pin, 50 ft.; CSU network receptacle to T1 carrier
15-pin CSU receptacle to spade terminals	7750	15 pin, 50 ft.; CSU network receptacle for T1 carrier

Table 4-6. Media Cables for Various Applications *(continued)*

Media Interface and Cable Type	Cable Number	Comments
Serial and Dial Services		
RS232, 15-pin to RS232 plug, WAN	7255	15 pin, 15 ft.; RS232, general WAN use
RS232, 15-pin to RS232 plug, synchronous pass-through	7218, 7941(F[5])	15 pin, 15 ft.; RS232, synchronous pass-through; use 7941 for female synchronous pass-through
RS232, 15-pin to RS232 plug, RaiseDTR	7118	15 pin, 15 ft.; RaiseDTR dial service
RS232, 15-pin to RS232 plug, V.25bis	7119	15 pin, 15 ft.; V.25bis dial service
RS232, 15-pin to RS232 plug, KG crypto resync	7225	15 pin, 15 ft.
RS232, 44-pin to RS232 plug, WAN	7826	44 pin, 15 ft.; RS232, general WAN use
RS232, 44-pin to RS232 plug, synchronous pass-through	7833, 7943(F)	44 pin, 15 ft.; RS232, synchronous pass-through; use 7943 for female synchronous pass-through
RS232, 44-pin to RS232 plug, RaiseDTR	7138	44 pin, 15 ft.; RaiseDTR dial service
RS232, 44-pin to RS232 plug, V.25bis	7826	44 pin, 15 ft.; V.25bis dial service
RS232, 50-pin to RS232 plug, WAN	7934	50 pin, 15 ft.; general WAN use
RS232, 50-pin to RS232 plug, RaiseDTR	7935	50 pin, 15 ft.; RaiseDTR dial service
RS232, 50-pin to RS232 plug, synchronous pass-through	7945(F)	50 pin, 15 ft.; female synchronous pass-through
RS422, 15-pin to RS422 plug, WAN	7315	15 pin, 15 ft.; RS422, general WAN use
RS422, 15-pin to RS422 plug, RaiseDTR	7117	15 pin, 15 ft.; RaiseDTR dial service

Table 4-6. Media Cables for Various Applications *(continued)*

Media Interface and Cable Type	Cable Number	Comments
RS422, 15-pin to RS422 plug, V.25bis	7116	15 pin, 15 ft.; V.25bis dial service
RS422, 15-pin to RS422 plug, KG crypto resync	7226	15 pin, 15 ft.
RS422, 44-pin to RS422 plug, WAN	7318	44 pin, 15 ft.; RS422, general WAN use
RS422, 44-pin to RS422 plug, RaiseDTR	7139	44 pin, 15 ft.; RaiseDTR dial service
RS422, 44-pin to RS422 plug, V.25bis	7318	44 pin, 15 ft.; V.25bis dial service
RS422, 50-pin to RS422 plug, WAN	7937	50 pin, 15 ft.; general WAN use
RS530, 15-pin to 25-pin, WAN	AA0018011	15 pin, 15 ft.; general WAN use
RS530, 44-pin to 25-pin, WAN	AA0018012	44 pin, 15 ft.; general WAN use
RS530, 50-pin to 25-pin, WAN	AA0018013	50 pin, 15 ft.; general WAN use
V.35, 15-pin to V.35 plug, WAN	7215, 7216	15 pin, 15 ft.; V.35, general WAN use; use 7215 for lines without DTR, 7216 with DTR
V.35, 15-pin to V.35 plug, synchronous pass-through	7219, 7942(F)	15 pin, 15 ft.; V.35, synchronous pass-through; use 7942 for female synchronous pass-through
V.35, 15-pin to V.35 plug, RaiseDTR	7121	15 pin, 15 ft.; RaiseDTR dial service
V.35, 15-pin to V.35 plug, V.25bis	7120	15 pin, 15 ft.; V.25bis dial service
V.35, 15-pin to V.35 plug, KG crypto resync	7227	15 pin, 15 ft.
V.35, 44-pin to V.35 plug, WAN	7220	44 pin, 15 ft.; V.35, general WAN use

Table 4-6. Media Cables for Various Applications *(continued)*

Media Interface and Cable Type	Cable Number	Comments
V.35, 44-pin to V.35 plug, synchronous pass-through	7834, 7944(F)	44 pin, 15 ft.; V.35, synchronous pass-through; use 7944 for female synchronous pass-through
V.35, 44-pin to V.35 plug, RaiseDTR	7137	44 pin, 15 ft.; RaiseDTR dial service
V.35, 44-pin to V.35 plug, V.25bis	7220	44 pin, 15 ft.; V.25bis dial service
V.35, 50-pin to V.35 plug, WAN	7932	50 pin, 15 ft.; V.35, general WAN use
V.35, 50-pin to V.35 plug, synchronous pass-through	7946(F)	50 pin, 15 ft.; V.35, female synchronous pass-through
V.35, 50-pin to V.35 plug, RaiseDTR	7933	50 pin, 15 ft.; RaiseDTR dial service
V.35, 50-pin to V.35 plug, V.25bis	7932	50 pin, 15 ft.; V.25bis dial service
V.28, 15-pin to 25-pin, WAN	7256	15 pin, 15 ft.; V.28, general WAN use
V.28, 44-pin to 25-pin, WAN	7837	44 pin, 15 ft.; V.28, general WAN use
X.21, 15-pin to X.21 plug, WAN	7221	15 pin, 15 ft.; X.21, general WAN use
X.21, 15-pin to X.21 receptacle, synchronous pass-through	AA0018003(F)	15 pin, 15 ft.; X.21, female synchronous pass-through
X.21, 44-pin to X.21 plug, WAN	7224	44 pin, 15 ft.; X.21, general WAN use
X.21, 44-pin to X.21 receptacle, synchronous pass-through	AA0018003(F)	44 pin, 15 ft.; X.21, female synchronous pass-through
X.21, 50-pin to X.21 plug, WAN	7936	50 pin, 15 ft.; X.21, general WAN use
X.21, 50-pin to X.21 receptacle, synchronous pass-through	AA0018005(F)	50 pin, 15 ft.; X.21, female synchronous pass-through

Table 4-6. Media Cables for Various Applications *(continued)*

Media Interface and Cable Type	Cable Number	Comments
Serial and HSSI Crossovers		
15-pin crossover	7260	15 ft.
44-pin crossover	7835	15 ft.
50-pin crossover	7938	15 ft.
15-pin to 44-pin crossover	7836	15 ft.
15-pin to 50-pin crossover	7939	15 ft.
44-pin to 50-pin crossover	7940	15 ft.
50-pin HSSI crossover	7832	50 pin, 20 ft.
Cable Adapters		
RS232, 44-pin to 15-pin	7122	0.5 ft.; RS232
RS422, 44-pin to 15-pin	7131	0.5 ft.; RS422
RS422, 50-pin to 15-pin	AA0018024	0.5 ft.; RS422
V.35, 44-pin to 15-pin	7124	0.5 ft.; V.35
V.35, 50-pin to 15-pin	7948	0.5 ft.; V.35
X.21, 44-pin to 15-pin	7123	0.5 ft.; X.21
50-pin to 44-pin	7947	0.5 ft.; interface dependent

[1] MCE1 cables in this matrix can be used with E1/G.703 interfaces.
[2] This cable is unterminated.
[3] This cable can be used only with COM1 ports for external timing.
[4] MCT1 cables in this matrix can be used with T1/DS1 interfaces.
[5] Female sync pass-through cables are marked with an F.

Table 4-6. Media Cables for Various Applications *(continued)*

Next, check the status LEDs on the connector or link module, if any. Connectors display a LINK or RLSD LED to indicate a valid cable connection and the detection of the downstream link. Hot-swapped link modules or slots that have been reset will have the red FAIL LED turned on. This does not necessarily indicate that the link module has failed diagnostics. Check the event log or rerun diagnostics on the link module to verify that the hardware is functioning properly.

Port Configuration and Statistics

With some connectivity issues, the media interface will be up and operational, but it may not appear to be passing traffic; connectivity may be intermittent or performance slow. In these cases, looking at the interface state alone may not provide the clues necessary to solve the problem, but looking at the port configuration and statistics may help reveal the cause. Also, performing loopbacks in a few installations may not be simple or straightforward, depending on the physical access to customer-premise CSU/DSU or other communications adapters. In these circumstances, too, port statistics are handy.

Earlier in this chapter, we talked about researching and evaluating recent configuration changes. The interface configuration should be reverified even if no changes have been made; this is because large networks tend to have many network managers, administrators, and operators, each of whom may have made a configuration change without considering its impact on the network.

Each medium has unique attributes that affect the port operation. In this section, we will look at some of the most common configuration attributes for various media. Notice that we do not discuss all attributes, only attributes that directly and significantly affect the operation of the physical layer.

Ethernet Interfaces

The Ethernet BCC object is `ethernet`, and the corresponding MIB object is `wfCSMACDEntry`. MIB instances have the format *slot.connector*, and BCC instances have the format *slot/connector*. Examine the following attributes when verifying an Ethernet interface configuration. The BCC attribute is listed first, followed by the MIB attribute in parentheses. The log entity for Ethernet is `CSMACD`.

bofl (`wfCSMACDBofl`) This attribute enables or disables the breath-of-life packets transmitted from the interface. The line driver uses the BOFL to determine the status of the link; periodic and successful transmission of the BOFL indicates to the line driver that the line is up. Disabling BOFL will disable the ability of the line driver to detect a link failure and will force the link to be up at all times. BOFL is enabled by default.

bofl-timeout (`wfCSMACDBoflTmo`) When bofl is enabled, the bofl-timeout (default: 5 seconds) controls how often a BOFL is transmitted. Any successful packet transmission restarts the bofl timer; sustained transmit traffic may not use any BOFL packets on the line. Shorter timeout periods will help you detect link failure faster, but will increase BOFL overhead on the line. Longer timeout periods are useful with high-throughput lines. BOFL transmission timeouts are reported as warnings in the log, followed by LLC service withdrawal.

```
#   97: 09/12/1999 04:09:16.791  WARNING  SLOT  9  CSMACD            Code:   4
Connector XCVR2 transmitter time-out.

#   98: 09/12/1999 04:09:17.019  INFO     SLOT  9  CSMACD            Code:  14
Connector XCVR2 LLC1 service withdrawn.
```

bofl-tmo-divisor (`wfCSMACDBoflTmoDivisor`) The BOFL timeout divisor can be used with bofl-timeout for rapid line-failure detection; the value is specified in milliseconds. The bofl-timeout value divided by the `bofl-tmo-divisor` equals the periodic BOFL transmission interval. For example, setting bofl-timeout to 2 and bofl-tmo-divisor to 100 will send BOFL packets every 20 ms (2s/100). The `bofl-tmo-divisor` attribute is used with interface redundancy and critical interfaces and is set to 1 by default.

hardware-filter (`wfCSMACDHardwareFilter`) Some high-density Ethernet link modules support hardware-based packet filtering. Using high-speed content addressable memory (CAM), Ethernet ports can filter out packets based on the outbound MAC address, without involving the bridging function on the processor module. This attribute does not affect network connectivity, but should be enabled on modules that have hardware filters and have bridging configured. By default, this attribute is disabled, and enabling it on unsupported hardware will generate a warning event message.

```
#  199: 03/02/1999 03:11:54.573  WARNING  SLOT  3  CSMACD            Code:  20
Connector XCVR1 failed to enable hardware filter.
```

fc-enable (`wfCSMACDFlowControlEnable`) Enables flow control on Gigabit and Fast Ethernet interfaces supported by the FRE4 processor. Flow control is not supported on 10-Mbps Ethernet interfaces or the dual 100-Mbps Ethernet modules. It is enabled by default on all supported interfaces.

fc-pause-time (`wfCSMACDFlowControlPauseTime`) This attribute specifies the flow control pause time, which dictates how long the transmitter pauses before attempting to retransmit. The value specified is multiplied by 512-bit times. For example on a Fast Ethernet interface, setting this attribute to 1950 will pause the transmitter 10 ms (1950 x 512/100E^6). The default is 65535.

pause-zero-enable (`wfCSMACDTxFlowControlPauseZeroEnable`) Enables the ability of a congested receiver to send a pause frame with a pause time of zero. When the receiver is no longer congested, sending a zero pause frame will allow the remote transmitter to resume sending packets. Sending zero pause frames is enabled by default.

Some Ethernet ports such as those on the dual and quad Fast Ethernet link modules, Gigabit link modules, and 10/100 ports on ARNs are capable of auto-negotiating speed and duplex setting. Using the `auto-neg` BCC context on these interfaces does this; the auto-neg context cannot be configured on interfaces that do not support auto-negotiation. An instance of the `wfCSMACDAutoNegEntry` MIB object is created to correspond with the `auto-neg` object.

```
#   89: 05/24/1999 18:49:08.663  WARNING  SLOT  5  CSMACD            Code:  36
Connector XCVR1: PHY device does not support Auto-Negotiation.  100Base-TX assumed.
```

Interfaces that do not support auto-negotiation will report a warning message like the one shown here. The following auto-negotiation attributes apply to interfaces and link modules that support auto-negotiation.

auto-neg/speed-select (`wfCSMACDAutoNegSpeedSelect`) By default, this attribute is set to auto-negotiation, which allows the port to negotiate speed and duplex settings.

```
#  132: 04/03/1999 12:00:11.314  INFO     SLOT  8  CSMACD              Code:  31
Connector XCVR2: Auto-Negotiation is enabled.

#  133: 04/03/1999 12:00:11.435  INFO     SLOT  8  CSMACD              Code:  33
Connector XCVR2: Auto-Negotiation started.

#  134: 04/03/1999 12:00:11.972  INFO     SLOT  8  CSMACD              Code:  34
Connector XCVR2: Auto-Negotiation completed.
Interface setting is 100Base-TX Full Duplex Flow Control.
```

Auto-negotiation can be overridden by setting the speed manually using the `speed-select` attribute when there are interoperability issues with the downstream switch or hub, or when auto-negotiation is not supported. When auto-negotiation is disabled, it must also be disabled on the downstream port.

```
#   98: 02/05/1999 07:14:09.240  INFO     SLOT  2  CSMACD              Code:  32
Connector XCVR1: Auto-Negotiation is disabled.
```

The speed and duplex mode operation values are limited to what the port can support. An error in the speed or duplex setting may prevent the port from initializing fully. Some common messages for unsupported configurations or capabilities are shown here.

```
#   13: 05/04/1999 11:41:28.667  WARNING  SLOT  1  CSMACD              Code:  37
Connector XCVR1: MAC device does not support 10Mb/s operation.  100Base-TX assumed.

#  115: 03/02/1999 09:11:10.817  WARNING  SLOT  3  CSMACD              Code:  43
Connector XCVR2: Unsupported option 100-base T4 (100Base-TX assumed).
```

If the directly attached switch, hub, or router does not support auto-negotiation, a common symptom of auto-negotiation timeout failure is the report in the log.

```
#  174: 02/01/1999 08:22:01.516  WARNING  SLOT  4  CSMACD              Code:  42
Connector XCVR2: Auto-Negotiation didn't negotiate within 2000ms.
```

auto-neg/advertised-capabilities (`wfCSMACDAutoNegAdvertisedCapability`) When the speed selection is configured to auto-negotiate, the capabilities of the port can be restricted to design limitations using the `advertised-capabilities` attribute. For example, a Fast Ethernet port may be restricted to operate in 10 M-half, 10 M-full, or 100 M-half, but not 100 M-full mode. When you suspect a problem related to this attribute, try manually forcing the interface speed and duplex setting on both ends. Attempting to advertise unsupported capabilities will generate a warning and may put the interface in an unsupported setting, thereby disabling it.

```
#  100: 04/07/1999 22:12:31.842  WARNING  SLOT 10  CSMACD                Code:  38
Connector XCVR2: Attempting to advertise an unsupported capability.
ADVERTISED CAPABILITIES = 0x1E0  LOCAL CAPABILITIES = 0x180
```

auto-neg/auto-neg-restart (`wfCSMACDAutoNegRestartAutoConfig`) If an Ethernet port fails auto-negotiation due to mismatched capabilities, you can change the advertised-capabilities attribute and attempt renegotiation using `auto-neg-restart`.

Token Ring Interfaces

The BCC Token Ring object is `token-ring`, and the MIB counterpart is `wfTokenRingEntry`. BCC instances of Token Ring interfaces have the format *slot/connector* (for example, 2/1), and MIB instances have the format *slot.connector* (for example, 2.1). Token Ring attributes are listed here, with the BCC attribute first and the MIB attribute in parentheses. Token Ring events are reported under the `TOKEN_RING` entity in the log.

speed (`wfTokenRingSpeed`) When you encounter a Token Ring problem, check the speed attribute to ensure that the ring is set for the correct operational speed. Older Token Ring interfaces (such as those from link modules from VME routers that can be used in the BN routers) may support only 4-Mbps operation; this includes interfaces on modules such as single token and single sync, single token. When using an earlier module version, setting the speed to 16 Mbps will generate the following event:

```
#  148: 04/03/1999 22:36:35.002  DEBUG   SLOT  8  TOKEN_RING          Code:  46
Connector MAU1 Bad Ring Speed Configured.
```

Most current Token Ring interfaces can operate either at 4 Mbps or 16 Mbps, and they default to the latter. If the ring speed is different than the configured interface speed, the interface will beacon upon insertion.

```
#   33: 09/08/1999 19:13:16.719  WARNING  SLOT 13  TOKEN_RING          Code:   3
Connector MAU2 Ring beaconing on insertion (Ring speed incorrect?).
```

The following MIB attributes currently do not have representative BCC attributes but can be accessed using the TI/BCC (using **get** or **mget**) or using Site Manager.

`wfTokenRingCfgMadr` Token Ring interfaces can be configured with a locally administered MAC address if the `wfTokenRingMadrSelect` attribute is set to manual. By default, `wfTokenRingMadrSelect` is set to prom, which allows the burned-in address to be used.

`wfTokenRingMadrSelect` The `wfTokenRingMadrSelect` attribute can also be set to boxwide, which allows a single MAC address to be used boxwide for all Token Ring interfaces.

Commonly, the default configuration prevents any duplication of MAC addresses by using the burned-in addresses. However, manual setting of the Token Ring address can cause a possible duplication in the MAC addresses if they are not administered properly. When locally administered MAC addresses are used, reverify that the address assigned to the interface is unique.

`wfTokenRingEarlyTokenRelease` Supported only on 16-Mbps rings, early token release (ETR) can be configured to improve ring utilization. If token-related problems arise, disable ETR and observe the results. Not all token issues are resolved by disabling ETR.

FDDI Interfaces

BCC refers to FDDI interfaces using the `fddi` object; `wfFddiEntry` is correspondingly used by the MIB. BCC instances of FDDI interfaces have the format *slot/connector* (for example, 2/1), and MIB instances have the format *slot.connector* (for example, 2.1). This section lists the FDDI attributes, with the BCC attribute first, followed by the MIB attribute in parentheses. Fiber Distributed Data Interface events are logged under the `FDDI` entity.

bofl (*`wfFDDIBofl`*) This attribute enables the use of BOFL frames to detect link failures, similar to the way BOFL frames are used on Ethernet and proprietary HDLC lines. If disabled, LLC service will not be withdrawn when the PHY goes down. The default enabled state is recommended, except in cases where it is appropriate to emulate the presence of the link for network stability.

bofl-timeout (*`wfFDDIBoflTmo`*) The bofl-timeout attribute for FDDI lines is identical in function to its Ethernet counterpart. The default timeout on FDDI interfaces, however, is 3 seconds.

reject-policy (`wfFDDISmtConnectionPolicy`) A FDDI interface can enforce an SMT connection policy, accepting only certain neighbors in certain port-type configurations. By default, the reject-policy attribute is set so that only A-B, B-A, A-M, and B-M connections are considered acceptable. For example, A-A, B-B, M-M connections will be rejected. A nonstandard connection, such as a daisy-chained peer or tree connection, will require that this policy be relaxed.

neighbor-notify-time (*`wfFDDISmtTNotify`*) The neighbor notification protocol used by SMT uses the T_Notify timer, which can be configured using the neighbor-notify-time attribute. The default timer is 22 seconds, and customized ring configurations may use a different T_Notify value—the valid range is 2 to 30 seconds. An incorrect configuration will result in the following message:

```
#   34: 10/17/1999 17:58:13.120  DEBUG    SLOT  8  FDDI                Code:  63
Node 1 illegal T_Notify value requested (range: 2 - 30).
```

token-rotation-time (`wfFDDIMacTReq`) This attribute directly affects the token rotation timer. It specifies the maximum period that can elapse without this local interface having access to the token; all devices negotiate the token rotation time during ring insertion or initialization. Changing this attribute to a lower value restricts token transit delays and extended token holding by some stations. Setting it to a higher value allows higher-traffic stations to transmit longer, but other stations in the ring may be starved of the token. The default value is recommended whenever possible, unless the value has to be customized to fit a specific ring configuration.

hardware-filter (`wfFDDIHardwareFilter`) As with Ethernet interfaces, certain FDDI modules have hardware content addressable memory (CAM), which can be used to filter local frames. Filtering local frames using hardware provides efficiency by allowing the CPU to process only data destined for remote destinations. If hardware filters are present on the module, it is recommended that they be enabled to improve performance.

```
#   30: 02/13/1999 14:18:21.780  WARNING  SLOT  8  FDDI                 Code:  65
Node 1 failed to enable hardware filter.
```

promiscuous (`wfFDDIPromiscuous`) The promiscuous attribute allows address filtering based on the destination MAC address of an inbound frame. When enabled (the default), all frames are received. Disabling this attribute will accept only frames destined for the local station address and functional group address.

Serial Interfaces

Most serial interfaces support both synchronous and asynchronous transmission circuits. They use different industry-standard interfaces, such as V.35 and RS232. They also use different connectors, such as 15-pin and 44-pin connectors terminating in another industry-standard connector type. Regardless of the serial protocol and connector type, serial interfaces are configured using the `wfSyncEntry` MIB object or the `serial` BCC object. Events from the serial driver are reported by the `SYNC` entity in the log.

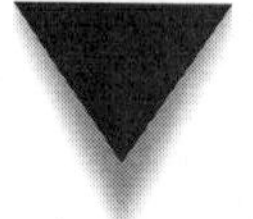

TIP: The term *sync interface* is sometimes used to refer to serial interfaces, even during asynchronous operation. Serial interfaces on older software usually support synchronous operation only, which is why the traditional `wfSyncEntry MIB` object and the term *sync driver/interface* are still used. When referring to a serial interface, always identify synchronous or asynchronous operation wherever appropriate.

bofl (`wfSyncBofl`) The `bofl` attribute enables or disables breath-of-life packets on serial interfaces as required by the data link protocol. This attribute is also enabled on loopback circuits to detect the presence of a loop. Certain data link protocols do not support the use of BOFL frames, in which case BOFL must be turned off. A warning message is printed in the log.

```
#   38: 10/13/1999 09:34:01.500  WARNING  SLOT 10  SYNC                 Code:  43
Connector COM2: can't run BOFL with chosen WAN protocol.
```

bofl-timeout (`wfSyncBoflTmo`) When BOFL is enabled, this attribute controls serial interface behavior in the same way as with Ethernet and FDDI interfaces. The only difference is that serial line drivers monitor the receive BOFL packets as well as the transmit BOFL packets. Timeouts must be matched on both ends; otherwise, one of the following events will occur, followed by LLC service withdrawal:

```
#  130: 06/05/1999 22:22:29.432  WARNING  SLOT 10  SYNC                 Code:   3
Connector COM3 transmitter timeout.

#  129: 09/16/1999 13:09:48.263  WARNING  SLOT 13  SYNC                 Code:   4
Connector COM1 receiver timeout.
```

async-baud-rate (`wfSyncAsyncBaudRate`) The `async-baud-rate` attribute is used on asynchronous connections, such as connection to an asynchronous modem for a dialup service. This attribute controls the baud rate of the data stream, which should be equal to or greater than the data rate at which the modem connects to the remote end. The default baud rate is 9600, and many V.34 modems support a modular speed of 28,800 kbps. With the use of V.42bis or MNP5 compression on advanced modems, data rates with high compression (4:1) ratios can reach as much as 115 kbps. The maximum supported asynchronous baud rate is 115,200 kbps.

media-type (`wfSyncMediaType`) As mentioned earlier, many media types can be connected to a serial interface, including T1/E1 lines, RaiseDTR and V.25bis lines, switched ISDN BRI lines, leased ISDN lines, and Hayes modem lines, to name a few. This attribute, when modified, must be set to the media type to which that interface is attached. If you leave the media-type attribute at the default setting, the router will automatically detect the media type whenever possible.

cable-type (`wfSyncCableType`) Similar to the media-type attribute, the cable-type attribute provides information to the line driver regarding the cable used. This attribute is crucial for the use of RaiseDTR and V.25bis dialup services, where the driver uses the cable interface to operate correctly, according to whether the cable type is RS232, RS422, X.21, or V.35. On normal leased circuits, this attribute can be left at the default setting to allow the router to automatically detect the cable type.

mtu (`wfSyncMtu`) The maximum transmission unit (MTU) is the largest data link frame size that this serial interface can support; the valid range is 3 to 4608 bytes, with 1600 bytes as the default size. This value must match the remote end on leased point-to-point lines and must be set to the lowest supported transmission unit on a multidevice link. The MTU directly affects higher-layer PDU sizes, which are calculated by subtracting the packet header overhead. The maximum value of 4608 may not be readily supported on Ethernet-based AN/ANH and ARN routers because of the global packet buffer size, in which case the buffer size must be increased first. Invalid MTU size configurations will result in a log message.

```
#   23: 03/02/1999 12:47:50.661  WARNING  SLOT  3  SYNC                 Code:  50
Connector COM4: Configured MTU of 5000 is too large, use 4608 or smaller.
```

promiscuous (`wfSyncPromiscuous`) Unlike the promiscuous attribute on FDDI interfaces, the serial promiscuous attribute is set to disabled by default. This is because most serial interfaces are deployed on point-to-point links, where acceptable data traffic always originates at the remote end. Data traffic coming from any other device can be considered inapplicable or bogus. This attribute must be set to enabled when performing loopback testing on PPP interfaces, so that PPP can receive and recognize its own frames. It should also be enabled when verifying the HDLC address of the remote device.

clock-source (`wfSyncClkSource`) Serial interfaces can be clocked either internally or externally. Most WAN applications require that the serial interface derive the clock from the network, which is accomplished using the received data signal. Hence, the clock-source attribute defaults to the external timing source. Occasionally, you may need the serial interface to use its own internal clock for data transmission. This is common practice when internetworking two serial devices back to back in a laboratory or a production shelf using a crossover cable. If another BayRS router is being used at the remote end, both ends must be set to internal clocking and to the same internal clock speed.

internal-clock-speed (`wfSyncClkSpeed`) The internal clock speed can range from 1200 bps to 5 Mbps, with the practical speed dependent on the media and interface type. The default value is 64 kbps. For example, the RS232 standard does not support high speeds such as 5 Mbps, but smaller (unsupported) speeds may be achieved using shorter, well-shielded cables and a matching clock at the remote end. This attribute has meaning only when the clock-source attribute is set to internal.

rts-enable (`wfSyncRtsEnable`) Some serial communication devices and links use signaling prior to actual data transmission—a request-to-send (RTS) signal is first sent, which is responded to by a clear-to-send (CTS) signal when the receiver is ready. Most modern serial devices do not require the use of RTS signals in full-duplex operation, so RTS is disabled by default. When connecting to devices using CTS/RTS and appropriate cables, this attribute must be enabled.

service (`wfSyncService`) This attribute controls the type of service that this interface provides. The default service is LLC1, which prefixes the HDLC address and control fields to the outbound frame and is normally used with HDLC-derived protocols. The transparent service is used with raw HDLC data. On X.25 lines, the service is set to LLC2, which provides LAPB services on the interface.

```
#   71: 03/02/1999 12:27:06.373  WARNING  SLOT  3  SYNC              Code:  44
Connector COM1: service must be LLC2 for chosen WAN protocol
```

minimum-frame-space (`wfSyncMinFrameSpace`) The flag (0x7E) character is used to synchronize and delimit HDLC frames and indicate the beginning of frames. By default, a minimum of 1 flag frame space is used between adjacent frames. This can be increased up to a maximum of 32 on lines where a single flag is not enough to delimit frames and indicate the beginning of the next frame.

passthru-local (`wfSyncPassThruLocalMadr`) The `passthru-local` and `passthru-remote` attributes are applicable in a synchronous pass-through environment. Frames accepted from this interface are encapsulated in a bridge frame, where the source MAC is set to emulate the address set in the passthru-local attribute. This attribute must match the passthru-remote value on the remote sync passthru interface.

passthru-remote (`wfSyncPassThruRemoteMadr`) This attribute indicates the emulated MAC address of the remote end; frames bridged to the remote device have the destination MAC address set to this value. This allows the remote receiver to accept the frame by comparing the destination MAC address to the frame's passthru-local address.

crc-size (`wfSyncCrcSize`) The cyclic redundancy check (CRC) is used to validate the integrity of received frames. Each transmitter computes the CRC, which is verified by the receiver upon receipt of the frame. By default, a 16-bit CRC is calculated as specified by the ITU-T. The CRC can be set to enhanced 32-bit, but this value must be matched on both ends. Mismatched CRC sizes will result in CRC errors on the line.

sync-line-coding (`wfSyncLineCoding`) Line coding is critical to proper operation. Coding types must be matched on all devices on the transmission circuit, unless intermediate coding conversion is performed. On supported serial interfaces, the sync-line-coding attribute can be set to nonreturn to zero (NRZ), nonreturn to zero inverted (NRZI), or NRZI mark. Mismatched line coding will result in receive errors.

external-clock-speed (`wfSyncExternalClkSpeed`) This attribute is set to reflect the data rate supplied by the external clocking device, usually the CSU/DSU. Usually, the external clock speed is not required for normal router operation. However, it is used by services such as priority queuing to determine queue depth, which ultimately affects data throughput and processing. This attribute does not have any meaning when internal clocking is configured.

bchannel-rate-adaption (`wfSyncBChRateAdaption`) When a serial line is used as an ISDN B-channel over an ISDN interface, this attribute specifies the rate adaption of the B-channel: 64 kbps or 128 kbps. This detail must match that of the service being provided by the ISDN carrier and will affect the data rate. Specifying 128 kbps on a 64-kbps (one B-channel) circuit will oversubscribe the line and cause data to be dropped by the ISDN device.

wan-type (`wfSyncWanType`) Another important attribute, `wan-type`, specifies whether this interface is operating as a synchronous, asynchronous, or polled-asynchronous interface. Incorrectly configured wan-type attributes will cause line errors, and circuit initialization will fail.

Bisync Interfaces

Bisynchronous communication services are provided to transport legacy BSC traffic with an IP backbone. Bisync over TCP/IP (BOT) is an encapsulation mechanism in which low-speed, high-cost, point-to-point links are terminated using a local bisync interface, and bisync data is transported over TCP/IP to a remote peer. BayRS supports the latest and concurrent version of BSC, called BSC3, which is polled and unbalanced.

Recall that when discussing media state attributes, we used the `wfBisyncEntry` MIB object. Bisync media interfaces, however, are not unique—traditional serial interfaces, when stripped of the serial driver and interfaced with the bisync driver, can handle BSC3 traffic. Hence, many of the common serial attributes are also used in bisync. Since bisync is a legacy protocol, currently configuration attributes are accessible only using Site Manager or directly using the MIB. Configuration attributes for verification and their corresponding MIB attributes are discussed here. The BCC-like attributes are represented to conform to the other sections of this chapter and may be equivalent if bisync support is added to BCC. Bisync events are logged under the `BISYNC` entity.

mtu (`wfBisyncMtu`) The default maximum transmission unit size is 1580 octets. The valid range is 1 through 4568. Higher-speed BSC3 lines can be configured to support larger data sizes for efficiency. Bisync is still a character-based transmission protocol. The MTU refers to the frame size delivered to and from the upper layers.

```
#  188: 12/23/1999 02:16:59.626  WARNING  SLOT  2  BISYNC              Code:  68
Connector COM2: Configured MTU of 5200 is too large, use 4568 or smaller.
```

media-type (`wfBisyncMediaType`) The default bisync media type assumes that a leased circuit is directly attached. If a RaiseDTR or V.25bis connection is used, the media type must be set appropriately. See also the cable-type attribute that follows.

cable-type (`wfBisyncCableType`) The bisync interface supports four cable types: RS232, RS422, X.21, and V.35, with RS232 being the default. As with the serial interface, RaiseDTR and V.25bis media require that the appropriate cable type be configured.

clock-source (`wfBisyncClkSource`) The bisync interface can be clocked either internally or externally. The default clock source is derived from the external device.

internal-clock-speed (`wfBisyncClkSpeed`) This attribute affects the data transmission rate when internal clocking is used. When set to internal clocking, the data rate can range from 1200 bps to 19.2 kbps (the default is 9600 bps). Invalid clock-source or clock-speed settings may generate receive or transmit errors on the bisync interface.

external-clock-speed (`wfBisyncExternalClkSpeed`) If an external clocking source (such as a DCE) is used, this attribute must be set close to the line speed for proper operation. The default value is 9600 bps.

character-mode (`wfBisyncCharMode`) The bisync interface supports both character modes used by BSC devices: EBCDIC and ASCII. Many strict BSC devices use EBCDIC characters for communication, but some BSC3 variants use the ASCII character set. When using a BSC3 variant, consult the protocol documentation to ensure that the correct character mode is used.

rts-enable (`wfBisyncRtsEnable`) This attribute has the same behavior as the corresponding serial attribute. It is most commonly used with bisync transmission devices.

High-Speed Serial Interfaces

High-speed serial interfaces (HSSI) are configured and managed using the `wfHssiEntry` MIB object. BCC represents HSSI interfaces using the `hssi` object, in the format *slot/connector* (for example, 6/1), and the MIB uses the format *slot.connector* (for example, 6.1). When you encounter a HSSI problem, check the following configuration attributes. The HSSI driver uses the entity `HSSI` to report events in the log.

bofl (`wfHssiBofl`) As we have seen before, this attribute enables or disables the transmission of BOFL frames. By default, it is enabled and should be used on Bay standard (proprietary HDLC) lines. The HSSI line driver monitors both transmit and receive BOFL packets on the line.

bofl-timeout (`wfHssiBoflTmo`) This attribute functions the same as the bofl-timeout attribute discussed in the "Serial Interfaces" section. On HSSI interfaces, the default BOFL timeout period is 1 second.

```
#  144: 08/03/1999 08:28:39.814  WARNING  SLOT  8  HSSI              Code:  2
Connector HSSI1 transmitter time-out.

#   76: 08/23/1999 09:36:11.370  WARNING  SLOT  8  HSSI              Code:  3
Connector HSSI1 receiver time-out.
```

mtu (`wfHssiMtu`) This attribute is identical in function and value range to the serial mtu attribute. The default value on HSSI interfaces is 4608 bytes.

media (`wfHssiTransmissionInterface`) HSSI interfaces can be used with two transmission interfaces: DS1 and DS3. This attribute value should be set according to the transmission circuit provided.

internal-clock-test (`wfHssiInternalClkTestMode`) By default, clocking is derived from the network. For loopback testing and back-to-back operation, internal clocking must be enabled.

external-clock-speed (`wfHssiExternalClkSpeed`) Defaulting to DS3 speed, the `external-clock-speed` attribute is necessary for proper operation of priority queuing. Normal data service is not affected by the external-clock-speed value when external clocking is used.

crc-size (`wfHssiCrcSize`) This attribute functions the same as the serial `crc-size` attribute.

Multichannel T1/E1 Interfaces

The MCT1 and MCE1 interfaces support channelized T1 and E1 services, along with ISDN PRI service. These interfaces provide a broad range of functionality, which in turn requires several attributes to be configured correctly for the interface to be provisioned properly.

Many attributes used by the MCT1 and MCE1 media interfaces are the same; this is because they share the DS1E1 driver and use the wfDs1E1*Entry objects. (The DS1E1 driver uses several MIB objects prefixed with wfDs1E1. Most configuration attributes are stored in the wfDs1E1PortEntry MIB. Some other configuration and action attributes are stored in wfDs1E1ConfigEntry and wfDs1E1ActionEntry, respectively.)

The BCC object for MCT1 is mct1, and the object for MCE1 is mce1. Both objects use the logical-line (wfLogicalLineEntry) child object for logical services and the pri object for ISDN PRI services. All attributes for the mct1 and mce1 objects are discussed together in this section, with attributes applicable to only one object detailed wherever appropriate. The log entity for the MCT1/MCE1 cards is DS1E1.

signal-level (wfDs1E1PortSignalLevel) This attribute is applicable for MCT1 interfaces only. It specifies the T1 transmit power level in decibels (dB). It is either specified by the carrier or configured according to the cable length. The default signal-level value, 0.0 dB, is appropriate for cable runs of 0 to 133 feet in length. The signal-level attribute can be set to a maximum of DSX1+1.5 dB, to allow cable runs of up to 655 feet. Having either a large or a very low line buildout can cause signal problems.

mtu (wfDs1E1PortMtu) This attribute is identical in function to the mtu attribute on serial and HSSI interfaces.

send-prm-cr-address-bit (wfDs1E1PortSendPrmCrAddressBit) The attribute applies only to QMCT1 modules. Performance messages sent from this interface can originate from either the customer installation (CI) or carrier (CR). Customer-premise installations must set this attribute to prm-ci (default); in a provider environment, this attribute must be set to prm-carrier. Incorrect settings may cause performance messages to be dropped at the remote end.

accept-prm-cr-address-bit (wfDs1E1PortAcceptPrmCrAddressBit) This attribute controls the acceptance of performance reports from the network. By default, the QMCT1 accepts performance messages only from the remote CI device.

primary-clock-source (wfDs1E1ModulePrimaryClock) This attribute is configured using the wfDs1E1ModuleEntry MIB object. Timing on T1/DS1 circuits can be obtained externally, internally, or using loop timing from any of the configured interfaces. By default, the module uses loop timing from the port. If the primary clock source fails, the router uses the secondary clock source.

```
#   32: 12/19/1999 21:56:35.398  WARNING  SLOT  8  DS1E1            Code:  76
Connector COM1, Primary clock lost.

#  114: 02/25/1999 23:10:49.488  INFO     SLOT  8  DS1E1            Code:  35
Connector COM1 Loop timing source lost.

#   49: 09/12/1999 23:45:32.287  INFO     SLOT  9  DS1E1            Code:  37
External Clock Source lost.
```

secondary-clock-source (`wfDs1E1ModuleSecondaryClock`) Like the attribute `primary-clock-source`, this attribute is configured using the `wfDs1E1ModuleEntry` object. If the primary clock source is not operational, the secondary clock source is used, which defaults to internal. If you have an accurate and reliable external timing source, you can use it as the secondary clock. If both the primary and secondary clock sources fail, the module will use the internal clock source.

```
#  103: 05/08/1999 09:09:56.231  WARNING  SLOT  9  DS1E1             Code:  78
Connector COM1, Secondary clock lost.

#  105: 05/08/1999 09:09:56.383  INFO     SLOT  9  DS1E1             Code:  23
Primary and Sec clocks unoperational - Switching to Internal Clock Source.
```

facilities-data-link (`wfDs1E1ConfigFdl`) The facilities data link (FDL) is used over ESF links to send performance and maintenance messages, when available. Two common FDL types are available: the ANSI T1.403 (ansi403) and AT&T 54016 (att54016) based on the respective specifications. ANSI T1.403 is used as the default, and FDL can be disabled when it is not available. If you disable FDL, then you will not be able to send out-of-band loop codes or maintain performance and error statistics.

fdl-target-hdlc-address (`wfDs1E1PortFdlTargetHdlcAddress`) If FDL is configured, then the target HDLC address must be set to AZ or BY. This specifies the address to which the near-end FDL will respond for FDL messages received from the far end.

line-coding (`wfDs1E1ConfigLineCoding`) Line coding and framing, as mentioned before, are crucial to the operation of the T1/DS1 circuit. Valid values include jbzs, b8zs, hdb3, zbtsi, and ami. The line coding must be matched on all devices in the transmission circuit. Certain loopback and BERT tests can help determine if all devices in the circuit are provisioned correctly using the same line coding. The DS1E1 driver can detect certain mismatched line codings.

```
#  149: 09/28/1999 24:09:00.251  INFO     SLOT  7  DS1E1             Code:  24
Connector COM2 B8ZS code received on port configured for AMI.
```

framing (`wfDs1E1ConfigLineType`) The framing type on the circuit also must match all other devices on the transmission circuit. Depending on the module type, several framing types, such as unframed T1, ESF, SF/D4, E1, and E1/CRC, can be chosen. Check with your provider to determine whether the frame type is set correctly.

international-bit (`wfDs1E1PortInternationalBit`) Supported on MCE1 interfaces only, the international-bit attribute, if enabled, sets the bit value in the E1 frame. Consult with your service provider to see if you need to enable the international bit on the E1 frames.

Fractional T1/E1 Interfaces

The fractional T1 and fractional E1 interfaces are currently available for the AN/ANH and ARN platforms. As mentioned earlier, the line service is provided by the serial driver using the `wfSyncEntry` MIB object, and the T1 digital service is provided by the DS1E1 driver using the `wfDs1E1*Entry` MIB object. The DS1E1 driver also uses the `wfDs1E1FracEntry` MIB object for the fractional interfaces. These drivers are managed using the `ft1` and `fe1` BCC objects, respectively. Event log entries for the FT1 and FE1 interfaces can be viewed under the `SYNC` and `DS1E1` entities.

The `logical-line` object (`wfLogicalLineEntry`) is also used with FT1/FE1 interfaces, as with the MCT1/MCE1 interfaces. Many attributes for the multichannel and fractional interfaces are the same, although the supported values for the fractional interfaces are limited. Attributes discussed in the previous section for the MCT1/MCE1 interfaces equally apply to FT1/FE1 ports, except the performance report attributes and the international bit attribute.

ISDN Basic and Primary Rate Interfaces

ISDN BRI interfaces are available on the AN/ANH, ARN, and ASN platforms. The BN family supports only PRI interfaces using MCT1/MCE1 modules. Both BRI and PRI interfaces share the `isdn-switch` object on a per slot basis; the corresponding MIB object is `wfIsdnSwitchCfgEntry`.

The `isdn-switch` object is instantiated using the number of the slot where it resides. Common attributes for this object are discussed here. Log events specific to the interaction with the ISDN switch can be seen using the `ISDN` and `SWSERV` entities.

switch-type (*`wfIsdnSwitchCfgType`*) The interaction with various ISDN switches depends on the supported national standard or the vendor's standard. This information is supplied by your ISDN provider, and this attribute has no default switch type. Switch-type values are prefixed with bri (for example, bridms100) or pri (for example, pri5ess), depending on the interface type. Make sure that you select the value based on both the switch and the interface type.

```
#  179: 11/10/1999 02:43:18.441  WARNING  SLOT  1  SWSERV            Code:  97
ISDN Unsupported switch type configured on slot 1.
```

incoming-filter (*`wfIsdnSwitchIncomingFilter`*) This attribute is responsible for screening inbound calls on ISDN interfaces on this slot. If incoming calls are being dropped, disable this attribute (the default). If outbound calls are being rejected, then the remote router may be performing call screening.

```
#  138: 02/21/1999 19:54:29.680  INFO    SLOT  2  SWSERV            Code:  91
Incoming ISDN call with screened Calling Party number 2249182 being dropped on DSL 1.
```

sending-complete (*wfIsdnSwitchSendingComplete*) If the ISDN network requires that complete information elements (IE) be sent in a setup message, this attribute must be enabled. By default, the sending of complete IEs is disabled. If call setup fails as a result of invalid IE contents, a partial or full resolution may be achieved by sending complete IEs. Certain switch types do not support the receipt of complete IEs, so consult your provider to ensure that this attribute is configured correctly.

global-rate-adaption (*wfIsdnSwitchGlobalRateAdaption*) Normally, calls are established with the default rate adaption of 64 Kbps. When one of the terminal endpoints or a transit link performs rate adaption at 56 Kbps (LSB), then you may need to set the `global-rate-adaption` attribute. Explicitly setting this attribute will force the slot to request the specified rate adaption when establishing calls.

x25-over-isdn (*wfIsdnSwitchX25D*) If your provider supports X.25 virtual calls over the D-channel, then this attribute must be enabled to send X.25 traffic.

ISDN BRI interfaces are configured using the `bri` object in BCC, and the configuration detail is stored in the `wfIsdnBriInterfaceEntry` MIB object. BRI object instances have the format *slot/connector* in BCC and the format *slot.connector* in the MIB. BRI-specific initialization messages can be viewed under the `ISDN_BRI` entity.

mode (wfIsdnBriMode) The mode attribute can be set to dialup, leased, or floatb; dialup is the default. The value of the leased mode also can be set in certain countries, such as Germany and Japan, where bearer channels are preestablished in a point-to-point leased configuration and do not require any call setup. Applicable to AN/ANH and ARN routers, the floating B-channel automatically selects an available serial communications controller when the COM1 port is in use. This configuration is further described in Chapter 5.

mtu (*wfIsdnBriMtu*) This attribute refers to the largest data unit that can be received on this interface and defaults to 400 bytes. Valid values range from 3 to 1024. Normally, this attribute does not need to be changed unless your provider explicitly instructs you to do so.

t3-timer (*wfIsdnBriTimer3Tmo*) The router uses the default 10-second T3 activation timer when it attempts to activate the S/T interface. During this time, the router will attempt to detect a signal on the line to activate the interface. When the timer expires, the interface is deactivated. Configure this timer as specified by your ISDN provider. If no value is provided, reset the timer to the default, 10 seconds.

t4-timer (*wfIsdnBriTimer4Tmo*) The default value of 750 milliseconds for the T4 debounce timer allows the BRI interface to automatically alleviate spurious problems at ISDN layers 1 and 2. Temporary problems are not reported to the upper ISDN layer until a problem persists for longer than T4 milliseconds. This attribute also does not require any changes unless specified by the provider.

line-type (*wfIsdnBriLineType*) The line-type attribute can be set as either ptp (point to point) or mtp (multipoint). Use multipoint if the interface is sharing the line with other devices on a passive bus, which eventually is connected to the switch. When there are multiple devices on the line, this attribute must be set to multipoint. If the router interface is the only device connected to the line, set the attribute to ptp.

x25-isdn-dchannel (*wfIsdnBriX25D*) If you are using X.25 virtual connections over the ISDN D-channel, then the `x25-isdn-dchannel` attribute must be enabled.

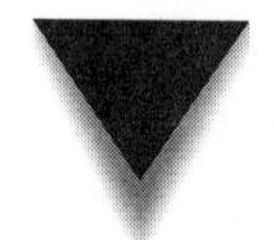

NOTE: ISDN PRI interfaces are configured under their respective `mct1` and `mce1` objects; the `pri` child object indicates that the MCT1/MCE1 port is to be used for the primary rate application. Using Site Manager, initial configuration of a multichannel T1/E1 interface requires that the port application mode be set to PRI for ISDN use.

ATM Interfaces

BayRS currently supports OC3 interfaces using SONET and SDH framing, and only asynchronous transfer mode (ATM) circuits are currently transported over SONET/SDH interfaces. ATM is also supported on DS3/E3 circuits. It is configured using the `atm` object in BCC, which uses several MIB objects, including `wfAtmizerCfgEntry`, `wfAtmInterfaceConfEntry`, `wfAtmAlcFrmConfEntry`, and `wfAtmizerDrvCfgEntry`. Certain MIB objects used by DS3/E3 and OC3 interfaces are unique. ATM BCC object instances have the format *slot/connector*, and MIB instances use both the physical line number and the *slot.connector* format.

Physical layer attributes pertaining to ATM interfaces are described here. Higher-layer issues, such as signaling, SSCOP, and ILMI issues, are discussed in Chapter 5. Log events for ATM interfaces can be viewed under the `ATMINTF` and `ATM` entities.

clipping (*wfAtmizerTxPerVcClipEnable*) An attribute of the `wfAtmizerCfgEntry` object, the per-VC clipping attribute enables the driver to clip packets at the VC level. Normally, the cell transmit driver discards cells prior to transmission if the queues are full. Loss of one cell indicates that a whole frame is lost, in which case an oversubscribed VC can be clipped at the frame level to improve efficiency.

mtu (*wfAtmizerDrvCfgMtu*) The Atmizer driver supports a maximum transmission unit of 9188 bytes, with a default of 4608. Applications such as LANE that require a larger frame size should be accommodated by setting the mtu attribute appropriately.

dp-notify (*wfAtmizerDrvCfgDpNotify*) This attribute controls whether the data path layer receives notifications of failures in lower-level functions. Disabling this attribute will keep all circuits and protocol interfaces in an operational state, even if the physical layer is unavailable. It is recommended that this attribute be left enabled by default.

dp-notify-timeout (*wfAtmizerDrvCfgDpNotifyTimeout*) After the specified number of seconds, the data path is notified that the medium is unavailable. This attribute is applicable only if dp-notify is enabled and the default timeout is 1 second.

vc-inactivity-control (*wfAtmizerDrvCfgVcInactEnable*) Switched virtual circuits, when inactive for an extended period of time, are automatically brought down if this attribute is enabled. Disabling this attribute minimizes call setup time but may consume resources for numerous destinations.

vc-inactivity-timeout (*wfAtmizerDrvCfgVcInactTimeout*) When vc-inactivity-control is enabled, the timeout value (the default is 1200 seconds) specifies the time period during which inactivity should be measured. This attribute must be set appropriately networkwide.

```
#   67: 03/10/1999 23:35:38.537  INFO     SLOT  3  ATMINTF               Code: 178
Port 1: inactivity timer expired for VC specified by VPI=0, VCI=90.
```

framing-mode (*wfAtmizerDrvCfgFramingMode*) The framing-mode value for an ATM interface depends on the medium. If the interface is connected to a public network, the framing detail is supplied by your provider. On private networks, it depends on the network provision. Some interfaces or lines that do not support a particular framing mode will automatically fall back to a supported one.

```
#   59: 03/10/1999 06:35:54.977  WARNING  SLOT  9  ATMINTF               Code: 184
Port 1: Falling back from DS3 Cbit mode to DS3 M23 mode.
```

clock-signal-source (*wfAtmizerDrvCfgClkSource*) The clocking source can be derived either internally or externally. Some networks have a stable, high-level stratum clock that can be used throughout the network. By default, the router uses internal clocking.

scrambling (*wfAtmizerDrvCfgDs3Scrambling*) Applicable to DS3 circuits, this attribute enables or disables scrambling on the line. DS3 scrambling helps randomize the data pattern in the cell payload so that cell synchronization is maintained. The setting of this attribute depends on the line provision; if your provider requires the use of scrambling, leave this attribute enabled with its default setting.

ds3-line-build-out (*wfAtmizerDrvCfgDs3LineBuildOut*) This attribute is applicable only to DS3 circuits. Lines less than 255 feet in length must be set to short, and longer lines should be set to long. By default, DS3 interfaces condition the signal for short distances.

atm-interface/use-hardware-mac (*wfAtmInterfaceUseHwMacAddr*) An attribute of `wfAtmInterfaceConfEntry`, the MAC address burned in the link module PROM is used by default. This can be disabled and a locally administered MAC address can be stored using the mac-override attribute.

atm-interface/mac-override (*wfAtmInterfaceHwMacOverride*) If the preceding attribute is disabled, the mac-override attribute specifies the MAC address used with the interface. This becomes part of the end-system identifier (ESI) when registering with an ATM switch.

V.34 Modem Interfaces

V.34 modems are available on the AN/ANH and ARN platforms for dialup services. The availability of a dialup interface eliminates the need for an external modem connected to a serial port. The `wfModemEntry` and `wfSyncEntry` MIB objects are used for configuration details, which are represented using the `modem` object in BCC. Event messages are logged under the `MODEM` entity.

The `modem` object is also used to configure attributes for externally attached modems (that is, those using a serial interface). This section describes common modem configuration attributes that require verification. Many other modem attributes are the same as those for serial interfaces and are used similarly.

async-baud-rate (*wfSyncAsyncBaudRate*) If the modem is used for asynchronous operation, the baud rate should be set appropriately for maximum throughput. It must be set to a value that is supported by the modem on the remote end.

modem-type (*wfModemType*) This attribute is normally crucial if an external modem is attached. Selectable values depend on the software version; BayRS will automatically send an appropriate initialization string based on the modem type. If you do not know your modem type, or if you want to set a custom initialization string, set the modem-type attribute to custom.

factory-defaults-load (*wfModemCfgFactoryDefaults*) When the modem-type attribute is set appropriately for the modem you are using, the router will automatically initialize the modem to the factory defaults using a factory default initialization string. Internal V.34 modems are also initialized to factory defaults.

```
#  141: 05/08/1999 06:33:08.947  DEBUG    SLOT  1  MODEM             Code:  17
Sending default string to modem on Slot1 COM3.

#  141: 05/08/1999 06:33:08.959  DEBUG    SLOT  1  MODEM             Code:  21
Command sent to modem on Slot1 COM3: AT&FE0Q0V1
```

init-string (*wfModemCfgInitString*) If you set the modem-type attribute to custom or require the use of a specific initialization string, you must set this attribute correctly. Every time the internal or external modem is initialized, this init-string value is sent to it.

```
#   62: 10/21/1999 12:02:41.980  DEBUG    SLOT  8  MODEM             Code:  18
Sending initialization string to modem on Slot8 COM1.

#   62: 10/21/1999 12:02:41.992  DEBUG    SLOT  8  MODEM             Code:  21
Command sent to modem on Slot8 COM1: AT&M2&Q2&D0&S1&R0S0=0M1L2T
```

country-code (*wfModemCfgCountry*) This attribute specifies the country where the modem is installed to operate. In certain public-switched networks, the router may need to know this information because it is not read directly from the modem. Internal V.34 modems must be configured initially with the country code.

phone-number (*wfModemPhoneNumber*) Applicable to RaiseDTR connections using an internal modem, the phone-number attribute specifies the outgoing phone number stored in the internal register. This number is not necessary for V.25bis applications or externally attached RaiseDTR devices. Upon initialization, the router transmits the configured phone-number attribute to the modem for storage. The modem does not dial the number until requested to do so.

```
#  128: 08/07/1999 17:56:51.454  DEBUG    SLOT  1  MODEM              Code:  19
Sending phone number to modem on Slot1 COM3.
```

unit-reset (*wfModemUnitReset*) This attribute is used to reset the internal modem or to send a reset code to an external modem. Resetting a modem will cause the router to send the initialization string; hence resetting it is essential if you change the initialization string.

```
#  109: 05/04/1999 02:05:16.459  DEBUG    SLOT  1  MODEM              Code:  15
Resetting modem on Slot1 COM3.
```

Loopbacks and Testing

A loopback is a common procedure used to validate the function of the media connector or the cable, and in most cases both. The process can be simply thought of as looping the transmit signal pin back to the receive signal pin at a designated point in the transmission link; multiple pins are used to emulate the presence of a cable link and a DCE.

LAN media interfaces such as Token Ring and Ethernet can be connected to loopback terminators, which loop back the signal so that the interface hardware can be verified. Looping port A to port B can test FDDI interfaces. If the Link light comes on in the case of Ethernet, or the Insert light in the case of Token Ring, or the Tx and Rx lights in the case of FDDI, then you can be assured that the interface is operating normally.

Wide-area interfaces and links can be tested using more extensive loopback methods. This allows testing of the media interface, the cable, the CSU/DSU, and the line. Figure 4-8 illustrates several different points in the link where a loop can be requested.

Most CSU/DSUs vary in configuration, but they can perform the same functions, including the accepting and enabling of loopbacks. If you are using an external CSU/DSU, consult the user manual prior to beginning a loopback test for steps to enable and disable

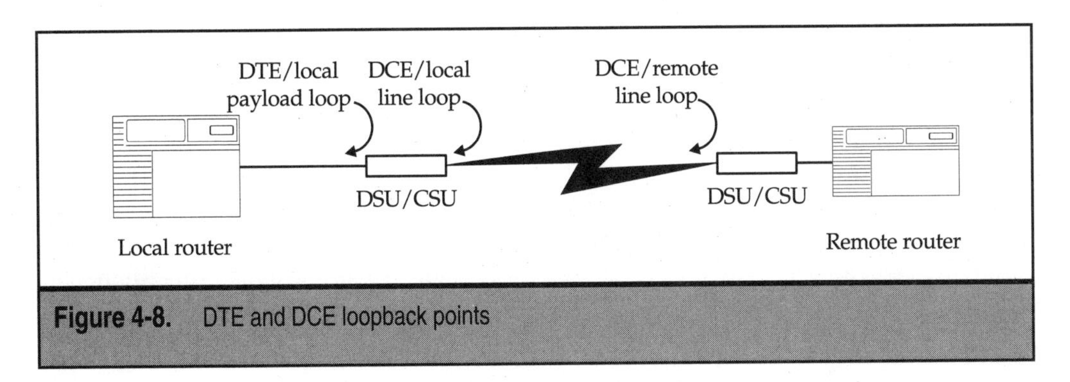

Figure 4-8. DTE and DCE loopback points

the type of loopback you want to perform. The following paragraphs assume that you know the steps to enable and disable loopbacks on the CSU/DSU.

CAUTION: When performing loopback testing on a port, disable bridging on the circuit if it is configured. Performing loopback tests on bridging ports will adversely affect network operation in a production environment.

We earlier mentioned loopback terminators for LAN interfaces. These can also be made to order or purchased for WAN interfaces so that the only target component for the loopback test is the media interface. For example, running a loopback test with a HSSI loopback terminator will help determine if the port is working properly. Special breakout boxes can also be used to manually interconnect pinouts to emulate a transmission link. When connecting the loopback terminators, the router interface must receive a valid clock signal for it to transmit data successfully.

During normal operation, WAN media chipsets automatically clock themselves based on the receive signal clock. Clocking to most DTE interfaces is provided by the CPE (customer premise equipment, usually the CSU/DSU), where the CSU/DSU is normally termed the *master* and the router the *slave*. The transmission speed is based on the receive signal clock, which will be absent when a downstream clocking device is unavailable. This will be the case when a loopback plug is put in place—the router's interface must be set to internal clocking so that the clocking is not lost. The absence of clocking will result in signal loss, which is undesirable in a loopback test scenario. In the example below, S42 is set to internal clocking before the start of a loopback test.

```
box->serial/4/2
mx2(13)# clock-source internal
```

Note that certain interfaces such as the MCT1, MCE1, FT1, and FE1 have built-in CSU/DSUs and usually require a framed carrier for proper testing of the CSU/DSU. Table 4-7 lists the clocking attributes that can be used to set internal clocking on various media interfaces.

NOTE: The secondary clock source for the DS1E1 driver is by default the internal clock, and the primary source is external. If the external (primary) clock source is unavailable, the driver automatically uses the secondary internal clock. You do not need to change the primary clock, but you can do so if necessary for immediate testing.

Some media interfaces do not support configurable clocking: for instance, V.34 modems (which use self-clocking with negotiated data rate) and ISDN BRI (whose data rates are fixed). ISDN PRI interfaces using MCT1/MCE1 modules can be tested using internal clocking, but only to test the digital carrier services and framing. (The MCE1-I module is not approved for use with ISDN PRI lines, although it can be used with them. The next-generation MCE1-II modules are approved for use with PRI lines.) Higher-layer testing for ISDN BRI and PRI interfaces requires the use of an external ISDN tester.

Media	Clocking Attribute (Attribute Number)	Internal Clocking Value
Serial	wfSyncEntry.wfSyncClkSource (13)	1
Bisync	wfBisyncEntry.wfBisyncClkSource (7)	1
T1/E1	wfT1Entry.wfT1ClockMode (10) wfE1Entry.wfE1ClockMode (8)	1
HSSI	wfHssiEntry.wfHssiInternalClkTestMode (16)	1
MCT1/MCE1	wfDs1E1ModuleEntry.wfDs1E1ModulePrimaryClock (4)	4
SONET/SDH	wfAtmizerDrvCfgEntry.wfAtmizerDrvCfgClkSource (18)	1
FT1/FE1	See serial and MCT1/MCE1[1]	
56/64K DSU/CSU	wfDsuCsuIfEntry.wfDsuCsuTxClkSelect (7)	2

[1] As mentioned earlier, fractional T1/E1 interfaces use the sync driver for the fractional line service and the DS1E1 driver for the digital service. When running a local DTE loopback, the sync driver must be set to internal clocking for the line to loop properly.

Table 4-7. Internal Clocking Attributes

When a port is put into loopback mode either using a loopback plug or the local or remote CSU/DSU, it will transition into either an Up state or Loopback state. The Loopback state is reported by those interfaces than can discern a loop, and the Up state is reported by all other interfaces. A loopback timeout points to a problem, which you can troubleshoot as follows:

- If you are using a loopback plug, use another loopback plug and test again. If the interface does not go into a Loopback or Up state, the port is defective.
- If you are using a DTE local loop, verify that the cable being used is correct. Replace the cable and try again.
- If you replace the cable and the interface still does not come up, reset the CSU/DSU or replace it and test again.

These steps will help you isolate a physical layer problem to the port, cable, or CSU/DSU.

Loopback Capable Interfaces

High-speed serial interfaces, fractional T1/E1 interfaces, MCT1/MCE1 interfaces, and ISDN basic rate interfaces all provide the ability to initiate and accept loopback requests. On dual and quad multichannel T1/E1 modules, individual channels can be looped up and down without affecting the remaining DS0s. These loopback-capable interfaces must be configured manually to send or accept specific loop codes to upstream devices.

Loopbacks can be enabled and disabled using Configuration Manager in dynamic mode, but they are commonly set using the TI or BCC. This section discusses loopbacks with the direct use of MIB/BCC objects.

HSSI Loopbacks

HSSI interfaces are available for the BN and ASN platforms. Figure 4-9 shows different points on the HSSI line where a loop can be requested. On HSSI interfaces, loopbacks can be enabled using the `wfHssiLoopback` MIB attribute or the `loopback` BCC attribute as shown here.

```
box->hssi/2/1
fp3# loopback ?
Current value: none
Legal values:  none,localdigital,localline,remoteline
Default value: none
box->hssi/2/1
fp3# loopback localdigital
```

Using the `wfHssiEntry.wfHssiLoopback` attribute directly, enter the following:

```
box->hssi/2/1
fp3# mset wfHssiEntry.wfHssiLoopback.6.1 X
```

When using the `wfHssiLoopback` attribute, set the value X as follows:

- Local digital loop (loop A): Set to 2.
- Local line loop (loop B): Set to 3.
- Remote line loop (loop C): Set to 4.
- Reset to normal: Set to 1.

Once testing has been completed, the loopback must be turned off. Local digital loops, as mentioned earlier, are useful for testing the HSSI port and the cable. A local line loop validates the integrity of the local CSU/DSU, as does the test performed by the local digital loop. The remote line loop verifies the operation of all components up to the remote CSU/DSU and back.

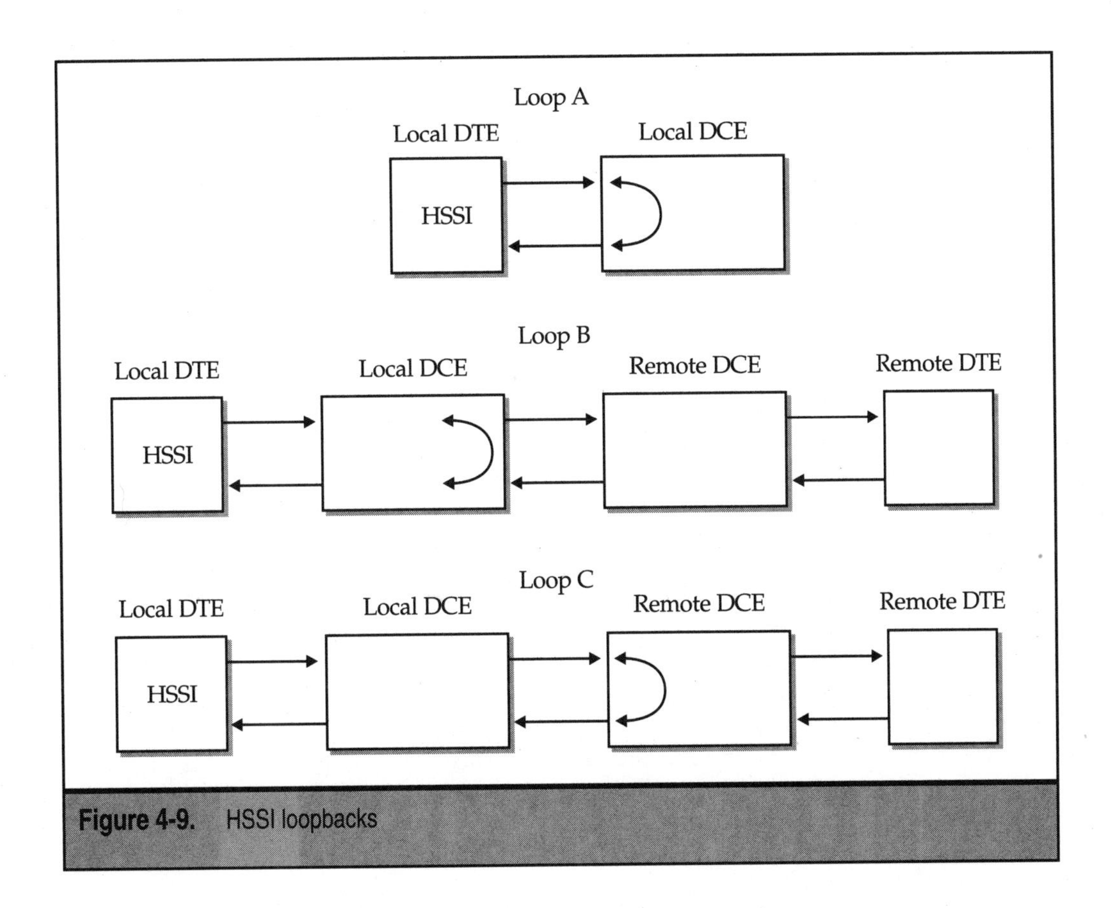

Figure 4-9. HSSI loopbacks

Digital Service Loopbacks

As mentioned earlier, the FE1, FT1, MCT1, and MCE1 interfaces contain their own CSU/DSUs for channel and digital service functions. The CSU/DSU by default is configured to accept loop codes from the network. For example, the remote end can request a remote line loop, which will be accepted and activated by the local CSU/DSU without any user intervention. The `wfDs1E1PortEntry.wfDs1E1PortAcceptLoopbackRequest` attribute can be set to disabled so that the loop up and down code detection circuitry is turned off. It is recommended that the loop code detection be left enabled.

The DS1 interface, however, can be put into a manual payload or line loop. In a payload loopback, the signal received is reframed using the T1 framer and retransmitted back out the interface. The line loop does not involve the framer, and data is looped before the framer. The line loop is seen as a remote line loop from the far-end device. Only the signal is regenerated using the line loop, which allows full testing of the transmission line. If a payload loop test fails following successful testing using a line loop, there is a problem with the framer on the interface, which may require a replacement. The following example illustrates the configuration of the loopback-mode attribute for MCT1 interfaces.

```
box->mct1/5/1
fp3(30)# loopback-mode payload-loop

box->mct1/5/1
fp3(31)# loopback-mode line-loop

box->mct1/5/1
fp3(31)# loopback-mode no-loop
```

Two of the most commonly used framing methods on MCT1 lines are ESF (Extended Super Frame) and SF/D4 (Super Frame). ESF is the default framing method used and is widely used by service providers, though D4 is still used in some remote environments. One of the most common advantages of ESF is support for facilities data link (FDL), a provisional service that operates out-of-band from user data to provide management and statistical functions. FDL is available only on MCT1 lines and is not used on MCE1 lines.

Manual loopbacks can be set on all digital service circuits regardless of the framing and fractional services. On D4 circuits, loop up and down codes can be sent in-band (`wfDs1E1ActionEntry.wfDs1E1ActionSendLoopCode`), and on ESF circuits, several FDL loop codes can be sent out-of-band (`wfDs1E1ActionEntry.wfDs1E1ActionSendFdlLoopbackCode`); these are listed in Table 4-8. Loop codes can be transmitted using the `port-line-test` object under the `mct1` context, whose attribute syntax is also listed in Table 4-8.

Bit Error Rate Testing

Digital data services are highly susceptible to failure from even small errors, on the order of one error every 10,000 bits. Bit error rate tests are standard in most digital circuits. The goal of BER tests is to measure the quality of service on the line using realistic data patterns, to detect malfunctioning equipment, and to detect incorrectly provisioned devices. Network devices with incorrectly set options can be detected using BER tests. Digital circuits are usable only if all devices along the transmission path support the same line coding.

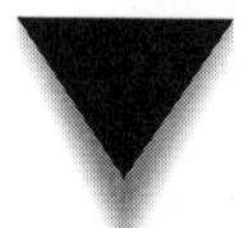

NOTE: On link modules and net modules with more than one MCT1/MCE1 port, only one port can be configured for BERT at any given time. Each module features only one BERT chipset that controls one DS1 interface.

Each specific BERT is characterized by a specific test pattern. A test pattern is a repeating sequence of logical ones and zeros that is sent by the test interface. The receiving test set (the local interface in the case of a loopback) will compare the received bit pattern against the expected pattern and determine whether any bits are in error. Various data patterns are used for stress testing and to simulate a dynamic data environment; stress testing exposes network components and interconnections to a wide range of pattern se-

Command	Description	MIB Attribute Setting	BCC Syntax
Loop Up	Sends a loop up code to the remote end to initiate a line loopback.	`wfDs1E1ActionSendLoopCode = 1`	`box->mct1/5/1->port-line-test/5/1 fp2(41)# send-loop-code loopup`
Loop Down	Sends the loop down code to terminate the remote line loop.	`wfDs1E1ActionSendLoopCode = 2`	`send-loop-code loopdown`
FDL Payload Loop	Sends a loop up code to the remote end to initiate a payload loopback.	`wfDs1E1ActionSendFdlLoopback Code = 5`	`send-fdl-loopback-code payloadloop`
Line Loop CI	Sends a customer installation (CI) loopback request.	`wfDs1E1ActionSendFdlLoopback Code = 1`	`send-fdl-loopback-code lineloopci`
Line Loop IA	Sends a CI loopback originating from the CSU.	`wfDs1E1ActionSendFdlLoopback Code = 2`	`send-fdl-loopback-code lineloopia`

Table 4-8. MCT1 Loopbacks and Configurations

Command	Description	MIB Attribute Setting	BCC Syntax
Line Loop IB	Sends a CI loopback originating from the DSU.	wfDs1E1ActionSendFdlLoopback Code = 3	send-fdl-loopback-code lineloopib
FDL Disable Payload Loop	Sends a loop down code to the remote end to terminate the payload loopback.	wfDs1E1ActionSendFdlLoopback Code = 6	send-fdl-loopback-code deactivatepl
FDL Disable Line Loop	Sends a loop down code to the remote end to terminate the line loop.	wfDs1E1ActionSendFdlLoopback Code = 4	send-fdl-loopback-code deactivatell
FDL Disable All Loops	Sends a loop down code to the remote end to terminate all loops.	wfDs1E1ActionSendFdlLoopback Code = 7	send-fdl-loopback-code deactivateall

Table 4-8. MCT1 Loopbacks and Configurations *(continued)*

quence density extremes and quick changes as is common with data traffic. If data patterns are received without any errors (or only insignificant errors), then the digital circuit can be presumed to operate without any errors in consumer use.

Several data test patterns available are summarized in Table 4-9. The table lists only BERT test patterns supported and available for use with the DS1E1 driver, although many others are commonly available with specialized digital service testers.

Pattern	Application
Zeros[1]	Use to verify that all network components in the digital circuit are provisioned with B8ZS for clear channel operation. A mismatched count of bipolar violations or bit errors indicates a line code optioning problem.
Ones[2]	Use to test power requirements for all components, particularly regenerators. Repeating devices will consume maximum power when they are forced to regenerate the mark (all ones) pattern. Repeaters with insufficient power may cause errors due to distorted pulses.
QRSS	The quasi-random signal source (QRSS) is an extended, but basic, pattern used to test performance and data continuity. It tests all line options and has an average ones density of 50%.
2E15	The 2E patterns are generally used in individual channel testing and are normally not used to test full T1 services. They violate either the ones density and/or consecutive zeros restrictions for digital signals. Commonly, they are used to test selected DS0s (see the discussion of logical line tests) in drop or insert mode or for gateway testing for E1. The 2E15 is a pseudo-random sequence based on a 15-bit shift register, and the 2E15Inv is an inverted sequence.
2E20	Similar to the 2E15, the 2E20 is a pseudo-random sequence based on a 20-bit shift register. An inverted sequence is also available.
2E23	Another pseudo-random sequence, the 2E23 is based on a 23-bit shift register. An inverted sequence is also available.

[1] A framed all zeros pattern may cause a false yellow alarm when used with D4 framing.
[2] The mark pattern, when unframed, may cause a false blue alarm (AIS) on DS1 circuits.

Table 4-9. BERT Test Patterns

Once a specific pattern has been selected, the bit error rate test can be started. During the test, deliberate errors can be introduced into the bit stream, and the return data can be examined for spurious errors. Using the `wfDs1E1ActionBertErrorInsert` attribute for the `wfDs1E1ActionEntry` instance, you can introduce BERT errors as described in Table 4-10.

In BERT mode, statistics on the bit stream are maintained based on the valid data bits and errored bits received. These statistics are maintained in an instance of `wfDs1E1BertStatsEntry` for each line that has undergone BERT. They can be viewed in BCC using the **show mct1e1 bert-stats-port** and **show mct1e1 bert-stats-ll**. The latter command is used to view BERT statistics on logical line (ll) tests. If a DS1 circuit has already undergone BERT testing, you must reset the counters on the BERT device by setting `wfDs1E1ActionEntry.wfDs1E1ActionBertReset` to 1.

An AIS or RAI can be sent while in BERT mode, to maintain continuity and to prevent downstream devices from generating unnecessary alarms during testing.

```
box->mct1/6/1->port-line-test/6/1
mx4(10)# bert-send-alarm ais-alarm
```

Valid BERT alarms are ais-alarm and yellow-alarm.

V.54 Loopbacks

Similar to MCT1/MCE1 interfaces, which can be set for digital service loopbacks, the 56/64K DSU/CSU can be set for V.54 loopbacks. Internal clocking is enabled by setting `wfDsuCsuIfEntry.wfDsuCsuTxClkSelect` to master (2), which defaults to slave mode and acquires clocking from the network.

Bit Error	Description	MIB Attribute	BCC Syntax
Single	A single nonrepeating error is introduced into the bit stream.	`wfDs1E1ActionBertErrorInsert = 1`	`bert-error-insert oneerror`
1 x 10^{-3}	One error is introduced every thousand bits.	`wfDs1E1ActionBertErrorInsert = 2`	`bert-error-insert errorperthousand`
1 x 10^{-6}	One error is introduced every million bits.	`wfDs1E1ActionBertErrorInsert = 3`	`bert-error-insert errorpermillion`
Disable	Stop all error insertions.	`wfDs1E1ActionBertErrorInsert = 4`	`bert-error-insert disable`

Table 4-10. Introducing Deliberate BERT Errors

Various loopback tests are available to test components or a section of the circuit. Table 4-11 summarizes the tests available and their purpose; these tests are illustrated in Figure 4-10. A loopback test of the CSU portion of the DCE involves testing the analog circuitry, and testing the DSU involves testing the digital circuitry:

```
box->dsucsu/1/1
an2(4)# loopback-test-timer 30

box->dsucsu/1/1
an2(5)# loopback-mode remDigLpbk
```

Test	Application
Digital Loopback	Tests the local DSU and the telecommunications circuit.
Remote Digital Loopback	Tests the local DSU/CSU, the remote DSU/CSU, and the telecommunications circuit.
Local Analog Loopback	Tests the local DSU/CSU. This is a local self-diagnostic test, where the CSU immediately loops the transmission data back to the receive path.
Remote Digital Loopback with Pattern	Same as the Remote Digital Loopback test, but the use of a test pattern allows the DSU/CSU to measure the number of errors in the received bit pattern. The error count is maintained in the attribute `wfDsuCsuIfEntry.wfDsuCsuV54Errors`.
Local Analog Loopback with Pattern	Same as the Local Analog Loopback test, but used with a test pattern. Bit errors are counted in `wfDsuCsuV54Errors`, which can provide stress results for the local DSU/CSU.
Telco-Initiated Loopback	This test is initiated by the telco provider, who can perform a CSU loopback (to test the circuit) or a DSU loopback (to test both the DSU/CSU and the telco circuit).
Pattern 2047 BERT	Strictly speaking, this test is not considered a loopback but is used to test the telco circuit. A BERT tester is attached to the remote end, and the received pattern-2047 is compared to the expected stream. Pattern-2047 is named after the length of the bit pattern.

Table 4-11. V.54 Loopback Tests

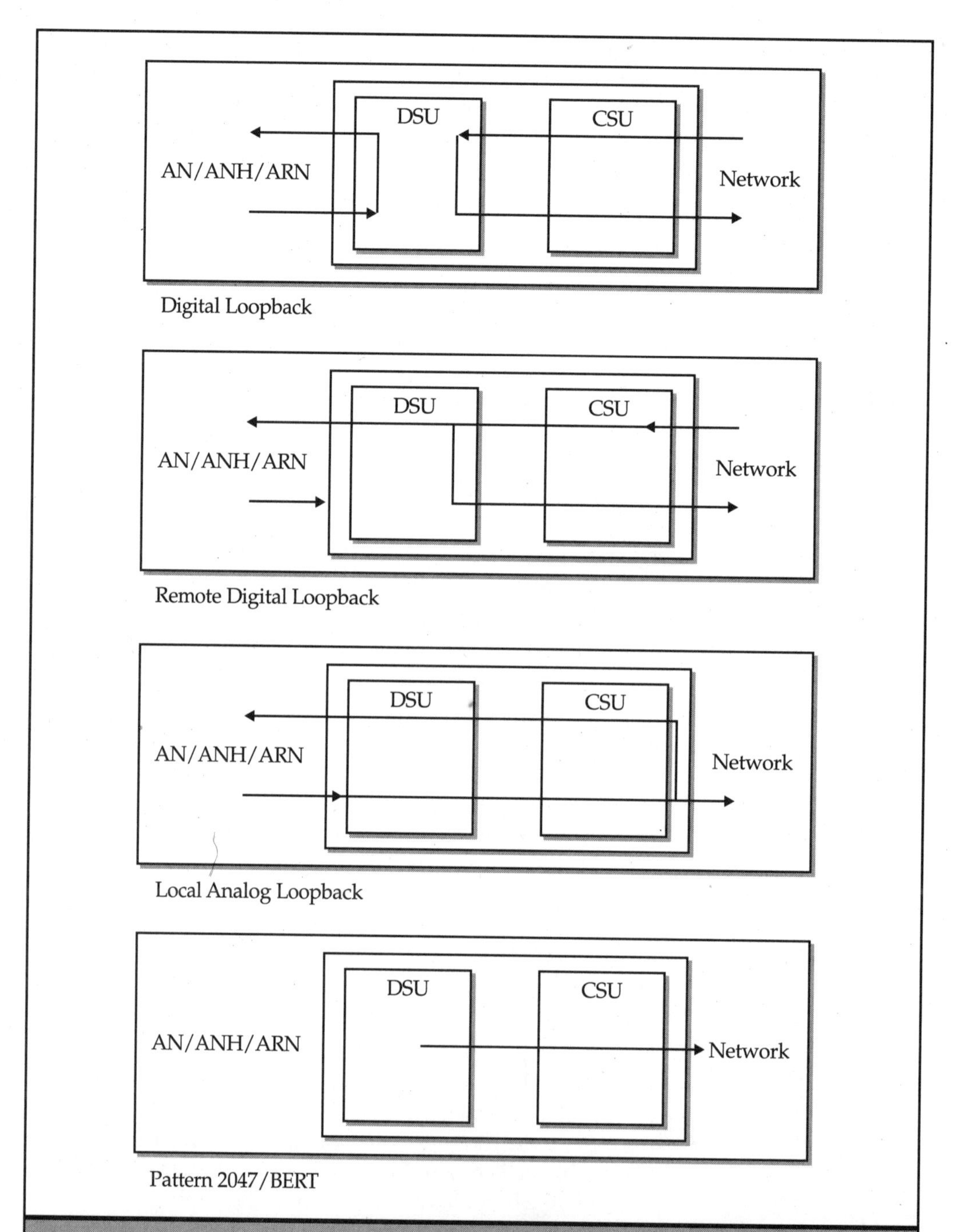

Figure 4-10. AN/ANH/ARN DSU/CSU loopback tests

Loopbacks are set with the `loopback-mode` (`wfDsuCsuV54Lpbk`) attribute in the `dsucsu` context. Valid tests are listed in Table 4-11; the values are noLoop, localAnlgLpbk, localDigLpbk, remDigLpbk, remDigLpbkWPattern, localAnlgLpbkWPattern, and pattern2047Gen. Normally, a test runs forever unless the test period is manually configured (in seconds) using the `loopback-test-timer` attribute. The corresponding MIB attributes for the loopback mode and timer are `wfDsuCsuV54Lpbk` and `wfDsuCsuV54Timer`, respectively.

ISDN BRI Loopbacks

BRI interfaces support bearer channel loopbacks for physical layer conformance testing. The `bchannel-loopback` (`wfIsdnBriBChanLoop`) attribute for the bri interface is enabled for testing the interface.

```
box->bri/1/1
arn(3)# bchannel-loopback enabled
```

Once loopback testing has been completed, the attribute must be disabled for normal operation of the `BRI` interface.

SUMMARY

This chapter covered many operational problems and physical layer issues, as well as configuration and loopbacks. The most important tasks before beginning to troubleshoot are gathering information, characterizing the symptoms, and determining the extent of the problem. In addition, once you log in to a router, you should always check to make sure that you're working with the correct device and that there are no critical faults.

This chapter also discussed various faults seen on BayRS routers, with a goal of helping you understand the scope of a particular problem; certain fault messages may indicate a critical condition that affects overall system operation and that must be resolved immediately to restore network stability. Other faults may indicate the onset of a serious problem. The chapter also discussed the use of local memory and global memory, along with the interpretation of memory consumption. Physical media and interface configuration issues for various media were also covered. The chapter concluded with a discussion of loopbacks; loopbacks can help you isolate a problem to a certain portion of the transmission line when the exact problem source is unknown.

In the next chapter, we will cover the troubleshooting of switched services and data link protocol issues.

REFERENCES

1. M. A. Miller, *Troubleshooting TCP/IP*, 3rd Edition, IDG Books, July 1999.
2. J. S. Haugdahl, *Network Analysis and Troubleshooting*, Addison-Wesley, December 1999.
3. *BLN and BCN Fuse Service Manual*, Publication 113911 Rev A.
4. *Cable Guide*, Publication 308604-14.10 Rev 01.
5. *Configuring Ethernet, FDDI and Token Ring Services*, Publication 308623-14.00 Rev 00.
6. C. F. Coombs Jr. and C. A. Coombs, *Communications Network Test and Measurement Handbook*, McGraw-Hill, August 1997.
7. R. W. Buchanan Jr., *The Art of Testing Network Systems*, John Wiley & Sons, April 1996.

CHAPTER 5

Troubleshooting Dial Services and Data Link Issues

It's a good thing to have all the props pulled out from under us occasionally. It gives us some sense of what is rock under our feet, and what is sand.

—Madeleine L'Engle, *The Summer of the Great-Grandmother*

The preceding chapter discussed basic troubleshooting concepts such as analyzing and characterizing network issues, understanding and resolving physical layer problems, and performing loopback testing. That chapter focused on layer 1 of the OSI reference model, the physical layer. Some network troubles are caused by issues with layer 2—the data link layer. As we will explore in this chapter, there are myriad data link protocols, and each requires a different troubleshooting approach.

Some data link protocols, such as Point-to-Point Protocol (PPP) and X.25 (see the following note), have a comprehensive design; they yield themselves well to troubleshooting because inherent in their operational scheme is exception handling and state management using finite-state machines. Others, such as HDLC (as used in the proprietary HDLC encapsulation called Wellfleet or Bay Standard), are simpler in design and operation. Each protocol has its pros and cons for any solution. The ideal data link protocol is one whose pros for a certain solution implementation far outweigh the cons.

NOTE: X.25 is not strictly a data link layer protocol alone. In conjunction with LAPB, X.25 features all the data link services and some common network layer services.

Two types of services are commonly used: switched services and leased services. Switched services, often referred to as dial services or dial solutions, provide on-demand functionality. This includes dial-on-demand service for temporary enterprise intranet or Internet connectivity, dial backup for primary leased-line redundancy, and bandwidth-on-demand service for incremental bandwidth. As the name indicates, switched services are established only when they are necessary. Physical layer and data link protocols are used only when a link is activated, an approach that differs from that of dedicated LAN interconnections and leased WAN links. PPP is the data link protocol most commonly used with switched services. Leased lines, on the other hand, are dedicated wide-area connections. A majority of regional and corporate WAN connections fall into this category.

After an initial section devoted to switched services, this chapter discusses data layer issues for both leased and dial services. Prior to troubleshooting an issue as a data link problem, you should look for any physical layer issues as discussed in Chapter 4.

SWITCHED SERVICES

Dial service configurations typically fall into three categories, each with different applications:

- **Dial-on-demand services** A dynamic connection is established between two routers only when data is to be sent between them. The connection may be established by either end, depending on where the data originates. Frequently, this configuration is used when a leased line is expensive or unnecessary because heavy traffic loads are temporary. It is also used for backing up circuits at the routing layer and in hot-standby mode.

- **Dial backup services** When a primary leased circuit fails, a master router dials out to the remote router to reestablish the same protocol services exactly as it was on the primary circuit. The backup line is thus an alternate link providing the same protocol services as the primary line. Dial backup services are automatically terminated when the primary line returns to normal operation.
- **Bandwidth-on-demand services** Bursty data can be accommodated by adding additional links to an existing line for more bandwidth. The primary line may be either a dial-on-demand service or a type of leased service.

Popularly used by remote offices for temporary connectivity during e-mail downloads or for Internet connectivity, the dial-on-demand service is also sometimes used for redundancy in backing up a primary link.

Switched services can be established using various media types and WAN connections, as shown in Figure 5-1. Links can be set up to PSTN, ISDN, or DDS (switched 56K) networks. A variety of interface connections can be used on the router, including the following:

- Serial connection to a dialup modem, connected to PSTN (analog)
- Serial connection to a terminal adapter, connected to ISDN (digital)
- Serial connection to a CSU/DSU, connected to a DDS network
- Native ISDN cloud connection using an ISDN BRI (S/T or U) or PRI interface
- Native PSTN connection using a V.34 modem

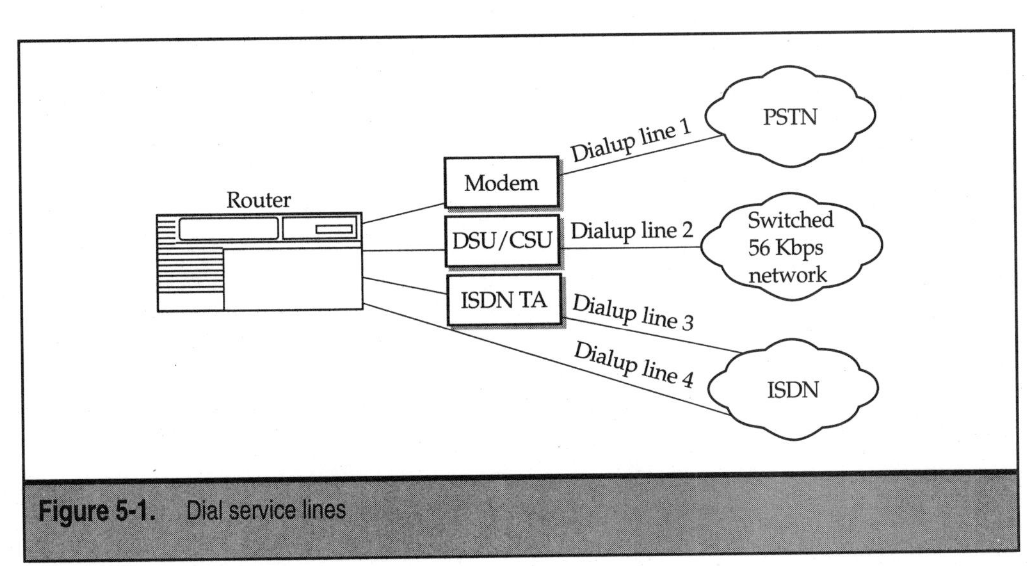

Figure 5-1. Dial service lines

Using interface types such as V.35, RS232, RS422, and X.21, switched service calls can be established using narrowband ISDN, V.25bis, RaiseDTR, and Hayes signaling methods. Each physical and logical line is added to a *pool* of available lines, from which the switched service application automatically selects an available line by interacting with the line manager. Three types of dial pools are available: dial-on-demand, dial backup, and bandwidth-on-demand. Dial pools do not span multiple slots, and lines in a pool are assigned from the same slot.

Each switched service circuit is configured to use a specific pool of the appropriate type. For example, a dial-on-demand circuit will use a demand pool, and a dial backup circuit will use a backup pool. Switched service circuits use the `wfSwservOptsEntry` MIB object for configuration, and the appropriate pool identifier is configured in `wfSwservOptsBackupPool`, `wfSwservOptsDemandPool`, or `wfSwservOptsBandwidthPool`. A pool may contain one or more lines, each of a different type. When a line is selected from the pool, the appropriate media and signaling method for the specified line is used.

Interface and Signaling Problems

The first step in establishing a call is signaling. Device misconfiguration related to signaling can cause unsuccessful call setup, resulting in incomplete establishment of dial service.

When you begin work on a dial service problem, first check to make sure that you are using the correct cable for both the interface type and the signaling type. Consult Table 4-6 to verify that the correct cable is being used on both ends of the dial service, particularly if this is the first time you are attempting to establish a connection. Use a breakout box if necessary to check that custom-made cables have all the pins connected correctly.

Once you verify that the cable is okay, check the media type and cable type configured on the port for serial interfaces, by checking the `media-type` and `cable-type` attributes. Although these attributes are not as significant on leased services, they are extremely important for dial services. First make sure that the media type is configured appropriately for the signaling method; then make sure that the correct cable type for the right standard is being used. The media type and cable type values default to RaiseDTR and RS232, respectively. For example, if you accept the default values for the RaiseDTR line but are using a V.35, the circuit may never be established.

```
#   51: 03/26/1999 13:11:26.489  WARNING  SLOT  5  MODEMIF               Code:   3
Connector COM2: modem not present, V.25bis mode
```

Warnings such as the preceding event message may indicate a cable-type or media-type setting problem. If you see this message even when a working modem is turned on and connected using the proper cable, then you may have an incorrect media-type or cable-type setting.

Some setups, such as those using RaiseDTR, may not be able to immediately see the modem when the interface initializes.

```
#   97: 01/08/1999 23:45:08.695  INFO     SLOT  9  MODEMIF              Code:   5
Connector COM1: Starting, raise dtr mode, is modem connected and turned on?
```

This event may not indicate a problem; check the interface state. When the modem asserts DSR (data set ready) on the line, the interface should be reported as up and should be added to the available list of lines in the dial service pool.

Signaling issues arise not only when the interface is plugged in or initialized. They can also arise when the interface is configured correctly for the media type and cable type, and the right cable is being used. This is because calls are established using a series of pin signal handshakes, which must be accomplished using the same signaling standard between the router and the modem or TA. In addition, a similar signaling handshake must occur on the remote (receiving) end before the two routers can successfully converse over the line.

For this to process to succeed, the data set also must be configured to support the same media and cable. Some modems are configured using dip switches; others are configured using a menu system or AT command sets. Configuration attributes specific to the device you are using must be set so that common signals such as DTR, DSR, RTS, CTS, and DCD are asserted correctly when they are needed. Table 5-1 shows common pin signals. On V.25bis connections, V.25bis commands such as CRN will not have any meaning if the modem is unable to understand them. The outbound number is stored on the router and sent to the V.25bis DCE using the CRN command.

Event messages such as the following indicate that the modem may be configured incorrectly because it asserted a signal when it should not have. Signaling messages are logged under the `MODEMIF` entity, which is the common software interface to the physical lines.

```
#  166: 02/25/1999 18:46:33.428  TRACE    SLOT 10  MODEMIF              Code:  18
Connector COM1: DCE set DSR while waiting for CTS

#   67: 11/02/1999 14:19:18.289  TRACE    SLOT 13  MODEMIF              Code:  19
Connector COM2: DCE set DSR while waiting for IND

#   46: 10/05/1999 07:14:13.572  TRACE    SLOT 12  MODEMIF              Code:  20
Connector COM1: DCE set DSR while sending DIC

#  167: 11/18/1999 13:43:54.605  TRACE    SLOT 11  MODEMIF              Code:  21
Connector COM1: DCE set DSR TRUE before DTE sent CIC
```

Pin	Description
DTR	Data terminal ready. Asserted by the router (DTE) to indicate availability.
DSR	Data set ready. Asserted by the modem/TA (DCE) to indicate availability.
RTS	Request to send. Indicates that the requesting terminal has data to send.
CTS	Clear to send. An acknowledgment from the device that receives the RTS signal, allowing transmission.

Table 5-1. Common Pin Signals

Consider the MODEMIF/18 event in the preceding example. Both RaiseDTR and V.25bis use the DSR and CTS signals during call setup and answering. On the call initiation side, a RaiseDTR DCE asserts DSR to raise the CTS signal (that is, the CTS signal is tied to the DSR signal), while CTS is asserted uniquely in the case of V.25bis. If a modem is configured as RaiseDTR and the router as V.25bis, the router will receive a DSR indication from the DCE (modem or TA) instead of a CTS signal. The DCE will expect the cable to have the CTS and DSR signals tied together in the case of RaiseDTR, but the V.25bis cable will not, and the configuration will not allow it. The call setup will stall repeatedly at this point as the router retries to dial.

You might see the same event message if the cable pinouts are incorrect instead of the DCE configuration. If the modem configuration is correct, check the cable or swap it with another one. Modem configuration errors are also indicated by messages like the one that follows, which shows an exception to the normal state transition typical of the underlying hardware or software signaling method.

```
#   77: 09/04/1999 12:13:04.075  WARNING  SLOT  5  MODEMIF              Code:  57
Connector COM1: Received packets while waiting for CTS.
```

```
#  166: 06/17/1999 01:42:17.332  WARNING  SLOT 12  MODEMIF              Code:  63
Connector COM2: received unknown indication -> 0x00af
```

Figures 5-2 through 5-5 show call setup and answering sequences for RaiseDTR and V.25bis. Knowledge of the call setup and answering process can help you establish a non-standard call setup to a remote router. Table 5-2 provides a list of V.25bis commands for reference and use with Figures 5-4 and 5-5.

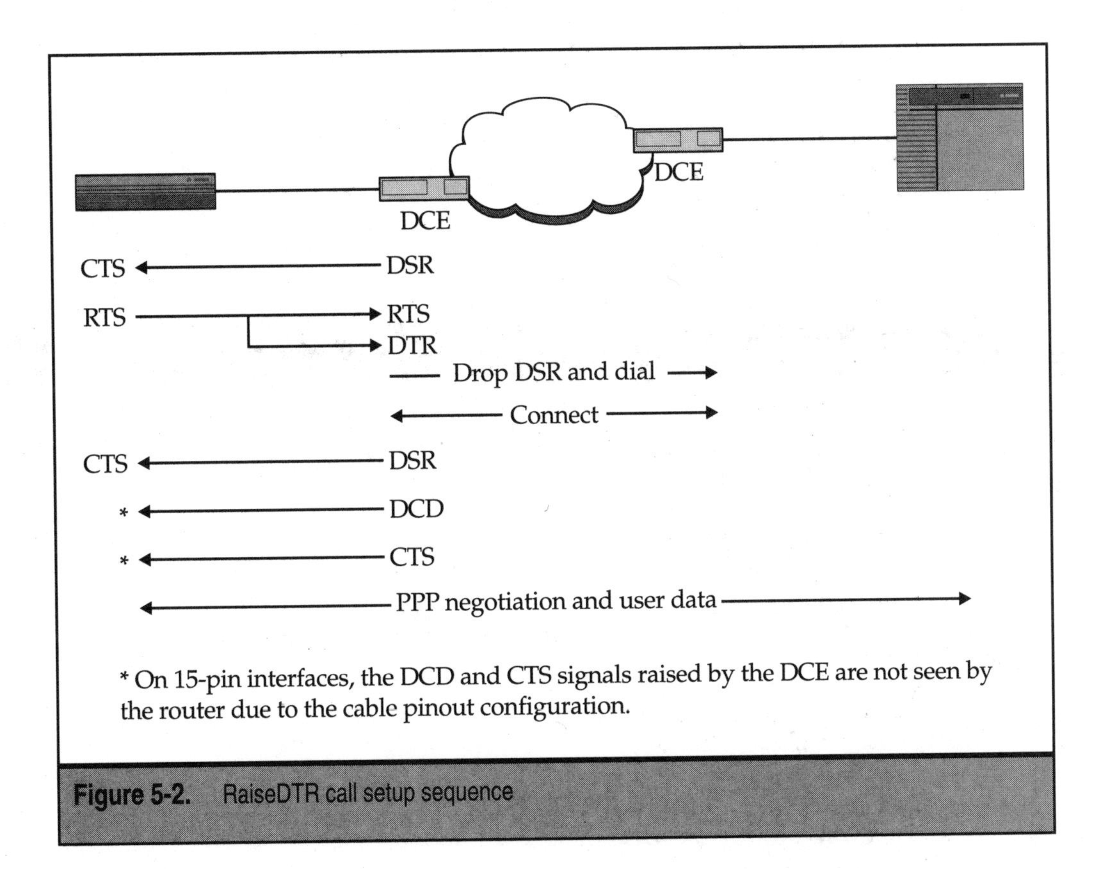

Figure 5-2. RaiseDTR call setup sequence

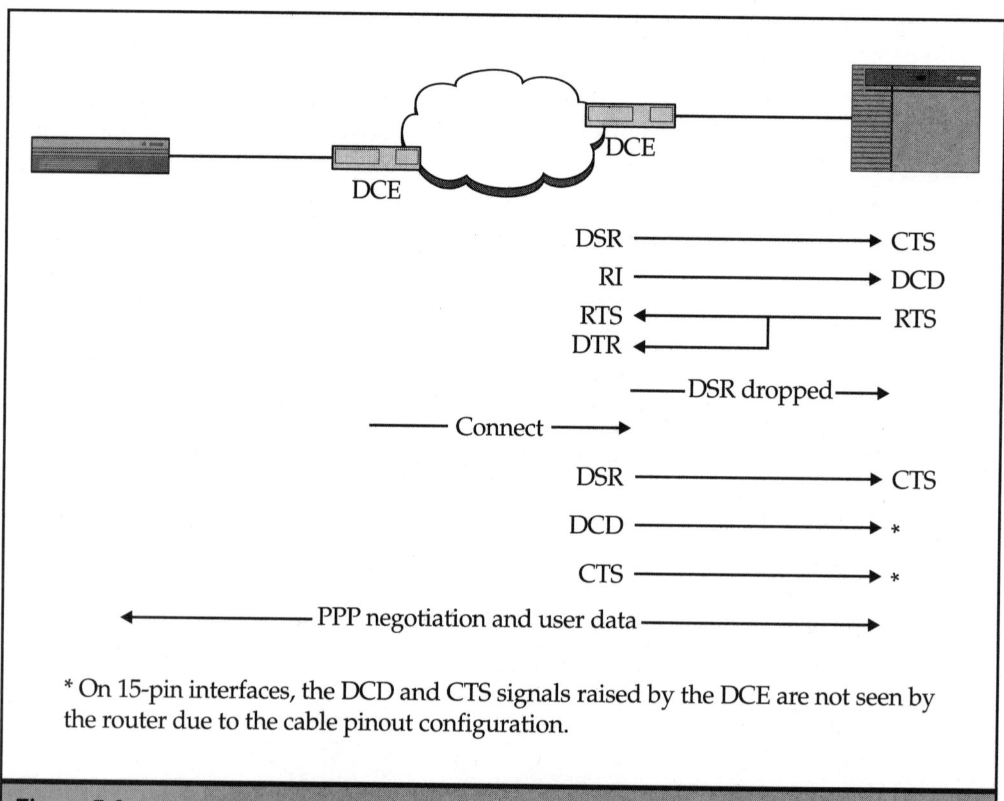

Figure 5-3. RaiseDTR call answering sequence

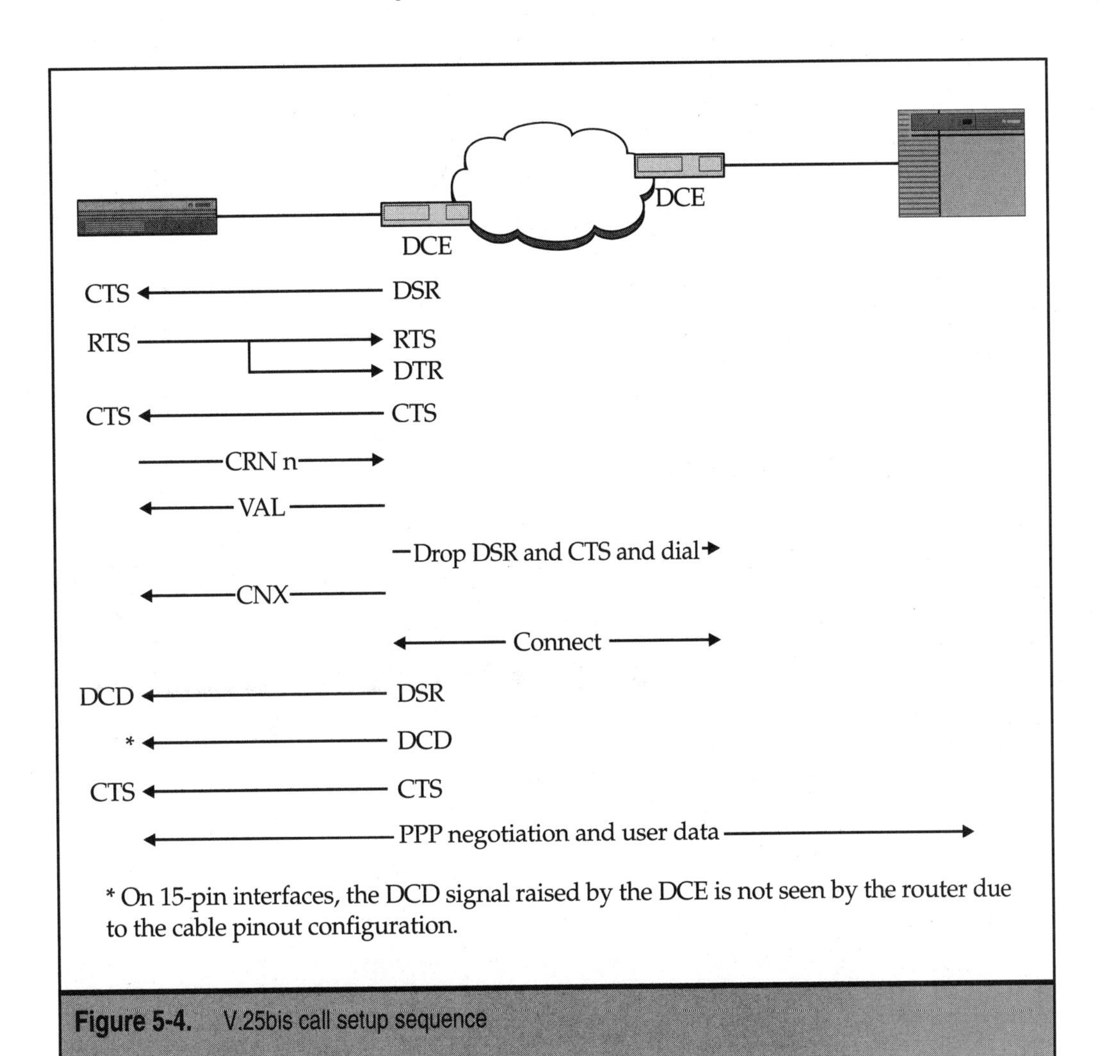

Figure 5-4. V.25bis call setup sequence

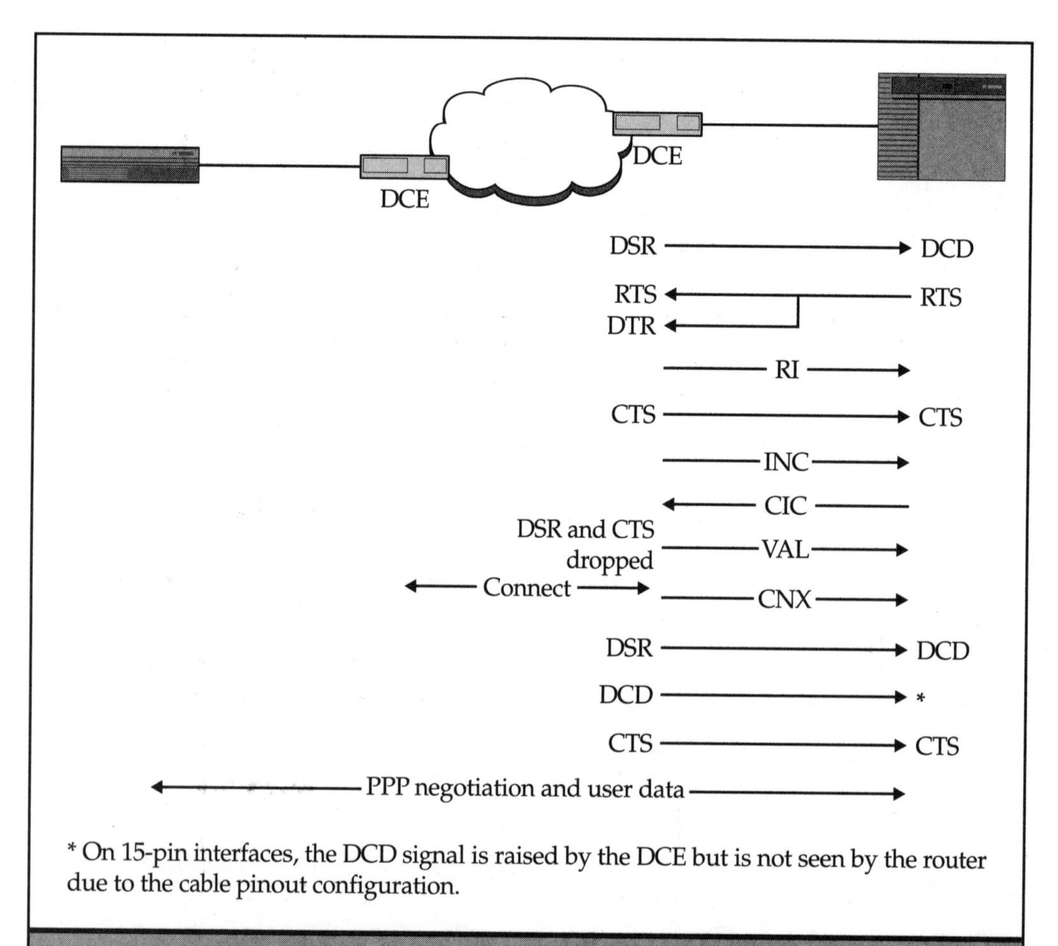

Figure 5-5. V.25bis call answering sequence

Command	Description
RI	Ring indication.
CRN*n*	Call requested number *n*. Used by the calling party to specify the remote router's phone number.
VAL/INV	Valid and invalid indications. Used to indicate that the V.25bis command issued was valid or invalid.
INC	Incoming call indication. Generated by the DCE to indicate an incoming call.
CIC	Connect incoming call. Used by the router in response to an INC command.
DIC	Disconnect call. Used by either router to terminate the call.
CNX	Connect indication. Generated by the DCE when an end-to-end session is established.
CFI	Connect failure indication. Generated when call setup fails.

Table 5-2. V.25bis Commands

Interface Initialization

Modems and ISDN terminal adapters attached to serial interfaces allow a line easily to be added to the dial-on-demand, backup, or bandwidth-on-demand pool. Once the DCE raises a DSR signal on the line, the router recognizes that the modem or TA is present and initializes the line driver.

```
#  101: 09/08/1999 14:37:48.627  INFO     SLOT  5  SWSERV              Code:   9
Backup Line 205101 available for backup pool 13.
```

```
#  128: 08/11/1999 07:32:43.558  INFO    SLOT  4  SWSERV              Code:  10
Demand Line 204102 available for dial on demand pool 2.

#   39: 03/18/1999 15:39:42.317  INFO    SLOT  7  SWSERV              Code: 141
Bandwidth Line 207104 available for bandwidth on demand pool 32.
```

If you see that the lines have not been added to the appropriate dial pool, then you may have a problem with the serial interface, cable, or DCE. Consult the previous section to resolve the problem.

ISDN BRI interfaces, on the other hand, follow a more detailed initialization process. ISDN BRI interfaces on an Access platform are S/T-type interfaces, which require an external NT1 to provide a U interface for attachment to the provider's ISDN switch. The exception is the ISDN adapter module for the ARN, which is available in S/T and U interface types. Figure 5-6 shows the S, T, and U reference points. PRI interfaces available on the ASN and BN platforms do not use a long initialization process.

Basic rate interfaces are specified in a service profile, which must be registered and activated by the ISDN network. Upon activation, each unit on a BRI line is dynamically assigned a terminal endpoint identifier (TEI) by the ISDN switch to uniquely identify it to the network. The directory numbers and service profile identifiers (SPIDs) are assigned by the provider; the SPID registration is crucial for the line to be usable.

```
#   75: 07/22/1999 14:19:14.985  INFO    SLOT  1  ISDN                Code:   4
TEI 81 assigned on DSL 0.
```

A successful TEI assignment indicates that the ISDN interface is successfully communicating with the ISDN switch. Following an activation request on the BRI interface, you will see an event indicating the switch type configured. If each DSL is not uniquely assigned a TEI, check the switch type (with `wfIsdnSwitchCfgEntry` or `wfIsdnSwitchCfgType` or with the **show isdn-switch** command) and set it accordingly. If you are still not communicating with the switch, make sure that the NT1 is working and contact the ISDN provider.

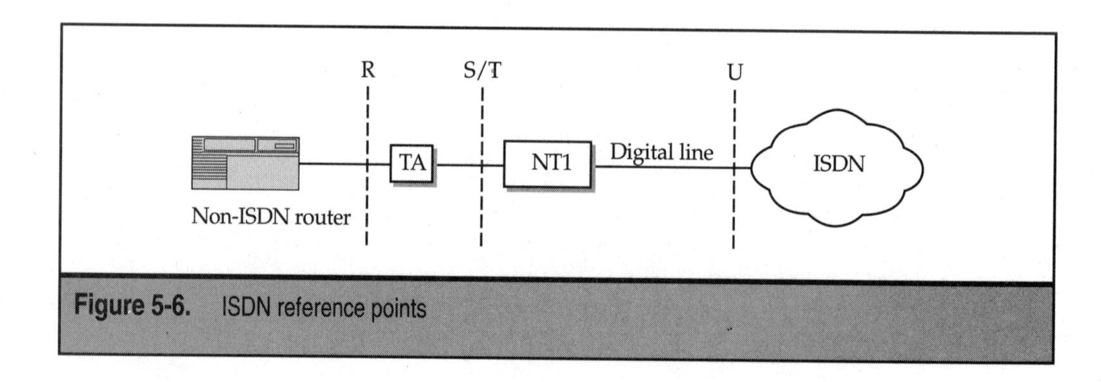

Figure 5-6. ISDN reference points

Following the TEI assignment, the BRI interface will attempt to register the configured SPIDs. In the following log events, we see that the registration for the first SPID is being NACK-ed repeatedly.

```
#  193: 01/16/1999 14:45:00.487  INFO     SLOT 1   SWSERV                Code:  96
Sending Registration for SPID1:97891628230101 on DSL 1.

#  194: 01/16/1999 14:45:00.803  DEBUG    SLOT 1   SWSERV                Code:  46
NACK Received(3 times) with cause 1 for SPID1:97891628230101 on DSL 1.

#  195: 01/16/1999 14:50:00.958  DEBUG    SLOT 1   SWSERV                Code:  46
SPID1 registration getting NACK'ed 5 times, stopping retries.
```

SPIDs are registered over the D-channel. To see the registration status of each SPID entry, you should use the **show isdn spid** (TI) command, or use `wfIsdnLocalPhoneNumEntry.wfIsdnLocalSpidStatus` if significant time has elapsed since the last reset and the log has wrapped. When a SPID is rejected, either an incorrect number is configured or the ISDN network is unable to accept the registration. An unsuccessful SPID registration may add the line to the pool, but the line cannot be used to make or accept a call.

Call Setup and Answering

In the preceding section, we briefly talked about call setup and answering on RaiseDTR and V.25bis lines (refer to Figures 5-2 through 5-5). Figure 5-7 depicts the call setup and answering behavior using a DSL on an ISDN BRI or PRI interface. ISDN, V.25bis, and RaiseDTR all have a state transition mechanism to indicate the current state of call establishment. If an ISDN call is made using V.25bis or RaiseDTR signaling to an external TA, you must check the ISDN call status on the terminal adapter. Certain ISDN terminal adapters are available with built-in display panels and out-of-band management access.

Several obstacles can arise during the call establishment phase. As you already know, an unsuccessful call is symptomatic of connectivity loss. The log will help in most cases to determine the cause of the problem. In the following event message, the V.25bis call has timed out, and the router is retrying the call.

```
#   79: 03/10/1999 00:47:42.973  TRACE    SLOT  7  MODEMIF               Code:  29
Connector COM2: Connection establishment timeout.

#   80: 03/10/1999 00:47:43.074  TRACE    SLOT  7  MODEMIF               Code:  28
Connector COM2: Connection retry in progress
```

If the call setup times out repeatedly, either the number being dialed is incorrect or the remote end is not accepting the call. Failing to set autoanswer on the remote DCE is a

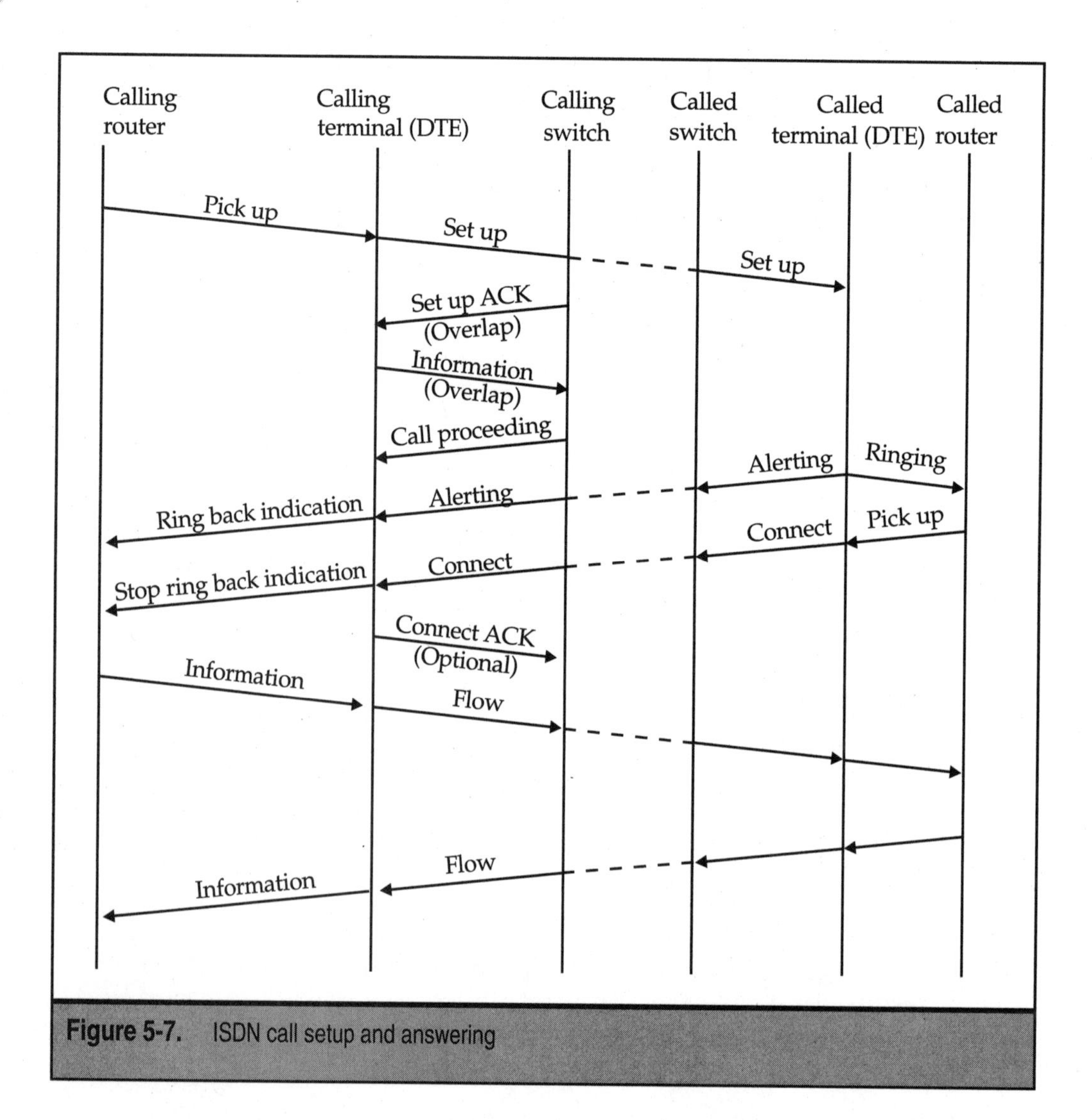

Figure 5-7. ISDN call setup and answering

common cause of connection establishment timeout. In the case of V.25bis, the DCE may reject the phone number if it is invalid or cannot accept it, as in the next example.

```
#   89: 05/04/1999 19:13:40.631  WARNING  SLOT  7  MODEMIF            Code:  41
Connector COM2: Adapter did not accept CRN command
```

V.25bis DCEs that accept the CRN*n* (call requested number *n*) command normally may be influenced by external call failures. In these cases, a connect fail indication (CFI) is transmitted by the DCE to the router indicating the cause.

```
#   75: 07/22/1999 00:15:14.769  TRACE    SLOT  7  MODEMIF              Code:  47
Connector COM2: Received a connect fail indication (CFI)

#   76: 07/22/1999 00:15:14.932  WARNING  SLOT  7  MODEMIF              Code:  48
Connector COM2: Adapter error = CFICB
```

Call failure indicators reported by V.25bis are listed in Table 5-3. In the preceding example, CFICB indicates that the call failed because of a communication busy (CB) signal. This can happen if the DCE is already connected to a call through an externally forced dial, or if the DCE has been busied out. Reset the modem or TA if it is erroneously in this state. Corrective actions for all V.25bis call failure indications are described in the table. Finite state tran-

Call Failure Indication	Description
CFI-CB	Call failure indicator, communication busy. The DCE is busy or has been busied out. Reset the unit and make sure that the line is not being shared by another dialup unit.
CFI-AB	Call failure indicator, abort. The DCE has aborted the call due to an unspecified error. Reset the unit. If the problem persists, replace the device.
CFI-ET	Call failure indicator, engaged tone. The DCE has detected that the line is engaged. This often occurs if a line is shared by more than one device. Disconnect the sharing device from the line or supply a dedicated line for the DCE.
CFI-FC	Call failure indicator, forbidden call. The number *n* provided to the DCE using the CRN*n* command is forbidden. Attempting to dial a long-distance number on a line restricted to local service only can cause this condition. Make sure that the line you are using is provisioned to dial the remote device.
CFI-NT	Call failure indicator, no tone. The DCE was answered by the remote device but did not receive an answer and negotiating tone. This is common if the number is miskeyed during configuration and rings a residential or office line.
CFI-RT	Call failure indicator, ring tone. The DCE was able to call the remote device but did not receive an answer. This can occur if the remote device is not configured to autoanswer or is not plugged in.

Table 5-3. V.25bis Call Failure Indications

sitions on RaiseDTR and V.25bis lines can be observed in detail by enabling debugging on the line; for this purpose, use the attribute `wfModemIfEntry.wfModemIfFsmDebug` for a particular line.

As with V.25bis call failure indications, ISDN calls are released with a clear cause code.

```
#   83: 03/06/1999 18:39:42.433  INFO     SLOT 13  SWSERV               Code:  95
ISDN Call ID 32801: Sending Setup to 5761023 from 7232972.

#   85: 03/06/1999 18:39:42.587  INFO     SLOT 13  SWSERV               Code:  78
ISDN Setup Request with Call ID 32801 on DSL 4.

#   86: 03/06/1999 18:39:42.979  INFO     SLOT 13  SWSERV               Code:  64
ISDN Clear Indication for Call ID 32801 with Cause 18 (NO USER RESPONDING) on DSL 4.
```

In this example, cause code 18 indicates that the remote end is not answering. The remote device did not respond with an alert or connect indication, which will occur if the remote ISDN device is not connected or has not successfully registered its SPIDs.

Call setups originated by the local interface can fail due to other reasons, as specified in Q.850. Table 5-4 summarizes the cause codes from Q.931. As when you use the debugging feature on RaiseDTR and V.25bis lines, you can observe ISDN layer-3 call control events by turning on the trace facility. You can enable the `trace-facility` attribute (`wfIsdnSwitchCfgEntry.wfIsdnSwitchTraceFacility`) for the `isdn-switch` object on a particular slot to see call events.

```
#   92: 12/23/1999 16:44:35.874  TRACE    SLOT 14  ISDN                  Code:  11
PRIMITIVE:
BufType: 36 Source_ID: 222 Dest_ID: 41392  Source State: 7
L2_Protocol: 0 Prim_ID 0x140 Sapi:0 Ces:1  LLI:0
Call_ID:0 DSL_ID:1  DSL_Chan_ID:24
```

ISDN layer-2 (LAPD, D-channel) and layer-3 (Q.931, call control) messages can be traced using packet capture or an ISDN sniffer. This is usually not necessary unless an extenuating circumstance requires the observation of call control frames. Packet captures are discussed in detail in Chapter 6.

Calls that are successfully connected must continue to remain established for uninterrupted data connectivity between the two routers. Misconfiguration or temporary connectivity issues can cause undesirable data interruption. In the following paragraphs, we look at a few issues that can arise with established calls.

Sometimes the router may abruptly drop dial sessions at 1800 hours and then not dial again, even if there is data to send. When established calls terminate at a precise time on the wall clock, this usually indicates that the dial service is configured with an availability schedule. Certain analog and digital lines are tariffed on a per-minute basis, and circuits that stay up unrestrictedly cost more to operate. Each demand or backup circuit may have a schedule when it is allowed to dial out; for example, some remote office installations may find it unnecessary to dial a corporate router at 0300 hours.

Code	Description
1	Unassigned number. This code indicates that the called number cannot be reached because it is currently not allocated, although it is in a valid format. The cause may be an incorrectly configured outgoing phone number or incomplete service provisioning for the destination number (contact the ISDN provider). See also cause code 22.
2	No route to specified transit network. This code indicates that a network device routing the call request does not recognize the particular transit network. This problem may occur if the transit network does not exist, or if it exists but does not serve the current ISDN network. You may see this code if you dial a long-distance number while connected through a LEC that automatically picks a long-distance carrier that does not have adequate service agreements to establish the call. You can pick a specific long-distance carrier by preceding your phone number with the appropriate access code. In the U.S., for example, use 10-288 for AT&T and 10-222 for MCI/WorldCom.
3	No route to destination. This code indicates that the called number cannot be reached because the network through which the call was routed does not serve the destination. This problem is similar to a code 2 problem, and you can try resolving it by using a specific long-distance carrier. If the call is an intra-LATA call, contact your ISDN provider.
4	Send special information tone. This code indicates that the called number cannot be reached due to unspecified but long-term reasons. This problem may occur if the called number is allocated but not provisioned yet or has been disconnected. Contact your ISDN provider for the remote end to verify that the destination can be called.
5	Misdialed trunk prefix. This code indicates that a trunk prefix has been incorrectly included in the called number. Check the number in the outgoing phone list and remove the area code if the call is local. If it is a long-distance call, some ISDN networks require the use of the country code (1 for the U.S.) before the area code. Change the phone number format and dial the service again.
6	Channel unacceptable. This code indicates that the DSL most recently identified to place the call to the remote destination is unacceptable. This problem can occur on BRI interfaces that have registered the SPID for only one B-channel but not the other. An attempt to send a call setup on an unregistered DSL will be rejected. The router will automatically try other lines in the pool after the retry attempts have been exhausted for the current channel.

Table 5-4. ISDN Clear Cause Codes

Code	Description
7	Call awarded; being delivered in an established channel. This code indicates that an incoming call has been awarded to the user and is being connected to a channel that is already established for similar calls. This code does not appear for regular ISDN calls, but is used with X.25 virtual calls established over the D-channel.
8	Preemption. This code indicates that this call is being preempted by a higher-priority or privileged call. If calls are repeatedly preempted, there may be a service problem. When the ISDN network is able to provide service, a retry will attempt to establish the call.
9	Preemption; circuit reserved for reuse. This code is returned by a device that preempts the call because the available circuit is reserved for reuse. This condition should last only until the network is capable of accepting and establishing the call. Persistent preemptions indicate a problem with the service provider's network.
16	Normal call clearing. This is the most common code; it indicates that the call has been explicitly cleared by one of the routers. Unlike most other codes, this one usually originates from a terminal endpoint and not the network. This code does not indicate a problem unless a higher-layer protocol issue causes the call to be terminated. This situation is described in detail in the discussion of PPP troubleshooting.
17	User busy. This code indicates that the called number is unable to accept the call because it is busy handling another call. This code can be generated either by the remote router or by the network itself. This problem occurs when multiple routers are configured to reach the remote end, and all available channels are assigned. If there are multiple numbers to the remote router, the local router will retry the call using alternates.
18	No user responding. This code indicates that the called number did not respond to the call setup request with either an alert indication or a connect indication. The problem may lie either with the remote switch or the terminal endpoint (router or ISDN TA). Disconnect and reconnect the router or TA on the remote end.

Table 5-4. ISDN Clear Cause Codes *(continued)*

Code	Description
19	No answer from user (user alerted). This code indicates that the called party has been alerted to the incoming call but did not respond with a connect indication. This situation usually occurs when the network is fully functional up to the terminating switch, but the terminal endpoint (router or ISDN TA) is either disconnected or does not answer. It may also occur if the timers on the network are set so low that the message returns faster than the called party can connect. In the case of an ISDN TA, make sure that autoanswer is set. Disconnect and reconnect the remote devices to allow reinitialization.
20	Subscriber absent. This code indicates that the remote mobile device has logged off the network. You will not ordinarily see this code on ISDN circuits, but if mobile stations are connected via a wireless cloud to the ISDN network, this code may be raised when the stations are out of range. Subscriber land lines will not generate this code; if they do, contact your ISDN provider.
21	Call rejected. This code indicates that the called number has processed the call setup request and has decided not to accept it. It does not indicate a defect on the part of the network or the called party, and the destination could have accepted the call because it is neither busy nor incompatible. Occasionally, this code is generated by the network when it is unable to service the request fully. A call rejection received on a BayRS router indicates that the remote party has screened the call, and the calling party's number is not in the incoming phone number list. If the call is valid, you must either disable call screening or add the calling party's number to the list of incoming callers.
22	Number changed. When a remote number has been changed to accommodate new provisions, your ISDN provider will notify you of the new number. If a router dials a destination number that has been changed, the ISDN network will return this code to the calling router. Sometimes the new number may be returned in the diagnostic field, but the router does not redial this number. Certain networks may return code 1 if this code is unsupported.
26	Nonselected user clearing. This code indicates that the call has been cleared by the network because the user was not awarded the incoming call. Contact your ISDN provider if you receive this code.

Table 5-4. ISDN Clear Cause Codes *(continued)*

Code	Description
27	Destination out of order. This code appears when the remote destination called cannot be reached because the interface to the destination is not functioning properly. This code usually indicates that the call setup message could not be delivered to the called party due to a physical or data link layer failure (at the remote end). The router at the remote end may also be offline.
28	Invalid number format (address incomplete). This code indicates that the remote called party cannot be contacted because the number is not in a valid format or is incomplete. This problem may occur if the called number needs to be prefixed with the area code or the country code.
29	Facility rejected. This code is returned when a supplementary facility service requested by the calling party cannot be provided by the ISDN network.
30	Response to status inquiry. This code is returned in a status message in response to a previous status inquiry. It is not used with the call setup process.
31	Normal; unspecified. This code is returned by the ISDN network when a normal event occurs, but none of the predefined normal cause codes apply. Contact the ISDN provider if this code is returned.
34	No circuit or channel available. This code is returned by the ISDN network when a usable circuit or bearer channel is unavailable for the call. This problem may be temporary; if it persists, contact the ISDN provider.
38	Network out of order. This code is returned when the ISDN network has an out-of-order condition that may last for a long period of time. The router's immediate retry attempts will fail with the same code until the allotted number of retry attempts are exhausted and another line is tried.
41	Temporary failure. This code is returned when the network is unable to establish a call due to a temporary failure. A retry attempt to establish the call will likely be successful. If the problem persists, try dialing out using another line, if one is available, or contact the ISDN provider.
42	Switching equipment congestion. This code is returned by a switch in the ISDN network when the network is experiencing congestion. Intermediate switches can perform crankbacks and reroute the call, although most devices return this code to the calling party.

Table 5-4. ISDN Clear Cause Codes *(continued)*

Code	Description
43	Access information discarded. This code is returned if the network is unable to deliver a certain access information detail to the remote called party, including user-to-user detail, lower- or higher-layer compatibility, and subaddressing. The network may return the discarded access information in the diagnostic message.
44	Requested circuit or channel not available. This code indicates that the switching equipment is unable to accept or establish the call on the requested circuit or channel. This situation can occur when only certain bearer channels are available on an interface and it is incorrectly configured with a larger count of available lines.
50	Requested facility not subscribed. This code is returned when the calling party requests a supplementary service that is implemented by the ISDN switch generating the cause, but which the caller is not authorized to use. Consult your ISDN provider to ensure that the router's circuit is provisioned with the required services, particularly on a BRI line.
57	Bearer capability not authorized. Similar to code 50, this code is returned when the calling party is refused the use of a bearer capability implemented by the ISDN switch because the caller is not authorized to use it. Contact your ISDN provider.
58	Bearer capability not presently available. This code is returned by the ISDN switch when the bearer capability requested by the calling party is implemented, but is currently unavailable. This issue must be addressed by the ISDN provider.
63	Service or option not available; unspecified. This code indicates that a specific service or option requested is unavailable. It is used by the ISDN network when no other code applies. Contact the ISDN provider if you receive this code.
65	Bearer capability not implemented. This code is reported by the ISDN switch when it does not support a specific bearer capability requested by the calling party.
66	Channel type not implemented. This code is returned if the ISDN equipment in the network does not support the channel type requested by the calling party.
69	Requested facility not implemented. When an ISDN switch does not support a specific supplementary service requested by the calling party, this code is returned.

Table 5-4. ISDN Clear Cause Codes *(continued)*

Code	Description
70	Only restricted digital information bearer capability is available. This code is returned when the calling party requests an unrestricted bearer capability, but the ISDN device generating this code supports only a restricted version of the requested bearer capability.
79	Service or option not implemented; unspecified. This code is returned by an ISDN device on the network that does not have the requested service or option implemented. It is generated when no other code for the service or option applies.
81	Invalid call reference value. This code is returned by an ISDN device on the network when it receives a message with a call reference number that is not currently used on the user-network interface. BayRS uses call reference values starting with 32768 for outgoing calls, and values starting at 1 for incoming calls.
82	Identified channel does not exist. This code is returned when the calling party requests use of a channel that has not been activated on the interface. For example, if you are subscribed to use four bearer channels on a PRI interface and the router attempts a call setup on an inactive channel, then this code is returned. When you configure the PRI, verify that only the provisioned channels are configured for the pool.
83	A suspended call exists, but this call identity does not. This code is returned by an ISDN device when the calling party attempts to resume a call with a call identity that differs from that in use for any presently suspended calls.
84	Call identity in use. This code indicates that the ISDN network has received a call suspended request containing a call identity that is already in use for a suspended call within the domain of interfaces on which the call may be resumed.
85	No call suspended. This code is returned by the ISDN network when it receives a call resume request for a call identity information element that presently does not point to any suspended call within the domain of interfaces over which the call may be resumed.
86	Call having the requested call identity has been cleared. This code is returned by the ISDN network when it receives a call resume request for a call identity information element indicating a suspended call that was cleared while suspended. A suspended call may be cleared by a network timeout or by the remote user.

Table 5-4. ISDN Clear Cause Codes *(continued)*

Code	Description
87	User not a member of CUG. This code is returned when the called party for the incoming closed user group (CUG) call is not a member of the specified CUG, or when the calling party is an ordinary subscriber calling a CUG subscriber. BayRS does not use closed user groups, and this code does not apply.
88	Incompatible destination. This code is generated by an ISDN device in the network when it receives a request to establish a call that has low-layer compatibility, high-layer compatibility, or other compatibility attributes such as data rate that cannot be accommodated. Contact your provider to ensure that the provision identifier in the diagnostics is available in the network.
90	Nonexistent CUG. This code indicates that a specified closed user group does not exist. If either this code or code 87 is returned, contact your ISDN provider and indicate that BayRS does not use CUGs.
91	Invalid transit network selection. This code is generated by the ISDN network to indicate that the received transit network selection is invalid. If you are specifying a long-distance access code, make sure it is prefixed correctly in the configuration.
95	Invalid message; unspecified. This code is returned by the ISDN network when it receives a message that it cannot process because the message is invalid. This code is generated only when no other cause codes apply for the invalid message. If this problem persists, run a trace on the ISDN D-channel to trace the invalid message and the returned diagnostics.
96	Mandatory information element is missing. This code is generated by an ISDN device when a mandatory IE is missing in a message from the originating device. This code appears when the ISDN device on the network requires mandatory information elements (IE) for each message to be processed completely. If this code appears, check the ISDN switch configuration and enable the sending of complete IEs in messages.
97	Message type nonexistent or not implemented. This code is generated by an ISDN device when it receives a message with a message type that is not recognizable because it is not defined (nonexistent) or not implemented. Check the switch type configuration if you see this code.

Table 5-4. ISDN Clear Cause Codes *(continued)*

Code	Description
98	Message not compatible with call state, or message type nonexistent or not implemented. This code is returned by an ISDN device when it receives a message stating that the message is not permissible in the current call state or when it receives a status message indicating an incompatible call state.
99	Information element/parameter nonexistent or not implemented. This code is returned when an ISDN switch receives a message with at least one IE or parameter that it either does not recognize or has not implemented. The device discards the IE or parameter and generates the code.
100	Invalid information element contents. This code is returned by an ISDN device on the network that receives a message with an IE that it has implemented but which contains invalid data that it cannot process. If this condition persists, run a trace on the D-channel to observe call control messages to evaluate the validity of the IE being sent to the network.
101	Message not compatible with call state. This code is generated by a network device that recognizes that the message received is incompatible with the current call state. This code may indicate a temporary network problem in the ISDN cloud. If the problem persists, contact the ISDN provider.
102	Recovery on timer expiry. This code is automatically generated by an ISDN device on the network when it initiates a recovery procedure following the expiration of a timer in association with defined error handling procedures. Check with your ISDN provider for the conditions that trigger recovery and for the timers used.
103	Parameter nonexistent or not implemented; passed on. This code is generated when an ISDN device receives a message that contains parameters that it does not recognize because they are not defined or are not implemented. The device ignores the parameter, but if it is an intermediate point, the device passes on the message with the parameters unchanged. Contact your ISDN provider if this code is followed by a call failure.
111	Protocol error; unspecified. This code is returned by a device in the ISDN network to report a protocol error only when no other protocol error code applies. Contact your ISDN service provider.
127	Interworking; unspecified. This cause code indicates that an interworking call to another network (such as a switched 56-Kbps service) has been completed and the call is released. Contact your provider if you receive this code.

Table 5-4. ISDN Clear Cause Codes *(continued)*

Schedules are configured either on an exclusive or inclusive basis. When configured on an individual basis, the schedule may affect the dial service in two ways. An existing dial service will terminate when the scheduled period ends. Also, new data, especially on a dial-on-demand circuit, will not bring up the circuit because it is scheduled as unavailable. The following is a list of conditions, including scheduled timeouts, that will bring down an existing call:

- **Maximum up time has been reached** By default, the `max-uptime` attribute (`wfSwservOptsEntry.wfSwservOptsMaxUpTime`) of a demand circuit or backup circuit is set to 60 minutes, but `max-uptime-termination` is disabled. When enabled, the circuit is torn down and may be reestablished if data is available. If the session keeps terminating every hour, disable the `max-uptime-termination` or set the `max-uptime` attribute to 999999, which sets the up time to unlimited.

```
#   24: 12/27/1999 19:52:43.918  INFO     SLOT  8  SWSERV              Code: 155
Maximum amount of up time exceeded for circuit 31.

#   25: 12/27/1999 19:52:43.929  DEBUG    SLOT  8  SWSERV              Code: 46
Max up termination enabled - killing demand cct
```

- **Inactivity time limit has been reached** On demand circuits, data inactivity in one or both directions (default) for 60 seconds will bring down the circuit. If you do not want inactivity to terminate the circuit, increase the inactivity-time value to a day or more to enable constant up time. The maximum up time should also be modified for this setting.

```
#  145: 07/10/1999 01:59:54.565  DEBUG    SLOT  3  SWSERV              Code:  46
Line Mgr received Inactivity Timeout on line 120903101 for cct 4

#  146: 07/10/1999 01:59:54.892  INFO     SLOT  3  SWSERV              Code:  55
No data received for demand circuit 4 for the configured inactivity period.
Demand circuit being brought down due to inactivity.
```

- **Scheduled availability period has ended or unavailability period begins** Check the circuit's schedule and make changes, if necessary. Schedules are maintained in an instance of `wfSwservTODEntry` for each circuit, if configured. In the following example, primary circuit 12 failed during a period when the corresponding backup circuit was scheduled as unavailable. At 0730 hours, the backup circuit is scheduled to be available, at which time it becomes active. Later in the day, when the availability interval ends, it is terminated.

```
#   26: 02/21/1999 07:30:00.036  INFO     SLOT 12  SWSERV              Code:  48
Backup start period reached, establishing failed primary circuit 12

#   27: 02/21/1999 18:00:00.123  INFO     SLOT 12  SWSERV              Code:  50
Primary circuit 12 in backup mode terminated, end time for backup interval reached.
```

- **Automatic demand termination** When there are two paths to the same destination, a failed demand circuit can be terminated in favor of an alternate path to the same destination. Normally, a demand circuit will be continuously retried after a brief delay following a maximum number of connection attempts. When the `auto-demand-termination` attribute is enabled for a demand circuit, the router can choose an alternate circuit, if one is available.
- **Forced take down** An operator can force a dial service to be taken down by enabling the `force-take-down` attribute or by disabling the circuit.

The opposite of an unexpected call teardown is a persistent call that will not go down or a condition in which the router repeatedly establishes a call. If the `max-uptime` attribute is set to a large value or if `max-uptime-termination` is disabled, constant data traffic will keep the line continuously operational. Data traffic need not consist only of user data; routing and service protocols configured on the dial circuit will generate periodic updates that will prevent the inactivity timer from expiring.

If a dial-on-demand call remains active due to routing protocol traffic, enable dial-optimized routing (DOR) so that route updates are exchanged over longer periods of time and a dial service is brought up, usually for user traffic. DOR circuits may also be brought up for exchange routes, but very infrequently—for example, RIP updates are exchanged every 3600 seconds (1 hour). Another alternate solution is to use static routes over the dial-on-demand circuit to help eliminate routing protocol overhead and ensure that only data traffic causes the line to timeout due to inactivity. And if demand data traffic is intermittent, causing frequent call setups and initial delays in session establishment, you can increase the minimum-call-duration value (default: 60 seconds) and the inactivity-time value so that calls remain established longer and inactivity is measured over a longer time interval.

Configuration Considerations

Two common configuration considerations must be noted. On the AN/ANH platform, configuring a BRI interface for dialup in 2B+D mode disallows configuration of COM1 for serial communications. This is because the motherboard can service only four channel controllers: one for the LAN, two for the two B-channels, and another for the serial port.

```
#  163: 11/10/1999 02:47:02.601  WARNING  SLOT  1  SYNC                  Code:  78
Cannot configure ISDN and SYNC COM1 together.
```

An alternate configuration is the dialup 1B+D or floating B setup, where both the serial ports are available if only one B-channel is required. The software will automatically make available a channel controller for the requisite bearer channel on the BRI interface. When you configure the platform initially, first configure both serial ports as desired. Then configure the BRI for floating B or dialup 1B+D application. A similar configuration consideration applies to the ARN platform, where two ISDN BRI adapter modules can be installed in the same unit. Of the four B-channels (two per adapter module), only three can be used simultaneously.

Another common configuration error occurs in the backup circuit configuration for primary frame relay circuits. Backup circuits are of two types: PPP and frame relay. The router can use the frame relay backup circuit to dial back into a frame relay network using an ISDN network. It is a proprietary solution available only if your frame relay provider and ISDN provider are the same, and the provider uses BayStream switches to provide these services. Normally, backup of a primary frame relay circuit is performed using a PPP circuit on a point-to-point dialed connection. When using Site Manager or BCC, always plan on using a PPP backup circuit for a primary frame relay circuit (in direct mode) unless your ISDN provider has provisioned frame relay access using dialup ISDN.

Figures 5-8 and 5-9 illustrate the BCC configuration objects associated with demand and backup circuits.

When there is ambiguity about a dial service configuration, run through the following checklist:

- Make sure that the desired lines are in the correct pool. At least one of these lines must be active and available for use by the switched circuit.
- ISDN interfaces must have the switch type configured correctly, and if you are using a BRI, one or more SPIDs must also be configured.
- Check the switched circuit configuration to ensure that it is configured on the correct pool. Circuits configured on the wrong pool will report a warning when attempting to receive a call.

```
#  181: 09/24/1999 14:37:32.171  WARNING  SLOT  3  SWSERV            Code: 159
Call for cct 12 received on Line not configured for pool 2.
```

- If this is a backup circuit, check to see that the leased circuit is configured as primary and set to use the correct backup pool.
- On demand circuits, the mode must be set to either collision master or collision slave. Backup circuits must be configured as master or slave. If the local circuit is configured as master, then the remote device must be configured as the slave. If one of the sides is configured as no dial, then the other must be configured as master or must use force dial to establish the call.
- Both backup master and demand circuits must have at least one outgoing phone number associated with the circuit. The phone number must have the appropriate area code and country code prefixes.
- CHAP or PAP must be configured on the circuit or the associated physical line. Called parties (slaves) must have the caller resolution table configured with the calling party and the circuit associated with it.
- Demand and standby circuits must have network layer protocols configured as required. Backup circuits will inherit the protocol interfaces from the primary circuit.

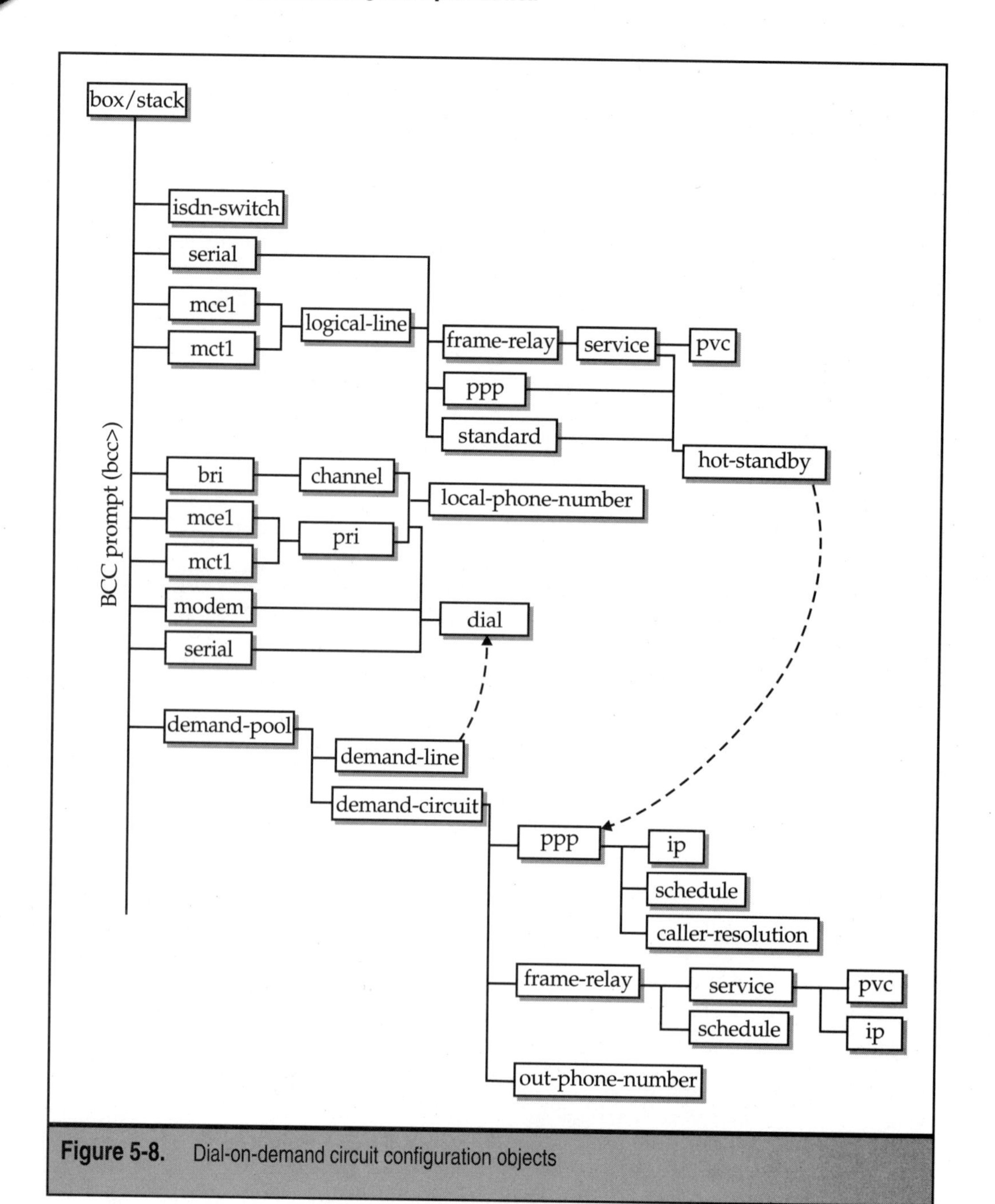

Figure 5-8. Dial-on-demand circuit configuration objects

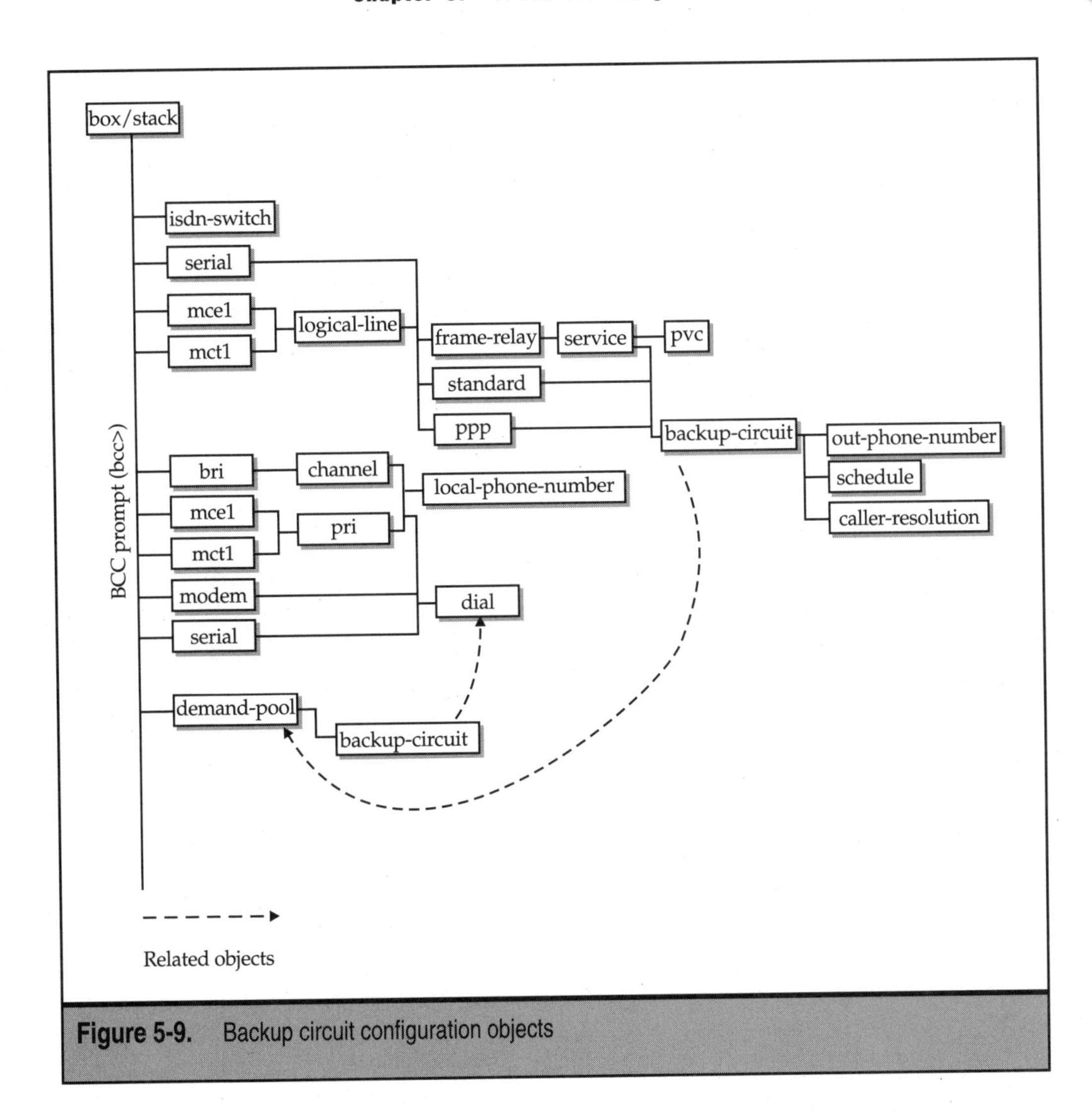

Figure 5-9. Backup circuit configuration objects

Security

When initially established, leased lines are tested to verify that both routers are connected to the desired destination or to multiple destinations (in the case of multi-access networks). Usually, you can be assured that data exiting your router's WAN interface will not be compromised and will not reach the wrong destination.

Dial services are a bit different than leased services in this regard. With the proliferation of data communication over analog phone lines, threats such as war dialing have allowed security compromises on insecure dialup lines. To combat threats such as these, security mechanisms have been developed for various dial services. These include user restrictions based on incoming numbers and destination authentication using PAP or CHAP protocols.

By default, BayRS does not filter inbound calls based on the caller. The router cannot distinguish callers on RaiseDTR lines, but can do so on ISDN and V.25bis lines. Implementing security may also generate problems if the configuration is incorrect. For example, if you incorrectly configure an acceptable inbound number, then valid incoming calls may be dropped.

```
#  137: 02/21/1999 19:54:29.253  DEBUG    SLOT  2  SWSERV              Code:  46
Calling party number is being validated for screening.

#  138: 02/21/1999 19:54:29.280  INFO     SLOT  2  SWSERV              Code:  91
Incoming ISDN call with screened Calling Party number 2249182 being dropped on DSL 1.
```

The preceding example shows a switched service message indicating that caller 224-9182 is not in the list of incoming phone numbers. All calls are screened on a slotwide basis when the `incoming-filter` attribute is enabled for the `isdn-switch` object. When configuring inbound call filtering, make sure that the caller's phone number is configured correctly. If the caller has more than one phone number, such as for multiple channels or inbound calls from one of several interfaces, you must configure all potential inbound numbers.

For calls originating in one area code and destined for another area code, the calling user and the ISDN network will automatically include the caller's area code. In these cases, inbound caller numbers must include the caller's full area code and phone number. If the same phone number is used by several routers, such as calls originating from a hunt group, then you must also configure the subaddress to uniquely identify the caller. Screened calls are counted with the attribute `wfIsdnSwitchCfgEntry.wfIsdnSwitchScreenedCalls`.

Enabling the `incoming-filter` attribute and configuring inbound phone numbers alone may sometimes be inadequate. The ISDN network also must support the propagation of the incoming phone number in an information element. In the following example, the number field for the calling party in the call setup message is incomplete, which will cause the router to drop the call. If this occurs, contact your ISDN provider to get the calling party's number (which is available with caller identification service) for the line.

```
#  186: 02/09/1999 21:34:33.960  INFO     SLOT 12  SWSERV              Code:  91
Incoming ISDN call with screened Calling Party number NONE being dropped on DSL 2.
```

On serially attached TAs and modems, the following message is logged:

```
#  126: 02/13/1999 07:50:05.964  TRACE    SLOT  6  MODEMIF             Code:  45
Connector COM2: Received call dropped (number not allowed)
```

Once a call is successfully established, protocols such as PAP and CHAP are used to authenticate the caller using a configured password or secret. Security issues with authentication protocols are discussed later in this chapter, after we discuss the LCP negotiation phase of PPP.

The callback feature for dial-on-demand circuits is used in conjunction with an authentication protocol both as a security mechanism and a connection cost-control mechanism. Remote sites initially call a central site, which is authenticated; then the call is dropped. The central site (the callback server) calls back the remote site (the callback client) and is authenticated, and services are established. This callback mechanism helps ensure that both the calling and called parties authenticate each other.

Design and External Problems

Most dial-on-demand, backup, and bandwidth-on-demand circuits, when designed properly, will accomplish the application's goals easily. However, for functional reasons or because of external factors, certain applications work best when you choose an alternative dial service.

Backup circuits are established when the switched service application recognizes that a primary circuit is down. In Chapter 2, we discussed the association between circuits and lines. A primary circuit is said to be down when its line goes down. On frame relay circuits, the associated line is the subcircuit.

Primary line failure is detected in various ways:

- **Bay Standard lines use BOFL packets** BOFL packets must be enabled on primary Bay Standard lines to enable switching to backup mode when necessary. Loss of DSR or clocking can also be used to detect line failure if an intelligent DCE is connected to the router.
- **PPP lines use echo requests or link-quality monitoring (LQM)** BOFL packets may also optionally be used if BayRS routers are installed at both ends. If the remote router does not support echo requests or LQR, line failure can also be detected by the loss of DSR or clocking.
- **Frame relay lines use the A-bit indicator for a specific DLCI in a status query** Primary circuits must be configured in direct mode, which can contain only one permanent virtual circuit (PVC) per service record. If a primary frame relay circuit does not go into backup mode when the PVC becomes inactive, consult your frame relay provider for information on the propagation of the active bit through the network.

Some backup circuits may not fully support the triggers required to indicate that the primary line has gone down. For example, if your frame relay provider is unable to propagate the A-bit (active bit) through the cloud, then a backup circuit may not be the best solution. Similarly, consider the case where a 512-Kbps leased PPP line is being backed up using a 64-Kbps ISDN dial circuit. The same protocol traffic and throughput that was on a

512-Kbps circuit cannot be transported on a 64-Kbps backup without losing some data. In this case, you may want to have only business-critical protocols or data transported over the backup circuit.

A regular dial-on-demand circuit can be triggered to behave as a backup using a higher-layer protocol. This is commonly called as routing-layer dial backup (RLDBU). With RLDBU, a lower-precedence route is installed for the demand circuit. When the primary circuit fails, all of its associated routes are removed from the routing table—the lower-precedence route from the demand circuit is installed in the routing table, and data traffic using the new circuit will automatically establish the dial service. RLDBU is appropriate when exacting functionality is required on the backup circuit, particularly the support of all protocols using similar addressing schemes.

Another common backup application is the hot-standby demand circuit. As mentioned earlier, it is impractical to immediately switch 512-Kbps of primary bandwidth to a 64-Kbps secondary line. A business decision will need to be made regarding whether to accept only data traffic critical to business operation. For example, you may want to include SNA traffic and database traffic over IP, and you may want to disallow DECnet and AppleTalk traffic on the secondary line, in which case selective protocol configuration is required. Using a hot-standby circuit on the master end and a standby on the slave, a backup mode connection can be established when a primary circuit goes down. Hot-standby is more flexible in rerouting because the circuit need not be established to the same remote router; the local router can dial into a different remote router to reestablish connectivity.

Troubleshooting Tips

When beginning to troubleshoot dial service issues, first start looking in the log under the appropriate entity.

- For ISDN BRI interface initialization and SPID registration, run **log -eISDN_BRI –eISDN**.
- For Raise DTR and V.25bis interface initialization, run **log –eMODEMIF**.
- For ISDN BRI/PRI, RaiseDTR, and V.25bis call setup and answering, run **log –eSWSERV**.

First check the interface initialization and signaling to ensure that the line and DCE are communicating properly. Then check SWSERV events to identify any problems. If you suspect that there may be a problem at a certain point in the log but are unable to understand or pinpoint them, check the debug severity events as well (log –ffwtid). If necessary, extended debugging can be enabled on certain objects to generate more debug events in the log for better troubleshooting. Extended debugging must be turned off once troubleshooting has been completed.

- For RaiseDTR and V.25bis DCEs, set `wfModemIfEntry.wfModemIfFsmDebug` to 1 (enabled) to view transitions in the RaiseDTR and V.25bis finite state machines (FSM). This will help identify signaling, configuration, and cabling problems.
- To debug ISDN BRI/PRI communication with the ISDN switch, enable `wfIsdnSwitchCfgEntry.wfIsdnSwitchTraceFacility` by setting it to 1. This will help identify TEI assignment, SPID registration, SAPI, and timer problems.
- For detailed messages on switched service operation for a particular circuit, set `wfSwservOptsEntry.wfSwservOptsDebugMsgLevel` to a value from 1 to 4, with 4 representing the most detailed messages. The instance identifier of the MIB object will be the same as the circuit number. Events with line management, dial service activation, call setup, and answering can be observed.

Sometimes, ISDN calls may disconnect with ambiguous cause codes. In particular, code 31 (normal; unspecified) does not provide much detail. Performing a packet capture on the D-channel in these cases is helpful in determining the cause of the problem.

DATA LINK ISSUES

Many data link protocols operate over various types of media. For example, PPP runs over leased serial (synchronous and asynchronous) lines, DS1/DS3 lines, and dial lines. Similarly, ATM/DXI can be configured to run over a DS3 line as well as a DS1 line, just to name two. However, all data link protocols exhibit the same functionality regardless of the media type they are provisioned for. In this section, we cover some common data link protocol issues and their symptoms and resolution.

Common Statistics

Media interfaces have both common statistical attributes and media-specific attributes. Chapter 4 discussed media and cable attributes that affect physical layer behavior. The following sections discuss statistical attributes related to physical and data link layers. Physical layer statistics are discussed here because they are often important in tracing data link layer symptoms. In many cases, even after you replace cables, ports, and modules, the same symptoms may still occur, leading you to the conclusion that the problem is with a higher layer.

Many attributes point to specific causes, and problems are thus easily resolved. Other parameters may not readily point to the cause of the problem, but they provide details that can help steer the troubleshooting. For example, if statistics reveal a port that appears to be successfully transmitting packets but not receiving any, the problem may be a defective cable or an issue related to the remote device.

Viewing Statistics

Chapter 2 discussed the syntax of the **show** command available using the TI and BCC. Appendix F lists the subcommands accepted by the **show** command. Most network administrators are familiar with the **show** command, which summarizes configuration and operational statistics. Three commonly used interface-related show subcommands are **errors**, **summary**, and **stats**. The **summary** subcommand displays comprehensive information for an interface or circuit, and **stats** displays dynamic operational statistics. A formatted display of transmit, receive, and internal errors is provided by **errors**. Not all objects support all three subcommands, and some of the information typically displayed by one subcommand may be provided by several interface- or protocol-specific subcommands.

Show commands for viewing summary and statistical detail use the following format:

```
box# show object summary|stats|errors
```

where *object* can be any one of the supported physical or data link values, including atm, atmdxi, bisync, csmacd, ds1e1, dsx3, ethernet, e1, fddi, frame-relay, ft1e1, hssi, isdn, mct1e1, modem, sdlc, serial, smds, sonet, sync, t1, token-ring, and x25. For the sake of simplicity, examples of **show** command output are condensed to highlight appropriate fields. Common statistical attributes returned by the **show** command are discussed in the next several sections. Each topic applies to several statistical attributes that are listed at the end of the corresponding section.

Runts

Runts are data frames that are smaller than the minimum length supported by the medium or data link protocol. If the minimum frame length is mandated by the physical layer, there may be a clocking issue at the transmitting station or a collision. Runts may be seen on asynchronous, bisynchronous, synchronous, and Ethernet lines. On Ethernet lines, the clocking speed is fixed, and runts are most likely to be caused by data collisions, which are discussed later. When runts occur, check the clock speed on all devices on the line.

```
show serial errors                                  Apr 03, 2000 16:08:16 [GMT-6]


RECEIVE ERRORS:

Slot/           Bad          Runt         Frame        Frames       Overflow
Conn  Circuit Frames         Frames       Rejects      Too Long     Frames
----- ------- ----------  ----------  ----------  ----------  ----------
3/7   S37_FR           0        9823           0           0           0
... ...
```

Runts can also appear on lines that frequently lose the data carrier, such as bouncy or noisy lines. Loss of the carrier during frame reception may result in a smaller-than-acceptable frame. These are considered inbound (receive) errors and may also be added to a cumulative receive error value.

Applies to: `wfAsyncRxRunts`, `wfBisyncRxRunts`, `wfSyncRuntsRx`

Receive Errors

Receive errors are counted whenever a received frame does not pass integrity checks or is not recognized. For example, receive errors occur if frames are received with an incorrect checksum or an invalid data symbol or are not octet aligned. Certain interfaces such as Ethernet separate these errors into different types, such as MAC receive errors and FCS errors. Some receive errors are caused by clocking or data errors on the line, which may occur if the line has more than one master clock source or the line coding is mismatched on the two ends.

```
show mct1e1 rx-errors                              Jan 13, 2000 12:10:49 [GMT-6]


Slot/                       Logical
<Module>/                   Line    Null   Short  CRC    Long   Abort
Conn      Circuit           Index   Frames Frames Errors Frames Frames
--------- ------------ ------- ------ ------ ------ ------ ------
1/2/1     MCT1_121_FR        1      0     92    982      0      0
```

On interfaces such as Ethernet and FDDI where the clock speed and coding are predefined by the underlying standard, a receive error may indicate a malfunctioning transmitter. Occasional receive errors, such as those that briefly occur on existing lines, may signal noise on the circuit. Check the clocking on the line and the cable and cable connectors.

Applies to: `wfAsyncRxErrors`, `wfBisyncRxErrors`, `wfCSMACDInternalMacRxErrors`, `wfCSMACDFcsErrorRx`, `wfLogicalLineRxErrors`, `wfDs1E1LineStatRxErrors`, `wfFddiXLineRxErrors`, `wfIsdnBriRxErrors`, `wfLnmInterfaceLrmRxErrors`, `wfSyncRxErrors`, `wfTokenRingInErrors`

Transmit Errors

Transmit errors may occur because of hardware problems, underflows, internal operational errors, or other errors specific to the medium. Unlike receive errors, which can be caused by external events, many transmit errors are local to the interface. Underflows can occur if the software is unable to deliver packets at a minimum rate to the line driver. Internal operation errors are unique to each interface and link module but generally indicate that the driver requires tuning to improve performance and resolve transmit errors. Transmit errors can also occur if the line is so noisy that the driver is unable to send frames successfully.

```
show hssi errors                                  Jan 14, 2000 06:17:28 [GMT-6]

... ...
TRANSMIT ERRORS:

Slot/
Conn  Aborts     Underruns
----- ---------- ----------
4/1            2      53871
```

If transmit errors accompany receive errors, first attempt to resolve the receive errors as this may eliminate the transmit errors as well. Continuous transmit errors may be addressed by running diagnostics and replacing the link module or interface, if necessary. If the errors persist, the line driver may require tuning, which is discussed later.

Applies to: `wfAsyncTxErrors`, `wfBisyncTxErrors`, `wfCSMACDInternalMacTxErrors`, `wfLogicalLineTxErrors`, `wfDs1E1LineStatTxErrors`, `wfFddiXLineTxErrors`, `wfIsdnBriTxErrors`, `wfSyncTxErrors`, `wfTokenRingOutErrors`

Lack of Resources

There are two types of lack-of-resource statistics: receive and transmit. Similarly, the line driver has two buffer queues (also called rings): one for receive data and another for transmit data. Received data frames are copied to the receive queue for processing, and the buffer is freed after the frames are processed. Likewise, the software places outgoing frames in the transmit ring (queue) for transmission by the line driver. The buffer is freed after the driver transmits the frames.

Queues are allocated on a per-interface basis, and frames received dynamically allocate more buffers to the queue as necessary. However, all processor modules have finite buffer resources, and on rare occasions, the line driver may be unable to obtain additional buffers. Frames that cannot be copied to a buffer on receipt will be dropped, and the receive lack-of-resource statistic will be incremented. Similarly, the driver will clip an outgoing packet if congestion prevents transmission and increment the transmit `lack-of-resource` attribute. Lack-of-resource statistics are not directly reported for all media and must be retrieved from the corresponding media MIB objects.

```
mx2(8)# mget wfCSMACDEntry.24.*
wfCSMACDEntry.wfCSMACDLackRescErrorRx.4.1 0
wfCSMACDEntry.wfCSMACDLackRescErrorRx.4.2 0
wfCSMACDEntry.wfCSMACDLackRescErrorRx.4.3 39722
wfCSMACDEntry.wfCSMACDLackRescErrorRx.4.4 0
```

When you observe lack-of-resource values, first check to see if any other interfaces on the same slot or link module are experiencing the same problem. If all other interfaces are experiencing the same problem, run show system buffers or look at `wfKernelBuffersFree` to see whether free buffers are available on the slot. You may want to monitor the buffers over a period of time to observe the low point of free buffers. If the buffer use fluctuates widely, then temporary periods of heavy buffer use may affect the line drivers. This problem can be resolved by adding more memory, allocating more memory to the global buffer pool, or statically defining a larger queue size for the receive or transmit driver. Temporary increments in the `lack-of-resource` attributes do not require any changes because higher-layer protocols usually adjust traffic windows during temporary congestion.

Next, check the dynamically allocated receive or transmit queue length for the interface driver. Currently allocated queue or ring length attributes usually end with a `TxQueueLength` or `RxQueueLength` suffix. The exception is HSSI, where the ring lengths can be obtained using `wfHssiRxRingLength` and `wfHssiTxRingLength` attributes. By default, buffers are fairly allocated among all interfaces in a slot or link module. Sustained lack of resources on one interface alone may indicate a higher density of data traffic on that interface and may require a larger allocation of buffers for that interface. Notice that lack of resources on the receive side is not necessarily paired with lack of resources on the transmit side.

```
mx2(18)# mget wfSyncEntry.57.*
wfSyncEntry.wfSyncTxQueueLength.8.1 64
mx2(19)# mget wfSyncEntry.58.*
wfSyncEntry.wfSyncRxQueueLength.8.1 64
```

You can increase the configurable queue lengths for the active receive and transmit queues using the `CfgRxQueueLength` and `cfgTxQueueLength` attributes, respectively, prefixed with `Cfg`. Increase the queue size in small (8 to 32 buffers) increments and observe the impact for several minutes. When you make a queue size change, recheck the active queue length to verify that the change took effect.

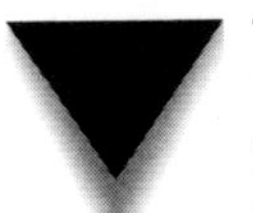

TIP: When you change a queue length, the line driver for the interface will be restarted, which will disrupt data traffic briefly. You thus should make the change after first disabling the interface or be prepared for the temporary disruption.

Applies to: `wfAsyncRxLackRescs`, `wfAsyncTxLackRescs`, `wfAtmizerIntfRxLackBufCredits`, `wfAtmizerIntfTxLackBufCredits`, `wfAtmizerVclRxLackBufCredits`, `wfAtmizerVclTxLackBufCredits`, `wfBisyncTxLackRescs`, `wfBisyncRxLackRescsChar`, `wfCSMACDLackRescErrorRx`, `wfLogicalLineLackRxResources`, `wfIsdbStatRxLacks`, `wfIsdbStatTxLacks`, `wfIsdnBriRxLackRescs`, `wfIsdnBriTxLackRescs`, `wfSyncLackRescRx`, `wfSyncLackRescTx`

Rejects

If a frame is not acceptable for processing by the line driver, it will be rejected, and a reject frame event will be sent to the remote device. On asynchronous interfaces, reject frame errors can be caused by data frames larger than the router's buffer size—this will occur if the remote device is using a larger frame size or a different stop data bit size.

```
show serial errors                                    Apr 01, 2000 10:12:19 [GMT-6]


RECEIVE ERRORS:

Slot/           Bad        Runt       Frame      Frames     Overflow
Conn  Circuit Frames       Frames     Rejects    Too Long   Frames
----- ------- ---------- ---------- ---------- ---------- ----------
2/4   S24              0          0         32          0          0
... ...
```

On serial lines, explicit mode addressing is used to filter frames destined for an incorrect address. If unique local and remote HDLC addresses are configured, a remote device will reject a frame if its local address does not match the received frame's address, which can occur due to misconfiguration of local or remote HDLC addresses. If this problem occurs only briefly, check the receive errors on the line. SDLC and LAPB lines can report both rejects and frame rejects (FRMR).

Applies to: `wfAsyncTxRejects`, `wfAsyncRxRejects`, `wfLapbRejectsTx`, `wfLapbRejectsRx`, `wfSyncRejectsTx`, `wfSyncRejectsRx`, `wfX25VcRejectTxs`, `wfX25VcRejectRxs`, `wfLapbPktStatsRejectTxs`, `wfLapbPktStatsRejectRxs`

Underruns

Underruns, sometimes called underflows, occur when the transmit driver empties the queue buffers faster than the higher-layer circuits deliver frames. This implies a temporary bottleneck between the protocol interfaces and the transmit driver. For example, high-speed interfaces (such as HSSI, Fast Ethernet, and FDDI) require that the transmit driver run more frequently than other, slower interfaces. If the transmit driver empties the buffer queue quickly, it may operate more efficiently if frames are delivered with a higher throughput.

```
show ethernet errors                                  Apr 04, 2000 10:14:27 [GMT-6]
... ...

TRANSMIT ERRORS:

Slot/ Frames     Underflow  Internal   Deadlock   Excessive  Late
Conn  Too Long   Errors     MAC Errors Errors     Collisions Collisions
----- ---------- ---------- ---------- ---------- ---------- ----------
 3/1           0          0          0          0          3          0
 3/2           0      45297          0          0          0          0
```

Transient underruns may not affect interface or link module performance significantly because of data traffic patterns. Underruns are most likely to occur on interfaces residing on link modules or net modules. When several interfaces within the same link module are operating at high speeds, the link interface can be tuned to operate more efficiently—data transfers between the processor and link modules can be delivered in larger chunks with a longer direct memory access (DMA) cycle.

Underflows with certain link modules using serial interfaces (using the MK5025 chipset) can be addressed by setting `wfSyncBurstCount` to unlimited. However, this setting does not have any impact on serial interfaces on Access routers. Ethernet interfaces on some link modules such as Quad Ethernet may report underflows as a result of unaligned frames (`wfCSMACDUnAlignedFrames`). In this case, all frames can be forcibly realigned, and the underflows may be eliminated by setting `wfCSMACDAlignmentMode` to all. In a similar fashion, underflows on HSSI interfaces can be addressed by increasing `wfHssiMemPageLength` to 256 bytes (the default is 32 bytes). ATM Adaptation Layer Coprocessor (ALC) underflows can be addressed by increasing `wfAtmAlcXmtQueueBurst` from its default value of 40 buffers, in increments of 10.

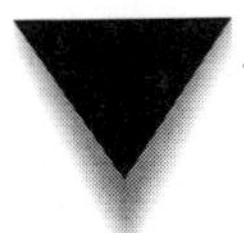

CAUTION: Tuning the link interface and coprocessor function will enhance interface performance, but it may also decrease performance if is overdone. Always observe the effect of a single change over a period of 5 to 10 minutes before making another change. To help you gauge the result, first monitor the underflow statistic to see the rate of change. Then make the next desired change and monitor the effect.

Applies to: `wfAsyncTxUnderFlows`, `wfAtmAlcCopDmaFifoUnderruns`, `wfBisyncTxUnderFlows`, `wfCSMACDUfloTx`, `wfLogicalLineTxUnderflows`, `wfDs1E1LineStatTxUnderflows`, `wfFDDIUnderrunTx`, `wfFddiXLineTxUnderruns`, `wfHssiTxUnderruns`, `wfIsdbStatTxUnderFlows`, `wfIsdnBriTxUnderFlows`, `wfSyncUnderFlowTx`

Overflows

The opposite of an underflow is an overflow, also sometimes called an overrun. Underflows occur on the transmit side; overflows happen on the receive side. If a driver is receiving frames from the medium faster than the processor can accept and process them, then the receive buffer ring (queue) will be overrun with frames. Like underflows, overflows may signify a condition where a link module is overloaded or there is temporary congestion preventing efficient operation. For example, a serial interface may be clocked to run at 6 Mbps, but not all eight ports on an octal-sync module may operate at the sustained rate (that is, 6 x 8 = 48 Mbps sustained rate) for an indefinite period.

```
show ethernet errors                          Apr 04, 2000 12:19:03 [GMT-6]


RECEIVE ERRORS:
```

Slot/ Conn	Circuit	CheckSum Errors	Alignment Errors	**Overflow Errors**	Frames Too Long	Symbol Errors	Internal MAC Errors	Late Collision
4/1	E41	0	0	**6**	0	0	0	0
4/2	E42	0	0	**8612**	0	0	0	0

Overflows can be resolved in the same fashion as underflows, by attempting to increase link interface communication between the processor and the link module. Overflows can happen, although rarely, as a result of a constraint imposed by the network design. If resolving overflows (or underflows) on one interface causes them on another interface, you should consult with Nortel Networks Customer Service to evaluate the implementation design of the interface or link module.

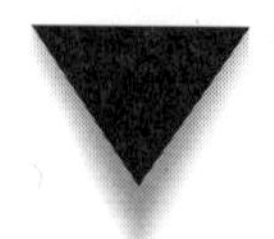

TIP: Both underruns and overflows are uncommon events; if they occur, check statistics for other errors (such as receive errors) and resolve them first. Several error statistics often increment simultaneously, giving the impression that each error is unique when, in fact, all are the result of the same problem.

Applies to: `wfAsyncRxOverFlows`, `wfAtmAlcCopBufOverflows`, `wfAtmAlcCopDmaFifoOverruns`, `wfBisyncRxOverFlows`, `wfFDDIOverrunRx`, `wfFDDIRingOverrunRx`, `wfFDDISmtRingOverrunRx`, `wfFddiXLineRxOverruns`, `wfHssiRxOverruns`, `wfHssiRxRingOverruns`, `wfIsdbStatRxOverFlows`, `wfIsdbPortStatOverrunErrors`, `wfLogicalLineRxOverflows`, `wfLogicalLineRxDescOverflows`, `wfDs1E1LineStatRxOverflows`, `wfSyncOverFlowRx`, `wfHwCompCompressionRingOverflows`, `wfHwCompDecompressionRingOverflows`

Queue Replenish Misses

In the discussion of lack of resources, we mentioned the use of the buffer queues for reception and transmission. When a frame is received, it is stored in a buffer and sent to the processor module for processing. The packet buffer becomes available again after the data frame is processed and forwarded. The receive line driver periodically replenishes its buffers. It will also attempt to replenish its queue buffers when a new data frame arrives but it does not have a free buffer for storage. On rare occasions, heavy buffer usage may cause the replenish attempts to fail, because there is no free buffer for storage, thereby temporarily starving the receive driver. A failed replenish attempt is called a miss.

Queue replenish misses occur only on the receive line driver—outgoing frames are delivered directly to the transmit line driver using a buffer. Replenish misses can occur if buffer use is so high or fluctuates so rapidly that buffers are unavailable during certain periods. If replenish misses occur periodically or frequently, then the total available buffer space must be increased to accommodate the high buffer use. Momentary and one-time replenish misses can be overlooked. The replenish misses statistic must be obtained directly from the corresponding MIB object.

```
fp3(3)# mget wfFddiEntry.41.*
wfFddiEntry.wfFDDIRxReplenMisses.4.1 0
wfFddiEntry.wfFDDIRxReplenMisses.12.1 723913
```

Total available buffer space can be increased in two ways. The existing memory can be recarved to contain more global memory and, therefore, more buffers, or the current total memory can be increased. Before upgrading your total memory, check the extent of buffer use. The kernel maintains the `wfKernelBallocFail` statistic, which indicates the number of times when buffer allocation requests could not be completed. A monotonous increase in this statistic indicates that even processor-based resources sometimes cannot allocate buffers; to resolve this problem, you will need to add more memory to the global buffer pool.

Applies to: `wfAsyncRxReplenMisses`, `wfAtmAlcRcvReplenMisses`, `wfBisyncRxReplenMisses`, `wfCSMACDRxReplenMisses`, `wfLogicalLineRxReplenMisses`, `wfFDDIRxReplenMisses`, `wfKernelReplenEmpty`, `wfIsdnBriRxReplenMisses`, `wfSyncRxReplenMisses`, `wfTokenRingRxReplenMisses`

Discards and Clips

When valid frames received on an interface cannot be copied to a packet buffer, such as when receive resources are not available, the frames are discarded, or dropped. Normally, these frames will already have passed integrity and checksum tests. Dropped frames usually are signaled by a lack-of-resources or other congestion statistic; addressing the lack of resources will usually eliminate the discards. If the value continues to increment on an interface that supports multiprotocol encapsulation (MPE), then the remote device may be sending frames for a protocol that the local interface is not configured to support. For example, IPX frames received on an IP-only service record will be discarded because the higher-layer protocol interface is not configured.

```
show frame-relay stats errors                        Apr 01, 2000 06:21:16 [GMT-6]


            -----------Last-Error-----------              Outbound Inbound
Interface     Type                Time           Faults  Drops    Discards
--------- ----------- -------------------- ------ -------- --------
4/3         reset       Mar 28 2000 18:12:23       0        18    232357
```

On the outbound side, the discarded frames are usually called clips. These frames are also valid and have passed integrity and checksum tests. Clips occur on the transmit side because of lack of transmission bandwidth on the line. For example, if frames are delivered to the line driver at a rate faster than the line rate, then the transmit driver must clip some packets. Broadcast storms or bridge loops can cause excessive use of transmission bandwidth, thereby resulting in clips.

Outbound frames can also be clipped by priority-queuing and traffic-shaping algorithms. If frames are being clipped or discarded at this level, check the configuration to ensure that the correct type of data packets is being clipped.

Applies to: wfAtmDxiDiscardedFrames, wfAtmAlcXmtPacketClips, wfAtmAlcCopRcvDropCells, wfAtmizerIntfTxDrvClipCount, wfAtmizerVclTxClipFrames, wfCSMACDTxClipFrames, wfFDDITxClipFrames, wfFrErrDiscards, wfFrCircuitDiscards, wfFrCctErrorDiscards, wfFrServiceRecordifInDiscards, wfFrServiceRecordifOutDiscards, wfFrCircuitShapedHiClippedPkts, wfFrCircuitShapedNormalClippedPkts, wfFrCircuitShapedLoClippedPkts, wfFrMlStatsPQHiClippedPkts, wfFrMlStatsPQNormalClippedPkts, wfFrMlStatsPQLoClippedPkts, wfHssiInDiscards, wfHssiTxClipFrames, wfHssiOutDiscards, wfIfInDiscards, wfIfOutDiscards, wfIsdbStatRxDataLinkDiscards, wfPppMlStatsPQHiClippedPkts, wfPppMlStatsPQNormalClippedPkts, wfPppMlStatsPQLoClippedPkts, wfSyncTxDropPackets, wfSyncRxDropPackets, wfTokenRingInDiscards, wfTokenRingOutDiscards, wfTokenRingTxClipFrames

Collisions

In BayRS, collisions are unique to Ethernet-based interfaces. Ethernet is based on the CSMA/CD specification, in which contention is inherent, which leads to the possibility of collisions. All stations are required to honor the interframe gap, during which they are required to ensure that no other device is attempting to transmit. Despite the interframe gap, collisions are still possible on CSMA/CD interfaces.

Collisions occur more frequently as data traffic increases on an Ethernet segment when more devices attempt to transmit. Occasional collisions are acceptable; however, a continuous increase in collisions indicates a potential problem. If collisions increase substantially, performance tends to degenerate. Single collisions indicate a successful frame transmission following a collision, which means that the router had fair access to the medium. A bit more severe are multiple collisions, indicating that several attempts to transmit resulted in continuous collisions. Late collisions indicate a condition where the collision was detected after the first 64 bytes of a data frame were transmitted (wfCSMACDLateCollnTx) or received (wfCSMACDLateCollnTx). This problem usually occurs because the cable length is longer than allowed by the appropriate specification. Check the fiber or UTP cable run and shorten it so that it is within the allowed limit. In some cases, collisions occur as a result of an excessive number of stations on the segment.

```
show ethernet errors                                  Apr 04, 2000 10:14:27 [GMT-6]


RECEIVE ERRORS:

Slot/        CheckSum Alignment Overflow Frames   Symbol   Internal   Late
Conn Circuit Errors   Errors    Errors   Too Long Errors   MAC Errors Collision
----- ------ -------- --------- -------- -------- -------- ---------- ---------
 3/1  E31           0         0        0        0        0          0         2
 3/2  E32           0         0        0        0        0          0      2812

TRANSMIT ERRORS:

Slot/ Frames     Underflow  Internal    Deadlock   Excessive   Late
Conn  Too Long   Errors     MAC Errors  Errors     Collisions  Collisions
----- ---------- ---------- ----------- ---------- ----------- -----------
 3/1           0          0           0          0           0           0
 3/2           0          0           0          0         176         251
```

Excessive collisions occur when the CSMA/CD interface cannot successfully transmit a frame because it encountered 16 consecutive collisions. Both multiple and excessive collisions indicate that data traffic on the segment is significantly high, such as when broadcast storms or bridge loops occur. If the router is attached to a repeater or shared hub segment, try to provide it with a dedicated switched port to deliver only unicast and broadcast/multicast data link frames. If the collisions persist, you must identify the source of the broadcast storm or loop and resolve it.

Applies to: `wfCSMACDSingleCollisionFrames`, `wfCSMACDMultipleCollisionFrames`, `wfCSMACDLateCollnTx`, `wfCSMACDLateCollnRx`, `wfCSMACDExcessvCollnTx`

When the term is applied as is, collisions (`wfIsdnBriTxDChanCollisions`) can also occur over an ISDN D-channel during call signaling and X.25 virtual calls. The collided frame is retransmitted, and it is unlikely that the D-channel will be congested with signaling or X.25 traffic. Collisions on ISDN/BRI interfaces will not significantly affect operation.

APPLICATIONS AND PROTOCOLS

When you begin to work on a data link issue, you must know the protocol you will be working with. Ethernet interfaces, for example, use only the CSMA/CD protocol, while serial interfaces can support a variety of data link protocols such as PPP, frame relay, and SMDS. Just as the line number is helpful in determining the media type, the WAN protocol attribute is helpful in determining the data link protocol if you're unsure. Usually, calls to the network operation center or ticketing system will yield only the circuit name or circuit number.

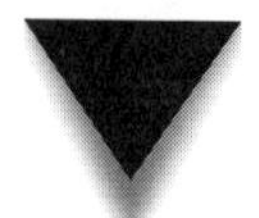

NOTE: The protocol used on a line cannot be modified by simply changing the WAN protocol attribute. Both existing protocol interfaces and the circuit must be deleted and reconfigured using the newly chosen WAN protocol. The WAN protocol attribute will automatically be set by BCC/Site Manager during reconfiguration.

The WAN protocol of an interface can be checked by running the **show media summary** command, where media represents the physical medium type.

```
show hssi summary                                  Apr 04, 2000 13:23:04 [EDT]

                                                         Total
Slot/          Admin                       WAN     Oper  Time
Conn  Circuit  State HW Address            Protocol State Up/Down     Errors MTU
----- -------  ----- ----------------- ---------- ------- ---------- ------ ----
 2/1  H21      up    00.00.A2.42.98.3D framerelay up      281h43m12s      0 4608


show mct1e1 summary                                Apr 04, 2000 13:25:15 [EDT]

Slot/                   Logical
<Module>/               Line    Line     Line                Wan
Conn      Circuit       Index   Type     Number State        Protocol
--------- ------------ ------- -------- ------ ----------   --------
3/1       MCT1_3_1_PPP 1       ds1e1    903101 up            ppp
```

Alternately, you can navigate to the appropriate media context and check the WAN-protocol attribute. Depending on the medium, performing a **get** or **mget** operation on the appropriate WAN protocol attribute also will help you determine the protocol being used. WAN protocol attributes include `wfSyncWanProtocol`, `wfHssiWanProtocol`, `wfLogicalLineWanProtocol`, `wfDs1E1LineCfgWanProtocol`, `wfIsdnBChanWanProtocol`, and `wfLapbWanProtocol`.

```
box(3)# mget wfHssiEntry.12.2.1
wfHssiEntry.wfHssiWanProtocol.2.1 framerelay
```

Once you determine that a particular line is running a certain protocol, you need to look at the WAN protocol state and the line driver state. The line driver state is the state of the physical or logical line being used. In the show listing, for example, the two line drivers are HSSI 2/1 and MCT1 3/1. Both lines show an up state, which indicates that the physical lines are operational. If either the line or the WAN protocol reports a status of notpres or notpresent, then the required line driver or protocol driver, respectively, is not loaded. Consult the "Media State" section in Chapter 4 to resolve the notpresent condition for media lines.

The notpresent condition for the protocol state is resolved by enabling the appropriate slot's bit in the protocol slot mask. The protocol MIB object is `wfProtocols` and must be accessed directly—there is no corresponding BCC object. Site Manager and BCC automatically load the appropriate drivers and protocols when an interface is initially configured. A subsequent discrepancy or configuration error will be the likely cause of a protocol's being unloaded from a slot.

Look at the PPP interface on MCT1_3_1 in the preceding example. The show mct1e1 summary listing indicates that the line driver is operational, but the PPP line protocol itself reports a notpresent condition.

```
show ppp interfaces                                    Mar 12, 2000 11:32:12 [EST]

... ...

PPP over Logical Line
---------------------
         Slot/
         <Module>/ Driver                                 Line
Circuit  Conn      State      Protocol   State   State
-------- --------- ---------- ---------- ------- ----------
MCT1_3_1 2/1       up         LCP        0       notpresent
                              IP         0
                              IPX        0
```

Notice that the Driver State column shows the actual line driver state, while the Line State column shows the protocol (PPP) line state.

The **loadmap** command displays a list of applications that are loaded in memory on each slot, which indicates that the ppp application is not loaded. Looking at the PPP protocol configuration for the dynamic loader, we can see that PPP is not loaded in any of the slots. There is one global instance of the `wfProtocols` object, with instance identifier 0.

```
fxc9(12)# mget wfProtocols.25.0
wfProtocols.wfPPPLoad.0 0
```

The PPP protocol is loaded onto slot 3, as discussed in Chapter 2, in the section "Slot Mask." The slot mask for multiple slots can be obtained using the slot mask calculator that you can download from the book's web site (http://www.mahalingam.com), or you can calculate the slot mask by adding slot values listed in Table 5-5.

TIP: Slot 1 on the BLN and slot 7 on the BCN are occupied by the system resource modules, which do not run any applications. To prevent confusion, these slot bits must not be enabled on the respective platforms. They can be enabled on VME-based LN and CN platforms.

```
fxc9(13)# mset wfProtocols.25.0 0x20000000
fxc9(14)# log -fd -eLOADER

#   113: 03/12/2000 11:58:32.039  DEBUG    SLOT  3  LOADER             Code:  43
Image ppp.exe loaded successfully from 3:bn.exe
```

In some situations, you might observe that the slot mask for the specific protocol is configured correctly, but the protocol still is not loaded. In the following example, check-

Slot Number	Slot Mask	Slot Number	Slot Mask
1	0x80000000	8	0x01000000
2	0x40000000	9	0x00800000
3	0x20000000	10	0x00400000
4	0x10000000	11	0x00200000
5	0x08000000	12	0x00100000
6	0x04000000	13	0x00080000
7	0x02000000	14	0x00040000

Table 5-5. Slot Masks

ing the log for LOADER events yields the cause of the problem—the ppp application has been removed from the runtime image archive and is not available. In this case, the runtime image must be customized (see "Customizing Images" in Chapter 3), or a new image including the PPP software must be transferred to flash.

```
#  142: 03/12/2000 03:17:05.959  DEBUG    SLOT  3  LOADER              Code:  17
Could not locate application ppp.exe in 3:bn.exe... not loaded
```

Circuits, Lines, and Protocols

In some cases, you will need to correlate the circuit name or number with the physical or logical line and determine the protocol or interfaces involved. In others, you will need to correlate the IP address or other network interfaces with the circuit number and the physical line. For example, if you have a circuit named S21_MCI_FR, first find the instance identifier of the circuit in the `wfCircuitNameEntry` object. (We will set aside the fact that we already know that the 'S21' portion of the circuit name indicates serial port 1 on slot 2.) Recall that in Chapter 2 we discussed circuit naming and why it is desirable to have circuit names contain the physical line information.

In this example, you will see how to identify the line and the protocol using the MIB; later we will discuss the easier-to-use **show circuit summary** and **show protocol interface** commands.

```
box(8)# mget wfCircuitNameEntry.3.*
wfCircuitNameEntry.wfCircuitName.1 E11_Net4
wfCircuitNameEntry.wfCircuitName.2 E12_Net5
wfCircuitNameEntry.wfCircuitName.3 S21_MCI_FR
wfCircuitNameEntry.wfCircuitName.4 S22_PPP2Macon
...
```

Circuit 3 with instance ID 3 represents the S21_MCI_FR circuit. There is a corresponding `wfLineMappingEntry` MIB object instance, which contains both the circuit number and the unique line number it is associated with.

```
box(9)# mget wfLineMappingEntry.3.*
wfLineMappingEntry.wfLineMappingCct.101101 1
wfLineMappingEntry.wfLineMappingCct.101102 2
wfLineMappingEntry.wfLineMappingCct.202101 3
wfLineMappingEntry.wfLineMappingCct.202102 4
...
```

Observe that the instance ID of the line mapping entry is the line number itself. The line number for this circuit can also be fetched from the `wfCircuitNameEntry` object itself, using the `wfCircuitLineList` attribute. However, the `wfCircuitLineList` attribute is an opaque (octet) string that should be converted to an ASCII display string first for viewing. The **octetfmt** command used in the following code is a TI command and must be called using **tic** if you are using BCC.

```
# octetfmt linenumber ASCII_STRING wfCircuitNameEntry.wfCircuitLineList.3
# echo $linenumber
202101:
```

Note that some circuits, such as multiline and multilink circuits, can have multiple physical or logical lines. In this case, the line list will contain multiple numbers, as in the case here for circuit S22_PPP2Macon, which uses the primary line 202102.

```
# octetfmt linenumber ASCII_STRING wfCircuitNameEntry.wfCircuitLineList.4
# echo $linenumber
202102:202103:202104
```

Protocols such as frame relay and ATM can contain multiple service records; we discussed the subcircuits in Chapter 2. Line numbers in these cases are mapped to the primary circuit—the one running the link management protocol. Individual virtual circuits, permanent and switched, are tied to their respective subcircuits. The ATM protocol represents the primary circuit number using the `wfAtmInterfaceConfEntry.wfAtmInterfaceCct` attribute, and frame relay represents it using `wfFrIfDlcmiEntry.wfFrDlcmiCircuit`. Subcircuit detail, as mentioned earlier, is stored in the appropriate VC: `wfAtmVclConfEntry.wfAtmVclCct` for ATM VCs, and `wfFrVCircuitEntry.wfFrCircuitSubCct` for FR VCs. The subcircuit information may be empty for a VC if it is part of group mode service record and is not in hybrid- or direct-access mode.

The **show *circuit* summary** or **show *protocol* interface** commands summarize the details reported earlier for easier interpretation, where *protocol* is the network layer protocol.

```
show ip interfaces                                    Apr 04, 2000 11:45:31 [GMT-6]

Circuit  Cct # State   IP Address       Mask              MAC Address
----------------------------------------------------------------------------
S21       2    up      80.23.10.1       255.255.255.252   00.00.A2.CB.B5.8D
E31       1    up      80.23.15.1       255.255.255.0     00.00.A2.CB.B5.8C

show circuit summary                                  Apr 04, 2000 11:46:32 [GMT-6]

               Interface  Circuit   IP                  Related
Cct#  Circuit  Type       Type      Address             Line List
----------------------------------------------------------------------------
1     E31      csmacd     normal    80.23.15.1          103101
2     S21      sync       master    80.23.10.1          202101
```

NOTE: The **show circuit** command is available from the book's Web site. It is an unsupported script in the public domain.

As always, it is most rudimentary to correlate the circuit and physical line by simply looking at the circuit name, if the circuit is named appropriately. In the case of the frame relay circuit in the preceding example, the name S22_MCI_FR tells us that the circuit resides on S22, the serial line on port 2, slot 2. Circuit names are unique, and unless changed from the default, they usually follow the format *MUSP*, where *M* represents the media type, *S* represents the slot number, and *P* equals the port number. The *U* represents the unit number in an ASN stack and is not used on other routers; on ASN routers, the *S* represents the net module number. Thus, you must be aware of the router type you are working with; O131 on an ASN represents Token Ring port 1 on net module 3, stack unit 1, but on a BCN it represents Token Ring port 1 on slot 13. Table 5-6 lists common circuit name prefixes and the media that they represent. Customizations to circuit names are

Prefix	Medium	Prefix	Medium
E	Ethernet (10/100/1000)	O	Token Ring (4/16)
F	FDDI	H	HSSI
T1	T1	E1	E1
FT1	Fractional T1	FE1	Fractional E1
S	Serial	A	ATM
MCT1	Multichannel T1	MCE1	Multichannel E1
PPP_ Demand	Variable[1]	FR_ Demand	ISDN

[1] Depends on the outbound line selected by the switched service application.

Table 5-6. Common Circuit Name Prefixes

Some references are frequently used and are very valuable in the troubleshooting process. In the following pages, we start with an architectural reference to the BayRS operating system. The loading of protocols, link module drivers and link controller drivers are configured using slotmasks—the representation of slotmasks is illustrated and a table of slotmask values is provided.

In Chapter 6, we discuss the troubleshooting of IP networks. The understanding of subnetting and classes is essential for administering IP networks. However, a handy reference is useful for computing, validating and analyzing an IP address and addressing schemes. An illustrated reference with IP tables is included for this purpose. The OSI stack is of common knowledge to readers; TCP/IP is the most dominant protocol suite today and the location and interaction of various transport and session layer protocols is also illustrated. Similar to TCP/IP, Ethernet is the dominant medium for network connectivity. The IEEE 802.3 Ethernet and the Ethernet Version 2 frame formats are illustrated in detail, with a reference table of common EtherType values.

Table of Contents

BayRS Architecture

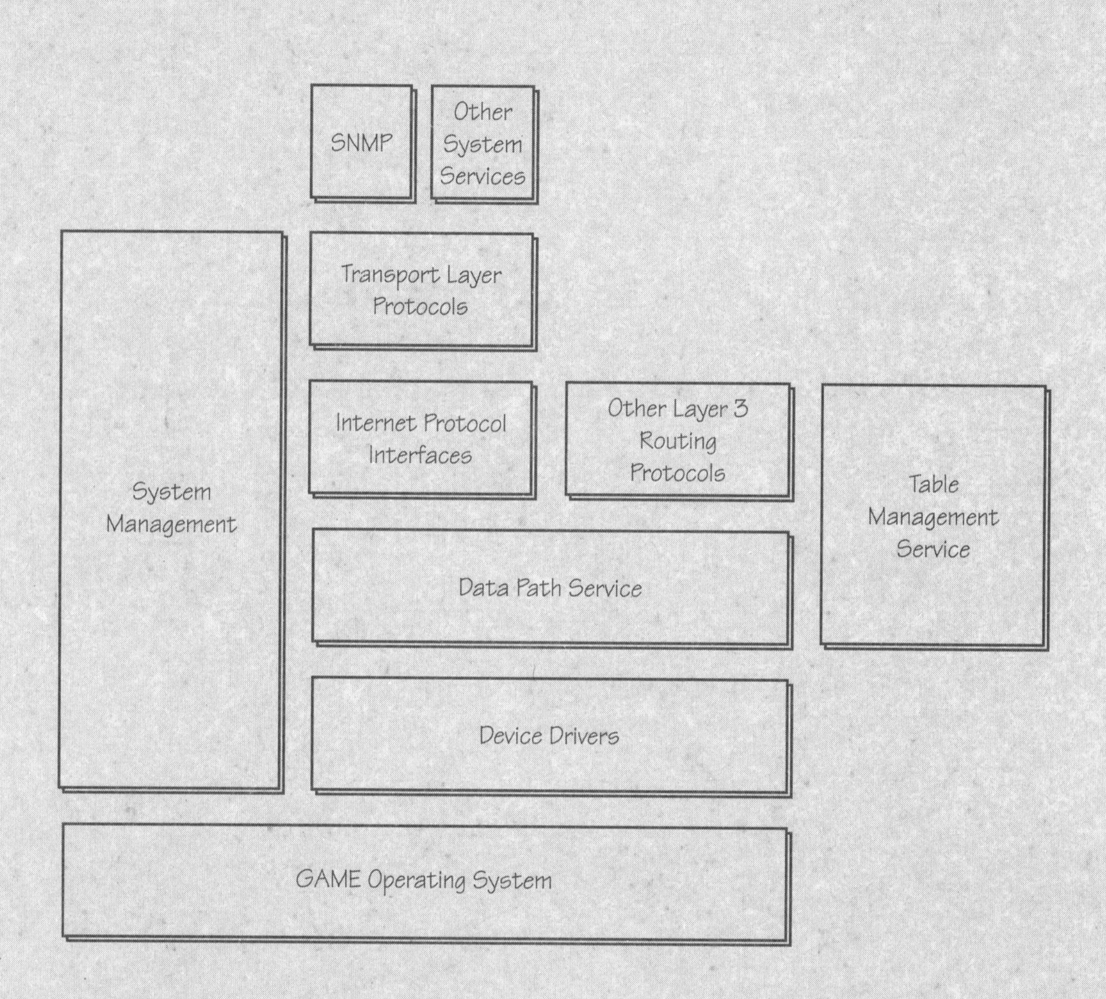
SNMP
Other System Services
Transport Layer Protocols
System Management
Internet Protocol Interfaces
Other Layer 3 Routing Protocols
Table Management Service
Data Path Service
Device Drivers
GAME Operating System

Slot Mask Reference

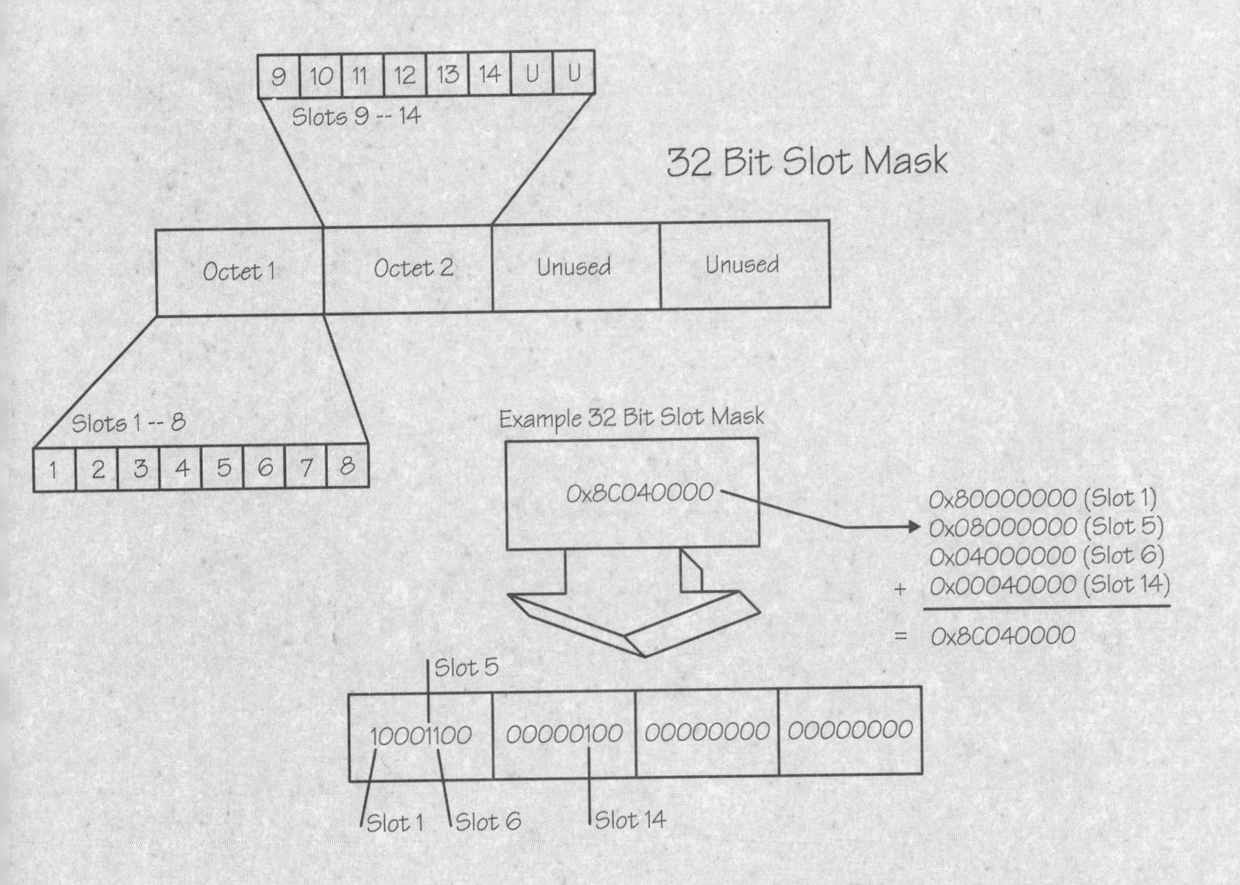

Slot Mask Table

Slot Number	Slot Mask	Slot Number	Slot Mask
1	0x80000000	8	0x01000000
2	0x40000000	9	0x00800000
3	0x20000000	10	0x00400000
4	0x10000000	11	0x00200000
5	0x08000000	12	0x00100000
6	0x04000000	13	0x00080000
7	0x02000000	14	0x00040000

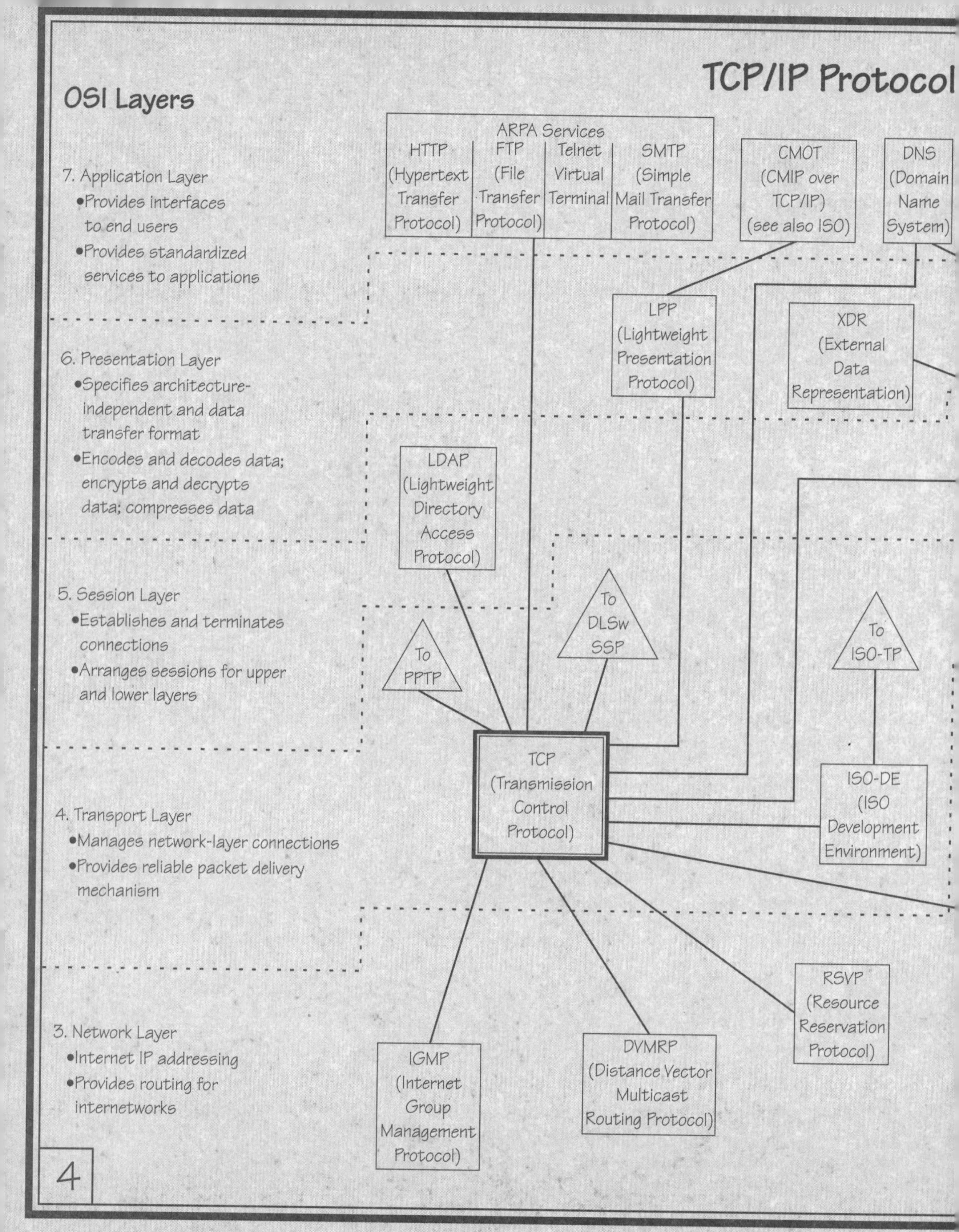

TCP/IP Protocol
OSI Layers
7. Application Layer
•Provides interfaces to end users
•Provides standardized services to applications
6. Presentation Layer
•Specifies architecture-independent and data transfer format
•Encodes and decodes data; encrypts and decrypts data; compresses data
5. Session Layer
•Establishes and terminates connections
•Arranges sessions for upper and lower layers
4. Transport Layer
•Manages network-layer connections
•Provides reliable packet delivery mechanism
3. Network Layer
•Internet IP addressing
•Provides routing for internetworks
ARPA Services
HTTP (Hypertext Transfer Protocol)
FTP (File Transfer Protocol)
Telnet Virtual Terminal
SMTP (Simple Mail Transfer Protocol)
CMOT (CMIP over TCP/IP) (see also ISO)
DNS (Domain Name System)
LPP (Lightweight Presentation Protocol)
XDR (External Data Representation)
LDAP (Lightweight Directory Access Protocol)
To DLSw SSP
To PPTP
To ISO-TP
TCP (Transmission Control Protocol)
ISO-DE (ISO Development Environment)
RSVP (Resource Reservation Protocol)
IGMP (Internet Group Management Protocol)
DVMRP (Distance Vector Multicast Routing Protocol)

Suite

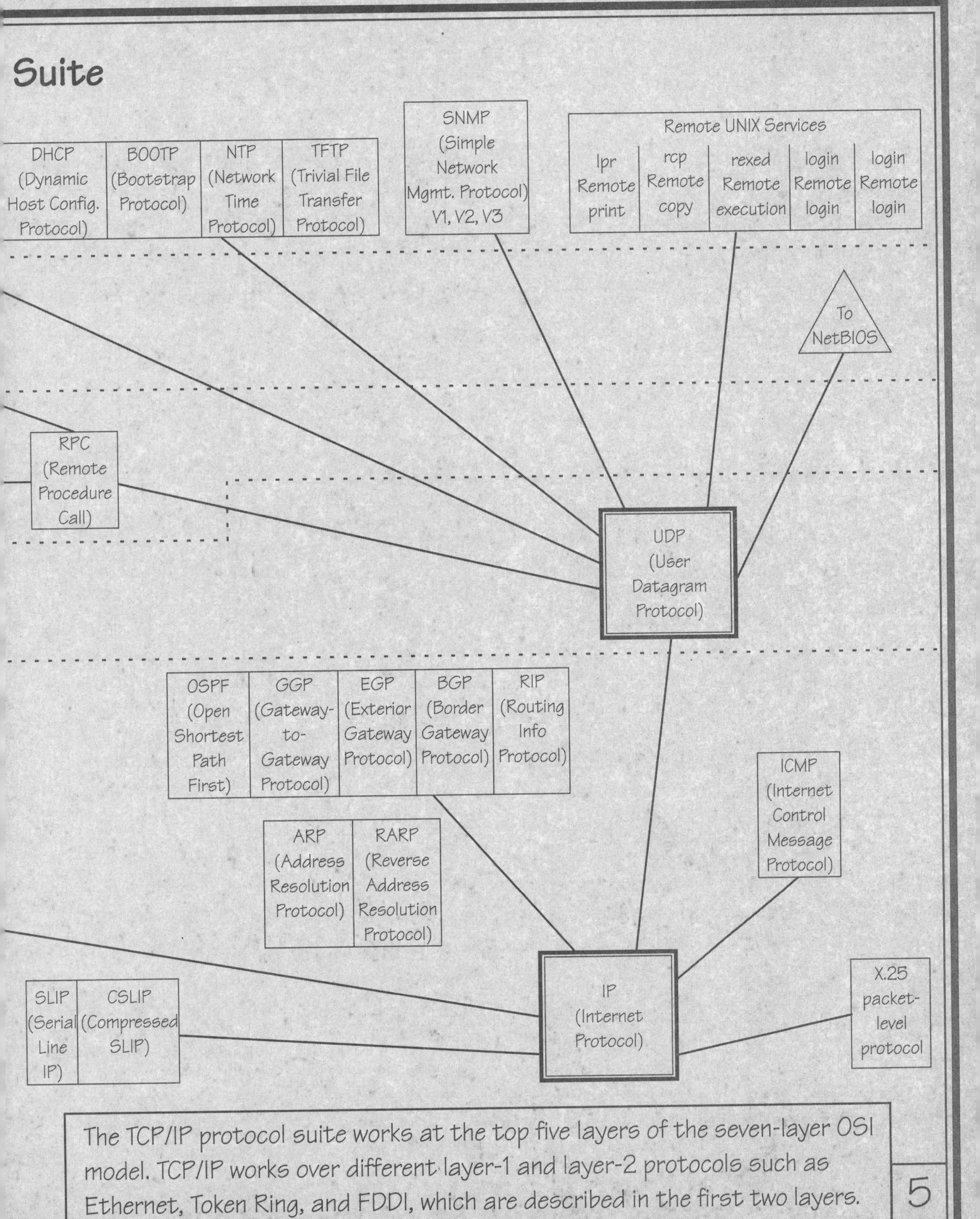

The TCP/IP protocol suite works at the top five layers of the seven-layer OSI model. TCP/IP works over different layer-1 and layer-2 protocols such as Ethernet, Token Ring, and FDDI, which are described in the first two layers.

Guide to IP

214.96.117.220

11011100 (220 in binary)

128 64 32 16 8 4 2 1 = 255

1 + 1 + 0 + 1 + 1 + 1 + 0 + 0 = 220

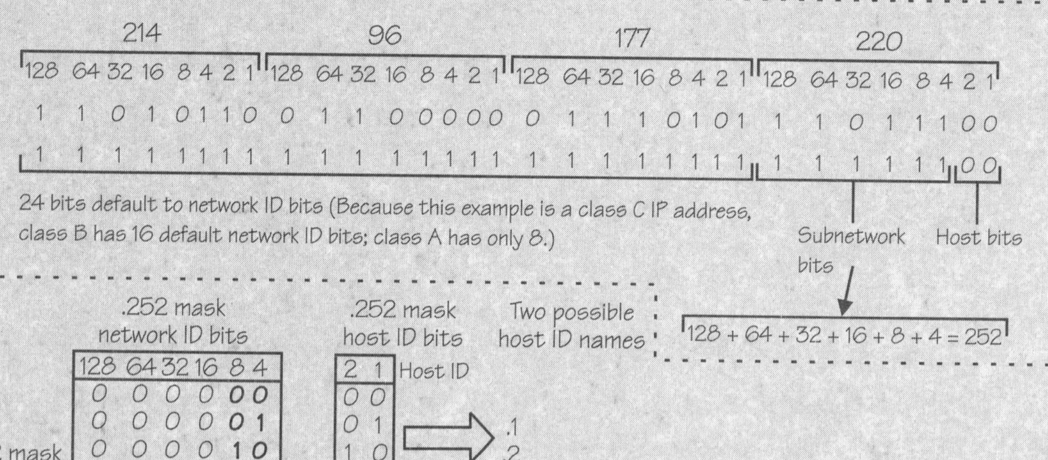

24 bits default to network ID bits (Because this example is a class C IP address, class B has 16 default network ID bits; class A has only 8.)

Host bits

How a .252 mask is applied

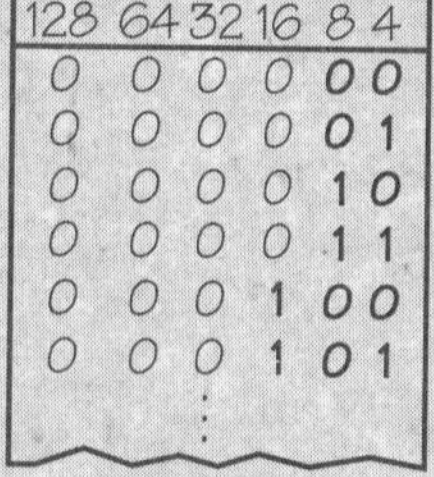

.252 mask network ID bits
128 64 32 16 8 4
0 0 0 0 0 0
0 0 0 0 0 1
0 0 0 0 1 0
0 0 0 0 1 1
0 0 0 1 0 0
0 0 0 1 0 1

A .252 mask has up to 64 subnet addresses

2	1	Host ID
0	0	
0	1	.1
1	0	.2
1	1	

.252 mask host ID bits — Two possible host ID names

Each subnet defined using the .252 mask can have up to two host addresses (addresses of all 0's or all 1's are reserved).

Using the .252 subnet mask, up to 64 LANs with up to two hosts each yields a maximum of 128 IP host addresses.

8 subnets

16 subnets

32 subnets

Destination IP Address	Subnet Mask	
11010110011000000111010111011100	11111111111111111111111111111100	← .252 mask
10101101100101110001001111001110	11111111111111111111111111110000	← .240 mask

Mask	Number of Network ID Bits	Example Notation	Number of Host ID Bits	Number of Subnets	Total Number of Host IP Addresses
.192	26	195.211.12.1/26	6	4	62
.224	27	195.211.12.1/27	5	8	30
.240	28	195.211.12.1/28	4	16	14
.248	29	195.211.12.1/29	3	32	6
.252	30	195.211.12.1/30	2	64	2

Addressing

This is an IP address shown in dotted decimal format. IP addresses are composed of four octets.

Machines work only with bits, not decimal numbers. Bits are set to either 0 or 1. An octet contains 8 bits, an IP address 32 bits.

Each column in an IP octet has a fixed value. A column value is counted only when the bit in the column is set to 1. The octet's decimal number is the total of all bits set to the value of 1; the maximum is 255.

A subnet mask tells the router which of the IP address's bits to use for subnetworks. The mask is indicated by a string of all 1-bits.

Usually, subnet masks end within an octet. The decimal total of all 1-bits in the subnet option gives it its number. The 1-bits in this example total to the .252 mask.

The subnet mask's length determines how many network and host addresses are possible. Each subnet mask has its use. For example, the .252 mask has two hosts per network, and therefore is used for point-to-point WAN connections.

Depending on which mask is used, between 4 and 64 subnets can be squeezed from a single class C block of addresses.

The router stores the mask alongside the IP address in its routing table. It tells which bits identify subnetworks instead of hosts.

There is a direct trade-off between subnets and hosts. The more you have of one, the fewer you can have of the other. As shown in this table, each subnet mask strikes a different trade-off.

Ethernet Frame Formats

IEEE 802.3 Frame

Preamble	SFD	DA	SA	Len	Data	Pad	FCS
7 bytes	1	6	6	2	← 46-1500 →		4

Preamble: A 7-octet stream with alternating 1s and 0s (101010....)

SFD: The start frame delimiter, 0 x AB

Destination & Source address:

I/G	U/L	46 bits

I/G: Represents an individual (0) or group (1) address
U/L: Represents a universal (0) or locally administered (1) address

Length: The length of the data and pad, ranging from 46-1500 bytes

FCS: The frame check sequence, a CRC-32 value

Ethernet V2 Frame

← 64-1518 → (DA through FCS)

Preamble	DA	SA	Type	Data	FCS
8 bytes	6	6	2	46-1500	4

Preamble: The preamble and SFD are integrated together from the 802.3 frame into an 8-byte preamble

Type: Otherwise called the Ethertype, indicates the higher layer protocol data carried in the payload

Hex Value	Description
2600	XNS IDP
0800	Internet Protocol
0805	X.25 Level 3
0806	ARP
0BAD	Banyan Vines
8035	Reverse ARP
809B	AppleTalk
80FF-8103	Nortel Networks
8137-8138	Novell
86DD	IP version 6
8847-8848	MPLS

usually made at the end of the circuit name for easy identification of the circuit (for example, O131 may be customized as O131_FEP rather than FEP or FEP_O131). Figure 5-10 illustrates the use of the default circuit name on various platforms.

Bay Standard

The Bay Standard protocol is commonly referred to as the Wellfleet Standard protocol. Other names used for this protocol are Bay/Wellfleet Proprietary HDLC and, simply, HDLC encapsulation. HDLC is a simple point-to-point link control protocol, and its legacy frame format was typically used for transporting protocol data belonging to a simple type.

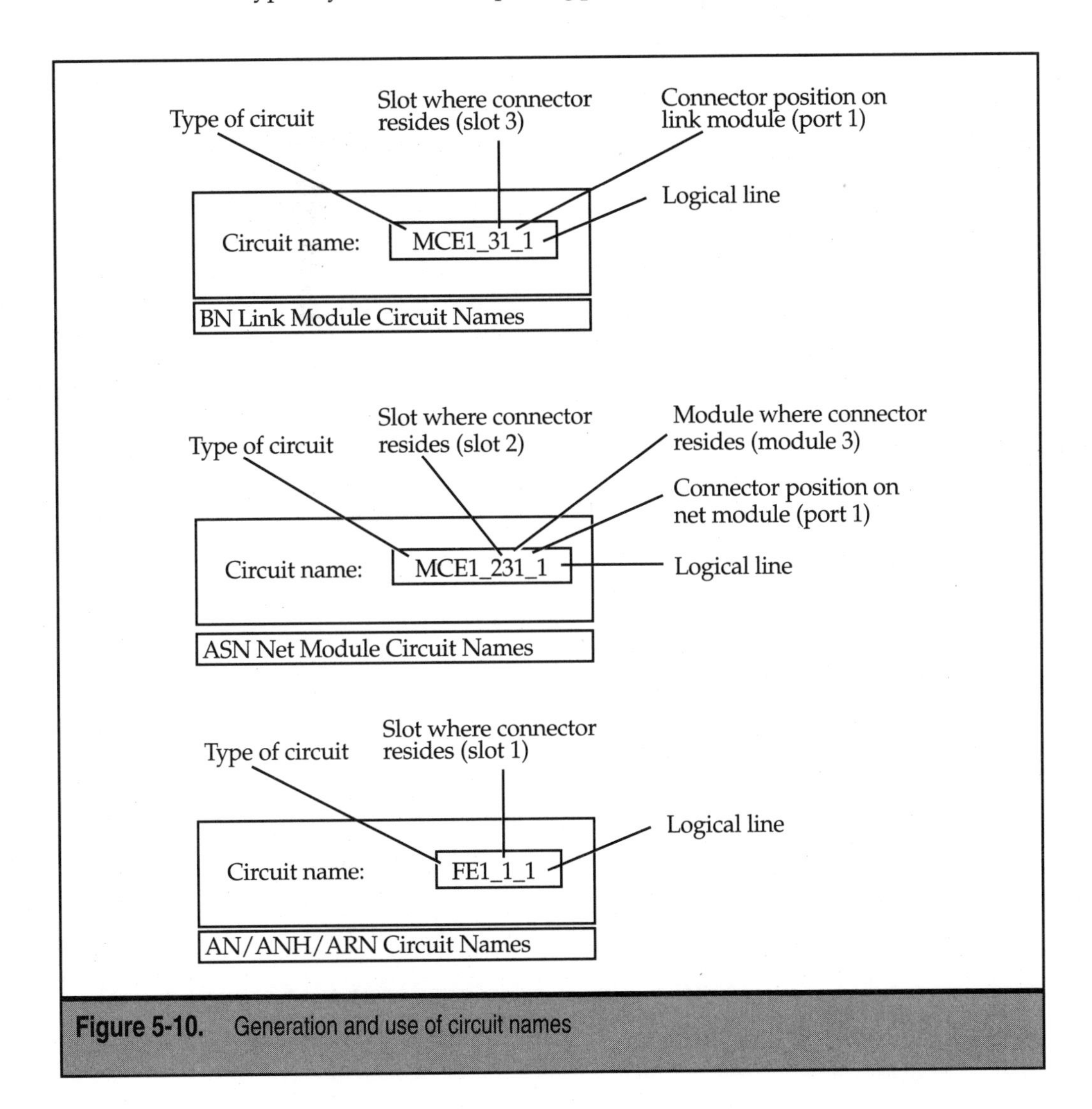

Figure 5-10. Generation and use of circuit names

Routers need to transport multiprotocol data across point-to-point links, and data link frames usually have a protocol type indicator to help the receiver deliver the traffic to the appropriate protocol stack (interface). HDLC's legacy frame format does not contain a protocol type identifier, but still its simple design is suitable for most point-to-point connections. Bay Standard capitalizes on the simple HDLC design and encapsulates a regular Ethernet II frame into the HDLC frame. Another Bay router on the remote end receives the HDLC frame, decapsulates the Ethernet II frame inside, and processes it based on the frame's protocol type. Figure 5-11 shows the HDLC and Bay Standard frame formats.

One of the most common errors made with Bay Standard protocol is the misconfiguration of BOFL and the BOFL timer. Each end of a Bay Standard circuit must receive BOFL packets every bofl-timeout period, and if five consecutive BOFL packets are missed, the line is brought down. On high-quality but low-throughput lines, BOFL packets can be disabled to conserve bandwidth; note that this must be done on both ends.

The service (`wfSyncService`) attribute also must be matched on both ends. By default, the router uses LLC1 service by including the address and control fields in the frame; these can be eliminated to improve bandwidth use by setting the service to transparent. However, if the service is mismatched between two routers, then frames may be rejected, and the routers will not communicate any higher-layer protocols. BayRS also uses explicit addressing 0x07 for both local and remote addresses, which must be correctly matched on both ends for received frames to be processed. Here also, mismatched HDLC addresses (the `local-address` attribute on router A must match the `remote-address` attribute on router B and vice versa) will cause the router to reject frames.

On Bay Standard lines, the higher-layer circuit and multiprotocol interfaces are automatically initialized and become operational when the LLC or transparent service be-

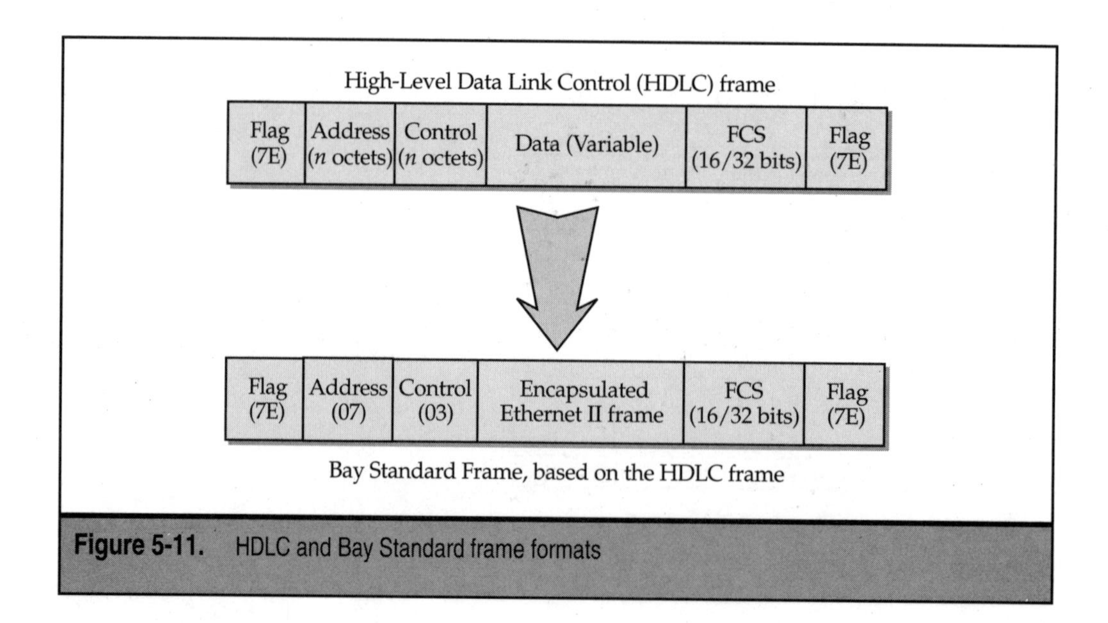

Figure 5-11. HDLC and Bay Standard frame formats

comes available, and the protocol interfaces will continue to be up and operational when there are small line level issues such as receive errors and runts. You should use BOFL packets because it is not possible to determine the actual functional status of protocols when data link layer issues could be causing problems.

The DSR signal is also crucial to enable the router to recognize the availability of the DCE. Occasionally, it may not be possible to deliver the DSR signal, or a custom cable may not provide the DSR pin. The router generates a warning when the DSR is lost and will subsequently disable the line driver.

```
#   93: 05/04/1999 13:01:04.683  WARNING  SLOT 11  SYNC                Code:  49
Connector COM2 Data Comms equipment (DCE) unavailable (connection indicator sig
lost).
```

You can force the Bay Standard interface (and all protocols using the serial driver) to ignore the DSR signal by disabling the `wfSyncPollingEnable` attribute. This should be done only in critical situations, however, and this attribute must not be disabled permanently. All HDLC and Bay Standard functionality is provided by the chipset driver (for example, the MK5025 HDLC controller used in many link modules) or the quicsync driver (`wfQsccSyncLoad`) on the Access platforms. Both drivers use the `wfSyncEntry` MIB object for configuration.

Synchronous Pass-through

The pass-through is an application where legacy traffic can be encapsulated in bridge frames for simple and efficient transport using a traditional bridged interface.

As we saw earlier in Figure 5-11, HDLC and other protocols such as SDLC using HDLC-derived frames pose problems in converting to Ethernet frame format. HDLC DTE/DCE addressing and SDLC link station addressing are unique and different from the MAC addressing used with Ethernet frames. Since the goal of the pass-through application is to establish a point-to-point connection between remote devices, virtual local and remote MAC addresses are assigned to each remote end point. Router A's local pass-through MAC address (`passthru-local`) must match router B's remote pass-through MAC address (`passthru-remote`) and vice versa.

This setup is simple to implement (see Figure 5-12), although configuration and connectivity issues occasionally cause problems. When you encounter problems with a synchronous pass-through circuit, check the following:

- The MAC addresses being used by the local and remote endpoints must be unique within the network. If you are in doubt, try generating new MAC addresses and remember to configure them on both ends.

- Make sure that at least one bridged interface is available on both routers. Bridged data originating from the pass-through interface must leave through another bridged interface on the router. Also, a continuous bridge path must be available from the local router to the remote router.
- Confirm that all serial interface parameters on each router match those of the device they are communicating with.
- Check the MTU value on both ends of the link. Certain devices transmit 4K blocks by default, which may not be supported by default on some routers. In this case, the MTU must be increased on the serial interface, and the buffer size may need to be increased as well.

Synchronous pass-through service is supported directly by the appropriate chipset driver and link module driver, similar to Bay Standard. Again, the `wfSyncEntry` object is used for configuration parameters.

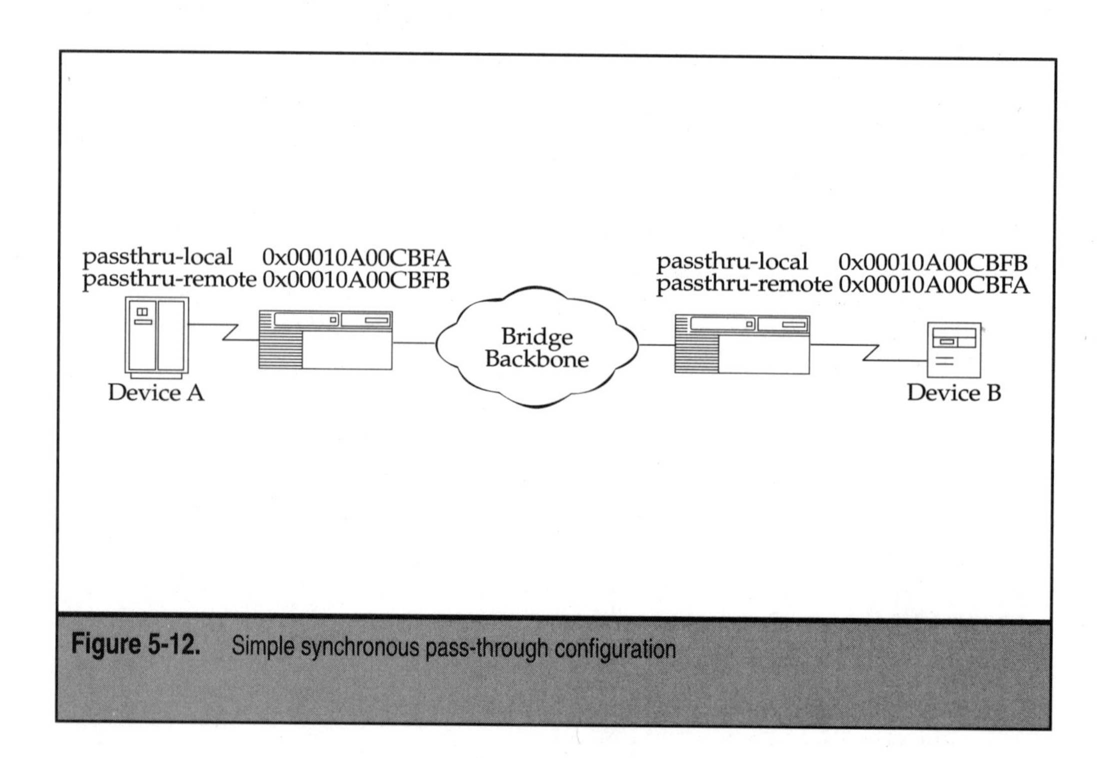

Figure 5-12. Simple synchronous pass-through configuration

ATM Data Exchange Interface

The Data Exchange Interface (DXI) can be used with serial and HSSI lines to connect to an ATM cloud using an ATM CSU/DSU. BayRS routers perform multiprotocol encapsulation and convergence sublayer (CS) functions above the DXI. The segmentation and reassembly (SAR) function and ATM layer function are performed by the ATM CSU/DSU.

DXI Modes

ATM DXI supports several modes of operation specifying the transfer of the service data unit (SDU) between the router and the DCE. The configured mode (`wfAtmInterfaceDxiMode`) must match that of the CSU/DSU for the interface to operate properly. BayRS supports modes 1A, 1B, 2, and a proprietary version of mode 2. A common configuration error is the incorrect selection of the AAL-type for a specified mode. For example, mode 1A can be operated with AAL-5 encapsulation only (see Figure 5-13); misconfiguration of either the service or an individual PVC will generate event messages such as the ones shown here.

```
#   82: 06/25/1999 23:10:09.616  WARNING  SLOT 14  ATMDXI            Code:  66
Line 214101 LLIndex 0: Illegal AAL configured for specified ATM DXI MODE.

#  174: 10/17/1999 02:22:53.108  WARNING  SLOT  2  ATMDXI            Code:  67
Line 202102 LLIndex 0 VPI/VCI 1/35: Illegal AAL configured for specified ATM DXI MODE.
```

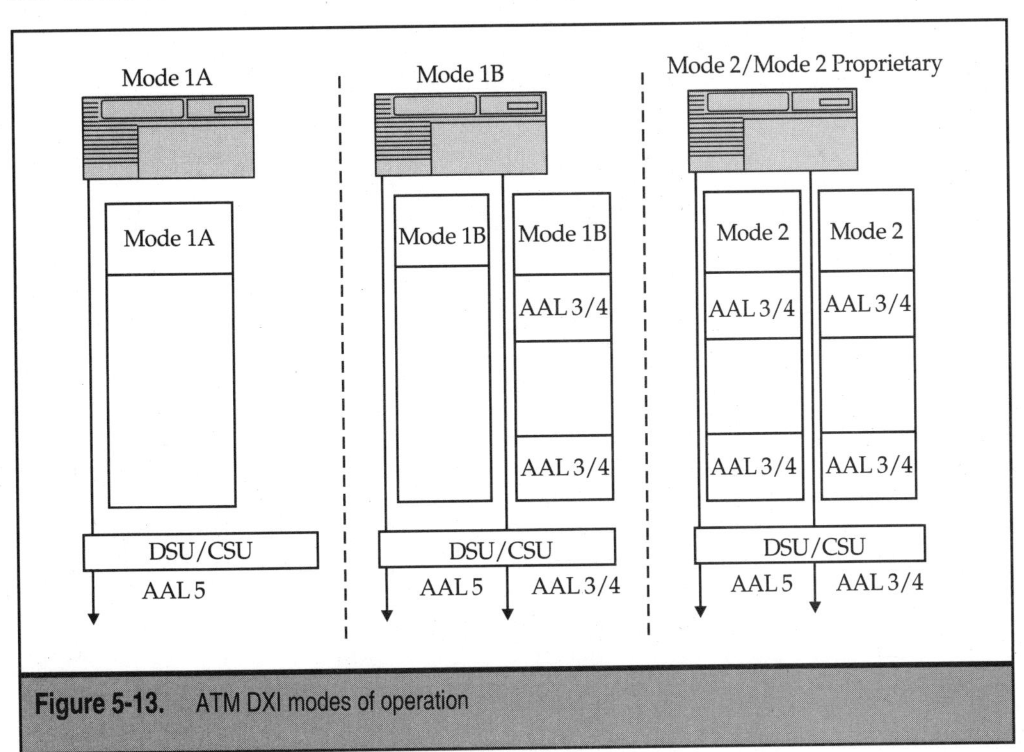

Figure 5-13. ATM DXI modes of operation

Link Management

The log entity for ATM/DXI is `ATMDXI`. BayRS supports a proprietary version of LMI on DXI interfaces. DSU/CSUs that do not support this will not be able to run LMI, and LMI will need to be disabled (default). If a device does not support LMI, the router will not receive an LMI response.

```
#  146: 06/13/1999 16:10:21.144  WARNING  SLOT 14  ATMDXI              Code:  35
LMI response from DSU timed out on Line 214101 LLIndex 0.
```

Invalid LMI messages received from a DSU/CSU indicate that the LMI version cannot be used on the line. In these situations, disable LMI on the router and the CSU/DSU to resolve the problem.

```
#  175: 11/10/1999 15:15:18.773  WARNING  SLOT 11  ATMDXI              Code:  22
Line 211104 LLIndex 0: ATM LMI received invalid message type 0x5e.
```

```
#   59: 11/02/1999 16:23:50.977  WARNING  SLOT  9  ATMDXI              Code:  23
Line 209101 LLIndex 0: ATM LMI address is of incorrect type.
```

Multiprotocol Encapsulation

In addition to the several CS encapsulations that BayRS supports, it uses LLC/SNAP (RFC 1483), NLPID (RFC 1490), or NULL encapsulation for multiprotocol support above the convergence sublayer. If null encapsulation is configured, the router operates as a VC-based multiplexer that disallows bridging on the circuit. Misconfiguration of a multiprotocol encapsulation type will be reported in the log with a debug message.

```
#  164: 08/23/1999 19:24:27.338  DEBUG    SLOT 10  ATMDXI              Code:  56
cct 4 : Waiting for MPE MIB record.
```

In addition to services provided by the underlying chipset and link module driver, the ATM DXI protocol service is available using the *atm_dxi* application (`wfATMLoad`). The *wan* application must also be running (`wfWPROXYLoad`) for link management to operate properly.

Frame Relay

Frame relay is a ubiquitous service that provides several advantages over traditional point-to-point leased lines, and BayRS provides extensive support for frame relay using both permanent and switched virtual circuits. It is a packet-switched protocol providing lower overhead yet higher reliability compared to older switched technologies such as X.25. Frame relay provides the ability to monitor congestion and shape traffic accordingly. Using bandwidth management, data traffic can be prioritized and queued based on required service guarantees.

Frame relay also provides the ability to set up the wide area network in a fully or partially meshed configuration for redundancy. In essence, it provides a more efficient transport than traditional technologies while providing a cost-effective and reliable solution.

The frame relay protocol requires the use of the *fr* application (`wfFRLoad`), and events are logged under the `FR` entity. When using Switched Virtual Circuits (SVCs), the *fr_svc* application (`wfFRSVCLoad`) must also be loaded; its events are logged under the `FR_SVC` entity. On the AN/ANH and ARN routers, the frame relay service is integral to the kernel and is not listed as a separate application in the loadmap.

Data Link Connection Management

A link management protocol is usually used and is necessary for proper frame relay operation, except in the case of statically defined PVCs. Use one of the following supported link management and signaling techniques on frame relay lines:

- ITU-T Q.933 Annex A
- ANSI T1.617 Annex D
- LMI

The ANSI T1.617 Annex D protocol is the default link management type. Annex A and Annex D use DLCMI 0 for link management, and LMI uses DLCMI 1023. The link access procedure for frame relay (LAPF) and Q.933 are used extensively for switched virtual circuits. Link management between the router and the FR switch is locally significant. Link management failure affects the frame relay interface globally and all virtual circuits carried over it. The fault status in a summary or in DLCMI statistics may indicate that the wrong type of DLCMI is configured for use with the switch. To see the overall DLCMI configuration and status, use the **show frame-relay summary** command.

```
show frame-relay summary                               Mar  2, 2000 21:15:39 [EST]

                    Management Interface                   Number-of-VCs
Interface Circuit     Type        Type     Status   Faults Total  Active Services
--------- ------- ---------- --------- ------- ------ ------ ------ --------
3/1         4       lmi        primary   running       0     24     22        6
```

The interface status will indicate a fault condition if the interface cannot communicate with the DCE. This problem often occurs when there is a DLCMI mismatch between the router and the frame relay switch. If the interface status is notpresent, then the frame relay application is not loaded either because it is unavailable in the image archive or because it has been disabled in the slot mask.

Detailed DLCMI statistics can be obtained by executing **show frame-relay stats dlcmi**. In the following example, the DCE (frame relay switch) has not responded to some of the router's polls, indicating that the switch has suddenly stopped responding. Polls can be missed as a result of either a poll interval configuration problem or data drops on the line.

```
show frame-relay stats dlcmi                         Mar  2, 2000 11:10:26 [EST]

          Maximum Current Sequence Received Passive  Passive        Polls
Interface Num VCs Num VCs Counter  Sequence Sequence Received Polls Missing
--------- ------- ------- -------- -------- -------- -------- ----- -------
3/2           100       3      872      831        1        1  2634       4

#   67: 03/02/2000 11:10:26.721  TRACE    SLOT  3  FR              Code:  83
Line 203102 LLIndex 0: Status message not received within time out.
```

Event messages such as the preceding indicate that DLCMI is not operating properly. BayRS sends status inquiry messages every `polling-interval` seconds and full status inquiry requests every `full-enquiry-interval` period; both attributes are configured for the `dlcmi` child object. These timers default to 10 seconds and 6 times (6 x polling-interval = 60 seconds), respectively. Failure to receive a response for `error-threshold` (default: 3) successful inquiries will bring down the line. Link management timers must be matched on both ends of a circuit, and the frame relay switch should respond to all status inquiries. The polling and full inquiry intervals are outbound timers only—BayRS does not support bidirectional LMI, and the switch should be configured so that status inquiry messages are not originated.

```
#   60: 08/03/1999 10:08:27.850  INFO     SLOT 10  FR              Code:  63
Service down on Line 210103 LLIndex 0.

#   61: 08/03/1999 10:08:29.887  INFO     SLOT 10  FR              Code:  62
Service up on Line 210103 LLIndex 0.

#   62: 08/03/1999 10:09:31.021  INFO     SLOT 10  FR              Code:  63
Service down on Line 210103 LLIndex 0.
```

If a service keeps bouncing as in the preceding example, the frame relay switch is likely bringing down the line because bidirectional LMI is configured. Bidirectional LMI normally is used on NNI connections and not on UNI connections and so must be disabled.

Data Link Connection Identifier

The data link connection identifier (DLCI), pronounced *del-see*, is a unique number used to represent each virtual circuit on the frame relay line. This number is only locally significant between the router (DTE) and the frame relay switch (DCE). The corresponding VC at the remote end may be represented by a different DLCI. When PVCs are referenced, they are usually indicated by their associated DLCI. Thus, it is important to know the frame of reference when referring to a DLCI number. Certain frame relay services provide enterprise networks with end-to-end VCs with the same DLCI. In this case, the DLCI is unique not only on the local interface but in the entire enterprise frame relay WAN. Such DLCIs are said to be globally significant, but their provisioning is rare.

Each virtual circuit (VC) represented by a DLCI can be in one of three states: active, inactive, or invalid. The active status indicates that the DCE is knowledgeable about the virtual circuit and that it is fully operational on an end-to-end basis. The inactive status indicates that the DCE is aware of the VC, usually due to the configuration, but end-to-end connectivity is unavailable. This condition will occur if the remote end is down or is unreachable due to an interswitch trunk loss. The invalid status indicates that the VC is statically defined on the DTE, but the DCE has no knowledge of its DLCI. This is usually the case where there is a misconfiguration or when the DLCI has been deleted from the network.

```
show frame-relay vcs                                   Apr 04, 2000 13:56:14 [EDT]


Interface   DLCI     State        Type        Hybrid     Service
--------- -------- -------- ------------ -------- -------------
2/1             80 active   pvc          disabled NYC_128k
2/1             88 active   pvc          disabled NYC_128k
2/1            100 active   pvc          disabled SLC_64k
2/1            112 active   pvc          disabled Philly_64k
2/1            204 inactiv  pvc          disabled 702101.0.1
2/1            205 invalid  pvc          disabled 702101.0.1
```

Virtual Circuits

When LMI is operational, the frame relay switch indicates the status of each VC as either active or inactive using the A-bit in the status report. A well-functioning virtual circuit will remain active and operational without any errors. However, a VC that bounces (turns inactive and becomes active again periodically) may indicate a problem with the remote frame relay DTE. If the DLCMI timers are misconfigured on the remote router, then the UNI may transition up and down, causing events on the local router.

```
#  152: 04/19/1999 18:40:07.470  INFO     SLOT  6  FR                Code:  70
Line 206104 LLIndex 0: VC 89 changed to INACTIVE.

#  153: 04/19/1999 18:40:08.687  INFO     SLOT  6  FR                Code:  70
Line 206104 LLIndex 0: VC 89 changed to ACTIVE.

#  154: 04/19/1999 18:40:09.482  INFO     SLOT  6  FR                Code:  70
Line 206104 LLIndex 0: VC 89 changed to INACTIVE.
```

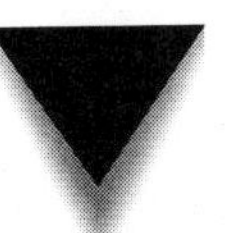

TIP: The full support of A-bit propagation between the two end points in a frame relay circuit is essential for backup circuit operation. If you have configured dial backup for a primary frame relay direct mode service record, the associated VC/DLCI must be marked inactive by the FR cloud when it becomes unavailable. If the backup circuit is not triggered automatically, check the log to see if the VC state changed to inactive.

Frame relay and LAPF derive their frame formats from HDLC and use a variable address length of from 2 to 4 bytes, with 2 bytes being the default (see Figure 5-14). Address formats follow Q.921, Q.922, or interim specifications (March 90 and November 90) of

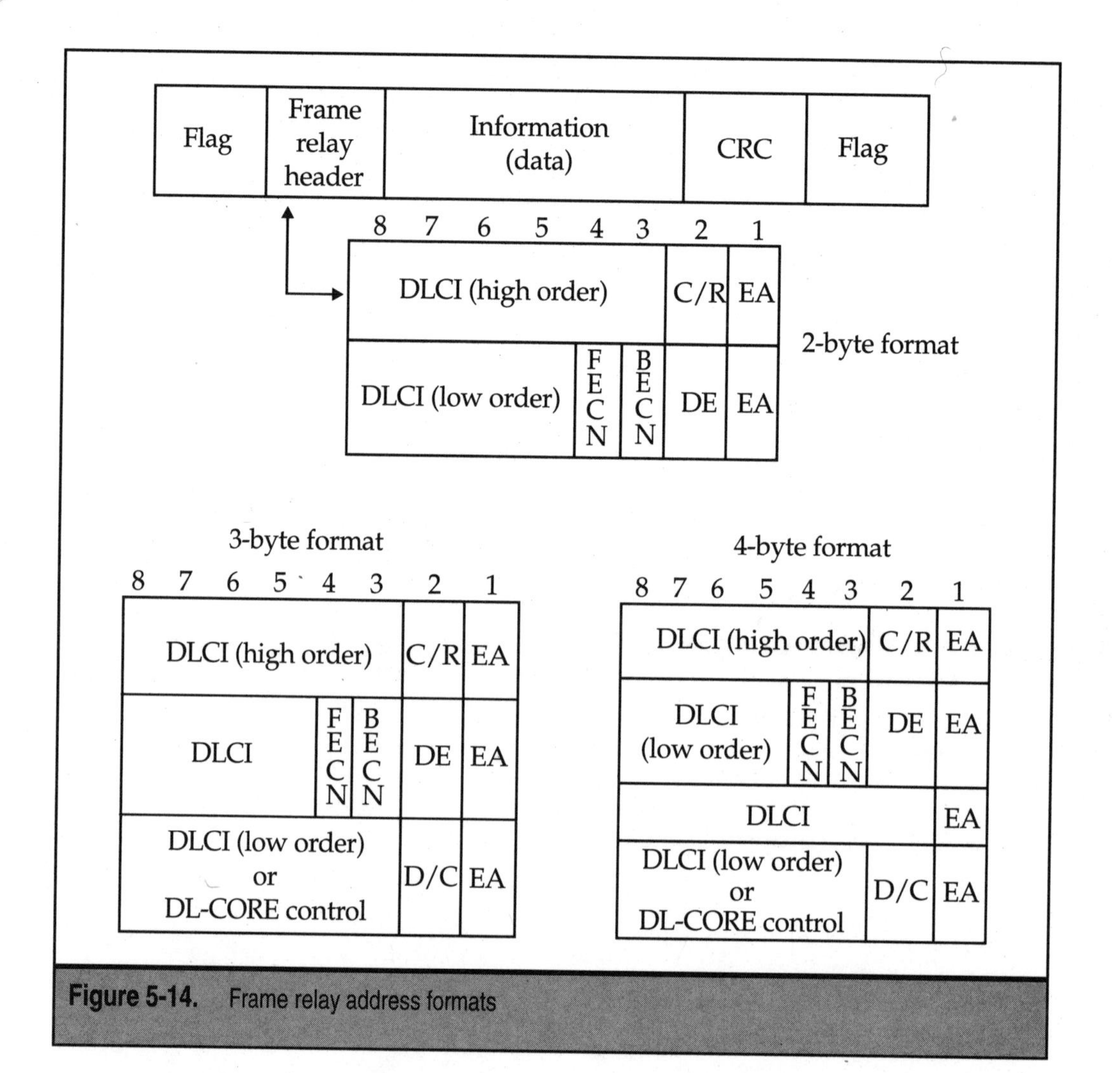

Figure 5-14. Frame relay address formats

Q.922. Most UNI interfaces are adequately provisioned with a 2-byte address field; however, your provider may require a configuration using a longer address (`address-length`) to support VC numbers (DLCI) in the higher range. If your provider uses an address length different from the one configured on the router, you will see symptoms related to DLCMI and address recognition.

```
#  108: 12/19/1999 07:20:31.746  TRACE    SLOT  6  FR                Code:  76
Line 206102 LLIndex 0: DLCI 1341 out of range - invalidated.

#  141: 09/04/1999 04:49:00.795  TRACE    SLOT  5  FR                Code:  80
DLCI extract failed for a DATA_PKT on Line 205101 LLIndex 0.

#   94: 06/01/1999 07:30:41.084  WARNING  SLOT 14  FR                Code: 115
Line 214102 LLIndex 0: Address length invalid for specified address type.
```

Service Records and Address Resolution

Frame relay and ATM share the concept of *service records*, which are in essence subcircuits on the physical interface. Multiple service records can be configured on an interface, providing various protocol services, including bridging. Each service record can contain one or more virtual circuits. When a service record contains only one VC, it is said to be in *direct* (point-to-point) mode. A service record containing multiple VCs is said to be in *group* (point-to-multipoint) mode. Protocol services on each service record are treated as if they each have their own physical interface. For example, if IP is required on two service records on the same physical interface, the IP interfaces must belong to two unique subnetworks.

```
show frame-relay services                              Mar 30, 2000 21:17:45 [EST]

                                        ---------Number-of-VCs----------
Interface       Service-Name       Total Static Dynamic SVCs Active SVC Support
------------ --------------- ----- ------ ------- ---- ------ -----------
2/1             NYC_128k               2      2       0    0      2 disabled
2/1             SLC_64k                1      1       0    0      1 disabled
2/1             Philly_64k             1      1       0    0      1 disabled
2/1             702101.0.1             2      1       1    0      0 disabled
...
```

The outbound DLCI used to reach a network address will serve as the data link address for the particular destination. For example, if a local service record's IP address is 10.23.82.1 and the neighbor 10.23.82.2 can be reached using DLCI 17, then 17 is the data link address. Frame relay service records automatically resolve the neighbor's address using inverse-ARP on each known DLCI. Occasionally, you may need to configure an adjacent host (static ARP) entry if the neighboring device does not support ARP or inverse-ARP. The physical address field in the **show ip arp** command output is equal to the DLCI, formatted as a hexadecimal number.

In Figure 5-15, router A acts as the central hub where routers B, C, and D converge to a single line. Data traffic between the hub and all the spoke routers, and among the spoke routers, will be transported normally. However, network management in the frame relay

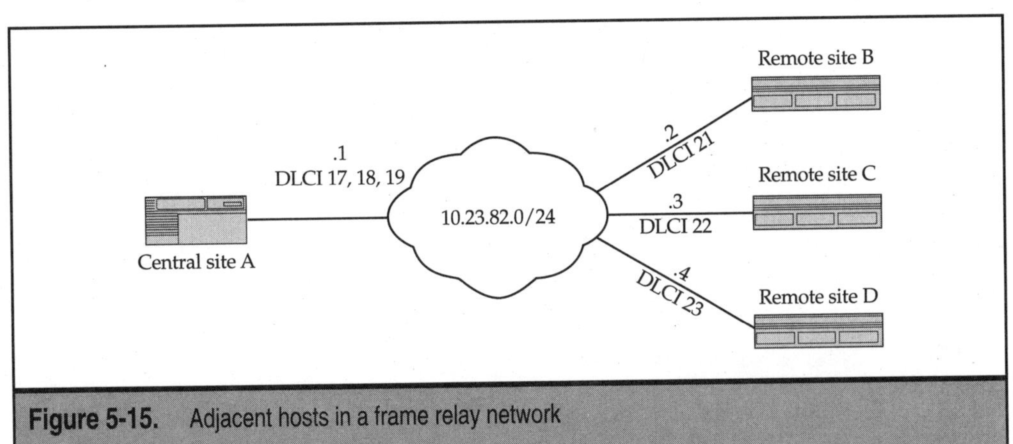

Figure 5-15. Adjacent hosts in a frame relay network

cloud may generate problems. The spoke routers will be able to ping the IP address of the hub router (10.23.82.1), but will be unable to ping other spokes. The reason for this is that frame relay is a nonbroadcast, multiaccess medium. Although the entire cloud represents one single subnetwork in this example, router B must pass through router A to reach routers C and D. For example, when router B pings the address of router D, it must first ARP out the local service record to determine the data link address of router D (10.23.82.4). This ARP request, however, reaches router A first, which will promptly drop the frame because it is not the intended recipient.

```
box->ip
router-B(4)# adjacent-host address 10.23.82.4 dlci 21

box->ip->adjacent-host/10.23.82.4
router-B(5)# show ip arp
show ip arp                                        Apr 06, 2000 18:14:08 [EDT]

IP Address        Physical Address      Type
------------------------------------------------
10.23.82.4        00-00-00-15-          static
```

Situations such as this are common in partially meshed frame relay networks. A partially meshed network is one where only a few of the routers in the cloud have VCs established to all other remote routers. The spoke routers are so called because they are tied together by means of the hub router, which interconnects them. In this scenario, spoke routers need to have an adjacent host entry configured so that the initial ARP broadcast is omitted. When router A receives the unicast ICMP echo request from router B, destined for router C, it will automatically forward the frame across VC 18 (see Figure 5-15).

Congestion and Traffic Shaping

One of the most beneficial aspects of frame relay is the ability to burst the data rate up to the line speed during a brief period of heavy use. The router exceeds the committed information rate (CIR) and transmits data at a higher rate during, say, a large file transfer. Naturally, the frame relay cloud may discard frames exceeding the data rate subscribed to by the subscriber, and the service provider is likely to filter frames, especially during a period of congestion.

Congestion is a normal occurrence that affects connectivity and delays data traffic. User and network interfaces rely on several congestion control mechanisms; the enabling feature of these mechanisms is congestion notification (CN). Two types of congestion notification are used, forward explicit congestion notification (FECN) and backward explicit congestion notification (BECN), as illustrated in Figure 5-16. A frame relay DTE or DCE monitors the FECN and BECN notifications carried on frames to evaluate congestion on the network. CNs are reported on one or more virtual circuits.

```
#   89: 05/24/1999 04:09:28.903  TRACE    SLOT  5  FR               Code: 116
Line 205104 LLIndex 0: VC 142 is experiencing congestion.
```

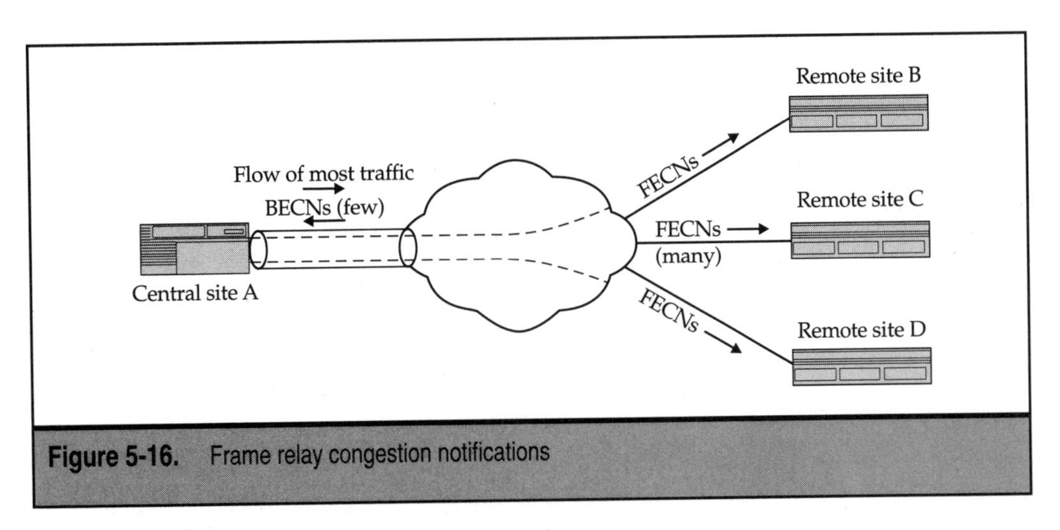

Figure 5-16. Frame relay congestion notifications

BayRS performs one of three functions when a certain number of congestion notifications (as specified by `congestion-counter`) are received over a specific period of time (as specified by `congestion-timer`):

- Shut down the VC (specify shutdown).
- Throttle outgoing data to the CIR and then shut down if CNs persist (specify throttleThenShutdown).
- Throttle back to the CIR (specify throttle).

The `congestion-method` parameter dictates the response to congestion on the line or VC using one of the preceding values; on the VC, the parameter can have the value *inherit*, so that it inherits the congestion method from the line. The first two actions will shut down the VC, which, if the system is operating in direct mode, will terminate all higher-layer protocol interfaces.

```
show frame-relay congestion                          Apr 04, 2000 14:02:31 [EDT]

                                                               --Received---
Interface   DLCI    Enable    State      Method   Timer Counter  FECN   BECN
--------- -------- ------- ---------- -------- ----- ------- ------ ------
2/1             80 enabled forwarding inherit    one      20      0      1
2/1             88 enabled forwarding inherit    one      20      0      0
2/1            100 inherit congested  throttle  five      20     12    197
2/1            112 disable forwarding inherit    one      20      0      0
2/1            204 inherit throttling throttle  five      20     23    325
2/1            205 inherit forwarding inherit    one      20      0      0
```

An occasional effect of rapid congestion control is the undesirable control of outbound traffic, as occurs when you use the shutdown method. It is usually desirable to throttle the traffic back to the CIR, although in some situations a complete shutdown is a better option than suffering data loss using the committed rate. Congestion can be managed easily by tuning and by implementing traffic shaping.

Sometimes congestion control may trigger too frequently and easily. This can occur if your congestion counter or congestion timer is set so low that the VC shuts down even with mild congestion. If you experience VC shutdowns frequently, check the congestion control parameters. You can also completely disable congestion control at the line level or the VC level. You can also completely disable the `congestion-control` attribute for testing purposes.

Configuration Considerations

Frame relay uses several MIB objects for configuration purposes:

- `wfFrIfDlcmiEntry` manages DLCMI on a per-interface basis.
- `wfFrVCircuitEntry` is used for individual PVCs and SVCs.
- `wfFrServiceRecordEntry` manages group mode, direct mode, and hybrid service records.
- `wfFrFRF4SigEntry` is used for FRF.4 signaling (signaling is mandatory for SVC applications but is not necessary for PVCs).
- `wfFrLapfEntry` manages LAPF (link access procedure, frame relay).
- `wfFrSVCOptionsEntry` manages SVC configuration options.

The *fr* (`wfFrLoad`) application must be running for frame relay operation, and SVCs require the *fr_svc* (`wfFRSVCLoad`) application. All frame relay lines also require that Line Manager (LM; `wfLMLoad`) be available on the slot.

Service records aggregate multiprotocol services into a single circuit that operates over one or more VCs. Site Manager and BCC automatically configure a default service record when defining a frame relay circuit; BayRS uses this service record to add dynamically learned PVCs. The default service record is generally used in group mode so that new virtual circuits are added automatically, without requiring any changes at a central site.

NOTE: BayRS inherently treats default service records as operating in group mode, even if they have only one PVC. This is because, if necessary, a new PVC will automatically be added to the default service record, and this addition would require group mode. This potential thus implies the group mode nature of the default service record.

There can be only one default service record per frame relay interface. Although you cannot create or delete a default service record, you can modify the `wfFrServiceRecordDefaultFlag` attribute for any given service. If you have previously configured the default service record with protocols but want to reconfigure another service record as the default, you simply set the default flag to true on the new service record and set it to false on the existing one. You must make this change directly by using the MIB attribute for the specific service records (`wfFrServiceRecordEntry`). Note again, however, that only one service record can be a default service record at any time.

```
#   12: 12/23/1999 04:52:15.914  TRACE    SLOT 12  FR                Code: 127
2 default service records found for configured Line 212102 LLIndex 0.
```

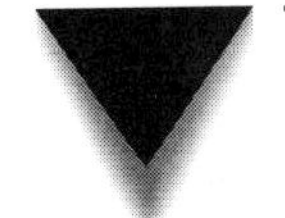

NOTE: As mentioned earlier, group mode operation is assumed for the default service record. Since group mode services have more than one VC, you create ambiguity if you define a single VC requiring backup because all associated protocol interfaces are automatically migrated to a backup circuit by definition. Doing so will leave other VCs without a controlling circuit and protocols. Thus, neither default service records nor any group mode service record can be configured as a primary for a backup circuit.

Routing Another common consideration is the application of the routing information protocol (RIP) to group mode services. RIP by default operates in split-horizon or poisoned-reverse mode, both of which help ensure that the route learned on an interface is not retransmitted as is, thereby preventing the countdown to infinity problem. However, use of either split horizon or poison reverse is not appropriate for a group mode service record. In Figure 5-15, for example, a route learned by router A from router B must be rebroadcast to router C using the same service record. Poisoning or split-horizoning the route will prevent router C from reaching the network at router B. Hence, BayRS implements the *actual* RIP mode, which retransmits the routes learned on an interface without applying the split horizon or poisoned-reverse algorithm. When configuring group mode service records with RIP, make sure that RIP mode is set to actual on the hub router.

Recall that we discussed the nonbroadcast multi-access (NBMA) nature of frame relay. The OSPF routing protocol defaults to multicast mode behavior, whereby it dynamically learns of adjacent neighbors. Frame relay service records are automatically aware of the nonbroadcast characteristic, and all broadcast and multicast packets are sent over all VCs on the appropriate subcircuit. Some partially meshed networks with numerous spoke routers may find it unnecessary to duplicate the multicast frames for each outbound VC. In other cases, sending numerous broadcast or multicast frames on VCs with low CIRs may reduce network performance. In situations such as these, OSPF interfaces on frame relay service records can be configured with NBMA neighbors, which are unicast neighbors that are static. When the OSPF interface initializes, it attempts to establish an adjacency with each of the configured NBMA neighbors directly. OSPF configuration considerations are discussed in Chapter 6.

Switched Multimegabit Data Service

Using commonly proven techniques from cell-relay technology and other principles from packet-based transport, Switched Multimegabit Data Service (SMDS) provides a public, high-speed, packet-switched data service. SMDS has many traits in common with ATM/DXI and frame relay; however, the application and management of this service are unique in several ways.

The log entity for Switched Multimegabit Data Service is `SMDS`. SMDS uses a link management protocol similar to ATM/DXI, which is operated by the same *wan* application (`wfWPROXYLoad`). Protocol services are rendered by the *smds* application, loaded using the `wfSMDSLoad` slot mask. SMDS interface configuration attributes are stored in an instance of `wfSmdsCircuitEntry` for use by the circuit.

Addressing

Each SMDS subscriber network interface (SNI) is assigned a set of data link addresses in E.164 format. An individual address uniquely identifies an SNI to the network, and a group address is similar to a broadcast address, to which several individual addresses belong. Broadcast frames originating from the router are sent to this E.164 group address. E.164 addresses are 64 bits (16 nibbles) long. The SMDS individual address can be compared to the MAC address on an IEEE 802.3 frame. Unlike PROM (hardware) derived MAC addresses, the individual (`wfSmdsCircuitIndivAddr`), group (`wfSmdsCircuitGroupAddr`), and ARP (`wfSmdsCircuitArpAddr`) addresses must be configured correctly on an SNI. The ARP address is usually the same as the group address, which is where broadcast frames are sent. Address configuration issues are commonly seen with SMDS circuits, as shown with the following event messages.

```
#   32: 12/07/1999 06:40:07.654  WARNING  SLOT  4  SMDS            Code:   4
Circuit 2 SMDS address is of incorrect length.

#  150: 06/09/1999 13:42:37.796  WARNING  SLOT  2  SMDS            Code:   5
Circuit 4 SMDS address is of incorrect type.
```

Individual and group addresses begin with 0xC and 0xE, respectively. If an address does not start with one of these prefixes, it is considered invalid by the E.164 addressing scheme. Also, each address must be 64 bits long. For example, a phone number (978) 555-1001 formatted into an E.164 individual address will be 0xC19785551001FFFF—note the 0xC prefix (indicating that this is an individual address) and the trailing 0xFFFF padding to meet the 8-byte address length.

Interfaces residing on an SMDS circuit automatically use the individual, group, and ARP addresses configured for the SNI. Certain protocols such as IP support the configuration of multiple interfaces (multinetting), which may use either the default interface E.164 addresses or several individual addresses. In the latter case, each IP interface must be configured with a unique SMDS individual address in the MAC address field. Both types of multinetting use one group address, whereas in a multigroup configuration, each IP interface has a unique individual and group address (see Figure 5-17).

SMDS circuits and interfaces running on BayRS will selectively accept only frames destined to configured individual addresses. Misdirected frames will be discarded and counted in the value of `wfSmdsCircuitAddrVerifyDiscards`, a discard statistic. If this occurs, check the interface and circuit configuration and work with your provider to verify the individual and group addresses.

Link Management and Heartbeat Polling

Link management on SMDS interfaces is performed using a protocol similar to SNMP. The SMDS DSU/CSU must support link management for you to use link management on the router interface. When enabled and working, the DSU will report warm and cold starts as well as link states changes.

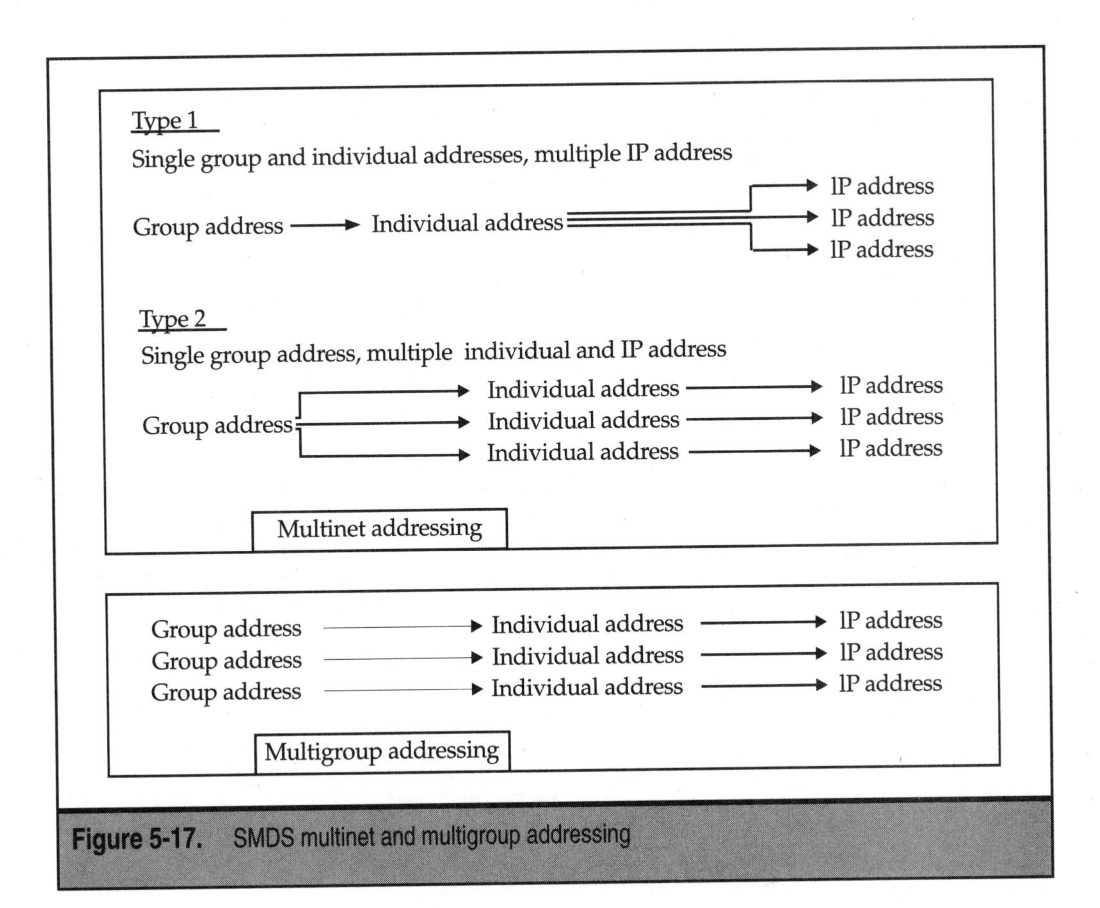

Figure 5-17. SMDS multinet and multigroup addressing

```
#   77: 05/28/1999 00:13:08.083  INFO     SLOT  1  SMDS                 Code:  19
Cold Start trap received from DSU on circuit 14.

#   78: 05/28/1999 00:13:08.098  INFO     SLOT  1  SMDS                 Code:  22
Link Up trap received from DSU on circuit 14.
```

The most common symptom of link management misconfiguration is an LMI time-out. If LMI on the DSU/CSU has been disabled or is an unsupported version, the router will report the event shown next. In these cases, check the LMI configuration on both the router and the DSU and, if necessary, disable LMI.

```
#  192: 08/23/1999 17:40:59.990  WARNING  SLOT 14  SMDS                 Code:  13
LMI response from DSU timed out on circuit 8.
```

The router uses a fundamental heartbeat polling method to ensure the availability of the DSU/CSU. Both LMI and heartbeat messages are prioritized over regular data traffic for management purposes.

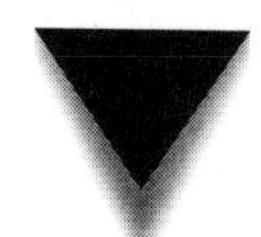

TIP: Both heartbeat polls and LMI should be disabled when you are connecting two routers back to back using a crossover cable.

```
# 171: 11/06/1999 18:07:26.665 WARNING SLOT 3 SMDS                Code: 10
No response to heartbeat poll on circuit 8.
```

Failure to receive `wfSmdsCircuitHrtbtPollDownCount` (default: 3) consecutive responses to heartbeat polls (sent every `wfSmdsCircuitHrtbtPollInterval` seconds; default: 10) will cause the router to shut down the SNI and attempt reinitialization. Some SMDS DSU/CSUs do not accept frequent heartbeat polls, and the router must be configured accordingly. If the heartbeat poll failure occurs soon after initialization, then the SMDS DSU/CSU may not support heartbeat polls, and you must disable it.

Point-to-Point Protocol

PPP, or Point-to-Point Protocol, is one of the most widely used WAN protocols on leased point-to-point lines. PPP is standards based, supported by numerous vendors, and enables communication among many vendors' hardware. PPP is a comprehensive protocol that not only provides data link connection and management but also certain fundamental functions of network layer connection and management. This latter aspect makes it especially desirable to many network managers because it enables easy integration for administration and troubleshooting.

Using a frame (see Figure 5-18) derived from the common HDLC format, PPP operates on both leased and dial lines. It requires the appropriate chipset and link module drivers, plus the *ppp* application (`wfPPPLoad`). Events are logged under the `PPP` entity, and the **show ppp** command can be used to display the configuration and statistics.

Protocol Negotiation

PPP circuits are initialized using negotiation in at least two phases: LCP and NCP. Link Control Protocol (LCP) negotiates data link layer parameters such as MRU, the authentication protocol, and the link quality protocol. An intermediate authentication phase may be

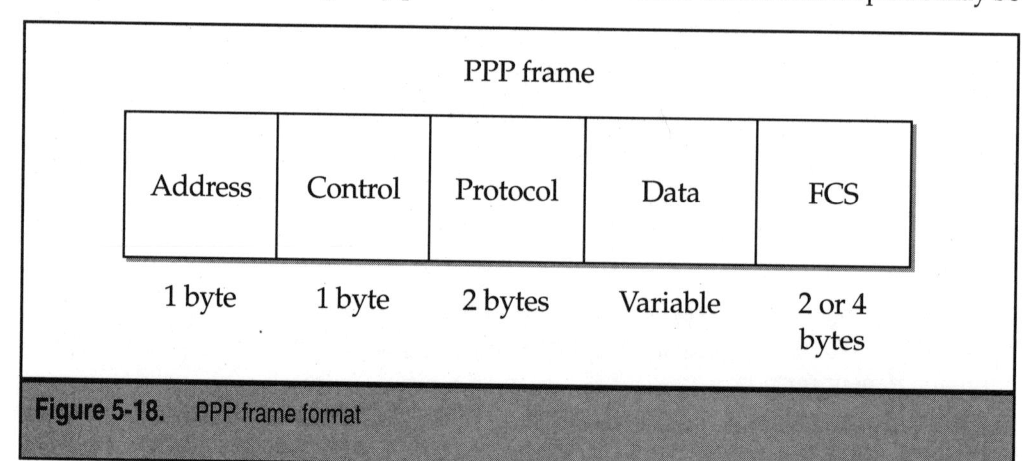

Figure 5-18. PPP frame format

applied in the case of dial lines for security and identification of the peer. Next, the Network Control Protocol (NCP) helps negotiate one of several higher-layer protocols between the peers. Several NCPs, such as IP and IPX, can be simultaneously negotiated and established. Once LCP or a particular NCP has successfully completed negotiation, it is said to be in an *open* state. PPP initialization between two peers is illustrated in Figure 5-19. In all, the three phases—LCP, authentication, and NCP—must complete for a network layer interface to become operational on a circuit.

```
show ppp interfaces                                   Mar 30, 2000 21:12:18 [EST]
... ...
        Slot/
        <Module>/ Driver                               Line
Circuit Conn      State      Protocol   State   State
------- --------- ---------- ---------- ------- ----------
H21     2/1       up         LCP        opened  up
                             IP         opened
                             IPX        reqsent
```

Each phase consists of several discrete negotiations, started by a request frame, which receives a response of ACK, NAK, or REJECT. ACK is an affirmative response acknowledging the configuration parameter request suggested by the peer.

```
#  108: 04/19/1999 19:08:51.194  TRACE    SLOT  8  PPP             Code:  44
Sending LCP Configure-Request on circuit 4.

#  109: 04/19/1999 19:08:51.201  TRACE    SLOT  8  PPP             Code:  56
Received LCP Configure-Ack on circuit 4.
```

NAK is sent as a negative response, but the message usually suggests an alternative value for the requested parameter. The peer then resends the request either as is or based on the suggested value; through repeated NAK exchanges, the peers arrive at a common value for a parameter.

```
#  187: 07/14/1999 10:15:22.177  TRACE    SLOT  5  PPP             Code:  55
Received LCP Configure-Request on circuit 8.

#  188: 07/14/1999 10:15:22.183  TRACE    SLOT  5  PPP             Code:  64
LCP Naking MRU option value 0xfa0 with value 0x5dc on circuit 8.

#  189: 07/14/1999 10:15:22.191  TRACE    SLOT  5  PPP             Code:  55
Received LCP Configure-Request on circuit 8.

#  190: 07/14/1999 10:15:22.198  TRACE    SLOT  5  PPP             Code:  45
Sending LCP Configure-Ack on circuit 8.
```

NAK exchanges do not always result in successful negotiation. Certain parameters such as network addresses (for example, IP address) are manually configured, and PPP may be unable to suggest alternate values. For example, if the IP address on one peer is

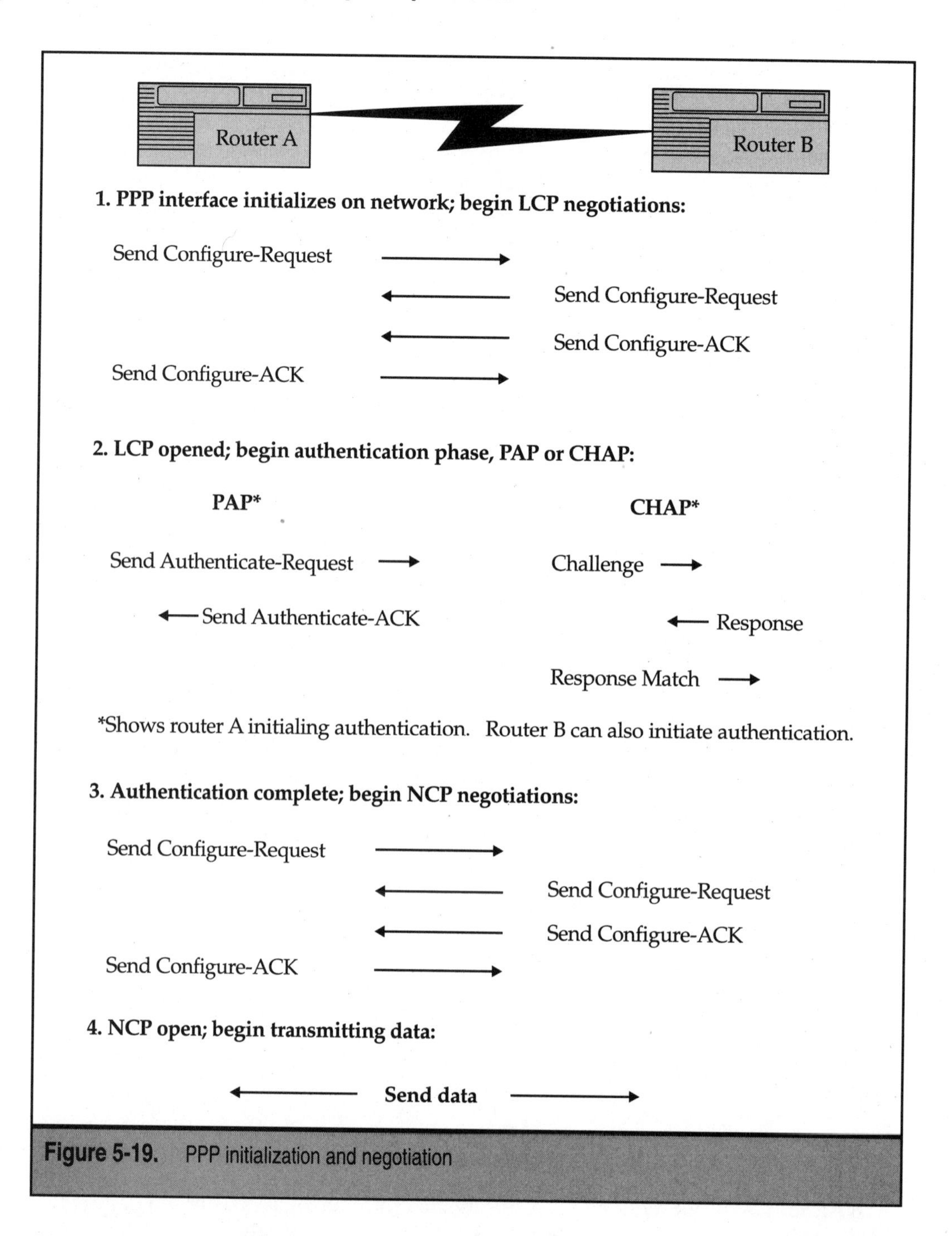

Figure 5-19. PPP initialization and negotiation

misconfigured and is the same as the other peer, then IPCP negotiation will never complete, as warned by IPCP in the following response.

```
#   57: 01/24/1999 09:45:52.439  WARNING  SLOT  1  PPP                Code:   6
Configuration not converging for IPCP on circuit 2.

#  140: 12/23/1999 19:48:31.706  WARNING  SLOT 10  PPP                Code: 231
Terminating Call because of unacceptable MRU/MRRU of 1800. Expected 1594 on cct 19
```

Configuration convergence problems can occur on both the LCP and NCP layers, as illustrated by the two preceding events. LCP negotiation failure will stop further progress; a convergence problem in one NCP (for example, IPCP) will still allow IPXCP to fully negotiate and enter the open state. Following successful negotiation of all mandatory LCP parameters, the LCP open state is entered. If the link establishment phase does not complete, match the circuit and line-level PPP parameters on the peers for the NAK attributes.

```
#    6: 02/13/1999 04:34:45.948  INFO     SLOT  8  PPP                Code:  38
Link Establishment Phase complete on circuit 12.
```

REJECT is a more severe form of the negative response; it indicates that the peer is unwilling or not capable of negotiating the specific attribute. For example, if the local peer has IPX configured and attempts to negotiate IPXCP, but the remote peer does not have IPX configured, then the remote peer will reject the request for IPX configuration. REJECT responses can occur on both LCP and NCP and usually indicate a mismatched configuration. You can easily identify this problem by looking at the code or protocol to see whether it is a valid value. If the code or protocol being rejected is undefined, it may be a proprietary protocol or code used by another vendor's PPP implementation. Consult the IANA-assigned numbers (STD 2) to see whether the code being rejected has been defined. Also consult the BayRS documentation for the version you're using to see if the particular protocol or feature is supported.

```
#   37: 01/16/1999 03:41:48.595  TRACE    SLOT  1  PPP                Code:  50
Sending LCP Code-Reject for code 100 on circuit 28.

#   12: 04/07/1999 00:52:23.738  TRACE    SLOT  8  PPP                Code:  51
Sending Protocol-Reject for protocol 0x8021 on circuit 35.
```

LCP code 100 has not been defined for use by any PPP RFC or standard; however, protocol 0x8021 (IP) is well known. This indicates that the peer sending the Protocol-Reject message does not have IP configured on the interface. You should pay attention to all NAK and REJECT responses during the initial configuration and during subsequent configuration changes.

Successful completion of authentication and NCP will be reported in the log.

```
#  132: 12/03/1999 03:40:43.466  INFO     SLOT  6  PPP                Code:  39
Authentication Phase complete on circuit 23.
... ...

#  140: 12/03/1999 03:40:43.982  INFO     SLOT  6  PPP                Code:  40
IPXCP up on circuit 23.
```

Network control protocols usually have several negotiable options, though in some cases, they have none. For example, IPCP (IP Control Protocol) will negotiate the IP Address (or IP Addresses, for backward compatibility) option, and ATCP (AppleTalk Control Protocol) will negotiate the network number, node number, and routing protocol; Vines and XNS do not have any negotiable NCP options.

Multilink PPP Negotiation

Multilink PPP (MLPPP) is based on RFC 1717 and is used to bundle multiple physical or logical lines into one single PPP circuit (see Figure 5-20). Multilink PPP has several advantages over proprietary multiline techniques:

- Multilink PPP is standards based.
- Multilink PPP supports multilink packet fragmentation and resequencing.
- Multilink PPP allows the use of lines with different data rates.
- Multilink PPP uses both leased and switched lines.

By default, PPP will attempt to negotiate multilink operation because of the possibility of using additional switched lines for bandwidth-on-demand during dynamic operation.

```
#  168: 08/19/1999 02:32:23.622  DEBUG    SLOT 14  PPP               Code: 101
By default, Line 214102 will attempt to negotiate PPP Multilink
```

One important attribute negotiated by MLPPP is the endpoint discriminator. BayRS routers automatically attempt to generate a unique endpoint discriminator, and this can be overridden through configuration (using `wfPppLineMyLinkDiscr`). All lines in a multilink bundle share this endpoint discriminator, which helps the router detect errors

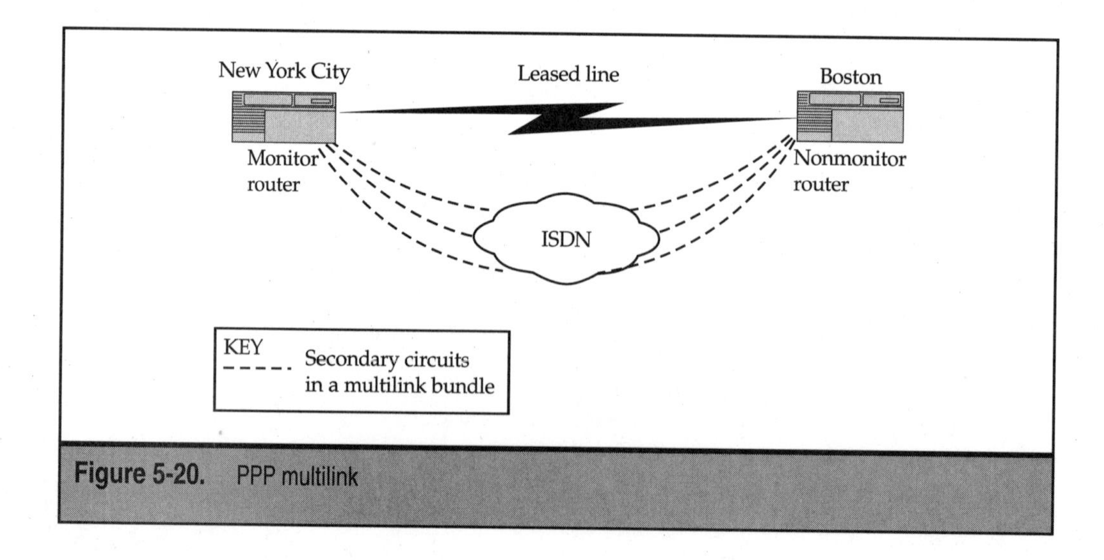

Figure 5-20. PPP multilink

such as an incorrectly dialed endpoint or a loop condition. If you observe errors with endpoint discriminators, check the configuration on both the local and remote ends (using `wfPppLinePeerLinkDiscr`). If the configuration is correct, check the outbound phone number dialed by the router, in the case of a switched line.

Although not specific to multilink, PPP uses magic numbers to detect loopback conditions. It will automatically bring the associated line down if it receives a frame with its own magic number. If you observe this condition, check the CSU/DSU or contact your provider to remove the loop.

```
#   68: 12/19/1999 09:44:27.914  WARNING  SLOT 10  PPP              Code:   5
Peer is using same Magic number, possible loopback condition on circuit 3.
```

Authentication

The optional authentication phase may follow the initial link establishment phase. Authentication is mandatory, however, on dial service lines. Authentication requests can be made by either peer on the link, and the receiving device will respond appropriately. If both peers request authentication, then the process is called two-way authentication (see Figure 5-21). One-way authentication is illustrated in Figure 5-22.

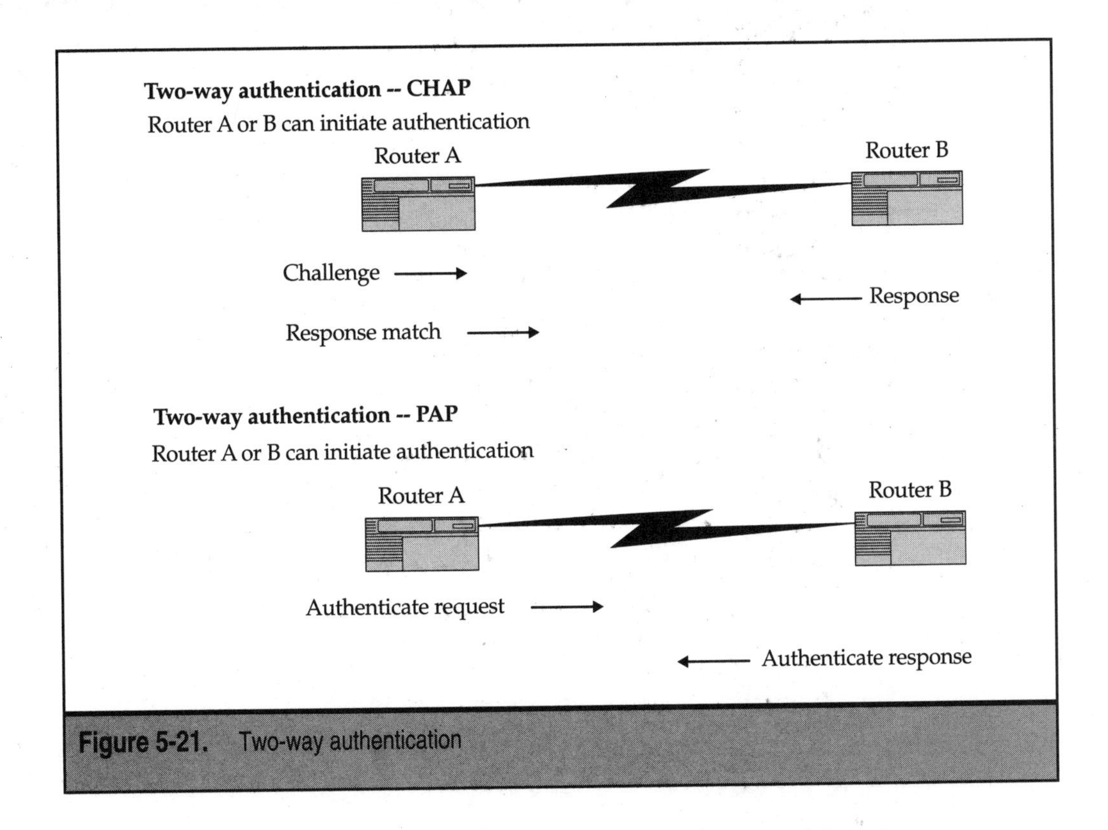

Figure 5-21. Two-way authentication

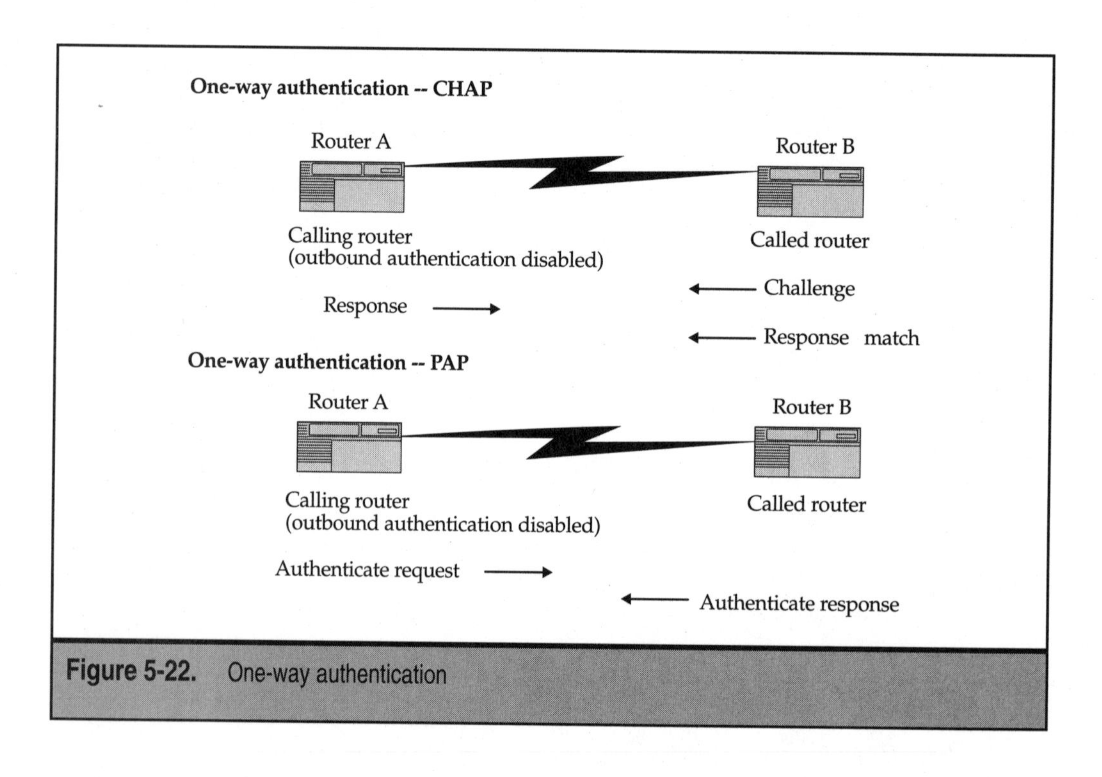

Figure 5-22. One-way authentication

Two authentication mechanisms are available: password authentication protocol (PAP) and challenge handshake authentication protocol (CHAP). PAP is the simpler of the two, which uses a simple plain-text identifier and a password to authenticate the caller.

```
#   96: 12/03/1999 10:52:47.070  TRACE    SLOT 14  PPP                Code:  71
Sending Authenticate-Request on circuit 24.

#  143: 03/14/1999 15:35:38.173  TRACE    SLOT  9  PPP                Code:  68
Received Authenticate-Request on circuit 19.
```

One PPP peer (usually the calling party on a dial service) sends a PAP authentication request to the remote peer (the called party), requesting the PAP identifier and the password. Again, note that authentication can also be requested by the calling party independent of the called party, a feature not necessary on a leased circuit.

Misconfiguration of either the `pap-id` or `pap-password` attribute will cause the authentication to fail and generate warning events. These can also occur if the wrong circuit is associated with the PAP detail.

```
#   14: 10/17/1999 10:54:41.100  WARNING  SLOT 12  PPP                Code:  10
Peer sent incorrect PAP ID 'arn4-miami' on circuit 6.

#    4: 08/15/1999 00:52:59.414  WARNING  SLOT  6  PPP                Code:  11
Peer sent incorrect PAP Password 'dingo99' on circuit 13.
```

After a PAP/CHAP failure, the origination sends an LCP termination request.

```
#   60: 08/11/1999 18:16:31.338  TRACE    SLOT  6  PPP                      Code:  48
Sending LCP Terminate-Request on circuit 4.
```

When PAP authentication fails, check the PPP warning messages to determine if the receiver is sending the wrong information. If the received PAP identifier and password are correct, then the local peer is misconfigured. Some vendors do not support PAP and will reject the request; in this case, you can configure lax authentication (`wfPppLineAllowPapReject`) and proceed to the NCP phase.

```
#  163: 11/18/1999 10:03:14.089  TRACE    SLOT  9  PPP                      Code:  65
Peer LCP Rejected PAP option on circuit 12.
```

A more secure mechanism is CHAP, where two PPP peers share a common *secret*. The secret is never sent across the line in the clear, which provides added security. One of the peers sends a CHAP challenge message using an incrementally changing challenge identifier and a variable challenge value. The receiving peer sends a CHAP response message, which is calculated using the previously known secret value and challenge identifier. This return value must match the value computed by the originating peer, thus providing authentication. Failure of the authentication process on the circuit prevents initialization of the network layer protocols. In this case, check the `chap-name` and `chap-secret` attributes.

```
#    4: 08/11/1999 07:56:59.090  WARNING  SLOT  4  PPP                      Code:  96
Received CHAP message with incorrect identifier on circuit 14.

#   47: 03/18/1999 13:39:34.421  WARNING  SLOT  1  PPP                      Code:  97
Received incorrect CHAP response from peer on circuit 12.
```

If CHAP authentication fails, PAP can be used as a fallback if it is configured (using `wfPppLinePapFallbackDisable`) on the line.

```
#  187: 07/22/1999 07:11:26.289  WARNING  SLOT  7  PPP                      Code:  82
Falling back to PAP on circuit 6.
```

Dial services use the caller resolution table, configured using the `wfPppWhoamiEntry` MIB object, to help associate a caller with a particular circuit. For example, a call received in a demand pool with several circuits needs to be resolved to one single circuit. Using the caller's PAP identifier or the CHAP name, the local router determines which circuit to use to negotiate NCP protocols. On the called side, LCP negotiations usually are associated with circuit 65535, which is a placeholder that will map to the actual circuit number (`wfPppWhoamiCircuit`) or circuit group (`wfPppWhoamiCctGrp`) once the authentication phase is complete.

The caller resolution table, sometimes called the *whoami* table, is used to identify the incoming caller and to associate the call with a particular circuit. Inbound calls, as mentioned earlier, initially are negotiated on the placeholder circuit 65535. Figure 5-23 shows a sample setup with multiple callers. When router A receives a call, it looks in the caller resolution table to determine the circuit with which to associate the call, based on the caller's CHAP name or PAP ID. NCP negotiations occur based on the protocols configured on the resolved circuit number.

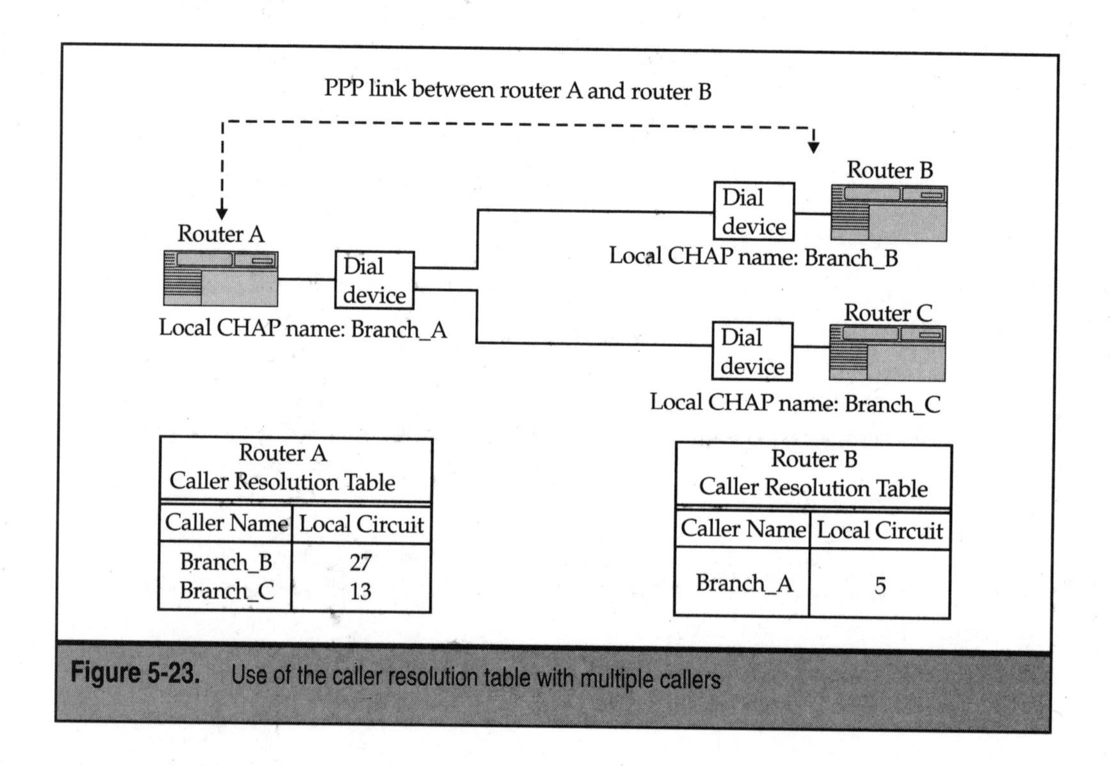

Figure 5-23. Use of the caller resolution table with multiple callers

```
#   76: 12/23/1999 05:36:07.794  WARNING  SLOT 14  PPP                  Code:  98
Failed to locate `la4-ttn1` in WHOAMI table.
```

Messages such as the preceding indicate that the caller resolution table does not know about the calling party. Caller resolution tables are important for dial-on-demand circuits. They are also essential for called parties in backup and bandwidth-on-demand circuits. Problems due to misconfiguration of the caller resolution table usually appear as authentication problems, discussed in the previous section.

Last Bad Packet

Most problems with PPP can be diagnosed and troubleshot using the **show ppp** display and the log. Occasionally when a protocol does not converge or repeated rejection leads to a termination request, the first 16 bytes of the last bad packet as perceived by the router will be stored in the `wfPppLineLastBadPacket` attribute.

```
fxc9(96)# show ppp bad-packets

show ppp bad-packets                                    Apr 07, 2000 13:27:56 [EDT]

Circuit Name     # Bad Packets   Last Bad Packet
-----------------------------------------------
S73_PPP          2               FF03C02101
```

This information may be sufficient for you to determine the cause of the problem without having to use a sniffer trace or a packet capture on the line. The PPP frame format illustrated in Figure 5-18 can be used along with PPP RFCs to help decode the packet. Alternately, the hexadecimal packet contents can also be entered in an expert window in a sniffer for interpretation as a PPP frame. In the preceding example, the protocol is identified as LCP (0xC021), which follows the initial address (0xFF) and control (0x03) values. The code indicates a configure request (0x01); however, there is no identifier or additional data following it. This indicates that the frame was either truncated in transmission or formatted incorrectly by the originating router.

Configuration Considerations

In software versions prior to 11.00, the PPP implementation and then-available standards did not support ARP over PPP lines. The function instead required configuration of an adjacent host (static ARP entry) on both ends of a link to allow IP connectivity, even after IPCP had successfully negotiated the IP addresses. In software versions subsequent to 11.00, this adjacent host entry is no longer required—it is, however, required if you are connecting to another PPP vendor's product that does not support ARP. If you are unable to ping the other end of the PPP circuit but can observe routing table exchanges, then an adjacent host entry is needed on both sides.

Certain vendors support the use of authentication only on inbound calls for dial services. For example, a third-party PPP device receiving a dial call will send an authentication request, but it may not accept an authentication request from the caller. BayRS routers can be configured so that outbound authentication is disabled when they are originating calls. By default, `outbound-authentication` (`wfSwservOptsOutboundAuth`) is enabled and must be disabled when the peer rejects the authentication request.

Link Quality Link-quality reports (LQRs) are useful in monitoring the link quality on a PPP line. If link quality falls below a configured threshold for either inbound or outbound traffic, LQM helps bring down the line. Peers send link-quality reports to each other to report successfully transmitted and received frames for each reporting period. A peer averages the values from five reporting periods and uses this metric in determining the acceptable configured thresholds.

```
show ppp lqr config                                  Mar 12, 2000 21:13:24 [EST]

        Slot/
Circuit <Mod>/  LQ       Remote      LQR Repeat Inbound  Outbound
        Conn    Protocol Timer       Period     Quality  Quality
------- ------- -------- ----------- ---------- -------- ---------
H21     2/1     linkqr   enabled              3      90%       90%
```

When the link quality falls below the acceptable thresholds, the NCPs are stopped but not closed. LCP continues to operate, and LQM continues to monitor the quality of the line.

```
#  163: 12/23/1999 17:48:15.253  WARNING  SLOT  8  PPP                Code: 121
Received LQR with inbound line errors on line 208101:0, circuit 14.
```

```
#  164: 12/23/1999 17:48:15.274  WARNING  SLOT  8  PPP                Code: 118
Inbound link quality at 92% on line 208101:0, circuit 14.

#  165: 12/23/1999 17:48:18.213  WARNING  SLOT  8  PPP                Code: 118
Inbound link quality at 88% on line 208101:0, circuit 14.

#  166: 12/23/1999 17:48:18.237  WARNING  SLOT  8  PPP                Code: 201
Link quality below configured minimum, killing line 208101:0, circuit 14.
```

When the quality of the line returns to normal, operation of NCPs resume.

```
#  180: 12/23/1999 17:48:23.834  INFO     SLOT  8  PPP                Code: 146
Inbound/Outbound link quality back to desired level on line 208101:0, circuit 14.
```

You should configure LQM on leased lines to help stop the operation of network protocols when quality degrades, because otherwise users will experience poor performance and response. On backup lines, however, you may configure lower LQM thresholds to allow some data loss due to noisy circuits but still allow user connectivity over the only available line.

```
show ppp lqr stats                                Mar 12, 2000 21:14:48 [EST]

        Slot/
        <Mod>/  LCP      LQR Repe-  Inbound  Outbound  LQR    LQR
Circuit Conn    State    at Period  Quality  Quality   In     Out
------- ------- -------- ---------- -------- --------- ------ ------
H21     2/1     opened            3      99%       99%   534    535
```

Synchronous Data Link Control

SDLC, or Synchronous Data Link Control, is a bit-oriented link protocol used in IBM subarea Systems Network Architecture (SNA) networks. It is used on point-to-point and point-to-multipoint lines to interconnect various types of SNA devices.

BayRS routers can behave as primary, secondary, or negotiable link stations on SDLC links. As a primary link station, the router can communicate with downstream nodes. When configured as a secondary link station, it can communicate with front-end processors and communications controllers. SDLC primary interfaces on BayRS routers control the data link, perform error recovery, and issue commands. SDLC and the established OSI standard, HDLC, have many common features.

The primary application required is *sdlc*, and events are logged under the `SDLC` entity.

Frame Formats and Frame Types

Figure 5-11 illustrates a typical HDLC frame, which is identical to an SDLC frame. The Address field is usually 1 octet long, followed by the Control field, which is 1 or 2 octets long; the Control field defines the function of the frame by indicating the frame format: whether it is a supervisory, information, or unnumbered frame. The Data field contains data, and the FCS field is usually 2 octets long. Flags (0x7E) delimit the frame at the beginning and the end.

Link Station Roles

An SDLC link station represents a logical connection between two adjacent devices. A link will have one primary link station and one or more secondary link stations. When using DLSw, you can configure a BayRS router as either a primary or secondary link station for PU 1.0, 2.0, and 2.1 nodes. For PU 4.0 devices in DLSw configurations and PU 2.1 devices in APPN configurations, the router can be configured as a negotiable link station.

```
show sdlc interface-operation                        Apr 07, 2000 13:27:14 [EDT]

Circuit
Name          Address         Role           Type           Topology
----------   ---------  -------------- ---------------   ----------
S41          C4        pri            leased           multipoint
```

A common error is the misconfiguration of the router's link station role. This can be rectified by setting the `linkstation-role` attribute for the sdlc interface or by setting the `mode` attribute for the `link-station` object for a specific link station.

```
fxc9(93)# show sdlc link-station-oper

show sdlc link-station-operation                     Apr 07, 2000 13:38:21 [EDT]

Circuit
Name        Address        Role      MAXDATA  MAXIN  MAXOUT MODULO
---------- ------- --------------- ------- ------- ------ ------
S12         A1      sec              pdu1033       7      7      8
```

Line Parameters

SDLC is supported over serial lines and MCT1 logical lines. The SDLC configuration objects in the BCC hierarchy are shown in Figure 5-24.

The following line parameters are crucial to the initialization of SDLC lines:

- **Clock source** The clocking source defaults to internal on serial lines and to external on MCT1 lines. This attribute must be set to external if clocking is provided by an external source. It is changed on the physical or logical line.
- **Internal clock speed** If the router is configured to use the internal clock source, this attribute must be configured properly. Only one device on the SDLC line must provide clocking, and this is set on the physical or logical line on the router.
- **Sync line coding** BayRS routers support NRZ (nonreturn to zero), NRZI (NRZ, inverted), and NRZI Mark (NRZI, mark) coding. The line coding must match that of all SDLC devices on the line.
- **Cable type** Although BayRS can automatically detect the type of cable plugged into the port, you may occasionally need to manually override the

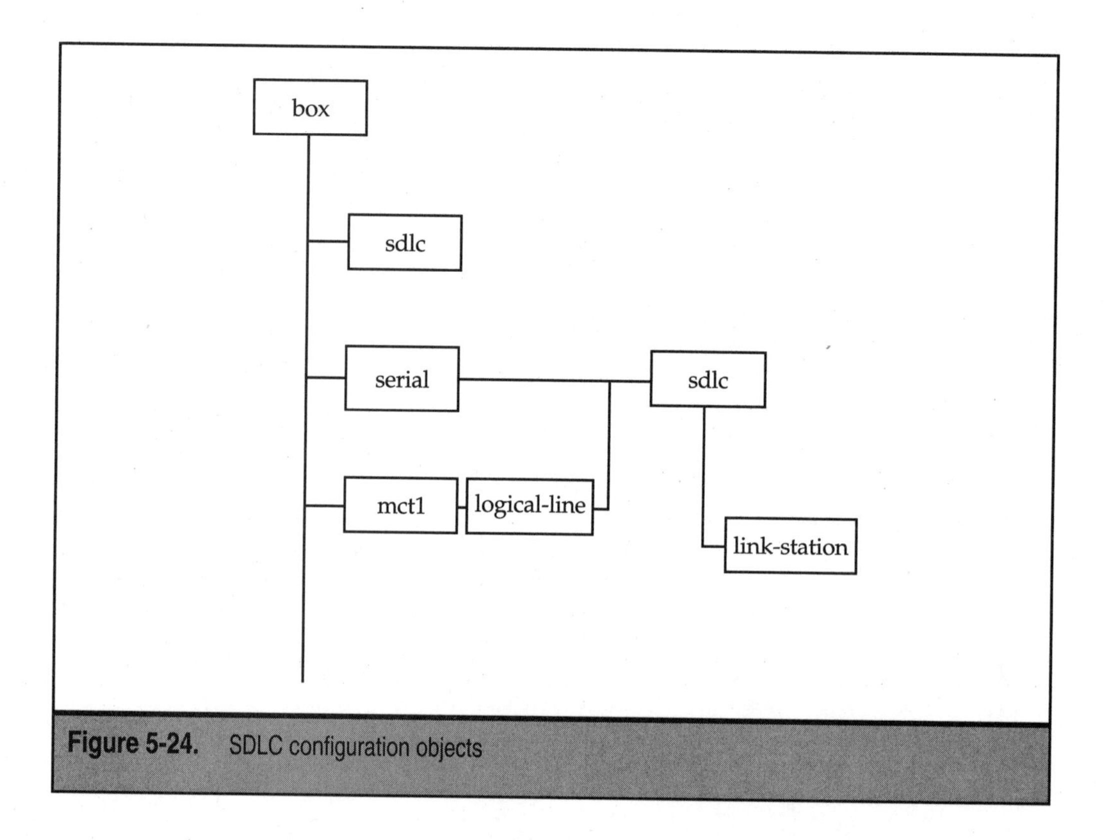

Figure 5-24. SDLC configuration objects

cable type: for example, if a custom cable is designed based on RS232, RS422, V.35, or X.21 cables. When using a particular cable type, make sure that it can support the line data rate.

- **RTS enable** The request-to-send (RTS) signal is disabled by default and must be enabled when configuring the router as a secondary interface on a multipoint line.
- **Interframe time fill** The IFTF character is the pattern that the router transmits when there is no user data across the line. It can be configured to use HDLC flags (0x7E) or IDLE flags (0xFF); by default, the IFTF character is based on the sync media type attribute. If the sync media type is set to ISDN leased, then IDLE flags will be used as the IFTF.

Reject Frames

The MODULO, a rotating acknowledgment window, keeps track of frames sent to and received from an adjacent SDLC node. It can be configured to take the value 8 or 128, providing maximum unacknowledged frame counts of 7 and 127, respectively. When an SDLC link station receives an information frame out of sequence, it may issue a reject (REJ) command. Some implementations may not support the reject command and instead use a sequence of RNR (receiver not ready) and RR (receiver ready) messages to request the retransmission of frames. Other implementations reset the link instead of rewinding the transmission window. In both cases, you can set the `reject-frame-allowed` attribute for the `sdlc` object to no.

```
box->serial/3/1->sdlc/serial/3/1
fxc9(5)#reject-frame-allowed no

show sdlc link-station-statistics                        Apr 07, 2000 13:27:37 [EDT]


                              I Frames  I Frames                    REJs   REJs
Circuit           Address    Ins       Outs       Retransmits  Ins    Outs
------------      -------    --------  --------   -----------  -----  -----
S31               C2              2341      1983            2      1      6
```

The reject command is most useful on full-duplex transmission lines because data can be retransmitted by the remote link station without affecting the outbound (local) transmission. A continuing increase in the number of reject frames indicates that frames are being lost on the line or dropped by the line driver. If receive errors occur on the line, the line driver may discard the frames before they can be delivered to SDLC. Check the serial or mct1e1 statistics to see if there are errors on the transmission line.

Preactivation Frame

The preactivation frame is used to determine whether connection is first established between DLSw and the host or with the SDLC station. SDLC configurations in BayRS typically employ Data Link Switching (DLSw) or Advanced Peer-to-Peer Networking (APPN). Five preactivation frame settings are available: XID (exchange identification), DISC (disconnect), SNRM (set normal response mode), SNRME (set normal response mode extended), and TEST (test). The default is XID, which indicates that DLSw should initiate the activation polling.

Occasionally, you may need to activate the link by sending one of the other frames to the SDLC station. Most SDLC devices use the SNRM, though certain configurations may require DISC, SNRME, or TEST. The `pre-activation-contact-frame` attribute is a commonly overlooked attribute that must be obtained from the IBM systems administrator and configured correctly for each link station.

```
link-station/12# pre-activation-contact-frame snrm
```

Configuration Considerations

SDLC is often used in conjunction with DLSw or APPN, as mentioned earlier. SDLC-to-LLC2 conversion is a common application whereby source-route-bridged networks are integrated to work with traditional SDLC links. APPN high-performance routing (HPR) and DLSw virtual ring configurations provide high reliability and require less overhead than traditional leased point-to-point SDLC lines.

BayRS routers normally perform SDLC/HDLC functions at the line driver level. On BN link modules, for example, the MK5025 chipset is responsible for most link control functions, and on the Access routers, the quicsync driver is responsible for these functions. Problems with basic SDLC functions are usually resolved by addressing the line (serial or mct1) issues and configuration parameters for the SDLC interface and the link stations. However, establishing LU-LU sessions end to end using DLSw or APPN may require you to troubleshoot higher-layer functionality, including SDLC-LLC2 conversion, DLSw, or APPN. Troubleshooting DLSw is discussed in detail in Chapter 6.

Asynchronous Transfer Mode

Asynchronous Transfer Mode, usually referred to as ATM, is a connection-oriented cell switching technology that runs over broadband ISDN networks. ATM provides several advantages over packet-switching technologies such as frame relay and X.25, including predictable cell delay variations and guaranteed quality of service (QoS). ATM is used in carrier networks to transport voice, video, and data. In enterprise data networks, ATM has found broad application, including in WAN interconnections and LAN backbones. ATM uses a fixed format of 53-byte cells, with a 5-byte header and a 48-byte payload (Figure 5-25).

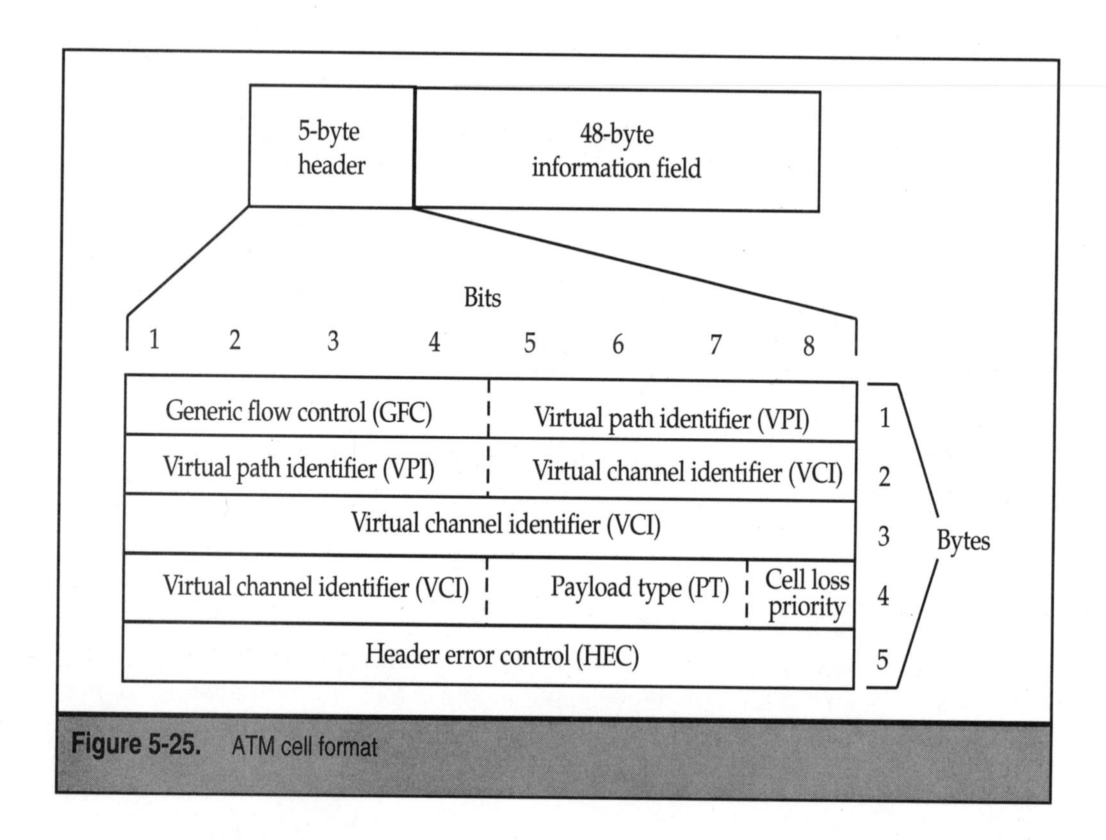

Figure 5-25. ATM cell format

The ATM service relies on several applications and drivers. The applications of most importance include the core *atm* application (`wfATMLoad`), the *atmsig* signaling application (`wfATMSigLoad`), and the *atm_le* LANE (LAN emulation) application (`wfATMLeLoad`). ATM interface-specific messages are logged under the `ATMINTF` entity, signaling messages under ATM_SIG, and LAN emulation messages under ATM_LE. General ATM messages are logged under the `ATM` entity. In recent software versions, the `ATM_SIG` entity has become obsolete, and signaling events are logged under the `ATM` entity.

Slot and Port Initialization

ATM lines are supported by multiple media types: OC3 SONET/SDH (multi-mode and single-mode), DS3 (C-bit and M23 framing), and E3 (G.832 and G.751 framing). OC3 link modules can be connected to both ARE and FRE2 processor modules, and DS3 and E3 modules use ARE processors exclusively. The image requirements for using ARE and FRE2 processors are unique and sometimes cause confusion.

BayRS software supports the ATM application in two suites: ATM and VNR (Virtual Network Router). The ATM suite (refer to Appendix A) provides full ATM functionality for permanent virtual circuits (PVCs), typically used with RFC 1577 applications, but does not support LANE, which requires switched VCs. The VNR suite supports both PVCs and SVCs and can be used for all applications, including LANE. When initially loading software, make sure that the suite being used supports all required applications. On 578x routers, only the VNR suite is available. If a Backbone router has ARE processor modules only and the incorrect suite is used, then all services may not be initialized.

Another common initialization problem occurs with the use of corporate suite images with routers installed with ARE processors. The PowerPC kernel and applications (.ppc) are available only in the ATM and VNR suites, which also contain the necessary images for the FRE processors. If the ARE processors fail to boot with a newly loaded image but all the FRE processors boot successfully, then run readexe on the image to see if the image contains the necessary .ppc applications for the ARE. When you are unsure of the image requirements, always use the VNR suite image to ensure that all the processors boot successfully and support all applications.

When an ATM interface initializes, the coprocessor, the SAR device, and the framer will initialize first. Following this, signal detection and framing will allow the port to enter service mode. Loss of cell delineation is the condition where the interface loses the cell boundaries, typically a framing problem on the circuit. A transceiver disconnect indicates that the device has lost the signal. Both conditions can occur if the circuit is bad or if a downstream path terminating device is malfunctioning.

```
#   14: 12/07/1999 23:24:31.308  WARNING  SLOT 10  ATMINTF           Code:  28
Port 1: framer out of cell delineation.

#   15: 12/07/1999 23:24:31.310  WARNING  SLOT 10  ATMINTF           Code:  27
Port 1: transceiver disconnected.
```

Service Records and Virtual Circuits

We discussed the concept of service records in the "Frame Relay" section earlier. You can view the service record as a logical interface providing the same set of higher-layer protocol services across multiple virtual circuits. Point-to-point and point-to-multipoint service records can be established over ATM interfaces.

```
show atm services                                    Apr 07, 2000 13:20:00 [EDT]

ATM Service Record Table
------------------------

Slot/
<Module>/
Conn       Service       Encaps     State  Type
--------- ------------ ---------- ------ ----------
4/1        ether_12      lane       up     svc
4/1        clip_net4     llcencaps  up     svc
... ...
```

The broadband ISDN (B-ISDN) protocol reference model is illustrated in Figure 5-26. BayRS routers perform all essential functions at the physical layer, ATM layer, and ATM adaptation layer (AAL). The AAL is further divided into the segmentation and reassembly (SAR) layer and the convergence sublayer (CS). At the higher layer are the service records, data encapsulation, protocols, and signaling functions. BayRS supports a CS function compatible with AAL 5. Data applications such as LANE are adequately serviced by AAL5; ATM termination devices requesting an unsupported AAL type must be configured to use AAL5 when connecting to a BayRS ATM/UNI.

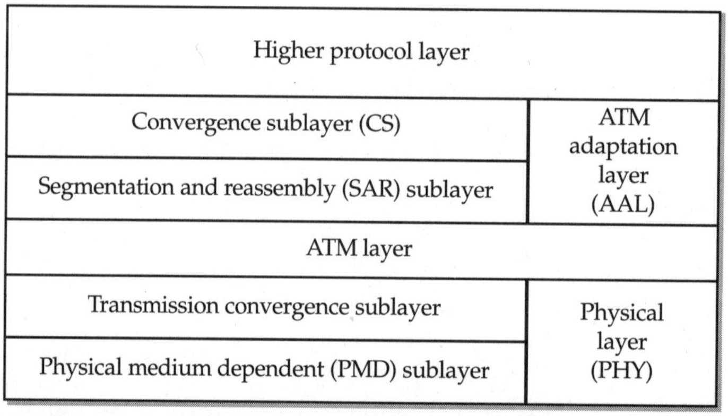

Higher protocol layer	
Convergence sublayer (CS)	ATM adaptation layer (AAL)
Segmentation and reassembly (SAR) sublayer	
ATM layer	
Transmission convergence sublayer	Physical layer (PHY)
Physical medium dependent (PMD) sublayer	

Figure 5-26. Broadband ISDN protocol reference model

```
# 173: 09/08/1999 10:25:32.555  WARNING  SLOT 11  ATMINTF          Code:  29
Port 1: activate virtual channel (call reference 12) failed;
        AAL type not supported (1).
```

ATM virtual circuits are identified using a virtual path identifier (VPI) and virtual channel identifier (VCI). Virtual circuits can be set up dynamically (switched, SVCs) or permanently (PVCs), depending on the application. SVCs and PVCs are commonly referred to as virtual channel links (VCLs). The virtual path and virtual channel components in an ATM transmission circuit are portrayed in Figure 5-27. Note that several virtual channels can be transported across a virtual path, and several virtual paths can exist in a single physical transmission line.

Encapsulation Methods

Above the ATM adaptation layer, multiprotocol data encapsulation functions must be performed to transport various protocols across the ATM network. Multiprotocol encapsulation (MPE) in frame relay service records is performed using a standard LLC/SNAP header. ATM service records, however, support multiple encapsulation options:

- **LANE** LAN emulation can be used to transport IEEE 802.3 (Ethernet) and IEEE 802.5 (Token Ring) encapsulated PDUs across an emulated LAN over the ATM network. Only switched virtual circuits can be used with LANE encapsulation.
- **LLC/SNAP** Defined by RFC 1483, the Logical Link Control/Subnetwork Access Protocol can be used to multiplex several protocols across a single virtual circuit. An 802.2 (LLC) header is used with each PDU, which can be transported across PVC service records, SVC service records, individual PVCs, and WAN SVC service records. When SVC service records are used with LLC/SNAP encapsulation, BayRS automatically uses Classical IP over ATM (RFC 1577) functions.

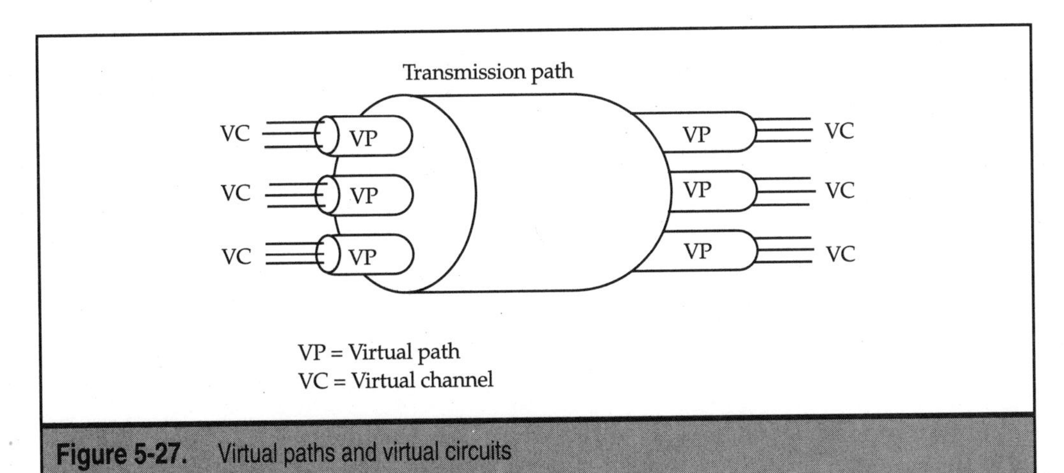

Figure 5-27. Virtual paths and virtual circuits

- **NLPID** Commonly used with frame relay–to–ATM interworking, the Network Layer Protocol ID (NLPID) encapsulation is defined by RFC 1490. NLPID encapsulation can be used with PVC service records only.
- **NULL** Often called VC-based multiplexing, this method multiplexes upper-layer protocols using virtual circuits. This encapsulation can be used with PVC service records, SVC service records, and individual PVCs that are also members of an LLC/SNAP service record. NULL encapsulation used on SVC service records automatically allows BayRS to perform Classical IP over ATM (RFC 1577) functions.

The use of a specific encapsulation with either SVCs or PVCs allows bridging and routing of certain protocols. A protocol set supported by one encapsulation may be different from that supported by another encapsulation; consult the release documentation for the BayRS version you are using. The data encapsulation type and the choice of PVC or SVC depend largely on your application. If a service record fails to become operational, check the virtual connection type used.

```
#   38: 02/01/1999 05:10:37.204  DEBUG    SLOT 12  ATM                Code: 231
Unsupported encaps type (5) for muxed SVCs.
```

Interim Local Management Interface

ILMI is used by the router for automatic link management. It is used for initializing the ATM address table, obtaining the network address prefix from the switch, and registering the interfaces' address (prefix plus end-system identifier, or ESI) with the switch. By default, the `ilmi` object is created as a child of the `signaling` object under the `atm` interface. ILMI is similar to other link management protocols discussed earlier and behaves like SNMP. ILMI uses the reserved VC 0/16 to communicate with the switch.

ATM uses a 20-byte address, with a 13-byte network prefix and a 7-byte user part. The user part is subdivided into a 6-byte end-system identifier (ESI) and one selector byte. The MAC address of a device usually is used in place of the ESI. The user part (`user-suffix`) can be manually set for each service record, or it can be automatically generated by the router by leaving the `autogenerate` attribute enabled (the default). The components of a typical ATM address are illustrated in Figure 5-28.

You should always use ILMI , except when the switch does not support ILMI. In that case, the ATM network prefix (`network-prefix`) must be configured manually for each service record. If the network prefix is configured and ILMI is left enabled, the interface will still attempt to register the address with the switch; however, the configured prefix will override the value assigned by the switch.

```
#   67: 11/22/1999 00:19:26.769  WARNING  SLOT 13  ATM                Code: 234
UME Line 1413101: Circuit 4: Failed to register address: 39004000 00000000
00000000 000000a2 82ac4501.
```

In both cases, registration of the ATM address for each service record must be successful. If the switch does not accept the address, then the service record will be unable to establish any SVCs for user data. The preceding message illustrates address registration failure.

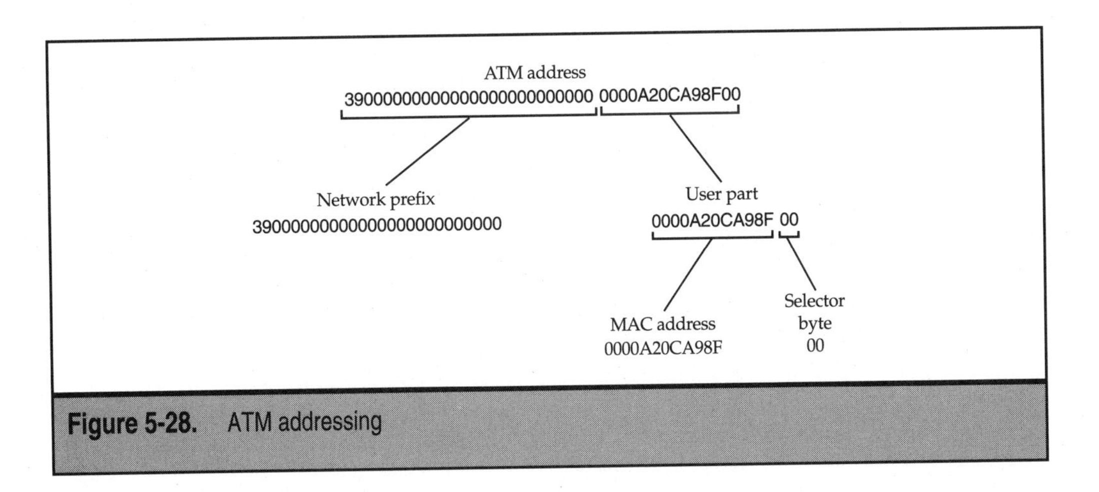

Figure 5-28. ATM addressing

Address registration commonly fails as a result of incorrect network prefix or user suffix configuration. If the user suffix is duplicated for the specific network prefix, the ATM switch may reject the registration.

```
#  138: 10/13/1999 16:26:21.512  INFO     SLOT 12  ATM                 Code:  35
Line 1412101 : ATM ILMI has been Disabled.

#  117: 05/16/1999 09:21:16.443  WARNING  SLOT 11  ATM                 Code:  60
Line 1411101: ILMI VC disabled.

#  164: 08/07/1999 12:56:59.426  WARNING  SLOT  8  ATM                 Code:  61
Line 1408101: ILMI VC not present.
```

Problems with ILMI also commonly occur because the service is disabled or the associated VC (0/16) cannot be established. The default ILMI VC 0/16 can be overridden under the ILMI configuration. The ATM switch must also be configured to use the same VC for ILMI if the network configuration requires it.

Signaling

Signaling over ATM circuits is used to dynamically establish, maintain, and clear switched virtual connections. BayRS supports Q.2931 signaling over ATM user-to-network (UNI) interfaces. UNI versions 3.0, 3.1, and 4.0 are supported. Signaling is not required if all ATM service records on an interface use only PVCs. The UNI version (set with the `uni-version` attribute) for the signaling object must correspond to the value set on the switch.

```
show atm signaling                                Apr 07, 2000 13:20:20 [EDT]

ATM Signaling Table
-------------------
```

```
Slot/
<Module>/
Conn      Signaling Sig Version
--------- --------- ------------
4/1       enabled   v30
8/1       enabled   v40
```

Most private data networks employ switched virtual connections for multiprotocol data traffic. In these scenarios, signaling is required, and its absence usually results in an inability to establish any dynamic VCs. You will see symptoms reported in the log, such as the absence of signaling configuration, the absence of the default signaling VC (0/5), or complete unavailability of the signaling service.

```
#   54: 02/17/1999 10:14:53.380  INFO     SLOT 12  ATM          Code:  33
Line 1412101 : ATM Signaling has been Disabled.

#   53: 05/28/1999 01:01:12.211  WARNING  SLOT  3  ATM          Code:  89
Line 1403101 : No ATM Signaling MIB entry found.

#    5: 01/08/1999 09:33:44.955  INFO     SLOT  7  ATM          Code:  20
Line 1407101: Signaling not loaded.

#   31: 11/10/1999 07:11:30.989  WARNING  SLOT  1  ATM          Code:  63
Line 1401101: SIG VC not present.
```

Another key aspect of ATM signaling is the Service-Specific Connection-Oriented Protocol (SSCOP). The connection-oriented nature of SSCOP allows Q.2931 signaling messages to be reliably transported across the ATM network. SSCOP resides above the AAL layer and below the Q.2931 signaling layer. SSCOP is sometimes referred to as Q.SAAL and is essential for signaling transport. The `sscop` child object is automatically created and enabled when the `signaling` attribute is enabled. If the SSCOP configuration object is either deleted or otherwise removed from the configuration, the following event message is logged.

```
#   44: 08/15/1999 23:36:43.114  WARNING  SLOT  8  ATM          Code:  90
Line 1408101 : No ATM SSCOP MIB entry found.
```

The UNI version helps you specify the format of the SSCOP frames as defined by the appropriate ATMForum standard. By default, the router uses UNI version 3.0 for interface signaling. The signaling protocol stack on ATM interfaces is shown in Figure 5-29.

Signaling services are provided to higher-layer service records that explicitly request them. Before a service record can establish SVCs, it must first bind with the signaling service. Each binding is called a service access point (SAP), and the maximum number of service records requiring these services is dictated by the `svc-applications-maximum` (`wfAtmSigMaxServiceUsers`) attribute. By default, 20 service records requesting the use of SVCs can be configured on an interface. If some service records can establish SVCs but others cannot, then check the log and configure svc-applications-maximum appro-

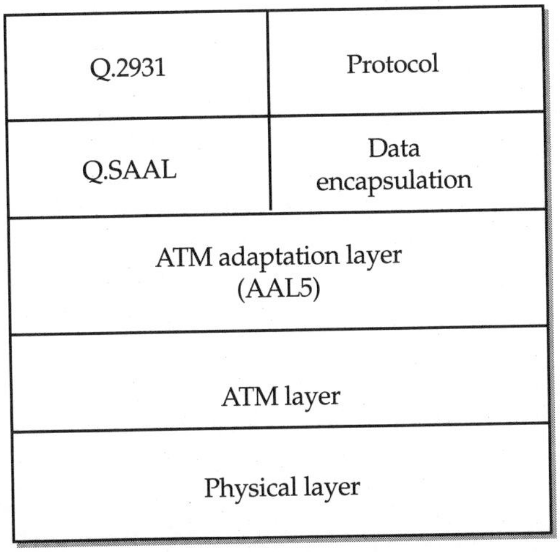

Q.2931	Protocol
Q.SAAL	Data encapsulation
ATM adaptation layer (AAL5)	
ATM layer	
Physical layer	

Figure 5-29. ATM signaling protocol stack

priately. Configure the maximum number of SVC applications based on the services that require SVC support; increasing this number will consume more memory resources on the processor module.

```
#  181: 09/08/1999 16:57:48.307  WARNING  SLOT  3  ATM             Code:  96
The maximum number of configured Service Users has been exceeded.

box->atm/3/1->signaling/3/1
fxc9(4)# svc-applications-maximum 50
```

Another signaling problem sometimes seen is the failure to establish SVCs on some service records after the interface has been operational for an extended period of time. This can occur if the maximum number of point-to-point SVCs are in use. This value is configured to limit the subscription of circuit resources. By default, the `point-to-point-maximum` (`wfAtmSigMaxPtPtConnections`) attribute for SVCs is set to 1000 simultaneous connections per port. It can be increased when the following symptom repeatedly appears:

```
#   49: 05/08/1999 10:57:16.895  WARNING  SLOT  9  ATM             Code:  95
A VC request for call_ref 2328 and SAP 13 has failed and is ignored.

box->atm/3/1->signaling/3/1
fxc9(5)# point-to-point-maximum 1800
```

The condition in the preceding example occurs when there are numerous simultaneous data sessions on an ATM interface. This can occur on service records with a significant number of routed or bridged devices.

Similarly, point-to-multipoint call requests may fail when the default limit of 40 connections is reached.

```
#   58: 10/09/1999 09:26:01.376  WARNING  SLOT  3  ATM                 Code:  97
A Point to Multipoint connection for call_ref 8723 and SAP 24 has failed.

box->atm/3/1->signaling/3/1
fxc9(6)# point-to-multipoint-maximum 1800
```

Point-to-multipoint connections are often used in LANE environments when establishing control distribute and multicast forward virtual circuits. The limit value typically is increased to double the number of maximum SVC applications. For example, if the `svc-applications-maximum` attribute is set to 50, then increase the value for the `point-to-multipoint-maximum` attribute from 40 to 100.

Signaling timers are often left at the default values. You can, though, modify the values to meet special network requirements. However, when changing signaling timers, be sure they are matched on the other end of the UNI connection, usually the ATM switch.

Classical IP and ARP over ATM

RFC 1577 defines the deployment of logical IP subnetworks (LIS) over ATM. This application was designed prior to the ATMForum work on LAN emulation and is still used in many IP-based networks. It is sometimes referred to as Classical IP (CLIP). Each LIS consists of several ATM devices including endstations and routers, which all have IP interfaces configured on the same IP subnet. All IP devices on the logical IP subnet communicate with each other using SVCs, and they communicate with other subnets using a router that resides in the LIS. In this topology, all devices can establish SVCs to each other using a meshed architecture.

A common requirement for all IP networks is the availability of an address resolution protocol (ARP). Ethernet and Token Ring networks use traditional IP ARP, and Classical IP defines the ATMARP scheme to resolve ATM addresses because of the nonbroadcast nature of the medium. In this scheme, each IP device opens a VC to an ATMARP server and registers its IP and ATM addresses. The ATMARP server uses this information to maintain a table of IP addresses and corresponding ATM addresses. When one IP device (ATMARP client) wants to communicate with another IP device, it uses the ATMARP server to resolve the ATM address of the destination station. Then it establishes an SVC to the destination.

BayRS routers can behave as both ATMARP clients and servers. When the router behaves as an ATMARP client, it must be configured with the ATM address of the ATMARP server. Misconfiguration and lack of connectivity to the ATMARP server are among the most common problems with Classical IP services.

If the log reports the following event, check the ip interface configuration under the classical-ip-service object. The arp-mode attribute defaults to client, which requires a valid ATM address in atmarp-server-address. If the service is enabled as ATMARP server, then you can override autogeneration of the user part by manually specifying the user-suffix attribute.

```
#  134: 10/17/1999 19:54:25.540  WARNING  SLOT  2  IP           Code:  87
ATMARP 10.93.81.1: Interface misconfiguration on circuit 14
```

In the following case, the classical IP server was unable to connect to the ATMARP server to perform the function. This problem can be caused by several factors, including an incorrectly configured ATMARP server address and the inability of the network to route the call because of invalid call routing tables.

```
#  112: 12/03/1999 03:36:31.302  WARNING  SLOT  6  IP           Code:  92
ATMARP 192.168.18.254: failed to contact ATMARP server
39001000:0000A000:00000000:000000A2:C54F5301 attempt #3
```

LAN Emulation

Emulation of traditional Ethernet and Token Ring LANs over an ATM network is a common application. Defined by the ATMForum, LANE (LAN emulation) allows the establishment of user-to-user virtual connections over the ATM network in a transparent fashion, emulating a traditional LAN. BayRS supports both IEEE 802.3 Ethernet and IEEE 802.5 Token Ring LAN emulation. BayRS supports ATM LANE versions 1.0 and 2.0; version 2.0 support is required to implement multiprotocol over ATM (MPOA) for layer-3 shortcuts.

Similar to frame relay, ATM is a nonbroadcast, multiple access medium. This nonbroadcast nature poses two considerations for LAN network emulation: broadcast capability and address resolution. Classical IP over ATM works around this problem by using the ATMARP scheme, because it solves not only address resolution problems, but also the broadcast problem, because broadcasting becomes unnecessary. However, LANE transports multiprotocol data over the ATM network by emulating a transparent LAN. It provides a broadcast and unknown server (BUS) for delivering broadcast frames, and a LAN emulation server (LES) for resolving ATM addresses based on MAC addresses.

A full-fledged emulated LAN (ELAN) has the following components:

- **LAN emulation configuration server (LECS)** The LECS is an ATM device that provides configuration parameters, particularly the address of the LES and the BUS. A LAN emulation client (LEC) will contact either the well-known LECS address (defined by the ATMForum, 47.00.79.00.00.00.00.00.00.00.00.00.00.A0.3E.00.00.01.00) or a predefined LECS address to obtain configuration parameters. Use of the LECS is optional, and the LES and BUS can be contacted directly.

- **LAN emulation server (LES)** The LES is the controlling entity for the ELAN. It maintains control direct and control distribute VCs to the LECs and is generally responsible for the LAN emulation-ARP (LE-ARP), which is used for MAC-to-ATM address resolution.
- **Broadcast and unknown server (BUS)** The primary purpose of the BUS is to forward broadcast traffic to all registered LECs and also frames destined to addresses that have not yet been resolved to a specific LEC in an ELAN. The BUS uses the multicast send and multicast forward VCs for this purpose.
- **LAN emulation client (LEC)** Several LECs connect to the ATM network, behaving as edge devices for the traditional LAN segments. They perform address resolutions, respond to resolution requests from the LES, forward broadcast data to the BUS, and establish point-to-point VCs for endstation connections.

An ATM service record configured to use LANE encapsulation will perform as a LAN emulation client (LEC). BayRS routers behave only as LECs, but will communicate with all the four LANE components in the preceding list. When an LEC initializes, it transitions through several states to reach the *operational* state. In the operational state, the LEC can communicate with the LES, BUS, and other LECs in the same ELAN. Each separate ELAN requires a unique service record for proper operation. The LEC establishes the following virtual circuits for ELAN operation and management:

- **Configure Direct** Established to the LECS, point-to-point. This VC is the first one brought up, except in the case of the manual LES configuration, in which case this VC is bypassed. The LES address is obtained from the LECS using this VC.
- **Control Direct** Established to the LES, point-to-point. This VC carries bidirectional traffic between the LEC and the LES.
- **Control Distribute** Established from the LES to the LECs, point-to-multipoint. The LES sends LE-ARP traffic and distributed traffic to all LECs over this VC.
- **Multicast Send** Established to the BUS, point-to-point. This VC carries bidirectional traffic between the LEC and the BUS.
- **Multicast Forward** Established from the BUS to the LECs, point-to-multipoint. Broadcast traffic sent by the BUS to all the LECs is sent over this VC.

```
show lane clients                              Apr 15, 2000 18:31:13 [EDT]

ATM LAN Emulation Client Status Info
------------------------------------
```

```
Slot/
<Module>/
Conn       Service        LecID State          FailCode      Cfg Src
--------- ------------ ------ ------------- ------------ --------
4/1        ether-12          92 operational    none          knownadr

  LanType  MaxDataFrameSize ElanName    Proxy
  -------- ---------------- ---------- -----
  ieee8023             1516 ether-12       0
... ...
```

The LAN emulation states for an LEC are illustrated in Figure 5-30. Once in the operational state, the LEC can communicate with other LECs in the ELAN, barring any higher-layer protocol issues. It is during the transition from the initial to the operational state that many issues arise, as a result of configuration errors, temporary call routing and signaling problems, and LES/BUS service errors. When a problem occurs, check the last failure code reported by the **show lane clients** command to better understand the cause of the problem.

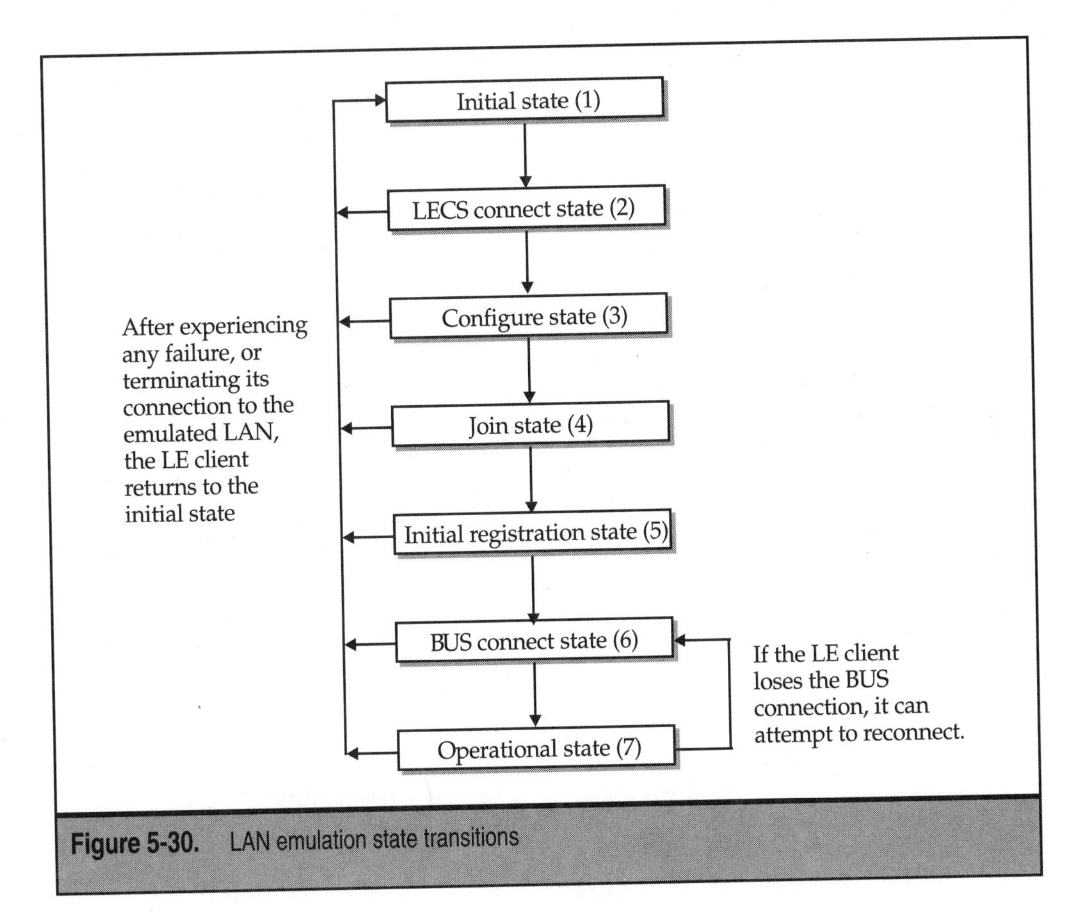

Figure 5-30. LAN emulation state transitions

The following are some common problems that may occur:

- If the LEC remains in the initial state, there may be an underlying signaling or interface problem.
- The LE client may cycle between the LECS connect and initial states, which typically indicates that the LECS is unavailable or is not reachable. Check the call setup failures on the LEC for confirmation of its inability to contact the LECS.
- The LEC may reach the configure state but may not proceed beyond that, which indicates that it was able to reach the LECS but unable to retrieve the necessary configuration for the ELAN. This problem may occur because of an incorrect configuration—ELAN names are case sensitive, and the ELAN type is also important. Check the ELAN name on the LEC and set the ELAN type to unspecified, which will allow the LEC to determine the type (802.3 or 802.5) from the LECS. In addition, on rare occasions, the MTU size on the LEC or the ATM interface may prevent the router from joining the ELAN.

```
#  198: 10/25/1999 08:10:33.820  WARNING  SLOT  4  ATM_LE              Code:  15
Line 1404101 : Circuit 22 : ATM LEC reconfiguration retry exhausted.
```

- The LEC may fail to proceed beyond the join state. During this state, the LEC attempts to establish control connections with the LES and join the specified ELAN. The LES assigns the LEC a unique identifier (LEC-ID). Failure to join the ELAN may indicate that the LEC cannot connect with the LES (see SVC establishment failures). The LES may also prevent the LEC from joining because it does not service the specified ELAN or the LEC specified an incorrect parameter. The LES will also be considered unreachable if it is manually configured but disabled.

```
#  182: 10/09/1999 17:38:49.540  INFO     SLOT  2  ATM_LE              Code:  51
Line 1402101 : Circuit 23 : Instance 1 : LES is disabled.

#  161: 05/12/1999 10:25:44.471  WARNING  SLOT  1  ATM_LE              Code:  52
Line 1401101 : Circuit 12 : Instance 1 : LES is unreachable.
```

If an alternate LES is available, the LEC will attempt to join it.

```
#  161: 01/16/1999 07:13:44.207  WARNING  SLOT  1  ATM_LE              Code:  54
Line 1401101 : Circuit 18 : ATM LEC now trying next le server.
```

- BayRS LECs do not enter the initial registration state because they do not register multiple MAC addresses. Third-party LECs that register multiple MAC addresses must enter the initial registration state.
- The router LEC enters the BUS connect state when attempting to establish a connection to the BUS. If this is the farthest state that the LEC reaches, then the BUS may be unreachable, down, or otherwise inoperable. If SVC failures increment at this state, the network cannot establish the connection. Check the state of the BUS, the BUS configuration, and the BUS address used by the LES.

```
# 113: 01/04/1999 00:09:56.343  WARNING  SLOT  9  ATM_LE          Code:  16
Line 1409101 : Circuit 57 : ATM LEC bus resolve retry exhausted.
```

- ▲ The LEC enters the operational state when it successfully establishes the control and multicast VCCs to the LES and BUS, respectively.

After the LEC is successfully registered and has entered the operational state, it can begin to resolve ATM addresses for MAC addresses residing on remote LE clients. Multiple LECs can be operational on a BayRS ATM interface, each residing in a single service record and joined to a unique ELAN. During normal operation, LE-ARP inconsistencies and errors are some of the common causes of disruption of user connectivity. The router maintains two types of tables: one for MAC address resolution and another for Token Ring route descriptor (RD) resolution. When there are user connectivity issues for a set of endstations, check the LE-ARP table to verify that the MAC-address–to–ATM-address mapping is correct. On Token Ring ELANs, the LE-ARP entry will be resolved for the MAC address if the destination station is on the current ring.

Operations and Management

When a PVC is inoperable because of a network failure, ATM endpoints do not receive any notification of the link failure at the ATM layer. Failure in this case is usually detected by a higher-layer protocol such as a routing protocol. The Operations, Administration, and Maintenance (OAM) feature available with SONET/SDH provides loopback cells and alarms for easier link failure detection. OAM loopbacks can be configured on all PVCs in a service record so that the failure to receive loopback cells will automatically bring down the service. Similarly, the router can be enabled to detect OAM alarms if the downstream switch is capable of generating them. When an AIS alarm is received (see "Loopbacks and Testing" in Chapter 4), the router will bring down the service immediately without waiting for a loopback response. Both conditions allow rapid PVC and link failure detection for data rerouting at higher layers.

SUMMARY

In the beginning of this chapter, we discussed the basics of dial services and the techniques useful in troubleshooting them. Data link issues are often observed by looking at common statistical values reported in the media and protocol statistics. The concepts of circuits, lines, and interfaces were introduced in Chapter 2. We took another look at this topic in the "Applications and Protocols" section of this chapter and discussed ways to correlate physical lines and logical lines based on circuit names and numbers and vice versa. Circuit, line, and interface values can be correlated easily when at least one of them is known. We also discussed some of the most common wide-area data link protocols. PPP, frame relay, and ATM are predominantly used in backbone enterprise data networks. SDLC is another common application, combined with DLSw or APPN to provide network connectivity for IBM SNA and APPN architectures.

In the next chapter, we discuss troubleshooting techniques for network layer protocols.

REFERENCES

1. *Configuring Dial Services*, Publication 308621-14.00 Rev 00.
2. W. Stallings, *ISDN and Broadband ISDN*, 4th Edition, Prentice Hall, 1998.
3. T. Bradley, C. Brown, and A. Malis, *Multiprotocol Interconnect Over Frame Relay*, RFC 1294, January 1992.
4. M. De Prycker, *Asynchronous Transfer Mode: Solution for Broadband ISDN*, Ellis Horwood, 1991.
5. *Configuring ATM DXI Services*, Publication 308608-14.00 Rev 00.
6. J. Heinanen, *Multiprotocol Encapsulation over ATM Adaptation Layer 5*, RFC 1483, July 1993.
7. J. T. Buckwalter, *Frame Relay: Technology and Practice*, Addison-Wesley, December 1999.
8. P. Smith, *Frame Relay: Principles and Applications*, Addison-Wesley, December 1993.
9. *Configuring Frame Relay Services*, Publication 308624-14.00 Rev 00.
10. *Configuring SMDS*, Publication 308643-14.00 Rev 00.
11. R. Klessig and K. Tesink, *SMDS: Wide-Area Data Networking with Switched Multi-megabit Data Service*, Prentice-Hall, 1995.
12. D. Piscitello and J. Lawrence, *The Transmission of IP Datagrams over the SMDS Service*, RFC 1209, March 1991.
13. J. Carlson, *PPP Design and Debugging*, Addison-Wesley, December 1997.
14. *Configuring PPP Services*, Publication 308639-14.00 Rev 00.
15. U. D. Black, *PPP and L2TP: Remote Access Communications*, Prentice-Hall, October 1999.
16. PPP RFCs 1332-1334, 1378, 1552, 1638, 1661, 1662, 1762-1764.
17. *Configuring SDLC Services*, 308642-14.00 Rev 00.
18. *IBM Synchronous Data Link Control: Concepts*, IBM Publication GA27-3093.
19. *IBM Systems Network Architecture: Technical Overview*, IBM Publication GC30-30723.
20. *IBM Systems Network Architecture: Concepts and Products*, IBM Publication GC30-3072.
21. *Configuring ATM Services*, Publication 308612-14.00 Rev 00.
22. M. Laubach, *Classical IP and ARP over ATM*, RFC 1577, January 1994.

23. R. Handel and M. Huber, *Integrated Broadband Networks: An Introduction to ATM-Based Networks*, Addison-Wesley, 1991.
24. *LAN Emulation Over ATM Version 1.0*, ATM Forum, January 1995.
25. J. Martin, J. Leben, and K. Chapman, *Asynchronous Transfer Mode: ATM Architecture and Implementation*, Prentice-Hall, November 1996.

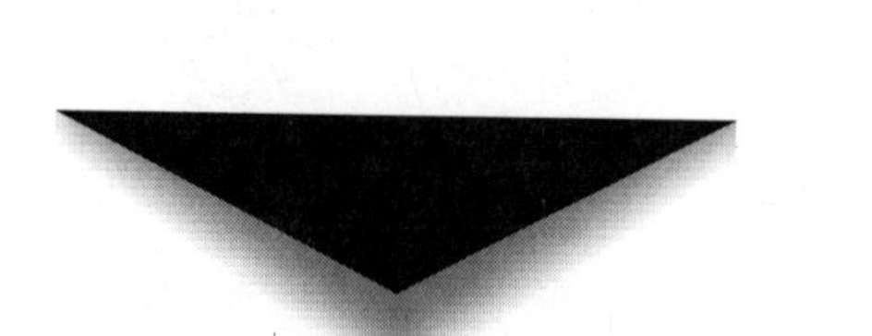

CHAPTER 6

Troubleshooting Network and Routing Protocols

The first rule of holes: when you're in one, stop digging.

—Molly Ivins, *Creators Syndicate*

The Open Systems Interconnect model locates network-level functionality between the data link layer (layer 2) and transport layer (layer 4). During the past two decades, numerous network layer protocols have been designed and implemented, providing a common set of network layer functions. Some historic networking models predate the OSI model and thus follow a different service and protocol hierarchy. At first glance, IBM Systems Network Architecture (SNA) may appear to follow a scheme similar to the OSI model, but the architectures are quite different.

Network control layer protocols generally provide a defined set of functions, but each implementation is geared toward one single network *architecture*. For example, the Internetwork Packet Exchange (IPX) protocol is used primarily in Novell NetWare internetworks, while AppleTalk Phase 1 and 2 protocols are used in Macintosh/Apple environments. Despite the architectural differences, the common functional structure allows methodical network analysis and troubleshooting. With the expansion of corporate intranets and the Internet, IP version 4 has emerged as one of the dominant network protocols. Commonly called simply IP, it has received widespread acceptance among most network vendors, including appliance manufacturers. The comprehensive set of features supported by IP has made it the choice for many higher-layer applications and services. Another significant protocol used in corporate intranets is IPX, for Novell networks. Both IP and IPX provide the backbone services for most enterprise networks and are discussed in detail in this chapter. IP has also been accepted as the de facto standard for Internet connectivity.

Subarea SNA and APPN is widely used in most mainframe environments and midrange server farms. BayRS routers are frequently used for interconnecting SDLC and LLC2 devices, using various methods including IP encapsulation, source route bridging, transparent bridging, SDLC pass-through, and DLSw.

Certain network problems may require a trace of packets transiting a LAN segment or WAN link. To trace packets, you can use the Packet Capture (PCAP) application on the BayRS router, or you can use an external snooping device, commonly called a sniffer. Performing a packet trace is useful for catching an "anomaly in the act," which may provide additional details on the source and cause of the problem. This investigative technique is commonly used to analyze protocol interaction problems and to detect errors in network provisioning. The configuration and analysis of traces is discussed later in this chapter.

INTERNET PROTOCOL

The concepts of *classes* and *subnets* are fundamental in working with IP networks. A 32-bit IP address is divided into *network* and *host* portions (in the form *network.host*), and the network portion may also contain a subnet portion (in the form *network.subnet.host*). Address classes, class A through class E, are recognized by InterNIC. Classes A through C maintain registered addresses. Class D addresses are used in multicasting, and class E addresses are used for experimental networks. Readers unfamiliar with IP classes or subnetting are encouraged to read the reference materials listed at the end of the chapter and use "Guide to IP Addressing" in the Blueprint section.

Ping

The Packet Internet Groper (ping) is a sophisticated yet simple-to-use troubleshooting tool for testing the reachability of remote devices. Ping is supported by all devices that support the TCP/IP protocol suite, including endstations. Internet Control Message Protocol (ICMP) echo requests are sent by the originating device (*source*) to the remote device (*destination*), which responds with an echo reply message. Several types of messages are returned by interim routers indicating different types of failures.

Syntax and Extended Pings

The **ping** command in BayRS supports reachability tests for various protocols; the syntax for IP pinging is shown here.

```
ping -ip <address|hostname> [-t<timeout>] [-r<repeat_count>] [-s<size>]
                            [-p] [-a <source address>] [-v] [-n]
```

The **–ip** option may be omitted because the **ping** command assumes the use of the IP protocol by default. A simple **ping** command uses the destination hostname or IP address as the only argument. DNS (see Chapter 3) must be configured for **ping** to resolve hostnames to IP addresses.

```
fp3# ping 80.23.10.2
IP ping: 80.23.10.2 is alive (size = 16 bytes)
fp3# ping dylan.infotech.mahalingam.com
IP ping: 80.23.10.2 is alive (size = 16 bytes)
```

Options for use with the **ping** command are described here. They are not necessary for basic connectivity tests but are useful in extended troubleshooting with customized ICMP packets.

- **Timeout** (-t) The router automatically considers the echo request to have failed if a response is not received within 5 seconds. Normally, a 5-second response delay is considered excessive because application performance tends to be sluggish or degraded at this level. The timeout value can be increased to determine whether the reachability problem is caused by excessive propagation delays or an overloaded destination.

  ```
  fp3# ping 192.168.10.14 -t1
  IP ping: 192.168.10.14 does not respond
  ```

```
IP fp3# ping 192.168.10.14 -t6
ping: 192.168.10.14 is alive (size = 16 bytes)
```

- **Repeat count** (-r) In a well-functioning network, packet loss is minimal, and ideally, nil. On certain occasions, heavily loaded segments or gateways may cause an echo request in transit to be dropped due to resource constraints. In situations like this, the single echo request sent by the router may be inadequate to determine the actual reachability of the remote device. To obtain details on the sustained reachability of the destination, the router can send a series of ICMP echo requests using a repeat count.

```
fp3# ping 80.23.10.2 -r4
IP ping: 80.23.10.2 is alive (size = 16 bytes)
IP ping: 80.23.10.2 does not respond
IP ping: 80.23.10.2 is alive (size = 16 bytes)
IP ping: 80.23.10.2 does not respond
```

- **Size** (-s) The default ICMP echo frame size is 16 bytes. This size can be increased and is commonly used in detecting misconfiguration of the maximum supported frame size along the path. It is also used to determine the maximum supported frame size on the destination device.

```
fp3# ping 192.168.32.11 -s1024
ping: 192.168.32.11 is alive (size = 1024 bytes)
```

- **Source address** (-a) BayRS routers typically have many IP interfaces and several circuits. Normally, when pinging a destination from the router, BayRS automatically uses the IP address of the outgoing interface as the source address in the ICMP echo request. This source IP address can be overridden by manually specifying one of the other configured (and operational) IP addresses on the router. This option is commonly used to test the destination's reachability from various networks supported by the router, including gateway and mask configurations.

```
fp3# ping 192.168.32.19 -a192.168.33.1
IP ping: 192.168.32.19 does not respond
fp3# ping 192.168.32.19 -a192.168.32.1
IP ping: 192.168.32.19 is alive (size = 16 bytes)
```

- **Path trace** (-p) Using a path trace, the local router will methodically analyze the reachability of the destination by checking the reachability on a hop-by-hop basis. Intervening routers are displayed one by one along with the round-trip time taken to reach each one. The path trace feature works similar to the traceroute function on many host-based operating systems, although the implementation is different. A path trace is frequently used to detect route table inconsistencies or the absence of a required route on an intervening router.

```
fxc9# ping 192.168.46.100 -p
traceroute to 192.168.46.100: 1-30 hops, 16 byte packets
1 192.168.32.5 3 ms 4 ms 3 ms
2 192.168.32.15 5 ms 5 ms 5 ms
3 192.168.49.1 6 ms 6 ms 6 ms
4 192.168.46.100 7 ms 7 ms 7 ms
```

- **Verbose** (-v) The verbose option generates more detail than usual, providing statistical values including success rates and round-trip times. This option is used with most extended pings.

```
fxc9# ping 172.16.14.125 -v
(172.16.14.125): icmp_seq=0, time= 1 ms
IP ping: 172.16.14.125 is alive (size = 16 bytes)
---- PING Statistics----
IP ping: 172.16.14.125 responded to 1 out of 1:
100% success.
round-trip (ms) min/avg/max = 1/1/1
```

Ping Problems

On rare occasions, two conditions may cause **ping** to return unsuccessfully without sending an ICMP echo request to the intended destination. These conditions are BayRS specific and represent a configuration or other provisioning error preventing the normal use of **ping**.

- **IP service is not running** This message is returned by the **ping** command if either the local slot or a remote outbound slot is the desired transit point, but it is not running IP. IP is considered not to be running if the IP Routing Table Manager (RTM) is not accessible on a slot, which typically occurs if the *ip* application is not loaded on the slot. Check the loader slotmask configuration for the IP protocol (`wfProtocols.wfIPROTOLoad.0`) and enable the appropriate slots.
- **Resource error** This message appears if either the TI/BCC or the IP protocol was unable to obtain a required resource, typically a buffer. This may be a transient condition brought about by a heavy load where traffic in the forwarding path is given greater priority. Wait a minute or more and try the ping again. If the problem persists, check the buffer utilization on all slots and see if more buffers need to be allocated.

ICMP Messages

When the network and the destination are in functioning order, the destination sends an ICMP echo reply in response to the ICMP echo request from the source. However, sometimes pings fail. In this case, other ICMP messages are returned indicating the cause of the failure. There are numerous types of ICMP messages, and they are very useful in understanding IP network problems.

The following ICMP messages are most frequently encountered during network problems:

- **Host unreachable from a.b.c.d** Generally categorized into the ICMP unreachable category, the host unreachable message is returned when the device returning the message cannot reach the specific destination IP address. ICMP unreachable messages generally indicate that the device originating the message does not have a usable path to the destination IP, network, protocol, or port. In this case, the specific IP destination cannot be reached by the returning device, normally a router. This condition can occur if the router originating the message has a route to the network on which the destination host resides (for example, a local interface), but the destination host does not respond to an ARP request. It can also occur if the last-hop router has an existing ARP entry, but the destination fails to respond due to a misconfigured network mask or gateway. On path traces, host unreachable messages returned by a router are marked with !H.

```
fp3# ping sniffles
IP ping: ICMP Host Unreachable from 172.16.94.1.
fp3# ping sniffles -p
traceroute to sniffles (172.16.94.27): 1-30 hops, 16 byte packets
 1 fcc4-e12  (172.16.85.1)   3 ms   3 ms   3 ms
 2 fxp2-f31  (172.16.89.2)   6 ms   5 ms   6 ms
 3 fxp2-o41  (172.16.94.1)   9 ms   9 ms   9 ms
 4 *   !H   *
fp3# ping sniffles
IP ping: 172.16.94.27 is unreachable.
```

  Notice that when the same destination is pinged again within 40 seconds, the command returns immediately with an unreachable message, formatted as shown in the last ping in the preceding example. This is because the router is already aware of the host unreachable message that it received previously, so it does not attempt the ICMP echo again. In this case, change the configuration parameters as appropriate and try to ping again after a minute to see if the changes have had any impact.

- **Net unreachable from a.b.c.d** Net unreachable messages are also classified in the ICMP unreachable message category. This message appears when the router originating the message does not have a valid route to the network on which the destination device resides. This occurs if the intervening router does not have a full routing table or does not have a default gateway to reach unknown networks. Route propagation problems commonly cause ICMP net unreachable messages that can be resolved by ensuring that all subnets within the enterprise can contact with each other at the network level. In some cases, net unreachable messages may be acceptable if certain routers and their respective subnets are disallowed by design from communication with the destination network. Network unreachable messages returned by intervening routers are marked with !N in path traces. The two types of network unreachable messages are illustrated here.

```
fp3# ping neutrino
IP ping: ICMP Net Unreachable from 172.16.11.2.
fp3# ping neutrino -p
traceroute to neutrino (172.16.14.12): 1-30 hops, 16 byte packets
 1 fcc4-e12 (172.16.85.1)  3 ms  3 ms  3 ms
 2 fxp2-f31 (172.16.89.2)  5 ms  5 ms  5 ms
 3 axp1-h61 (172.16.11.1)  9 ms  8 ms  9 ms
 4 bxp2-h41 (172.16.11.2)  12 ms  12 ms  12 ms
 5 bxp2-h41  !N  !N  !N
```

- **ICMP TTL exceeded from a.b.c.d** IP frames contain a Time to Live (TTL) field that specifies the maximum number of hops that the frame is allowed to traverse through the network before it is discarded. The device originating the frame sets the TTL to a nonzero positive value, which is decreased by one when the frame passes through a router. When the TTL reaches zero, the router receiving the frame will drop it and send an ICMP TTL exceeded message to the source. By default, BayRS routers use a TTL of 30 on locally originated packets. The goal of the TTL mechanism is to prevent frames from endlessly traversing the network, especially in the case of routing loops. TTL exceeded messages may indicate either that a routing loop is present in the network, or that the destination indeed lies too many hops away from the source. The ICMP TTL exceeded message is returned by the intervening router that observes the zero TTL and also discards the packet. Routing loops are easily identified by a path trace, which will report repeating gateways in the path as illustrated here.

```
fp3# ping jellybelly -p
traceroute to jellybelly (80.23.43.10): 1-30 hops, 16 byte packets
 1 m4des-an1 (80.23.10.2)  2 ms  2 ms  2 ms
 2 arn1-mktg (80.23.46.2)  4 ms  4 ms  4 ms
 3 blnB-core (80.23.45.2)  6 ms  6 ms  6 ms
 4 blnA-core (80.23.45.1)  7 ms  7 ms  7 ms
 5 blnB-core (80.23.45.2)  11 ms  11 ms  11 ms
 6 blnA-core (80.23.45.1)  14 ms  14 ms  14 ms
 7 blnB-core (80.23.45.2)  18 ms  18 ms  18 ms
...  ...
```

Many high-speed networks can support higher TTL values without compromising round-trip times. If no routing loop is present (that is, if all gateways in the path trace are unique), the reachability problem may be the result of the hop count limit imposed by the TTL. The `ttl` parameter under the global `ip` object can be increased to support additional hops.

CAUTION: Changing an IP parameter globally requires an IP restart that will temporarily affect IP forwarding on the router. Unless temporary disruption is acceptable, do not make parameter changes that affect any protocol globally.

- **Administratively prohibited from a.b.c.d** This self-explanatory message is returned when the intervening router has an administrative (configuration) directive preventing access to the subnet, the specific host, or the intended mode of communication. Some networks implement access policies and traffic filters prohibiting certain traffic for security reasons. This traffic may occasionally include pings or communication to a specific network. When this message is returned, try alternative data patterns (such as telnetting to the destination) to determine whether all communication is prohibited or only ICMP. Check the configuration on the intervening gateway or ask the administrator to relax the prohibition. On path traces, !U marks an administrative prohibition returned by a gateway.

```
fp3# ping bison-bob
IP ping: ICMP Administratively Prohibited from 10.48.118.1.
fp3# ping bison-bob -p
traceroute to bison-bob (10.48.119.13): 1-30 hops, 16 byte packets
 1 rtrtjn4-e82 (10.48.142.1)  3 ms  3 ms  3 ms
 2 rtrxpe1-f11 (10.48.190.2)  5 ms  5 ms  5 ms
 3 rtrcal2-f71 (10.48.176.1)  9 ms  8 ms  9 ms
 4 rtramp5-e44 (10.48.118.1)  12 ms  12 ms  12 ms
 5 *  !U
```

- **ICMP redirect** Although not directly seen in a **ping** command, ICMP redirect messages play a key background role that must be understood to troubleshoot certain problems. ICMP redirect messages are sent by intervening routers when they determine that the outbound route exists on the same interface on which the packet was received. This indicates that a shorter path can be used by the incoming packet, and the appropriate gateway is returned in an ICMP redirect message so that other routers can update their routing tables. BayRS routers automatically process ICMP redirects in the background and generate ICMP redirects whenever necessary, counted into `wfIpIntfStatsIcmpInRedirects` and `wfIpIntfStatsIcmpOutRedirects`, respectively.

ICMP statistics on the router can be retrieved on a per-interface basis using the **show ip icmp** command. This command is useful for quickly identifying network conditions that may portend a problem, such as a routing loop. As an ICMP client, an interface receives inbound ICMP datagrams and processes them. This may include unsolicited echo requests and echo responses to requests that were sent previously by the server.

```
show ip icmp client                                   Apr 27, 2000 06:39:59 [GMT-6]

Circuit  IP Address          Echo Re- Echo Re- Timesta- Timesta- Address  Address
                             quests   plies    mp Reqs  mp Repls Mask Re- Mask Re-
                                                                 quests   plies
------------------------------------------------------------------------------------
E32      172.16.14.101       0        10       0        0        0        0
E31      172.16.19.100       11       6        0        0        0        0
```

```
show ip icmp server                                   Apr 27, 2000 06:40:14 [GMT-6]

Circuit  IP Address          Echo Re- Echo Re- Timesta- Timesta- Address  Address
                             quests   plies    mp Reqs  mp Repls Mask Re- Mask Re-
                                                                 quests   plies
------------------------------------------------------------------------------------
E32      172.16.14.101       212      0        0        0        0        0
E31      172.16.19.100       47       11       0        0        0        0
```

As an ICMP server, each interface originates outbound ICMP datagrams either for internal use or as a response to a previous ICMP datagram. For example, consider a sequence of 10 ICMP echoes directed to an IP interface on the router from a workstation. The ICMP echo requests are received and processed by the client, while the corresponding echo responses are sent by the server. Both the client and server statistics should ideally be looked at simultaneously to correlate the In and Out frames. In the preceding example, the E32 circuit sent 212 echo requests from the server, but only 10 responses were received by the client.

The following commands illustrate the ICMP redirect statistics and other ICMP messages received and sent by each IP interface.

```
show ip icmp misc                                     Apr 27, 2000 06:40:21 [GMT-6]


                             SrcQunch Messages Redirect Messages    Prohibit
Circuit  Ip Address             In       Out      In       Out      In       Out
-------- ---------------- ----------------- ----------------- -----------------
E32      172.16.14.101          0        0       2639     1274        0        0
E31      172.16.19.100          0        0          0        0        0        0
```

```
show ip icmp in                                       Apr 27, 2000 06:40:41 [GMT-6]

Circuit  IP Address          ICMP Re- ICMP In  Destn.   Received Received
                             ceived   Errors   Unrea-
                                               chable   Time     Param
                                                        Exceeded Problem
------------------------------------------------------------------------
```

```
E32      172.16.14.101    2848     0       12        187       0
E31      172.16.19.100    65232    0       65203     12        0

show ip icmp out                              Apr 27, 2000 06:43:29 [GMT-6]

Circuit  IP Address       ICMP     ICMP    Destn.    Sent      Sent
                          Sent     In      Unrea-    Time      Param
                                   Errors  chable    Exceeded  Problem
------------------------------------------------------------------------
E32      172.16.14.101    1581     0       4         91        0
E31      172.16.19.100    69       0       11        0         0
```

Does Not Respond Problems

Network administrators and engineers commonly encounter two messages: a.b.c.d is alive and a.b.c.d does not respond. The "alive" message indicates that a successful ICMP echo reply was received from the destination device that was pinged, and the "does not respond" message indicates that an echo reply was not received within the timeout period. As mentioned earlier, ping is also supported by endstations, and the "does not respond" message often appears on a source workstation from which connectivity tests are run. This may indicate one of several issues:

- The TCP/IP stack on the source station or the destination station is not operating properly. If the router is the source station, the message "IP service is not running" is displayed.
- The default gateway or the subnet mask is incorrectly configured on the source station or the destination station.
- Either the source station or the destination station is connected incorrectly to the network

Figure 6-1 shows a simple flow chart to help troubleshoot a problem when a "does not respond" message appears. When troubleshooting a problem such as this, do not overlook a step because you believe that the configuration *ought* to be correct because it was always operational before. When many of these problems occur in existing networks, the cause is often a configuration change that was incorrectly implemented without the knowledge of the network administrator.

When troubleshooting the problem from the perspective of an endstation, most ping implementations return messages as just described. If a default gateway is absent in the configuration, then the ping will return a network unreachable message. The network and subnet portions of the IP address must match between all stations on the same subnet, and particular attention must be paid to the endstation and the router.

IP ADDRESSING, NAME AND ADDRESS RESOLUTION

IP network problems can be characterized into two broad categories: addressing and resolution issues and route management issues. Addressing and resolution issues primarily

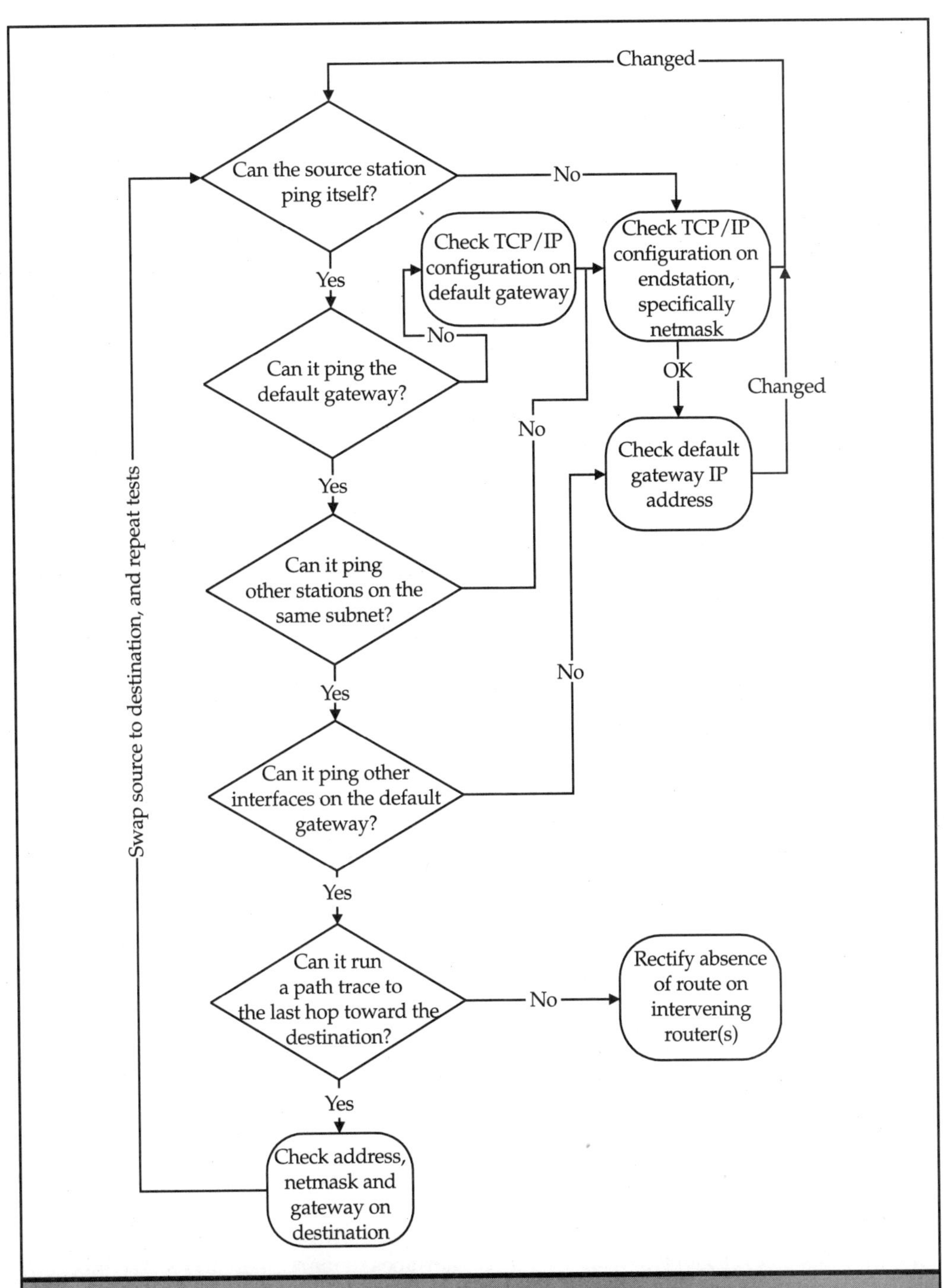

Figure 6-1. Troubleshooting a simple IP reachability problem

involve the endstation and server configuration, although the router configuration also may be involved. In this section, we consider the most common problems related to addressing and address resolution and describe methods for resolving them. Route management issues are discussed later in this chapter.

IP Addressing

In the preceding section, we recommended checking the IP address and subnet configuration on a workstation several times. IP addressing in large networks requires diligent management to ensure uninterrupted connectivity. Many enterprise networks use dynamic addressing, using the Dynamic Host Configuration Protocol (DHCP) for mobile stations such as laptops and for desktop systems. Some diskless clients use the Bootstrap Protocol (BOOTP) to obtain IP addresses. Most servers and dedicated workstations are configured with a static IP address and network mask.

Static Addressing

Because static IP addresses require manual administration on workstations and servers, they may be misconfigured due to a typographical error or incorrect provisioning detail. When there are problems with IP connectivity on static endstations, always check the configuration regardless of the connectivity information provided prior to the connectivity loss.

Reverse Address Resolution Protocol

Although the Reverse Address Resolution Protocol (RARP) provides an older, traditional method of dynamically obtaining an IP address, it is still widely used with devices that do not support a more advanced protocol such as DHCP. The endstation sends a reverse-ARP frame on the local LAN, which is responded to by a RARP server (typically a BayRS router), which returns the IP address. This IP address is returned based on a mapping table configured on the router, using the MAC address and IP address of the stations. The most frequent problem in this setup is the inability of the endstation to reach the RARP server (due to an intervening gateway) or the absence of an IP mapping for the station's MAC address.

Bootstrap Protocol and Dynamic Host Configuration Protocol

BOOTP and DHCP are closely related, although DHCP uses advanced extensions to support additional configuration parameters for endstations. Both protocols use the User Datagram Protocol (UDP), which is automatically enabled when IP is configured on the router. However, the most important requirement is the transport of the broadcast BOOTP request and DHCP discover frames through all routers between the workstation and the BOOTP/DHCP server. Since these frames are addressed as either 0.0.0.0 or 255.255.255.255, routers will typically drop them because their inherent design goal is to limit broadcast traffic from incessantly traversing the network. BayRS routers must be configured as BOOTP/DHCP *relay agents* on interfaces that are likely to receive inbound BOOTP requests or DHCP discover packets. All routers along the path, when properly configured as relay agents, will relay these frames correctly across other interfaces also configured to support relaying. This helps the initial broadcast request or discover frame reach the BOOTP or DHCP server, as appropriate.

If all BOOTP or DHCP stations in a subnet are unable to obtain IP addresses, check the reachability of the BOOTP or DHCP server from the local router using a ping. If it is reachable, then look at the relay agent configuration and statistics on all intervening routers to ensure that they are configured correctly and operating properly. The **show bootp stats** command provides a summary of statistics for all BOOTP and DHCP interfaces. The dropped packets are the sum of several drop attributes that can be individually fetched from the `wfBootpRelayIntfEntry` object.

```
# show bootp stats

Bootp Statistics
----------------

   Interface        Dropped     Requests    Replies
    Address         Packets     Received    Received
 --------------- ---------- ---------- ----------
 172.16.119.1     0            14          14
 172.16.124.1     12           110         65
...
# get wfBootpRelayIntfEntry.*.172.16.124.1
...
wfBootpRelayIntfEntry.wfBootpRelayIntfOpDrops.172.16.124.1 = 0
...
wfBootpRelayIntfEntry.wfBootpRelayIntfHopsDrops.172.16.124.1 = 7
wfBootpRelayIntfEntry.wfBootpRelayIntfBcastDrops.172.16.124.1 = 1
wfBootpRelayIntfEntry.wfBootpRelayIntfSecDrops.172.16.124.1 = 0
wfBootpRelayIntfEntry.wfBootpRelayIntfGiaddrDrops.172.16.124.1 = 4
wfBootpRelayIntfEntry.wfBootpRelayIntfResrcDrops.172.16.124.1 = 0
...
```

The individual drop counts refer to the following:

- **OpDrops** This value indicates the packets dropped due to an invalid value in the operation field. Packets may be dropped if an interface receives nonstandard BOOTP or DHCP frames from proprietary implementations.
- **HopsDrops** By default, BOOTP and DHCP frames are allowed to pass through four hops, and the hops field is decremented at each hop. When the hop count reaches zero, the frame is discarded. And increase in this value indicates that multiple hops are necessary to reach the server, and the maximum allowed hops must be appropriately increased.

- **BcastDrops** BOOTP and DHCP frames must be addressed to the IP broadcast address (all zeros or all ones). Frames received on an interface without this broadcast destination are automatically dropped.
- **SecDrops** Frames with a very small seconds field are dropped.
- **GiaddrDrops** Reply frames received by a relay agent interface are compared to see if the giaddr field matches the interface IP address. If the values do not match, the server's reply may be malformed, and the frame is dropped.
- ▲ **ResrcDrops** On rare occasions, frames may be dropped if sufficient resources are not available. An increase in this value indicates that additional resources (buffers) must be allocated to the slot's protocols.

If the DHCP server is reachable via a unicast and all relay agents are configured correctly, then the server likely ran out of IP addresses to assign to the endstations, or the provisioning of address ranges or the scope on the DHCP server may be incorrect. Note that BOOTP servers have a one-to-one mapping between an IP address and the source's MAC address; thus, this condition usually implies that the MAC address of the source is not configured correctly in the server's configuration. In the case of both BOOTP and DHCP, the nearest server that processes the request may not be provisioned appropriately, and the preferred server may reside elsewhere in the network. In this case, the first relay agent receiving the request must be configured with a preferred server so that the frame is directly unicast to the preferred server.

BOOTP/DHCP Events BOOTP/DHCP relay agent events are logged under the `BOOTP` entity. Some problems may not fall into the common categories discussed earlier and are reported as different events in the log. The following event indicates that the relay agent was unable to obtain required resources, which may indicate a temporary condition or a persistent lack of resources, typically buffers. This problem can be alleviated by adding more buffers, either by carving out more global memory or increasing the total memory.

```
#  135: 11/14/1999 01:15:02.717  WARNING  SLOT  7  BOOTP              Code:   2
Dropping BOOTREPLYs due to resource constraints.
```

Another common problem occurs when the BOOTP/DHCP request traverses the maximum number of hops before reaching the server. This problem can occur because the hop count is too small (default 4, maximum 16) or the server is too many hops away. In these cases, increase the hop count so that the request or discover frame can travel additional hops to reach the server.

```
#   78: 06/17/1999 20:50:37.076  INFO     SLOT  2  BOOTP              Code:   5
BOOTREQUEST received on Interface 172.16.67.1 dropped - hop count exceeded.
```

Router IP Addressing

IP interfaces on LAN and WAN routers are usually statically defined, because in addition to the default gateway assigned using DHCP extensions, gateways need to be specifically configured on many endstations. IP interface configuration usually does not require changes on an existing network, unless a broad network or subnet change is being implemented. Regardless, in large networks, the IP interface configuration may be changed accidentally, potentially affecting IP connectivity for numerous workstations. Configuration errors and other IP interface problems are logged under the `IP` entity, which must be looked at when all devices in a subnet are affected. First, check the status of the IP interface using the **show ip interfaces** command to verify that it is operational.

```
fxc9(34)# show ip interfaces
show ip interfaces                                    Apr 27, 2000 06:49:41 [GMT-6]

Circuit  Cct # State    IP Address     Mask            MAC Address
------------------------------------------------------------------------------
E21      1     up       172.16.17.1    255.255.255.240 00.00.A2.CB.B5.8C
E22      2     up       80.23.197.1    255.255.255.0   00.00.A2.CB.B5.8D
H41      3     invalid  10.119.0.1     255.255.255.0   00.00.A2.C9.FC.45
E23      4     down     172.16.29.1    255.255.255.0   00.00.A2.CB.B5.8E
```

The following three states are undesirable and must be resolved promptly:

- **Down** This indicates that the interface is operationally down, either because it is administratively down (that is, disabled) or the underlying circuit is not operational. The underlying circuit may not be operational for various reasons, including cable disconnects, loss of signal, and loss of framing.
- **Invalid** This state indicates that one of the configuration parameters is incorrectly configured. Parameters such as network masks and the IP address may be misconfigured, or one of the required features may be disabled, preventing proper operation. This state is discussed in detail later in this section.
- **Not Present** This state has been discussed several times previously, in the context of media states and protocol states. In particular, refer to "Applications and Protocols" in Chapter 5. The Not Present condition occurs if the IP protocol is not loaded on the appropriate slot, making only the MIB skeleton information available from the configuration. The IP protocol (`wfProtocols.wfIPROTOLoad.0`) needs to be added using the specific slotmask.

The Invalid state is sometimes difficult to understand because it can be caused by a variety of configuration errors. The following event is logged to indicate a configuration error.

```
#   49: 01/04/1999 05:13:04.399  WARNING  SLOT  4  IP                Code:   7
Interface 10.119.0.1 Misconfigured - Disabled
```

Misconfiguration can be due to one of the following:

- The network mask of the interface is smaller than 255.0.0.0 or larger than 255.255.255.252. Currently, 32-bit network masks are not supported.

- The IP address either has zero or all ones (broadcast) for the host number portion.
- The IP address resides in the first (zeros) subnet of the specific network, but the `all-subnets` parameter for the interface has not been enabled. By default, the first (all zeros) and last (all ones) subnets are disallowed.

```
box->ip
fxc9(94)# all-subnets enabled

box->ip
fxc9(95)# show ip interfaces 10.119.0.1
show ip interfaces                                    Apr 27, 2000 06:49:41 [GMT-6]

Circuit  Cct # State    IP Address       Mask              MAC Address
----------------------------------------------------------------------
H41        3     up     10.119.0.1       255.255.255.0     00.00.A2.C9.FC.45
```

- The network mask is noncontiguous. The ones portion of the netmask must not be interleaved with zeros for an IP configuration.
- The IP interface is configured as an unnumbered interface, but the associated address is not configured.
- Another IP address matching the current one or matching the same subnet is already up and operational. Multiple IP interfaces in the same subnet are not allowed.

Name Resolution

All enterprise networks and the Internet use a name resolution scheme to map IP addresses to logical, meaningful names. For example, the IP address 134.177.3.26 is currently mapped to the name ftp-support.baynetworks.com. Using logical names simplifies node management and provides many benefits. Different name resolution schemes are used by different intranets. This includes static methods such as the use of the /etc/hosts file on UNIX systems and the lmhosts file on Windows systems. Static methods use local files for name-to–IP-address mapping, and the common problem is that the mapping is misconfigured.

IP connectivity tests using pings can be performed using the logical name of the destination or using the IP address.

NetBIOS over TCP/IP

Commonly called NBIP, the NetBIOS over TCP/IP protocol has four methods for resolving NetBIOS names to IP addresses: p-node, b-node, h-node, and m-node. In simple terms, the methods differ primarily in whether they use a unicast or a broadcast frame to reach the name server. A p-node unicast can contact either a NetBIOS Name Server (NBNS) or a Windows Internet Name Service (WINS), and b-node broadcasts can contact only NBNSs. For proper use of NBNS, NetBIOS over TCP/IP must be configured and enabled on the IP interfaces that need to support it. Nonstandard implementations of NBIP may generate invalid packets or invalid names, which are promptly observed and reported in the log under the `NBIP` entity.

```
#  192: 08/15/1999 16:04:03.846  WARNING  SLOT 10  NBIP              Code:   2
invalid NetBIOS over IP packet received

#   22: 06/13/1999 24:22:33.292  WARNING  SLOT  2  NBIP              Code:   3
invalid NetBIOS over IP name
```

BayRS routers also can cache NetBIOS names, which can help reduce broadcast traffic on the network. By default, NBIP caches 100 names on a global basis. Certain large networks may find this inadequate, and symptoms of this condition are usually seen in the log.

```
#  143: 07/06/1999 15:23:54.421  DEBUG    SLOT  5  NBIP              Code:  11
NetBIOS name cache is full

#  144: 07/06/1999 15:23:55.049  DEBUG    SLOT  5  NBIP              Code:  12
aging from NetBIOS name cache
```

To fix this problem, try increasing the NetBIOS name cache to improve performance. Note that when newer name entries are learned from the NBIP interface, the earliest entries are aged out and replaced if the cache is full.

Windows Internet Name Service

The WINS protocol is a variant of the NBNS protocol and is implemented by Microsoft using p-node unicasts to the server. Since the protocol requires unicast mode operation, at least one server IP address must be configured on the workstation. Problems with WINS often occur because the client lacks a WINS server IP address or the WINS server is unreachable from the client subnet. In these cases, you need to perform connectivity tests using the IP address of the WINS server to check reachability. Multiple WINS servers can be configured for redundancy. Troubleshooting IP connectivity usually must precede WINS testing.

Domain Name Service

The Domain Name Service (DNS) is the backbone name service for the Internet, providing hierarchical name-to-address mappings. DNS is natively supported on the BayRS router, and like WINS, DNS requires at least one name server to contact for address resolution. The same is true of DNS clients, such as workstations, which must be configured with a DNS server manually or must obtain DNS servers dynamically from the DHCP server using extensions. Multiple name servers can be configured on the router for name resolution.

On the BayRS router, various DNS issues may surface during pings. The **ping** command automatically attempts to resolve IP addresses to names, for efficient and logical display of gateways and destinations. The following messages may appear:

```
#   89: 09/24/1999 01:01:48.439   DEBUG    SLOT  5  DNS                 Code:  11
DNS error, server failure (SERVFAIL).

#   88: 12/07/1999 06:52:39.542   DEBUG    SLOT 14  DNS                 Code:  12
DNS error, non existent domain (NXDOMAIN).

#  187: 07/26/1999 20:43:26.425   DEBUG    SLOT  1  DNS                 Code:  14
DNS error, query refused (REFUSED).

#   61: 01/12/1999 07:25:44.083   DEBUG    SLOT 13  DNS                 Code:  13
DNS error, not implemented (DNS_NOTIMP).

#  101: 09/16/1999 15:13:04.803   DEBUG    SLOT  1  DNS                 Code:  15
DNS error, no data.

#   56: 08/27/1999 00:36:43.774   DEBUG    SLOT 12  DNS                 Code:  19
DNS query timed out.

#   85: 01/24/1999 01:41:32.051   DEBUG    SLOT  7  DNS                 Code:  21
Can not get to server.
```

Most DNS events reported under the DNS entity are self-explanatory and occur because the router cannot reach the DNS server, a query failed due to a nonexistent name or domain, or a provisioning problem occurred on the DNS server.

TIP: When performing a ping or path trace, use the –n option to turn off name resolution so that only true IP connectivity is tested. If DNS lookups timeout or fail, pings may appear to take longer.

Address Resolution

The Internet Protocol uses address mapping to map IP addresses to unique physical addresses. On LAN networks using media access control (MAC) addressing, the Address Resolution Protocol (ARP) is used to map IP addresses to MAC addresses. ARP is automatically enabled when an IP interface is initially configured. If the ARP protocol is absent from an endstation, an adjacent host (static ARP entry) must be configured on the router. The endstation also requires a static ARP entry for the router that acts as the default gateway.

While name resolution is not mandatory for IP connectivity, ARP is essential for IP communication, particularly on broadcast LAN networks. Address resolution is performed by the IP stack dynamically on the router, as necessary. You can use the **show ip arp** command to view the current ARP cache on the router.

```
box
fxc9(15)# show ip arp
show ip arp                                              Apr 27, 2000 06:52:03 [GMT-6]

IP Address          Physical Address       Type
-----------------------------------------------
172.16.29.2         00.00.A2.0B.31.C9      dynamic
172.16.29.14        00.04.C9.45.7D.E2      dynamic
172.15.29.17        00.04.C9.DC.6F.4E      dynamic
... ...
```

A frequent problem in address resolution is the appearance of duplicate IP addresses. The IP address and the MAC address both must be unique in the network, and sharing the same IP address is disallowed. The user symptom of a duplicate IP address is intermittent connectivity to network devices and servers. BayRS routers automatically update the ARP cache to reflect a new MAC address for an IP address, a change common with dynamically leased IP addresses. This behavior will periodically display alternating MAC addresses for an IP address in the ARP cache. To force the router to relearn the MAC address, first clear the specific ARP entry using the **clear ip arp a.b.c.d** command. Then issue a ping from the TI/BCC to the suspected IP address and verify the ARP entry. If the MAC address for the specific ARP entry changes, then one of the stations must be reconfigured with a new IP address. An example of duplicate address detection is illustrated here.

```
box
fxc9(93)# show ip arp
show ip arp                                              Apr 27, 2000 06:54:03 [GMT-6]

IP Address          Physical Address       Type
-----------------------------------------------
...
172.16.29.14        00.04.C9.45.7D.E2      dynamic
...

box
fxc9(94)# clear ip arp 172.16.29.14

box
fxc9(95)# ping 172.16.29.14
ping: 172.16.29.14 is alive (size = 16 bytes)

box
```

```
fxc9(96)# show ip arp
show ip arp                                           Apr 27, 2000 06:54:03 [GMT-6]

IP Address          Physical Address       Type
----------------------------------------------
...
172.16.29.14        00.04.C9.45.96.49      dynamic
...
```

Certain ARP mismatches on an IP interface are reported by the `ARP` entity and can be readily observed in the log. ARP uses broadcast frames at the data link layer, which is seen by all devices on a broadcast LAN segment or WAN cloud. When the router detects an ARP frame originating from a source that does not reside in the appropriate subnet, the following message is logged:

```
#  155: 11/06/1999 15:23:30.449   INFO     SLOT  1  ARP              Code:   5
Invalid ARP Source: 172.16.79.137 MAC: 004e9c462a9d on 172.16.47.1
```

The problem may be a station that is configured with an incorrect IP address or a backdoor bridge on the network that is transparently bridging frames between different subnets. If several different invalid ARP sources are reported, there is likely a backdoor bridge on the subnet. If the router must accept nonlocal ARP source frames, then the global `arp` parameter `nonlocal-source` must be set to `accept`.

```
#    2: 02/17/1999 20:14:21.552   INFO     SLOT 10  ARP              Code:   6
Invalid ARP Request: Src: 172.16.97.81 MAC: 004e9c923a4c &
Dest: 172.16.99.47 on diff nets
```

The preceding event message is logged when the source and destination devices appear to be on different subnets. Normally, the router drops the ARP request in both cases. The case of a nonlocal destination can occur if endstations do not fully support subnet masks and instead use their natural class masks. In this case, the `nonlocal-destination` parameter for the global `arp` object must be set to `accept`. In other cases, use the MAC address to find the device originating the invalid ARP requests and reconfigure it correctly.

Clearing ARP Entries

In the preceding section, we briefly discussed the use of the **clear ip arp** command. The argument to the command can be of the form *network/mask*, or it can be a specific 32-bit host address. The ARP table is called the *host cache* in BayRS, and by default, local destinations in this table are not aged periodically. Historically, most IP endstations have supported static addressing, and this was the prevalent method. In statically addressed networks, aging the ARP entries periodically and relearning them is unnecessary because the entries do not change. This is true also for devices using RARP or BOOTP. However, the advent of DHCP and the use of dynamic stations brought MAC addresses that change periodically for an IP address depending on the lessee station. BayRS routers can be configured to periodically age out and relearn ARP entries and keep the host cache updated.

BayRS routers automatically update the host cache with a newer MAC address when a packet is received from a previously known IP station with a different MAC address. Many troubleshooting scenarios with address resolution require the clearing of a range or a specific host cache entry. For example, host cache entries in an entire subnet can be cleared using the *network/mask* argument of **clear ip arp**. A specific host address can be specified, as is often needed when troubleshooting duplicate IP addresses. When the mask is not specified, a 32-bit host mask (255.255.255.255) is automatically assumed.

```
box
fxc9(14)# clear ip arp 172.16.92.132

box
fxc9(15)# clear ip arp 172.16.78.0/24
```

The preceding examples illustrate the two methods used to clear ARP entries from the table. Using wildcards for the IP address in the form 0.0.0.0 is not allowed, because doing so would clear the entire ARP table.

IP ROUTING AND FORWARDING

The IP router is required to maintain a routing table for the various networks it can forward traffic to and from. On BayRS routers, this function is performed by the IP Routing Table Manager (RTM). The RTM receives routes from one or more static or dynamic routing protocols, such as RIP, OSPF, and BGP. Each routing protocol may have multiple routes to the same destination, but each submits the best possible route to the RTM. The RTM then takes all the routes submitted by the various protocols and recalculates the best route for use on a boxwide basis.

On multi-slot routers, each slot runs an instance of the RTM, which maintains a local copy of the routing table. Each RTM communicates across the backplane and sends updates to other RTMs. A slot is said to *own* a specific route if the route was received and processed on an IP interface residing in it. This owning RTM is responsible for updating or removing the route as the network topology changes. The RTM dynamically increases the size of the routing table as more entries are received from the routing protocols.

A *forwarding table* is used for individual interfaces. While the routing table is maintained on a slotwide basis, the forwarding table is maintained on a per-interface basis. It uses a preallocated amount of memory as a route cache for frequently used routes on the interface. On enterprise networks, data flows tend to use the same forwarding path during a presentation or application layer session, and using the forwarding table to cache repeatedly used routes reduces the number of routing table lookups. Lookups in a large routing table can consume a significant number of CPU cycles, depending on the size of the table. The forwarding table is usually disabled on core routers and ISP routers, where numerous and variegated data flows are unlikely to benefit from a forwarding cache that changes frequently. The routing and forwarding tables are illustrated in Figure 6-2.

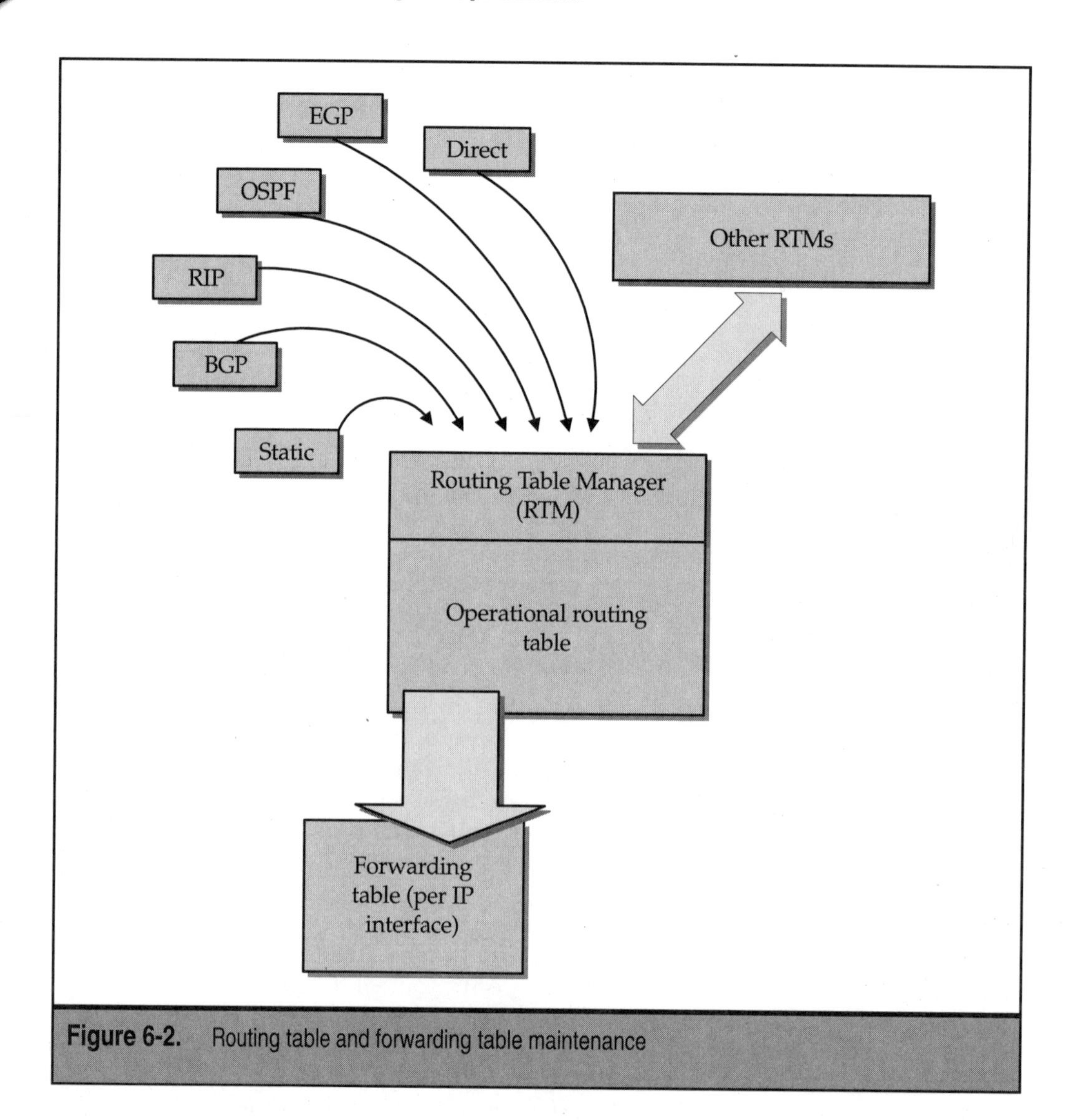

Figure 6-2. Routing table and forwarding table maintenance

We mentioned that the same route can be submitted by different protocol subsystems. These routes must again be considered by the RTM to determine the best possible route. The primary criterion for the best route is the use of a *preference* value: the route with the highest preference value wins. If the preference value is the same for two or more routes, an internal weight value is used to determine the best route. Table 6-1 summarizes the route selection criteria used to determine the best route.

Route Type	Preference Value	Weight Value
Directly attached route	16	0
OSPF intra-area route	16	0
OSPF inter-area route	16	1
OSPF Type-1 external route	1	2
BGP route	1	32
RIP route	1	34
EGP route	1	35
Static route	1 (see note below)	36
OSPF Type-2 external route	1	Uses weight of the external route origin

Table 6-1. IP Route Selection Criteria

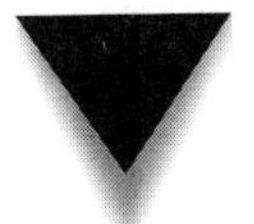

NOTE: Static routes have a default preference value of 16. If other preferable routes are learned via routing protocols, it is usually recommended that the static route be configured with a lower preference value. This allows the static route to behave as the route of last resort when other routes become unusable.

Directly connected OSPF intra-area and inter-area routes have a preference value of 16, which cannot be changed. Static routes by default have a preference value of 16 and will automatically be entered into the routing table in the absence of a route with a higher preference value. Table 6-1 assumes that static routes are configured with a preference value of 1. All other routes with a preference value of 1 in Table 6-1 are assigned that value by the protocol when they are initially received. These route preference values can be altered using a route accept policy for the specific protocol when it is initially processed. Understanding the route preference and weight values is key to analyzing route propagation and forwarding problems.

Classless Interdomain Routing

Classless interdomain routing, commonly called CIDR, eliminates the traditional restrictions that apply to class-based routing networks. A destination is routed based on the network address and mask, without regard to the natural class type (A, B, C) that it belongs to. Unlike regular class-based networks, CIDR can support nonstandard network masks. In particular, CIDR allows supernetting, which can be used in large corporate networks and on the Internet to aggregate multiple networks into a single route with a supernet mask. This reduces the overall number of routes in the routing table, for efficient routing. The Open Shortest Path First (OSPF) protocol and Border Gateway Protocol (BGP) are examples of protocols that use CIDR.

BayRS routers support CIDR and employ various protocols that natively use CIDR. Classless routing is disabled by default, because most traditional corporate intranets have a design using class-based subnets within the corporation's assigned class address. Historically, these networks did not need a default path for unknown networks because all well-known routes were propagated internally. If a destination network was absent in the routing table, the inherent implication was that the network was not reachable. However, this is not the case any more because more and more enterprise intranets are connected to the Internet, making a default path for unknown networks a necessity. If a default route is configured on the router but it does not appear to be used for unknown subnets within the intranet, then most likely the `classless` attribute is disabled. This condition is illustrated in the following listing.

```
box->ip
mx2(2)# ping godzilla
IP ping: 132.245.155.134 is unreachable

box->ip
mx2(3)# show ip routes 132.245.155.134
show ip routes                                     Apr 21, 2000 15:49:30 [GMT-6]

Network/Mask        Proto        Age Slot      Cost  NextHop Address     AS
------------------- ------ --------- ---- --------- ---------------- -----
0.0.0.0/0           Static   1131196    3         1 132.245.152.1

box->ip
mx2(4)# show ip summary
show ip summary                                    Apr 27, 2000 06:50:07 [GMT-6]

Configured State     : enabled
Current State        : up
All Subnets          : enabled
Number of Routes     : 6
Number of Hosts      : 1
Time-To-Live         : 30
Maximum Policy Rules: 32
RIP Diameter         : 15
Route Cache Interval: 60
Estimated Networks   : 0
Estimated Hosts      : 0
Classless            : disabled
Forwarding Mode      : forwarding
```

```
Route Filters       : disabled

box->ip
mx2(5)# classless enabled

box->ip
mx2(6)# ping godzilla
ping: 132.245.155.134 is alive (size = 16 bytes)
```

Routing Entries

The current routing table can be viewed using the **show ip routes** command. The syntax of the command is as follows:

```
show ip routes [address[/prefix]] [-A] [-s slot]
```

This command supports several options. The –A option displays all entries, including forwarding entries. The type specifier is used to list routes from a specific routing protocol. Directly attached routes and static routes are considered to be from the *local* routing protocol. The –A option returns the consolidated weight of each route entry, which takes into account the preference, metric, source, and normal weighting values. Entries marked with an asterisk (*) are currently being used in the normal routing table.

```
box
fp6(31)# show ip routes -A
show ip routes -A                              Apr 27, 2000 06:57:11 [GMT-6]

 Network/Mask         Proto       Age Sl      Cost NextHop Address     AS  Weight
 ------------------- ------ -------- -- -------- ---------------- ----- --------
*0.0.0.0/0           Static  1617662  3         1 192.168.155.129        03c00001
 0.0.0.0/0           Direct  1617700  0    131071 Unreachable            ffffffff
 0.0.0.0/32          F-Host      N/A  0         0 un# IP cct 0           00000000
 80.0.0.0/8          Direct  1617662  3    131071 Unreachable            ffffffff
 80.0.0.0/32         Host        N/A  3         0 ASB Broadcast          00000000
*80.23.10.0/24       Direct  1617662  3         0 80.23.10.1             00000000
 80.23.10.0/24       RIP           5  3         2 80.23.10.2             7b9e0002
 80.23.10.0/32       Host        N/A  3         0 Broadcast              00000000
 80.23.10.1/32       Host        N/A  3         0 This Router            00000000
 80.23.10.2/32       F-Host      N/A  3         0 80.23.10.2             00000000
 80.23.10.255/32     Host        N/A  3         0 Broadcast              00000000
*80.23.43.0/24       Static   691951  3         1 80.23.10.2             03c00001
*80.23.45.0/24       RIP           5  3         2 80.23.10.2             7b9e0002
 80.255.255.255/32   Host        N/A  3         0 ASB Broadcast          00000000
 192.168.0.0/16      Direct  1617662  3    131071 Unreachable            ffffffff
 192.168.0.0/32      Host        N/A  3         0 ASB Broadcast          00000000
*192.168.152.0/24    Static  1617662  3         1 192.168.155.129        03c00001
*192.168.155.128/28  Direct  1617662  3         0 192.168.155.134        00000000
 192.168.155.128/32  Host        N/A  3         0 Broadcast              00000000
```

```
192.168.155.129/32  Host       N/A  3       0 192.168.155.129       00000000
192.168.155.134/32  Host       N/A  3       0 This Router           00000000
192.168.155.143/32  Host       N/A  3       0 Broadcast             00000000
192.168.255.255/32  Host       N/A  3       0 ASB Broadcast         00000000
255.255.255.255/32  Host       N/A  0       0 All Ones              00000000

  Total Networks on Slot 3 = 6

  Total Hosts on Slot 3 = 14
```

Each line entry in the routing table is usually self-explanatory. In troubleshooting, you should pay particular attention to the routing protocol source and the next hop for the destination. If the destination route is using the correct next hop and is from the desired protocol source, then the route can generally be considered *usable*. The route is considered *reachable* if the metric (cost) is valid for the specific protocol source. The metric value is applied differently by various protocols. For example, RIP considers a destination to be unreachable if the metric exceeds the RIP diameter (default: 15). Directly connected routes are reported with a metric of 0 in the routing table. The age of the route also plays a role, where the loss of periodic updates may cause the route to be marked for removal because an update has not been received within the specified interval.

To look at a specific routing entry or the route used for a specific destination, you can use an argument of the form *network/mask*, where the mask value is optional. The following examples illustrate the available methods.

```
fp6(23)# show ip routes 192.168.152.9
show ip routes 192.168.152.9                          Apr 27, 2000 06:55:14 [GMT-6]

Network/Mask        Proto        Age Slot      Cost  NextHop Address     AS
------------------- ------ --------- ---- --------- ---------------- -----
192.168.152.0/24    Static   1617545    3         1 192.168.155.129

fp6(24)# show ip routes 192.168.152.0/27
show ip routes 192.168.152.0/27                       Apr 27, 2000 06:55:44 [GMT-6]

Network/Mask        Proto        Age Slot      Cost  NextHop Address     AS
------------------- ------ --------- ---- --------- ---------------- -----
192.168.152.0/24    Static   1617576    3         1 192.168.155.129

fp6(25)# show ip routes 192.168.190.0
show ip routes 192.168.190.0                          Apr 27, 2000 06:55:27 [GMT-6]

Network/Mask        Proto        Age Slot      Cost  NextHop Address     AS
------------------- ------ --------- ---- --------- ---------------- -----
0.0.0.0/0           Static   1617558    3         1 192.168.155.129
```

The second example uses a specific host as the argument, and the route that will be taken to reach that destination is returned. This is useful for locating the specific route that will be taken to reach a destination. Notice that no network-specific route is available for the third destination; for this destination, the router will use the default route returned.

If you use the –A option, all network routes and host routes are returned. A host route is one that matches a full 32-bit mask, but it does not necessarily indicate a unicast destination. For example, host routes are used for local subnet-broadcast addresses, to specially handle broadcast destinations. Host routes are not considered to be part of the network routing table, which normally displays reachability information only for the network layer.

Routing Information Protocol

RIP was one of the first routing protocols used in IP networks. It has a simple design and was the first protocol to be widely implemented for use by most networking vendors. RIP uses periodic updates between routing nodes, which learn which other networks can be reached via adjacent routers. The simplicity of RIP and the ease of routing table maintenance it allows have made it the choice for many small- to medium-sized networks. Even networks using a protocol like OSPF sometimes use RIP due to its easy maintenance and interoperability. However, the elementary design of RIP poses several limitations for evolving networks. The most common concerns about traditional RIP routing are these:

- **Network diameter** RIP uses a maximum network diameter of 15 hops. In today's high-speed networks, which span numerous hops, this limitation can significantly constrain the size of the overall network.
- **Bandwidth use** RIP does not maintain any state information about its adjacent routers. It simply relies on periodic refreshment updates, which is an inelegant way of using bandwidth. Bandwidth use becomes a crucial concern on low-speed WAN links.
- **Hop count** RIP uses a single route metric, the hop count, which is simply a measure of the number of routing hops taken to reach a destination. Occasionally, a higher-speed path with a larger hop count may be replaced with a nonoptimal route that uses a slower path with a smaller hop count.
- **VLSM** RIP was designed for use with classful networks, where nonstandard variable-length subnet masks (VLSM) were not used. Because it does not propagate network masks, a nonstandard VLSM route may not be entered and used correctly by other routers. Moreover, class-based addressing restricts the network administrator to a nonoptimal use of the network addressing space.

These functional limitations led to the development of RIP version 2, an enhancement that provided many improvements. RIP2 supports VLSM and propagates the network mask information along with the destination network, thereby permitting classless route maintenance. Also, RIP2 can perform route aggregation for a range of networks, which reduces the number of routing table entries in the overall network. A general understanding of the differences between RIP1 and RIP2 is necessary when troubleshooting RIP routing problems.

You can use the **show ip rip** command to view the status and configuration of RIP interfaces. Particularly useful are the **summary** and **timers** subcommands, which display various parameters, including the RIP mode, interface state, broadcast timer, and hold down timer.

```
box
fxc9(32)# show ip rip summary
show ip rip summary                                   Jun 06, 2000 11:20:06 [GMT-6]


                                              Def.
                                     RIP      Rt.     Poison   RIP   Trig.
IP Interface    Cct# State           Sup/Lis Sup/Lis Reverse  Mode  Updates  TTL
--------------- ---- ---------- ------- ------- -------- ----- -------- ---

80.23.10.1      2    up              ena/ena dis/dis poisoned rip2  disabled 1
172.16.99.1     3    up              ena/ena dis/dis poisoned rip1  disabled 1
172.16.47.1     6    up              ena/ena ena/dis actual   rip2  enabled  1
80.23.150.1     7    up              ena/ena dis/ena split    aggr  enabled  1
```

The RIP mode parameter (the `version` attribute of the `rip` object) indicates the protocol version being used on each interface. The RIP Time to Live (TTL) value usually should be left at the default setting of 1, which indicates that RIP updates must not be propagated to other networks. This value needs to be changed only in the case of unusual topologies with varying levels of broadcast support. If the RIP TTL value is set to a higher value, RIP route broadcasts may be propagated to other networks if the IP interface's broadcast address is set to the all-subnets broadcast (ASB) address. In certain implementations, you may have to manually configure the `broadcast` attribute for the IP interface so that routers exchange RIP routes with other implementations. This is most common in server and workstation environments that support RIP1 minimally and require that RIP broadcasts be addressed to 0.0.0.0. While these configurations are acceptable, care should be taken to avoid undesirable circulation of RIP broadcasts. IP/RIP will detect unacceptable RIP frames from routers that are not from the same network.

```
#  177: 05/08/1999 09:13:00.247  DEBUG    SLOT  9  IP                Code:  84
RIP packet received on interface 172.16.92.1 (cct# 52) dropped,
  bad source 172.16.169.10
```

Countdown to Infinity

RIP networks sometimes suffer from the limitations imposed by the use of periodic updates based on timers. Because updates are periodic, network failures may propagate slowly, resulting in a slow convergence.

Routes learned on one interface are readvertised on other interfaces, including the interface the route came in on. The advertisement on the interface the route was received on is usually *poisoned*, with the metric set to 16 (infinity, indicating unreachability). This is done so that the originating router does not use a reverse route.

The problem that arises from the timed nature of periodic updates is called the *countdown to infinity*. A simple network using RIP with actual advertisements is shown in Figure 6-3. Router 1 advertises the network 172.16.95.0/24 to router 2, which advertises it to router 3 with a higher metric. Now suppose that the interface connected to the 172.16.95.0/24 network goes down on router 1. Since the network is directly attached, the local route is immediately removed from the routing table. However, in the absence of triggered updates, the change in the network condition is not immediately propagated to router 2, and subsequently router 3, until the next scheduled periodic update. In the next update, the 172.16.95.0/24 network is simply missing or is poisoned in the advertisement from router 1. At this time, both router 2 and router 3 still have the route to the 172.16.95.0/24 network, pointing toward router 1. However, when router 2 receives the periodic update from router 3, it replaces the existing route (pointing toward router 1) with the one pointing to router 3. If the route loss was indicated by router 1 using an unreachable route, then the old route may have already been removed. In either case, when router 2 learns the route from router 3, the route is invalid because the 172.16.95.0/24 network is no longer available.

Now routers 2 and 3 both have an invalid route to the 172.16.95.0/24 network. The route is periodically readvertised between the two routers, each time with an increased metric. When the hop count metric eventually reaches the infinity (16) value, the route is removed from both routing tables. It may take a significant amount of time for the invalid route to be successfully removed from both routers. On large networks, the route may remain in the network for several minutes. This is undesirable because packets will traverse the network several hops before receiving a network unreachable message.

Two solutions are available to prevent the countdown to infinity problem:

- Use triggered updates, enabled using the `triggered-updates` parameter for the RIP interface. As soon as a local route becomes unreachable, router 1 will send

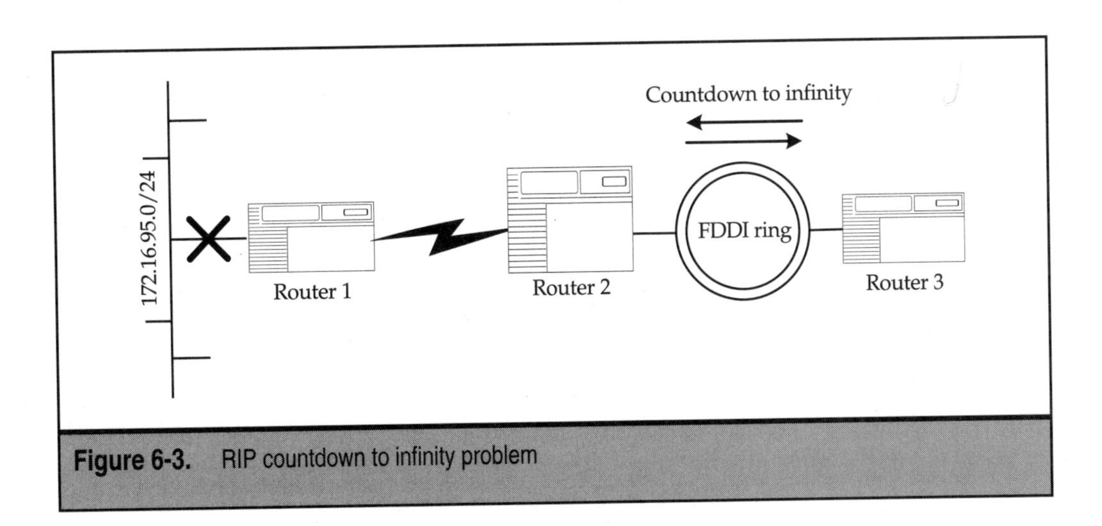

Figure 6-3. RIP countdown to infinity problem

an update indicating the unreachability of the network. Router 2 will similarly send a triggered update as soon as it updates its routing table. For this mechanism to work efficiently, all routers in the network must support triggered updates.

- Use either the poisoned-reverse mechanism or split horizon mechanism to reduce the likelihood that this condition will occur. The split horizon mechanism prevents RIP from readvertising a route out on the same interface on which it was learned.

Route Propagation over a Noncontiguous Network

Many intranets use some form of private addressing along with a combination of one or more IANA assigned network address ranges. This sometimes results in a noncontiguous network, where subnetworks belonging to the same class of network are separated by an intervening network belonging to another natural class. Figure 6-4 shows a noncontiguous network.

In Figure 6-4, router 1 advertises the network 172.16.45.0/24 to router 2. Similarly, router 3 advertises the network 172.16.47.0/24 to router 2. Both subnets (45 and 47) belong to the natural class B network, 172.16.0.0/16. Initially, assume that RIP1 is being

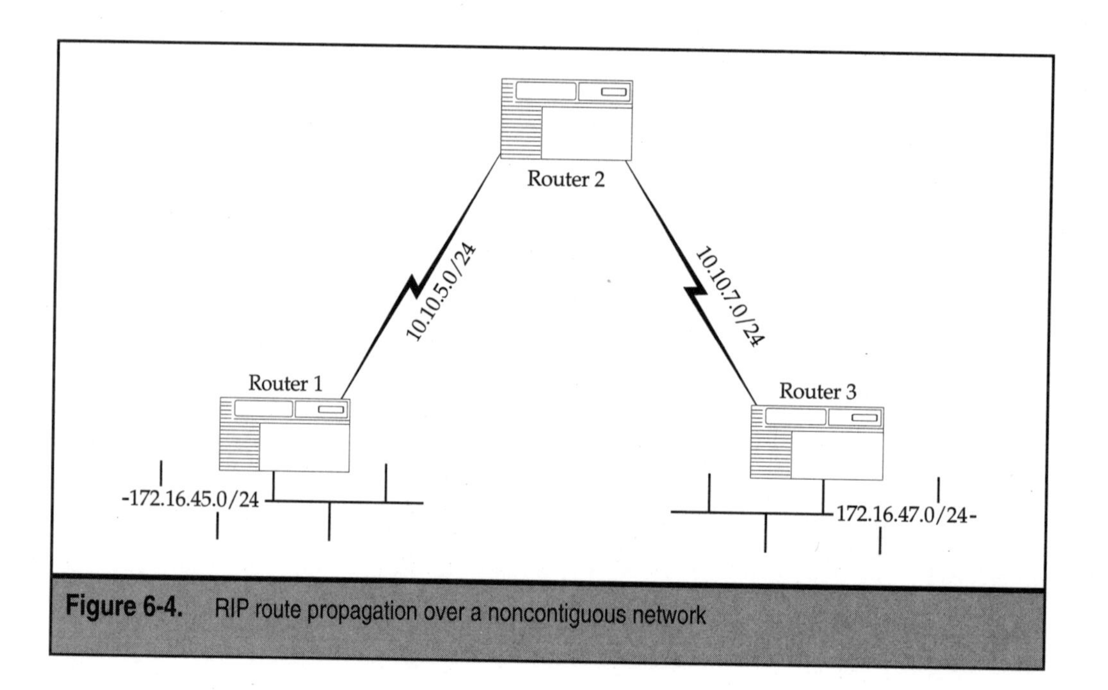

Figure 6-4. RIP route propagation over a noncontiguous network

used between all routers. When router 2 receives the respective subnet routes from router 1 and router 3, it does the following:

- ▼ It first checks the route to see if it belongs in the same network as the interface it was received on. If it does, then the network mask of the inbound interface is applied to the route.
- ▲ If the route received is from a different class of network, then the natural class mask of the inbound route is applied. In this example, the natural class B mask (255.255.0.0) is applied to both subnets, 172.16.45.0 and 172.16.47.0, yielding the same destination network, 172.16.0.0!.

You may begin to see the problem here. Router 2 now has two routes to 172.16.0.0, from which it chooses one. Assuming that all metrics are the same, the first learned route will be entered into the routing table. Subnets 172.16.45.0 and 172.16.47.0 will not be able to communicate through router 2 because router 2 does not know that this a noncontiguous network.

This is one of the problems rectified by RIP2. If all RIP interfaces have been migrated to RIP2, then the network mask will be propagated in addition to the destination network, which will allow our example network to operate as desired. The default RIP version on BayRS routers is version 1.

The same noncontiguous network problem will occur if RIP2 is used with auto-aggregation. The automatic aggregation feature attempts to summarize the 45 and 47 subnets into their natural class network, 172.16.0.0, producing the same routing problem as before. The RIP version is configured using the `version` parameter for the `rip` object, a child object of the relevant IP interface.

Open Shortest Path First

The OSPF protocol is the most widely deployed routing protocol in large IP networks today. Unlike RIP, which uses a distance vector algorithm, OSPF uses a link state algorithm and exchanges link state information with other routers in the autonomous system. During initialization, all routers exchange topology databases and synchronize link state information. A routing table is built from the link state database (LSDB) using the shortest-path-first algorithm. Subsequently, OSPF routers do not exchange full topology information and send link state updates only when a topology change occurs. OSPF routers need to periodically test only the integrity of the links to other neighboring routers to which they are adjacent. The protocol supports variable-length subnet masks in link state information and routing tables.

The extensive benefits of OSPF do not come without added complexity, however. OSPF routers need to maintain state information on their neighbors by forming an adjacency with each of them. The link state database (LSDB) is the supporting database used to run the SPF algorithm, with the local router as the root of the tree. This algorithm computes the shortest and best path, which usually offers the least-cost metric delay. The resulting paths are entered into the OSPF routing table. The LSDB and OSPF routing tables are separate from the comprehensive routing table maintained by the RTM. Maintaining these tables requires more memory than a protocol like RIP. The SPF algorithm, while simple to understand and implement, consumes significant processing resources when topology changes occur. The processing requirements increase with an increase in adjacencies, areas, or link state entries.

This section discusses several common OSPF problems and their symptoms, causes, and resolution.

Neighbors and Adjacencies

Routers that have an interface to the same network are generally said to be OSPF *neighbors*. A true neighbor relationship depends on the acceptability of various parameters such as the area ID, hello and dead intervals, authentication mechanism, and external and NSSA options. Neighbors communicate with each other using OSPF hello packets. The hello packets are used for discovering neighbors and establishing adjacencies with them. An OSPF router is said to have an adjacency with another router when the two exchange routing information with each other. The hello protocol is used by the OSPF interfaces for periodic bidirectional communication. Its most important role is in the formation of the adjacency, using logical state transformations. On broadcast networks such as Ethernet and Token Ring LANs, the hello protocol is used to discover neighbors dynamically. On NBMA networks such as frame relay, you may need to manually configure NBMA neighbors to form proper adjacencies, especially on partially meshed networks.

Figure 6-5 depicts a simple OSPF hello exchange between two routers on the same network. Adjacency establishment uses the following state transitions for each neighbor:

- **Down** The Down state indicates that the specific neighbor has had no activity on the network. This state usually follows an adjacency loss; configured NBMA neighbors also begin in this state.
- **Attempt** The Attempt state indicates that the local OSPF interface has sent hello packets out from the interface, but has not received any communication from the neighbor.
- **Init** The Init state indicates that at least one hello packet has been received from the neighbor. This is the start of bidirectional communication.
- **Two-Way** When both neighbors are aware of each other, the Two-Way (bidirectional) neighbor state has been reached. Most neighbor relationships on broadcast networks will be in the Two-Way state, except those with a designated or a backup designated router.
- **ExStart** The ExStart (Exchange-Start) state indicates that both neighbors are beginning the process of exchanging database details.

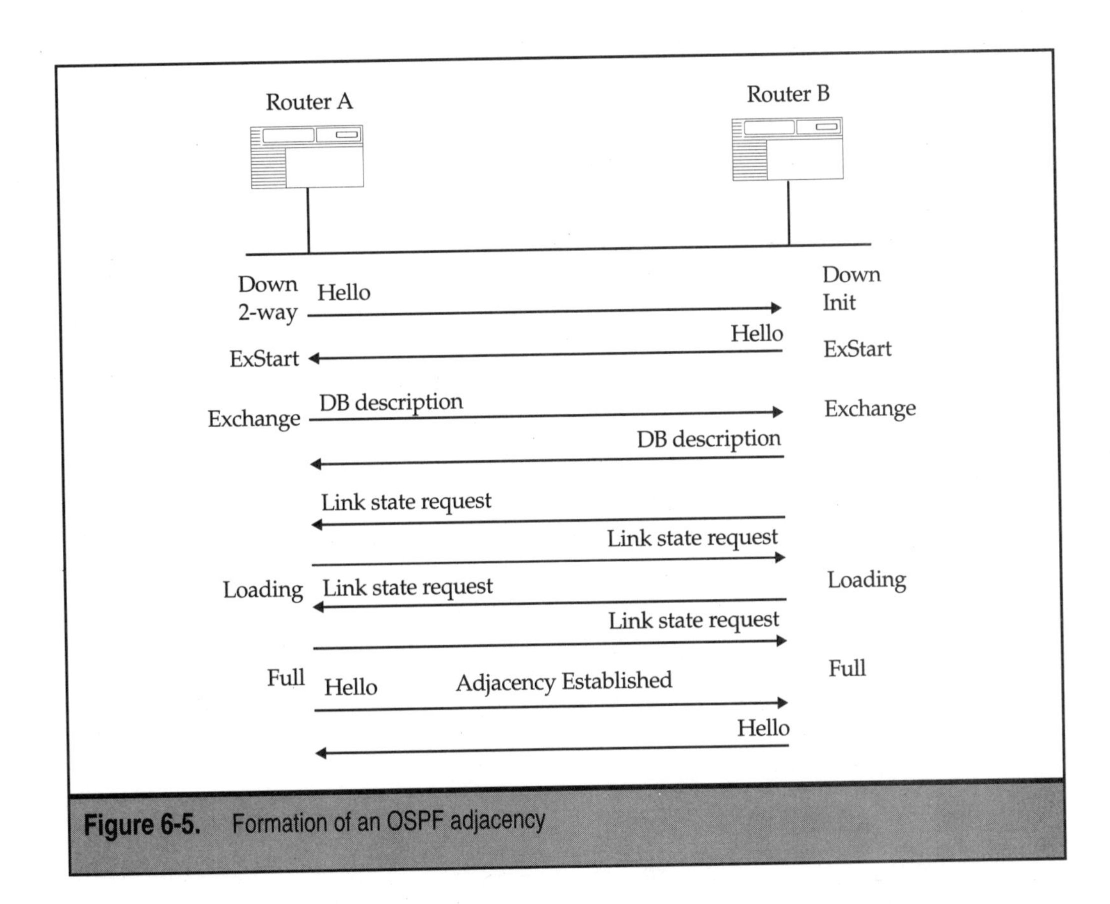

Figure 6-5. Formation of an OSPF adjacency

- **Exchange** To prevent deadlocks, a role selection takes place in the ExStart state. Following this, the Exchange state is entered, during which the neighbors exchange database description (DD) packets. The DD packets briefly describe link state details and sequence numbers.
- **Loading** After the exchange of DD packets, each neighbor specifically requests LSAs that it does not currently have in its LSDB. The Loading state is entered while LSAs are being loaded by each neighbor.
- **Full** The Full state indicates that a steady-state neighbor relationship has been established. Subsequent to this, neighbors update each other using link state updates.

Two routers are considered to be in a Full adjacency state when they are fully synchronized with each other's link state database contents. Many OSPF issues arise due to problems forming adjacencies. When two neighbors discover each other on a broadcast network, they observe several details in the hello packets, which must be agreeable for an adjacency to form.

The **show ospf neighbors** command displays the dynamic and configured neighbors.

```
fp3(48)# show ospf neigh
show ospf neighbors                                    Apr 27, 2000 07:00:17 [GMT-6]

OSPF Dynamic Neighbors
----------------------

                                  Neighbor
IP Interface     Router ID        IP Address       State      Type
---------------  ---------------  ---------------  ---------- ----------
172.16.218.10    172.16.218.14    172.16.218.14    full       Dynamic
172.16.218.10    172.16.218.11    172.16.218.11    full       Dynamic
172.16.218.10    172.16.218.19    172.16.218.19    twoway     Dynamic
172.16.220.10    172.16.220.101   172.16.220.101   full       Dynamic
172.16.220.10    172.16.220.102   172.16.220.102   full       Dynamic
... ...

OSPF Configured Neighbors
-------------------------

                                  Neighbor
IP Interface     Router ID        IP Address       State      Type
---------------  ---------------  ---------------  ---------- -----------
172.16.52.1      172.16.52.2      172.16.52.2      full       Config
172.16.52.1      172.16.52.3      172.16.52.3      full       Config
172.16.52.1      172.16.52.4      172.16.52.4      full       Config
... ...
```

If a router is neither the DR nor the BDR, it will have only two full adjacencies on an interface, regardless of the number of neighbors on the segment. For example, the neighbor with router ID 172.16.218.19 is in the Two-way state because OSPF requires adjacency formation only to the DR (172.16.218.14) and BDR (172.16.218.11). Virtual neighbors, if configured, are also displayed by **show ospf neighbors**. The following parameters are exchanged in the hello packets and can prevent adjacencies from forming:

- **Network address and network mask** The hello packet received from a neighbor must be sourced from the same network on which the OSPF interface resides. Since the network address is unique and is complemented by the network mask, the net mask must also match. When mismatches occur, OSPF logs a warning message. In the following example, although the network number is supposedly correct, the network masks used by the two routers are different—22 bits and 24 bits—thus making each network address unique.

```
# 169: 05/04/1999 22:05:04.815  WARNING  SLOT  3  OSPF              Code:  53
C2: Hello Rejected: NETMASK MISMATCH
    src 80.129.98.101:255.255.252.0 interface 80.129.98.100:255.255.255.0
```

- **Router priority** On broadcast networks, OSPF uses the router priority value to elect a designated router (DR) and backup designated router (BDR). Misconfiguration of the router priority may cause the incorrect router to be elected as the DR or BDR. For example, on a backbone FDDI ring, it may be preferable to have a BCN perform the DR function than an ASN. Most important, if the priority value is set to 0, then the interface becomes ineligible to operate as a DR or BDR. If all routers in a network are configured with zero priority values, then they will never form adjacencies because DR/BDR election will not take place. When a router fails to assume the DR/BDR responsibilities, check the `priority` attribute and compare it to other priority values for the `ospf` interface object, a subcontext of the IP interface. In the following example, the OSPF interface is in the waiting state and has not seen a DR in the network. The priority of the interface is set to zero, preventing it from becoming the DR.

```
box->ethernet/9/1->ip/10.17.143.200/24->ospf/10.17.143.200
fxc9# show ospf interface
show ospf interface                          Apr 27, 2000 06:59:54 [GMT-6]

                                 Interface Interface Metric Prio- Designated
IP Address      Area ID          Type      State     Cost   rity  Router
--------------- ---------------  --------- --------- ------ ----- ---------------
10.17.143.200   0.0.0.0          broadcast waiting       10     0 0.0.0.0

box->ethernet/9/1->ip/10.17.143.200/24->ospf/10.17.143.200
fxc9# priority 10
```

- **Hello and dead intervals** The hello interval determines the frequency of the periodic hello packets originated by each router. All routers must agree on the same hello interval. The dead interval is the period that must elapse from the last hello packet for an OSPF router to declare that a neighbor is down. By default, the hello interval is set to 10 seconds, and the dead interval is set to 40 seconds. These interval values are sent by each router in the hello packet during the initial formation of the adjacency, and they must match. Mismatch in these values will cause repeated hello rejections in the log and prevent the formation of the adjacency.

```
#  193: 01/28/1999 13:45:52.399  WARNING  SLOT 13  OSPF              Code:  54
C2: Hello Rejected: HELLO INTERVAL MISMATCH
    src 172.16.85.2(15)   interface 172.16.85.1(10)

#  194: 01/28/1999 13:46:02.403  WARNING  SLOT 13  OSPF              Code:  54
C2: Hello Rejected: HELLO INTERVAL MISMATCH
    src 172.16.85.2(15)   interface 172.16.85.1(10)
```

- **External (E) and NSSA (N) options** The Options field in the hello packet has two important bit values: for the E (external) bit and the N (Not So Stubby Area, or NSSA) bit. These capabilities are optional for OSPF, but neighbors forming an adjacency must agree on the use of these options. The external routing capability indicates whether the interface is operating in a *stub* area, and the NSSA option indicates whether the area is a Not So Stubby Area. Both of these options are rejected at the OSPF interface level, but are attributes of the area to which the interface belongs. The area parameters must be configured correctly to resolve for the following types of warnings:

```
#  120: 08/15/1999 06:44:23.318  WARNING  SLOT 14  OSPF                Code:  57
C2: Hello Rejected: N-bit MISMATCH
    src 10.119.92.10(0) interface 10.119.92.20(1)

#   74: 06/09/1999 09:54:13.768  WARNING  SLOT  2  OSPF                Code:  56
C2: Hello Rejected: EXTERN OPTION MISMATCH
    src 172.16.148.1(1) interface 172.16.148.2(0)
```

NBMA Neighbors On nonbroadcast multi-access (NBMA) networks, neighbors usually can discover others and form adjacencies in fully meshed environments. However, this may be undesirable in certain cases, especially when the WAN bandwidth is limited. On partially meshed NBMA networks, NBMA neighbors may have to be statically configured for adjacencies to form successfully. Most point-to-multipoint topologies do not support broadcast capabilities, as shown in the frame relay network in Figure 6-6.

NOTE: Do not assume that neighbors will automatically establish full adjacencies over NBMA networks such as frame relay and X.25. The default broadcast setting for the OSPF interface will not be appropriate in all instances, especially for services where the multiplexing method does not provide broadcast-type functionality.

When NBMA neighbors are configured, unicast hellos are sent periodically to the neighbors. The neighbors must be configured on both sides of the adjacency; otherwise the following warning message is reported:

```
#   31: 11/02/1999 06:03:26.765  WARNING  SLOT  5  OSPF                Code:  59
C2: Hello Rejected: UNKNOWN NBMA NBR
    src 172.16.213.14 interface 172.16.213.1
```

The type of an OSPF interface also may be misconfigured, which may stop its NBMA neighbors from establishing adjacencies. In this case, the error is diagnosed as an invalid NBMA neighbor. For example, the following warning is taken from a frame relay service record whose OSPF interface is incorrectly configured as the broadcast type, but the NBMA neighbors are configured correctly.

```
#  164: 08/27/1999 03:24:59.010  WARNING  SLOT 12  OSPF                Code: 103
intf not NBMA or Point-to-multipoint 172.16.15.1.0
config'd nbr ip addr: 172.16.15.7
```

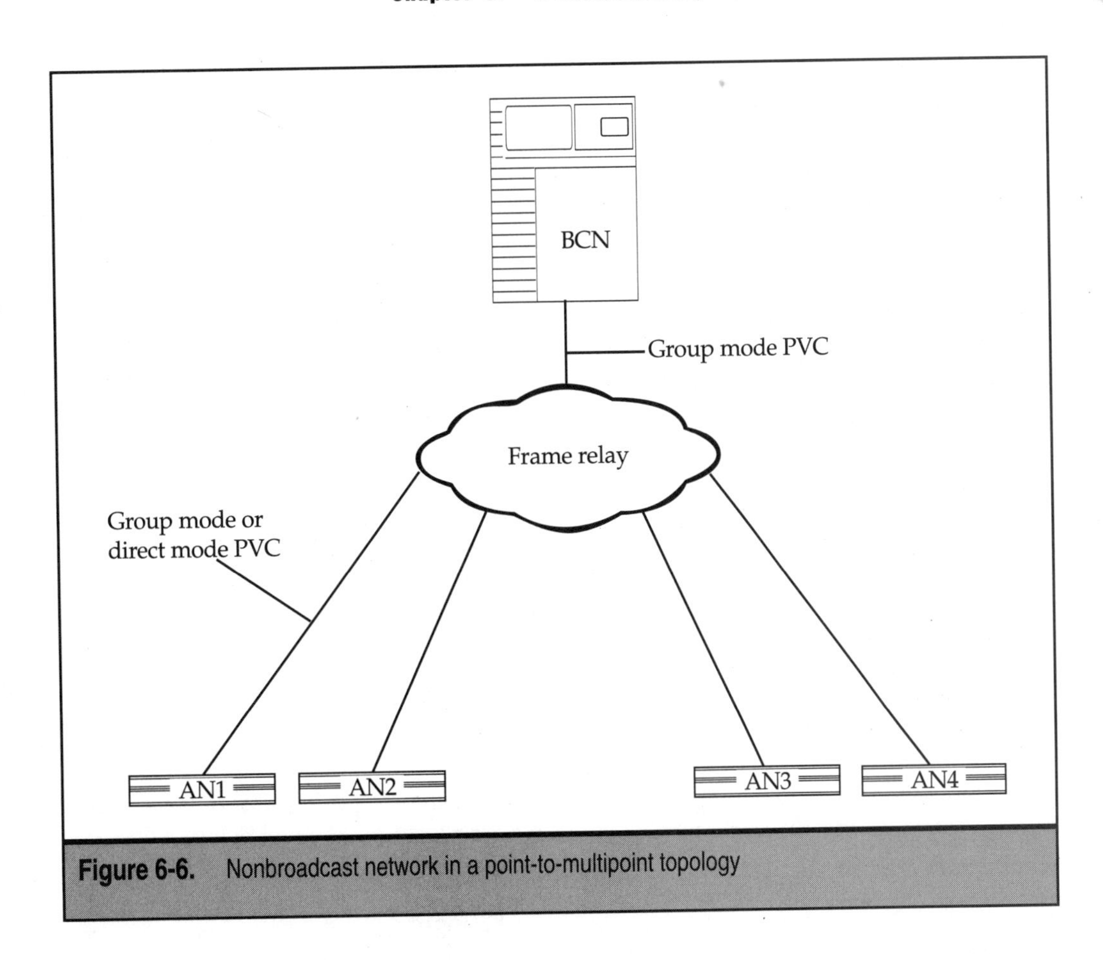

Figure 6-6. Nonbroadcast network in a point-to-multipoint topology

Adjacency Loss Occasionally, network hiccups such as a temporary PVC loss or a physical link reset will cause a circuit to reset. For example, if a downstream switch is reset, the circuit and all protocol interfaces on a router attached directly to the switch will also be reset. This will cause the OSPF interface to reinitialize and reestablish adjacencies. In the meantime, adjacent routers on the same network will use the dead interval to declare the loss of the adjacency.

However, in certain situations a transient network problem can result in an extended adjacency problem. Consider the network illustrated in Figure 6-7. Router A is connected to its local switch, and router B is connected to another switch. Both switches are interconnected using another high-speed Ethernet connection. Switch A is reset due to a power problem, which causes the Ethernet interface and the circuit to go down on router A.

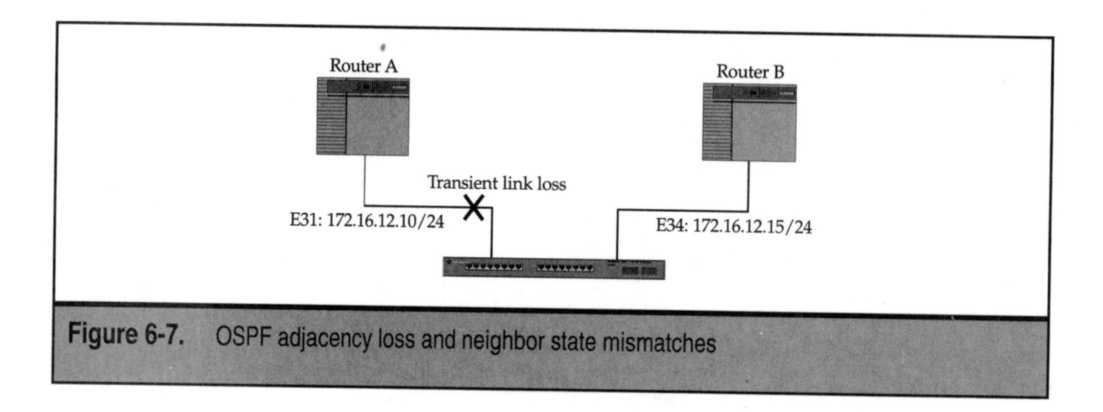

Figure 6-7. OSPF adjacency loss and neighbor state mismatches

Router B's circuit and protocol interfaces are unaffected at this time. Five seconds later, switch A is operational again, and router A reinitializes all protocols, including OSPF. At this time, router A does not have an adjacency to router B, but router B still believes that it has a full adjacency with router A. Router A reports the following warning when it receives a frame from router B:

```
#  128: 08/15/1999 13:00:27.198  WARNING  SLOT  2  OSPF                Code:  61
C3: Packet Rejected: SOURCE NEIGHBOR IN WRONG STATE
    src 172.16.12.15 state INIT type Link State Update
```

The message repeats a few times, until router B also recognizes that router A is in a different state. Both routers then reset their adjacencies and start the adjacency process again. In some cases, when many routers are involved, the state mismatch can cause an unstable condition, thereby preventing a quick recovery. If the adjacency does not recover within the dead interval period, disable the OSPF interface that was just reset. Leave the OSPF interface disabled at least for the duration of the dead interval (default: 40 seconds) so that all adjacent neighbors will declare it lost. Then reenable the OSPF interface to restart the adjacency establishment process.

Areas

The OSPF protocol considers the SPF domain to be an *autonomous system* (AS). The OSPF autonomous system is usually divided into *areas*, which are usually connected by a *backbone* area. Each OSPF area has a unique area ID, 32 bits in length and represented in a format similar to that of an IP address. The backbone area has the area ID 0.0.0.0. In addition to the adjacency problems discussed in the previous section, areas are another source of OSPF routing problems.

Each OSPF interface on a router is associated with a specific area. A single router can have multiple OSPF interfaces, each participating in a different area. Often, a router has several interfaces participating in the same area, but with different network numbers. When an OSPF interface detects a hello packet from a neighbor in a different area, it will be unable to establish an adjacency. This can be due to either misconfiguration or a backdoor bridge between two networks.

```
#   42: 06/17/1999 13:30:49.088  WARNING  SLOT 12  OSPF                Code:  49
C1: Packet Rejected: AREA MISMATCH (0.0.0.9)
    src 10.87.113.10 dst 224.0.0.5 routerid 10.87.113.10
```

In cases like the preceding one, check the `area` attribute of the OSPF interface. The OSPF interface must be deleted and re-created with the correct area ID, if it is incorrect.

You can view the state of all OSPF areas configured on a BayRS router with the **show ospf area** command. Use an optional area ID after the command to list details specific to that area only. A configured area is considered to be down when none of its associated OSPF interfaces are in an operational state. If an area is down, investigate the status of all interfaces in that area.

```
fxc9(38)# show ospf area
show ospf area                                  Apr 27, 2000 06:59:23 [GMT-6]

Non-NSSA Areas:
---------------

Area ID         Area State Area Type Authentication
--------------- ---------- --------- --------------
0.0.0.0         up         normal    none
0.0.0.1         up         normal    none
0.0.0.2         up         stub      none
0.0.0.3         down       normal    none

NSSA Areas:
-----------

Area ID         Area State Area Type Authentication
--------------- ---------- --------- --------------
0.0.0.7         up         nssa      none
... ...
```

Routers with multiple areas configured are said to be area border routers (ABRs). If a router is an ABR, then it must participate in the backbone (0.0.0.0) area. OSPF routers that have multiple areas configured without the backbone area will report that all areas are down. OSPF does not follow a hierarchy; however, all areas must be tied together using the backbone area. In addition to all the area states being down, the following event message is logged:

```
#   58: 10/01/1999 24:10:49.472  WARNING  SLOT  6  OSPF                Code:  90
No Backbone area configured for area border router
```

A simple OSPF autonomous system is illustrated in Figure 6-8. Observe that the backbone area interconnects all other areas. Occasionally, certain design changes and network evolution may prevent the addition of a new area to the backbone area directly using a physical network interface. In this case, you can configure a virtual link to the backbone area using a *transit* area. As shown in Figure 6-9, border router 1 is connected to both areas A and B. However, it does not have a physical connection to the backbone area, a condition it resolves by using area B as a transit to the backbone area. Virtual links are often considered poor OSPF design, and whenever possible, preliminary network designs should take into account the future evolution of the OSPF AS into multiple areas.

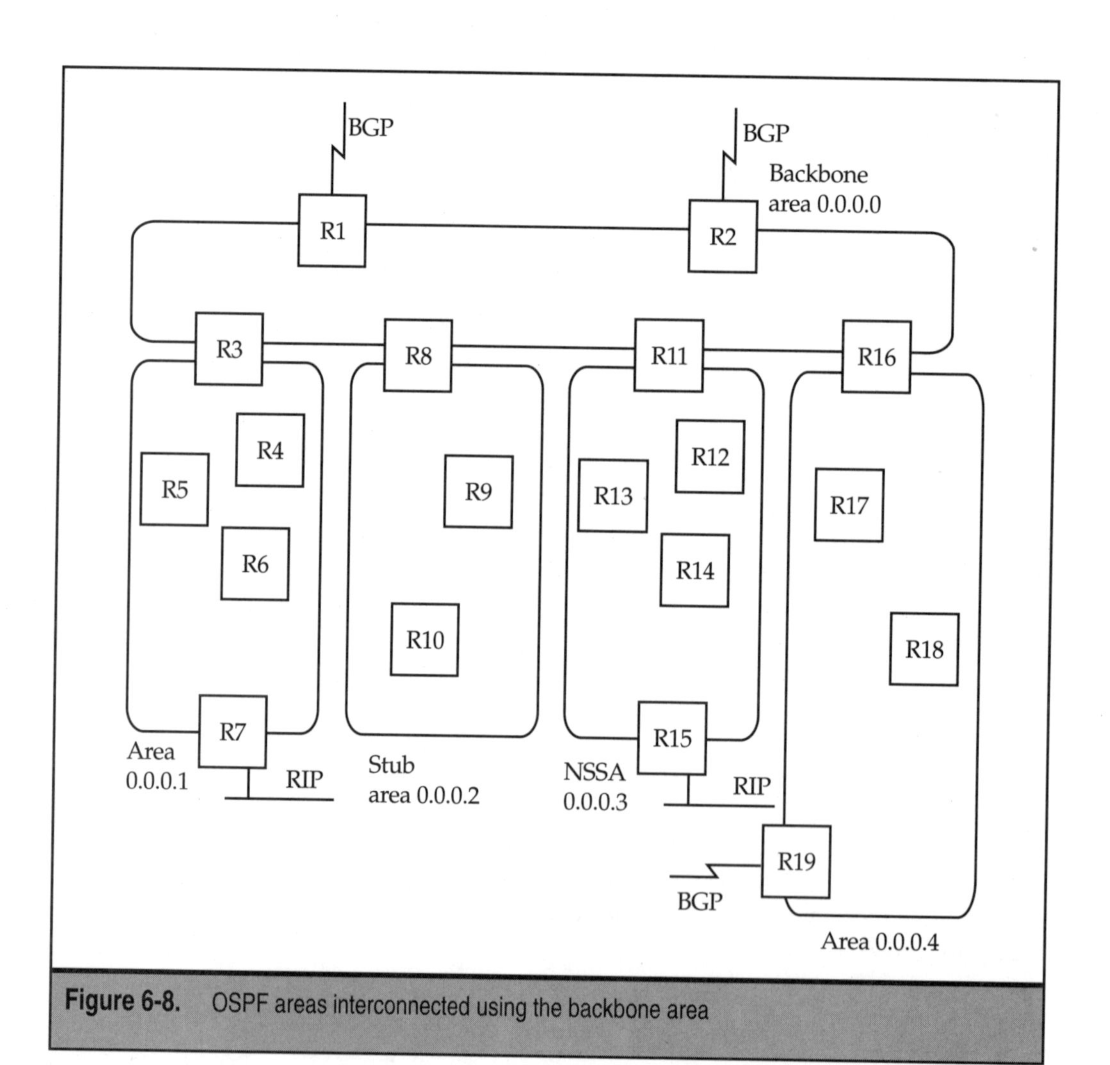

Figure 6-8. OSPF areas interconnected using the backbone area

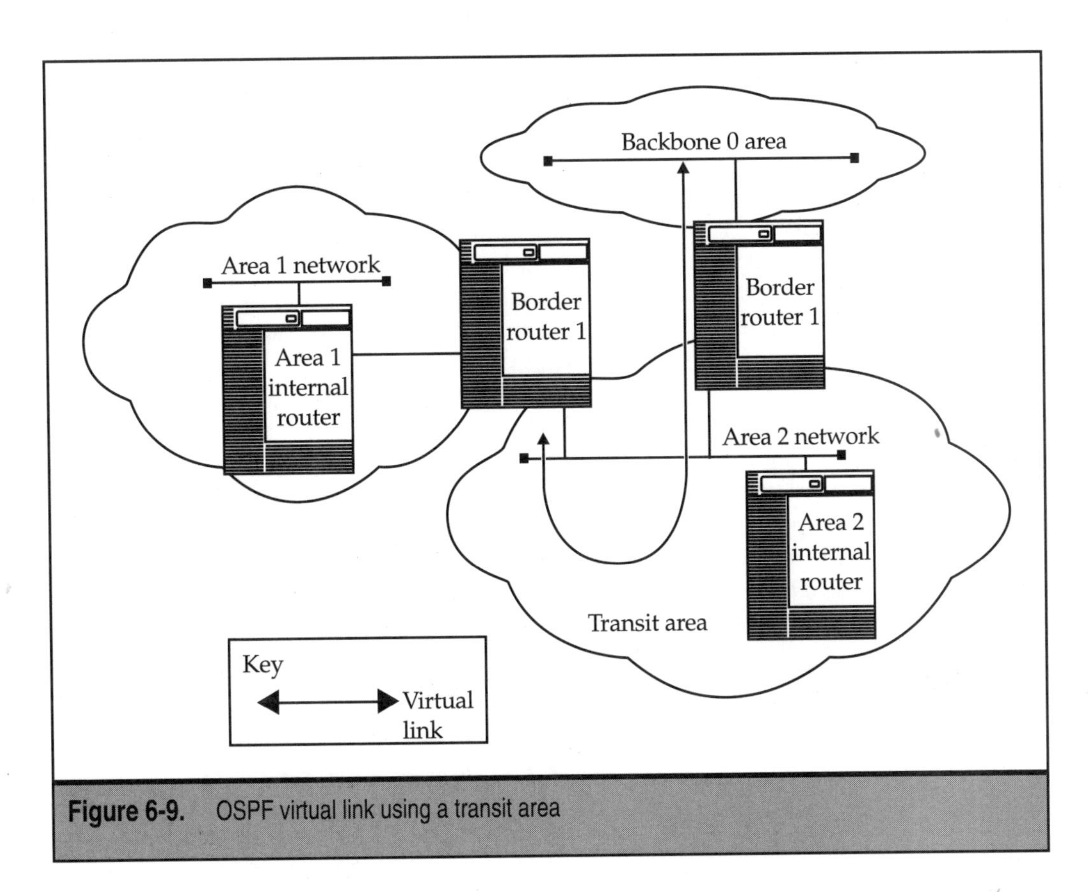

Figure 6-9. OSPF virtual link using a transit area

Virtual link configurations are easily set up simply by configuring the transit area and the router ID of the backbone router to which the link is to be established. Ideally, the router ID of the backbone router should be a valid operational IP address. Misconfiguration of a virtual link may make the OSPF area unreachable from the backbone.

```
#   28: 08/15/1999 21:16:59.946  WARNING  SLOT 10  OSPF              Code:  58
C2: Hello Rejected: UNKNOWN VIRTUAL NBR
    src 172.16.217.101 interface 172.16.19.1

#  192: 08/27/1999 22:48:39.022  WARNING  SLOT  6  OSPF              Code:  99
No transit area with ID 0.0.0.6 for virt intf/nbr 192.168.77.120

#  110: 02/01/1999 02:58:01.812  WARNING  SLOT  4  OSPF              Code: 100
Virtual interface/nbr 80.47.112.1 config'd, router NOT Area Border

#    8: 04/23/1999 22:04:35.782  WARNING  SLOT  6  OSPF              Code: 101
Area 0.0.0.3 is stub for virtual interface/nbr 172.16.57.101
```

The warnings in the preceding example indicate various configuration errors. The virtual link must have a valid transit area to pass through, which must be up and operational. If a router does not border multiple areas and contains only one area, then it is said to be an *internal* router for that area. Internal routers cannot be configured as virtual neighbors, and this task is usually relegated to the router bordering another area, usually the transit area. The last event message in the example illustrates that stub areas, by virtue of their design, usually are situated at the periphery of the autonomous system. They do not qualify for use as a transit area into the backbone, which would imply that the stub becomes a bordered area and must import summaries. These considerations must be applied during virtual link establishment.

Link State Database

The LSDB is the working source for the Dijkstra SPF algorithm, which eventually finds the shortest paths to other networks. Reachable networks can be advertised by various routers as different types. For example, a network may be advertised using a type 2 (network link) LSA by one router, and also advertised by another router using a type 5 (external link) LSA. The LSDB contains all LSAs that were received by the router and is stored on a per-area basis. The following LSA types are propagated with the local area only:

- Router (type 1) LSA
- Network (type 2) LSA

On multi-area OSPF autonomous systems, the following LSA types are passed between areas:

- Summary (type 3) LSA
- AS summary (type 4) LSA
- AS external (type 5) LSA

Each LSA type has several subtypes indicating the specific type of advertisement. For example, a type-1 LSA usually specifies the link type as point to point, transit, stub, or virtual. Other LSA types such as stub (type 0), multicast (type 6), NSSA (type 7), opaque (type 15), and resource (type 16) are also used within the OSPF AS for stub areas and QOSPF/MOSPF functionality. You can use the **show ospf lsdb** command to look at the LSDB contents to find discrepancies. You will sometimes need to do this because an incorrect OSPF route in the routing table is usually a result of incorrect entries in the LSDB. An optional address/prefix value will retrieve only entries that match the specified range.

```
fp3(14)# show ospf lsdb
show ospf lsdb                                        Jun 07, 2000 20:49:27 [GMT-5]


Ospf Area: 192.168.128.0
```

```
LS Type  Link State ID    Adv Router       Metric  ASE Fwd Addr    Age  Seq Nbr
-------  ---------------  ---------------  --------  ---------------  ----  --------
Stub     57.16.33.0       57.16.1.60       N/A                        0        0
Stub     57.16.34.0       57.16.1.60       N/A                        0        0
Stub     57.16.35.0       57.16.1.60       N/A                        0        0
Stub     57.16.36.0       57.16.1.60       N/A                        0        0
Stub     57.16.37.0       57.16.1.60       N/A                        0        0
Router   192.168.135.3    192.168.135.3    N/A                     1740 80001835
Router   192.168.135.213  192.168.135.213  N/A                      113 8000a9fe
Router   192.168.135.221  192.168.135.221  N/A                      753 8000ab17
Router   192.168.135.225  192.168.135.225  N/A                      634 8000004c
Router   192.168.135.245  192.168.135.245  N/A                     1155 800084fb
Router   192.168.136.8    192.168.136.8    N/A                      303 800010b8
Router   192.168.136.31   192.168.136.31   N/A                      303 8000196c
Network  192.168.135.13   192.168.136.130  N/A                      861 80000bcd
Network  192.168.136.5    152.130.1.133    N/A                       11 80000074
Network  192.168.136.130  192.168.141.1    N/A                      255 8000065b
Network  192.168.155.2    192.168.176.145  N/A                     1378 80000b08
Sum Net  57.81.80.0       152.130.1.129     11                     1631 800018e8
Sum Net  57.81.80.0       152.130.1.133      9                     1532 800016bd
Sum Net  57.81.82.0       152.130.1.129     21                     1631 80000d79
Sum Asb  192.177.13.71    152.130.1.133      9                     1532 8000181a
Sum Asb  152.130.1.137    152.130.1.129     12                      731 800018f6
Sum Asb  152.130.1.137    152.130.1.133     10                      933 800017ab
Sum Asb  152.130.1.141    152.130.1.129     10                     1660 800018ae
Sum Asb  152.130.1.141    152.130.1.133      8                     1561 80001793
Extrn 2  0.0.0.0          192.177.13.24      1 192.177.4.1          411 80000064
Extrn 1  0.0.0.0          152.130.1.129      1 192.168.136.5        760 800008f9
Extrn 2  0.0.0.0          152.130.1.133      4                      962 800008f8
... ...
```

OSPF gives precedence to an intra-area route over an inter-area route. This precedence is a result of a route derived from an LSA that originated in the same area. Both intra-area and inter-area routes have precedence over external routes that are generated as a result of type-5 LSAs.

Many times, LSDB contents may appear acceptable, and all entries will look valid. Hence, baselining LSDB contents is very important in a fully operational environment. When the OSPF AS is functioning normally, take a snapshot of the LSDB contents from all routers for reference. Large LSDBs with multiple areas should be perused for a minimal understanding of the area contents and LSA types. During a network problem, the reference LSDB will be very useful in pinpointing incorrect LSAs and LSAs that are missing from the LSDB.

External Routes

Recall that we mentioned that OSPF behaves as an autonomous system (AS). It operates as an independent system and interacts uniquely with other routing protocols. Any route derived from a non-OSPF source is said to be an *external* route. OSPF considers itself the primary routing source and assumes that a majority of the routes in the routing table originate from within the AS. On large networks, routes may also originate from protocols such as RIP and BGP. The following types of routes are considered external to the OSPF AS:

- RIP1 and RIP2 routes
- Static routes
- BGP and EGP routes
- Directly attached routes from interfaces not running OSPF
- Circuitless IP routes from virtual interfaces not running OSPF

By default, BayRS routers do not import external routes into the OSPF LSDB. In this default configuration, non-OSPF routes will not be propagated within the OSPF AS. To allow external routes to be incorporated into the LSDB and announced to OSPF neighbors, the autonomous system boundary router (ASBR) attribute must be enabled on any router that must propagate external routes. Often, network engineers overlook this attribute, resulting in the omission of external routes in OSPF LSDB. The `as-boundary-router` attribute resides in the global `ospf` object.

```
box->ip->ospf
lm5(54)# as-boundary-router true
```

Route interaction with external routes should be considered when running OSPF and other routing protocols on the same IP interface with ASBR enabled. When ASBR is enabled, the same set of routes will be advertised by both OSPF and an external routing protocol (say, RIP) out of the same interface. If the receiving router is also configured as an ASBR, then this may lead to a protocol source loop and generate unneeded LSAs. When routers are configured as ASBRs, make sure that neighboring routers are not running multiple routing protocols on the same interface.

Summaries and External Route Policies

Areas are used to optimize and limit the number of LSAs and routes maintained by various routers within the OSPF AS. Routes from one area can be summarized into the backbone using OSPF area ranges. If a range of networks resides within a specific area, it can be summarized using a single route with a larger network prefix. This reduces the number of both the routing table entries and link state database entries. This process is called *area summarization* and is comparable to RIP2 aggregation. In RIP2 with aggregation, networks are automatically aggregated using the appropriate network class. With OSPF, however, summary ranges must be manually configured with a specific network address and prefix. Two problems that commonly occur when OSPF summaries are used relate to the following:

- Routes in the backbone area cannot be summarized. Since the backbone area is the interconnection among other areas, backbone LSAs must be propagated in their original form.
- Network ranges are calculated with invalid or broad prefixes. The summary range must encompass only networks that truly reside in the specific area. If multiple areas attempt to summarize the same range, one of the following conflict messages will appear:

```
#  134: 10/21/1999 16:02:37.196  WARNING  SLOT 12  OSPF              Code: 108
Src Area 0.0.0.4 can't export SUM 10.10.0.0/255.255.0.0 TO: 0.0.0.0

#  174: 06/25/1999 24:42:57.916  WARNING  SLOT 14  OSPF              Code: 109
Dst Area 0.0.0.12 already has SUM 172.16.0.0/255.255.128.0
```

Area summarization is also called *inter-area summarization* and is configured using one or more `summary` objects under the appropriate `ospf` area. This is in contrast to the aggregation of external routes, which is sometimes called *external route summarization*. External route summaries are configured using an OSPF announce policy, which is applied to the entire set of self-originated external routes. OSPF announce policies are configured only on AS boundary routers. An announce policy can affect an existing route so that it is advertised as is, or advertised in a modified fashion, or removed from the announce pool. When configuring area summaries and policies, keep in mind that summaries affect inter-area routes, and policies affect external routes.

Configuration Considerations

OSPF is a robust protocol but generally requires additional resources as the network grows. Network growth occurs in three ways: new subnets or supernets are added to an existing area, new routers with new adjacencies are added, and new routers with new areas are added. Of these, an increase in the number of adjacencies and areas typically generates greater use of memory and CPU resources. When expanding a network, consider how OSPF operation throughout the network will be affected.

For example, an OSPF AS with 100 areas and 10 networks per area is considered to be a weak design. On the other hand, placing the same 1,000 networks in a single backbone area is also considered to be a deficient design. As a rule of thumb, an equivalent number of adjacencies and LSAs should be present in each area in the AS. If there are four OSPF areas, for example, there should not be an inordinate number of adjacencies and LSAs in one area and only a handful of adjacencies and LSAs in the other three areas. The exception to this rule is the design of the backbone area and stub areas.

The OSPF router ID (`router-id`) uniquely identifies the router to other OSPF neighbors in the AS. By default, the first configured IP interface address on the router is used as the OSPF router ID, though this setting is not mandatory and can be changed to any 32-bit number. When interacting with other routing protocols, specifically BGP, the router ID must be a valid IP interface address on the router. It is generally recommended that a critical IP interface on the router be used as the OSPF router ID. In most cases, if a virtual (circuitless) IP address is configured, it should be used as the router ID unless a different ID scheme is being employed.

Debugging

Many OSPF interactions between neighbors are logged in the event log. This includes messages such as those indicating adjacency state changes and general OSPF interface operation. OSPF also reports bad frames received, including link state requests, acknowledgments, and updates. Extended debugging can be enabled for OSPF, using the global `log-mask` (`wfOspfPrimaryLogMask`) attribute.

```
box->ip->ospf
lm4(31)# log-mask 0xfff
```

Setting the `log-mask` attribute to 0xfff will display additional messages, including messages indicating self-originated LSAs, receipt of new LSAs, changes to the OSPF routing table, receipt of less recent LSAs, receipt of more recent self-originated LSAs, and receipt of MaxAge LSAs. These event messages are useful in understanding the source of some OSPF problems without having to use packet captures.

Some network problems may cause OSPF interfaces to bounce frequently. For example, a defective external transceiver attached to an Ethernet interface may cause frequent signal loss, which effectively resets the OSPF interface and triggers OSPF to respin the Dijkstra algorithm repeatedly. This is undesirable because it consumes CPU resources and may disrupt traffic to the intended destination. By default, BayRS imposes a 1-second SPF hold-down, during which time only one SPF calculation may run. While this is adequate for most problems, the timer value may need to be increased if the network problem occurs occasionally but is significant enough to degrade performance. In the following example, the hold-down timer value is increased to 3 seconds. Increasing this value significantly, on the other hand, will affect OSPF convergence using recalculated routes.

```
box->ip->ospf
lm4(35)# holddown 3
```

Border Gateway Protocol

The Border Gateway Protocol is the successor protocol to the Exterior Gateway Protocol (EGP), which historically was used to interconnect traditional Internet core gateway routers. BGP has a complex, scalable design that allows it to be used in an exclusive application: routing between multiple autonomous systems. It is the preferred routing protocol in the Internet backbone and is used by all Internet service providers (ISPs). Most large enterprise networks connecting to the Internet using a high-speed connection use BGP as the routing protocol. When enterprise networks are merged, such as after a business merger, BGP interconnects the existing autonomous systems without requiring significant network changes, yet still providing an efficient routing solution. Figure 6-10 illustrates the connection of two OSPF autonomous systems using BGP.

To overcome some of the serious shortcomings of EGP, BGP-3 was introduced with two enhancements: reliable transport using TCP and loop detection using AS paths. Further improvements were made in BGP-4, which includes classless interdomain routing (CIDR) support and policy-based routing functionality. The support for CIDR has helped to reduce the growth of Internet routing tables by supporting supernet routes. Policy-based routing also helps in the propagation of selective routes.

Peers

Similar to the way OSPF uses neighbor adjacencies, BGP uses *peers*. OSPF can dynamically discover neighbors in broadcast networks, but BGP peers must be statically configured to establish neighbor relationships. BGP uses the transmission control protocol (TCP) for reliable transport between peers. An initial TCP session is established to the well-known TCP port 179 when a BGP connection is opened. The TCP session must be established successfully for the BGP connection to be successful.

A typical BGP connection establishment is illustrated in Figure 6-11. From an idle state, the connection eventually transitions to an established state. Subsequently, network layer reachability information (NLRI) is exchanged among the BGP peers. A router

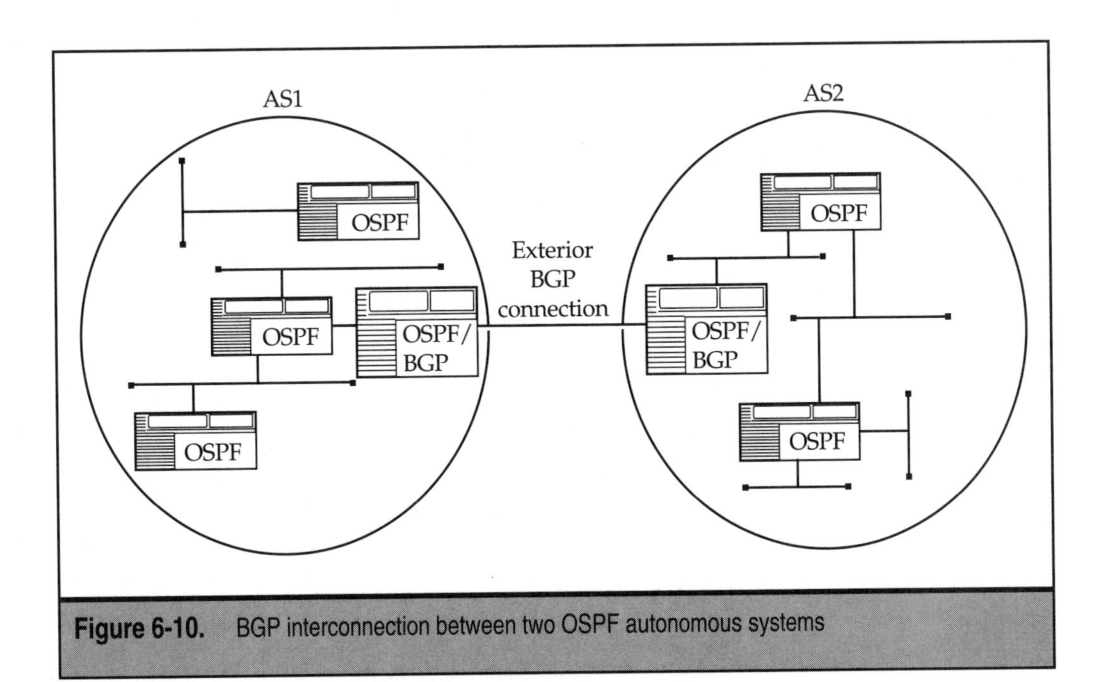

Figure 6-10. BGP interconnection between two OSPF autonomous systems

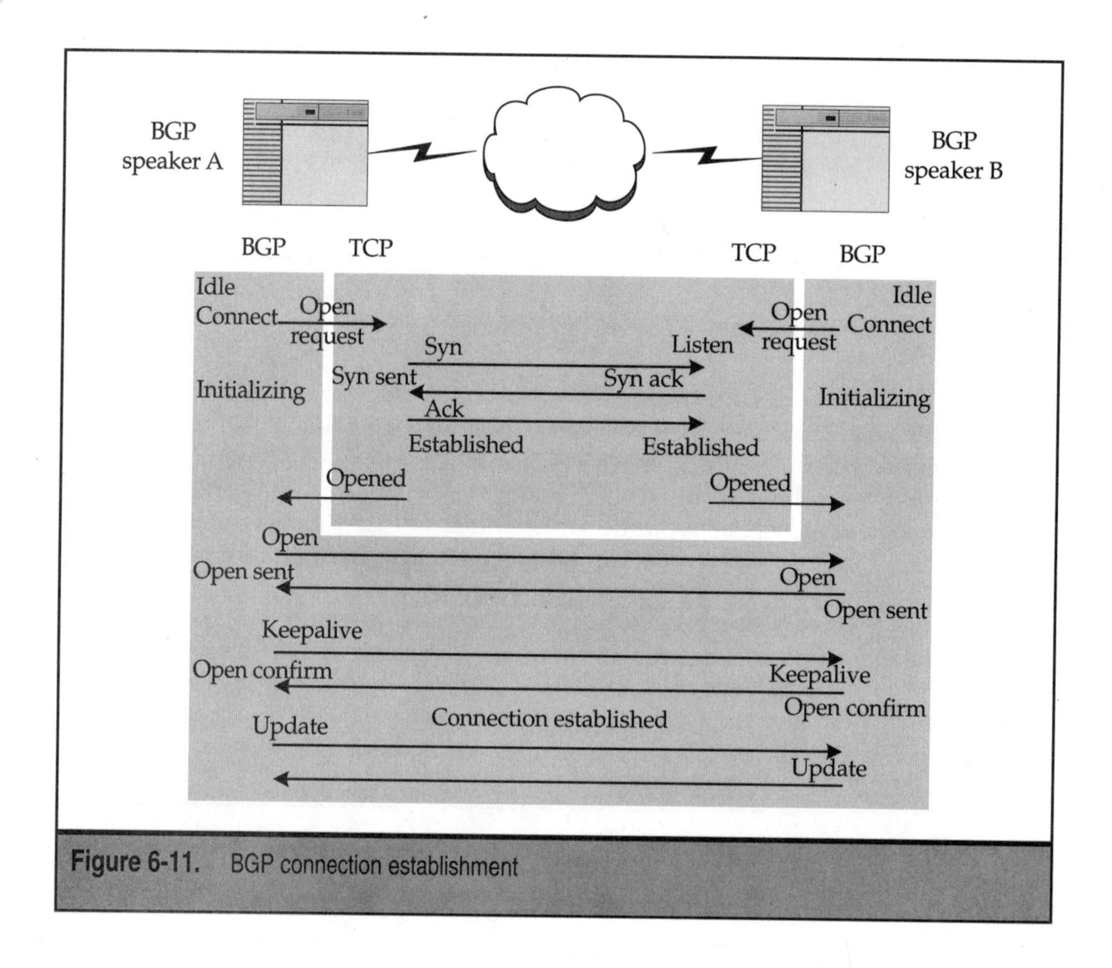

Figure 6-11. BGP connection establishment

running BGP is called a BGP speaker and is uniquely identified by a BGP router identifier (`router-id`) and a local AS number (`local-as`).

A connection is uniquely identified by the following three parameters:

- **Peer address** The IP address of the remote peer, used to establish the BGP connection. This is equal to the local address on the remote router. Misconfiguration of the remote peer address is a common error.

```
#   28: 08/23/1999 12:48:19.722  WARNING  SLOT  6  BGP               Code:  20
Invalid Remote IP address configured in wfBgpPeerEntry: 135.19.47.11

#    8: 04/19/1999 02:40:27.182  WARNING  SLOT  6  BGP               Code:  21
Local and Remote IP addresses on different subnets
in wfBgpPeerEntry: Local: 172.16.94.1, Remote: 172.16.104.1
```

Normally, BGP sessions are established over point-to-point connections. However, some BGP connections are established over multiple hops. In this case, the `multi-hop` attribute for the global `bgp` object must be enabled. Otherwise, the BGP/21 warning is reported as shown in the preceding example.

- **Local address** The IP address of the local peer, used to establish the BGP connection. This address is equal to the remote peer address configured on the remote router. If the local IP address is absent or otherwise invalid, BGP reports a warning, and the connection is not established.

```
# 185: 05/16/1999 21:25:04.527 WARNING SLOT 7 BGP                Code: 19
Invalid Local IP address configured in wfBgpPeerEntry: 190.85.57.1
```

- **Peer AS** Identifies the number of the autonomous system where the remote peer resides. When the peer AS and the local AS are the same, the connection is said to be an IBGP (internal) connection. If the BGP speakers reside in different autonomous systems, then the connection is called an EBGP (external) connection. The peer AS number is obtained from the administrator of the remote BGP speaker. Invalid AS numbers cause a BGP warning.

```
# 198: 10/05/1999 08:58:29.052 WARNING SLOT 8 BGP                Code: 28
Invalid remote AS configured in wfBgpPeerEntry: 298570
```

BGP, as mentioned, relies on the TCP protocol for reliable transport. The TCP system service must be running on the slot along with IP for BGP to attempt a connection.

```
#  61: 09/04/1999 06:17:12.459 DEBUG   SLOT 9 BGP               Code: 101
BGP connection aborted because TCP is not active
```

The preceding events are reported on the local BGP speaker when the configuration is incorrect. BGP connection establishment can also be affected when the remote BGP speaker is misconfigured or when values are mismatched. BayRS routers must be configured to require BGPv4 support on the remote BGP peer, by setting the `min-version` and `max-version` of the `peer` object to *bgp4*. BGPv4 is mandatory on many networks due to the need to support CIDR. However, when establishing a peer relationship with a BGPv3 speaker, the `min-version` attribute must be set to *bgp3*. A mismatch is reported in the log if the router has not been provisioned to support the specified version.

```
# 118: 06/09/1999 15:02:25.092 WARNING SLOT 14 BGP               Code: 37
Connection 172.16.72.1 - 172.16.72.2: OPEN Message Error,
Unsupported Version Number
```

Several other peer values during the open process may raise a flag in the local BGP speaker.

```
# 151: 03/18/1999 01:47:14.133 WARNING SLOT 7 BGP                Code: 38
Connection 135.19.144.5 - 135.19.144.6: OPEN Message Error, Bad Peer AS

# 198: 02/01/1999 04:30:05.236 WARNING SLOT 2 BGP                Code: 39
Connection 172.16.147.100 - 172.16.147.101: OPEN Message Error,
```

```
Bad BGP Identifier

#   96: 04/23/1999 24:44:39.854  WARNING  SLOT  2  BGP                Code:  40
Connection 80.23.154.1 - 80.23.167.10: OPEN Message Error,.
Unsupported Authentication Code
```

The peer must reside in the AS that the local BGP peer configuration indicates to the local BGP speaker. For example, if the peer AS is configured as AS 20 on the local router but the remote peer actually resides in AS 21, then the BGP peer connection will never be established. The BGP identifier uniquely identifies each BGP speaker and must be set to a valid IP address on the router, typically the virtual (circuitless) IP address. If the remote peer supplies an invalid BGP identifier in the open message, then the connection is rejected. BayRS supports TCP MD5 authentication only and rejects other authentication methods.

```
#   44: 04/19/1999 12:48:39.322  WARNING  SLOT  8  BGP                Code:  62
Connection 192.168.149.10 - 192.168.149.9: Hold timers do not match
Local: 180  Remote 90   Using 90
```

A common problem is the mismatch of BGP hold timers between two peers. BGP speakers running BayRS automatically select the smaller of the two proposed values, although a BGP speaker may reject a connection on the basis of the hold time. The hold time is the maximum period that can elapse between keepalive or update messages. If the hold time is less than 3 seconds but not zero, then the connection may be rejected. In the preceding case, the router chooses the lesser value proposed by the remote peer. If the hold timer causes the connection to fail, you must match the values between the peers by reconfiguring the `holddown` attribute for the `peer` object.

Peer Debugging The BGP protocol facilitates extended debugging operations by logging detailed messages, enabling the monitoring of additional messages pertaining to various connection states and protocol activity. In particular, you can explicitly track the four defined types of BGP messages—open, update, notification, and keepalive messages—in detail by creating a debug control object in the appropriate BGP peer configuration. An example of a BGP peer debug configuration is shown here.

```
 ...peer/132.245.155.134/132.245.155.34
lxc4(10)# debug-control

 ...debug-control/132.245.155.134/132.245.155.34
lxc4(11)# info
  log-message-type {debug info warning fault trace}
  log-packet-type disabled
  bgp-debug-codes {}

 ...debug-control/132.245.155.134/132.245.155.34
lxc4(12)# log-packet-type {open update notification keepalive}
```

```
...debug-control/132.245.155.134/132.245.155.34
lxc4(13)# info
  log-message-type {debug info warning fault trace}
  log-packet-type {open update notification keepalive}
  bgp-debug-codes {}
```

You can monitor specific message types by setting the appropriate value in the `log-packet-type` attribute. The peer control automatically updates the `bgp-debug-codes` attribute to reflect the event message and packet type configuration. Peer debugging generates a significant number of event messages in the log and should be used for troubleshooting only. It should not be left enabled during normal operation.

```
...debug-control/132.245.155.134/132.245.155.34
lxc4(21)# delete

...peer/132.245.155.134/132.245.155.34
lxc4(22)#
```

BGP Routing Entries

Network layer reachability information (NLRI) exchanged between BGP speakers is maintained in two different tables. The announce pool is the NLRI that the local speaker will send to a remote peer, and the accept pool is the NLRI that was received from a remote peer. In the next section, we discuss routing policies, which play a critical role in the acceptance and announcement of routes to other BGP speakers.

BGP employs a common set of path attributes, of which some are mandatory and others are optional. The well-known attributes are recognized by all conforming BGP implementations, including BayRS. Of the optional attributes, some are transitive (those that are propagated) and others are nontransitive. The following path attributes are commonly used in BGP environments:

- **Origin** This attribute specifies the source of the path information and is well-known and mandatory. A value of IGP indicates that the NLRI originated within the originating AS; a value of EGP indicates that the NLRI was learned using EGP. A value of INCOMPLETE indicates an external route source such as a static route.
- **AS Path** This well-known, mandatory attribute specifies a set or a sequence of autonomous systems that the update message has traversed. Sequenced paths are commonly used to detect routing loops.
- **Next Hop** This is also a well-known, mandatory attribute. It indicates the IP address of the border gateway that will carry data traffic for destinations listed in the update message.
- **Multi Exit Discriminator** The MED is an optional nontransitive attribute. When multiple exit or entry points are available to a neighboring (external) AS, this 32-bit unsigned integer is used to selectively prefer the connection with the

lowest metric. The MED is optionally propagated within the AS to internal BGP speakers.

- **Local Preference** This well-known, optional attribute is a 32-bit unsigned integer, like the MED. It indicates the degree of preference allocated to the specified route by the BGP speaker and propagated to its internal peers. The local preference value is not propagated to peers in neighboring autonomous systems.
- **Atomic Aggregate** This is another well-known, optional attribute. The local BGP speaker indicates to other BGP speakers that the local router has chosen a less specific route to the destination.
- **Aggregator** The aggregator is an optional, transitive attribute. A 48-bit value, it contains the last AS (16 bits) that aggregated the destination route, and the BGP router ID (32 bits) in the AS that aggregated the route.

BGP routes received from a remote peer are first processed using a set of configured accept policies. These accept policies specify which routes may be accepted by the RTM into the final routing table, and applies modifiers, if there are any. For example, suppose the local BGP speaker receives 1,000 routes from one remote peer. These 1,000 routes are stored in the accept pool for that peer, and the RTM enters all or a selected number of routes from this pool into the final routing table.

BGP uses classless routes and relies on special attributes not otherwise seen in other protocols. You can use the **show bgp routes** command to display the BGP routing table that is submitted to the RTM. From the TI, enter the **ip bgp_routes** command to display the same BGP routing table. When you use the **show ip routes** command, the AS column indicates the next hop BGP AS. The **show bgp routes** command supports various options, depending on the software version. An example of the command output is shown here.

```
fxc9(12)# show bgp routes
show bgp routes                                      Jun 08, 2000 03:17:41 [GMT]

Network/Mask         Peer Rem Addr    NextHop Address Org Loc Pref   B/U Sl
------------------ --------------- --------------- --- ---------- --- --
0.0.0.0/0            152.43.245.102  152.43.245.102  INC       4079 B/U  9
        AS Path: <764 762>
10.0.0.0/8           152.43.245.102  152.43.245.102  INC       4079 B/U  9
        AS Path: <764 762>
10.1.1.0/24          152.43.245.102  152.43.245.102  INC       4087 B    9
        AS Path: <764>
10.10.10.0/24        152.43.245.102  152.43.245.102  INC       4087 B    9
        AS Path: <764>
11.11.11.0/24        152.43.245.102  152.43.245.102  INC       4087 B    9
        AS Path: <764>
12.12.12.0/24        152.43.245.102  152.43.245.102  INC       4087 B    9
        AS Path: <764>
57.0.0.0/8           152.43.245.102  152.43.245.102  INC       4079 B/U  9
```

```
        AS Path: <764 762>
57.17.20.8/30      152.43.245.102  152.43.245.102  IGP       8183 B     9
        AS Path: <764>
57.17.20.12/30     152.43.245.102  152.43.245.102  IGP       8183 B     9
        AS Path: <764>
57.17.20.16/30     152.43.245.102  152.43.245.102  IGP       8183 B     9
        AS Path: <764>
57.17.20.20/30     152.43.245.102  152.43.245.102  IGP       8183 B     9
        AS Path: <764>
57.17.20.24/30     152.43.245.102  152.43.245.102  IGP       8183 B     9
        AS Path: <764>
57.17.20.28/30     152.43.245.102  152.43.245.102  IGP       8183 B     9
        AS Path: <764>
57.17.180.0/24     152.43.245.102  152.43.245.102  IGP       8183 B     9
        AS Path: <764>
57.17.181.0/24     152.43.245.102  152.43.245.102  IGP       8183 B     9
        AS Path: <764>
57.17.183.0/24     152.43.245.102  152.43.245.102  IGP       8183 B/U   9
        AS Path: <764>
57.17.184.0/24     152.43.245.102  152.43.245.102  IGP       8183 B/U   9
        AS Path: <764>
57.17.185.0/24     152.43.245.102  152.43.245.102  IGP       8183 B/U   9
        AS Path: <764>
57.17.186.0/24     152.43.245.102  152.43.245.102  IGP       8183 B/U   9
        AS Path: <764>
57.17.187.0/24     152.43.245.102  152.43.245.102  IGP       8183 B/U   9
        AS Path: <764>
57.17.188.0/24     152.43.245.102  152.43.245.102  IGP       8183 B/U   9
        AS Path: <764>
152.43.5.16/28     152.43.245.102  152.43.245.102  IGP       8183 B     9
        AS Path: <764>
152.43.5.32/28     152.43.245.102  152.43.245.102  IGP       8183 B     9
        AS Path: <764>
152.43.5.48/28     152.43.245.102  152.43.245.102  IGP       8183 B     9
        AS Path: <764>
152.43.5.64/28     152.43.245.102  152.43.245.102  IGP       8183 B     9
        AS Path: <764>
152.43.5.80/28     152.43.245.102  152.43.245.102  IGP       8183 B     9
        AS Path: <764>
152.43.5.96/28     152.43.245.102  152.43.245.102  IGP       8183 B     9
        AS Path: <764>
152.43.7.16/28     152.43.245.102  152.43.245.102  IGP       8183 B     9
        AS Path: <764>
152.43.14.0/24     152.43.245.102  152.43.245.102  INC       4087 B     9
        AS Path: <764>
152.43.15.0/24     152.43.245.102  152.43.245.102  INC       4087 B/U   9
        AS Path: <764>
... ...
```

Routing Policies

BGP is a complex protocol, and several components have to interact properly for its successful operation. BGP uses policy-based routing, employing specific policies to accept and announce routes from peers. Unlike protocols such as RIP and OSPF that implicitly accept routes from their own autonomous systems, simply establishing a peer session is not sufficient for BGP operation; specific policies must be configured to allow the BGP protocol to interact fully with the IP routing table manager.

Figure 6-12 shows the roles of BGP accept and announce policies. BGP does not have any default accept or announce policies. In the default configuration without an accept policy, routes received from a specific BGP peer (`show bgp routes -i peer_router_id`) will not be entered in the IP routing table (**show ip routes**) because no accept policy is configured. Similarly, the BGP announce pool (`show bgp routes -N`) containing the NLRI being propagated to a remote peer will be empty, because there is no announce policy configured by default. The lack of a proper announce policy is one of the most frequent oversights in BGP configuration.

```
#  147: 07/02/1999 00:11:10.097  DEBUG    SLOT  3  BGP                    Code: 185
bgp_update: net (204.14.4.0/22) failed policy check
```

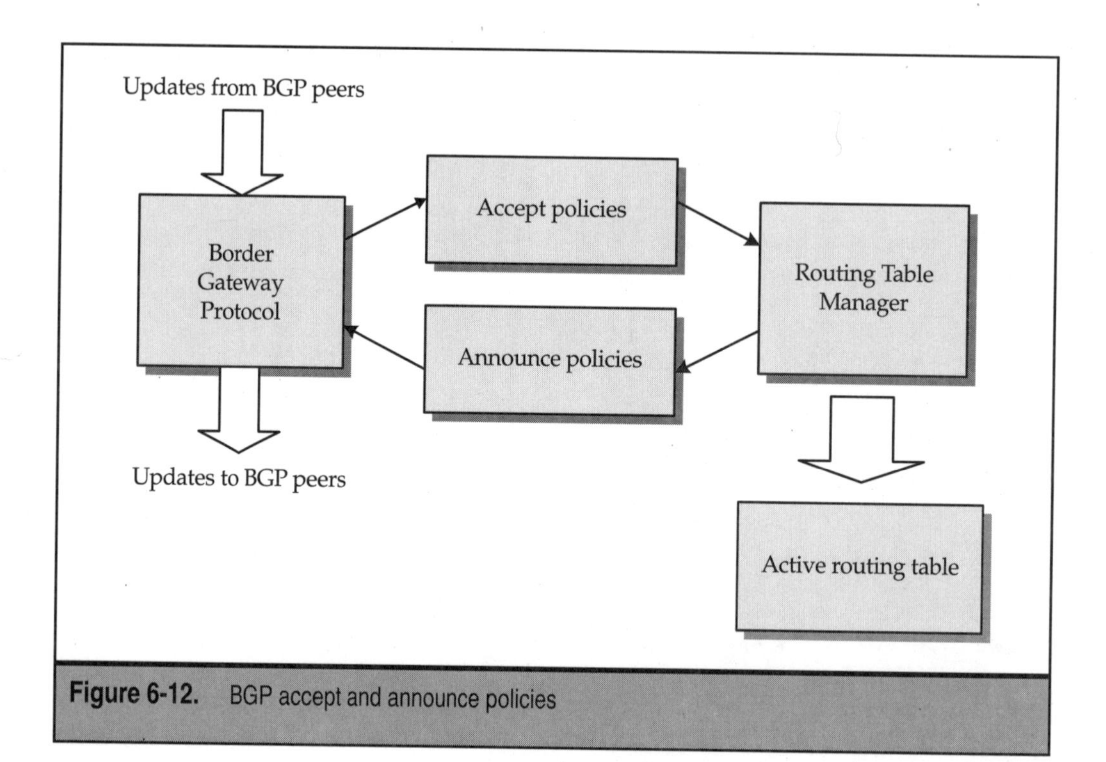

Figure 6-12. BGP accept and announce policies

Another common problem is the misconfiguration of accept or announce policies, preventing them from working correctly. An event message such as the preceding example is logged when BGP peer debugging is enabled for update messages. Each network in an update message is analyzed against all applicable accept or announce policies—it is reported to have failed the policy check when either no policies apply or the corresponding action is inapplicable.

BGP and OSPF Interaction

As we discussed earlier, BGP interacts with other autonomous routing systems, and one of the most frequently used interior gateway protocols (IGPs) is OSPF. OSPF also naturally supports CIDR, and the interaction of BGP and OSPF is defined in RFC 1403. Using both OSPF and BGP on a router requires certain considerations:

- Both the OSPF router ID and the BGP identifier must be valid IP addresses on the router.
- The OSPF router ID must be equal to the BGP identifier. It is common practice to use the circuitless (virtual) IP address on the router as this value.

Although primarily applied to OSPF and BGP, the concept of accept and announce policies generally also applies to other routing protocols. BGP accept policies affect the entry of BGP routes in the IP routing table maintained by the RTM. However, by default, routes learned via BGP are not advertised to the IGP system. Even though the RTM, using a BGP accept policy, accepts the BGP routes into the routing table, the BGP global intra–AS routing parameter (`intra-as-routing`) must be disabled for the route to be subsequently entered in the IGP.

```
box->ip->bgp
fxc9(14)# intra-as-routing disabled
```

When the IGP used is OSPF, recall that OSPF by itself behaves as an autonomous system. OSPF treats BGP routes as external routes. Hence, in addition to disabling the BGP intra-AS routing parameter, the OSPF global AS boundary router attribute (`as-boundary-router`) must be set to true.

```
box->ip->ospf
fxc9(20)# as-boundary-router true
```

Static Routing

Using static routes is often preferable to using a routing protocol on remote office routers. For example, an Access router such as an AN installed in a branch office may be configured to support two IP subnets. Running a routing protocol such as RIP over a PPP serial connection may be unnecessary because the reachable destinations can be statically defined at the hub router connected to the AN. On the branch office AN, you can configure a default route that is a comprehensive static route applicable to all destinations.

Static routes are defined using the `static-route` object under the global `ip` context. Misconfiguration of a static route is often the cause of routing problems, as the following event message indicates. Such messages usually indicate that the next hop address resides in a subnet different from the one that the local router resides in, or the network mask is not applicable to the destination.

```
#  136: 04/19/1999 14:40:39.238  WARNING  SLOT  6  IP                Code:  56
Error when configuring static route 172.16.147.0/255.255.255.0/4
```

In some cases, the global IP attribute `all-subnets` must be enabled if the destination is a nonstandard subnet using the all-zeros or all-ones subnet number.

```
#  173: 05/12/1999 19:01:48.067  WARNING  SLOT  5  IP                Code:  57
Subnet zero disabled, not allowed for this static route
172.16.0.0/255.255.255.0/12
```

The default preference for static routes is 16. This normally makes the IP RTM by default prefer a static route over other route sources that have a lower preference value. In most cases, static routes are configured as access routes of the last resort. In these configurations, the `static-route` objects must be configured with a preference of 1 so that routes learned via routing protocols are preferred. However, in the event that routes from routing protocols disappear from the routing table, the static route will automatically assume the role of the preferred route.

Equal-Cost Multiple Paths

Route preferences and weights play a key role in the best route selection process by the IP Routing Table Manager. In many redundant environments, multiple routes with the same metric can be learned for the same destination. If all other attributes are equal, these routes are called equal-cost multiple paths (ECMPs). Normally, the router chooses the first ECMP route that was learned.

The redundant ECMP routes can be used for load balancing by enabling ECMP routing on the router. The global IP parameter `ecmp-method` is disabled by default and must be set to one of the following values for ECMP routing to take place:

- **round-robin** All available routes to the given destination are used in a round-robin fashion. This option allows greater usage of all routes; however, it increases the likelihood that packets will arrive out of order.
- **src-dest-hash** A hash value computed using the source and destination IP addresses yields a single next hop that is used for all communication between the two devices. This option allows packet delivery in order, but may not fully use the network resources.

- **dest-hash** The path is selected using the destination IP address only, and all data destined for a specific host uses the same next hop. This option is compatible with RSVP.

ECMP routes are entered into the routing table on a per-protocol basis, and the maximum number of routes submitted by RIP is set using `rip-max-paths`, and the maximum number of routes submitted by OSPF is set using `ospf-max-paths`. Both attributes are configured under the global IP context. When multiple equal-cost paths are available, they are entered in the routing table up to the count limit imposed by the RTM.

The ECMP routing technique can also be applied to BGP connections, whether IBGP or EBGP. The BGP global parameter `ibgp-ecmp-method` and the peer parameter `ebgp-ecmp-method` can take one of three values: route-balance, traffic-balance, or disabled (default). Route balancing distributes a group of destination networks that share the same next hop among all available ECMP routes. In this case, only one route-balanced next hop is entered into the routing table. With traffic balancing, all available routes to the same destination are simultaneously used to share load across all routes. In this case, multiple routes appear in the routing table. When you use traffic balancing, a significant number of ECMP routes may be entered in the routing table, so you need to take care to manage them.

INTERNETWORK PACKET EXCHANGE

The Internetwork Packet Exchange (IPX) protocol is the basic connectionless service used by the Novell NetWare network operating system. The NetWare model has some of the features of the Xerox Networking System (XNS). It is both similar to and different from the OSI model, which it predates. The IPX protocol layer in NetWare can be considered the equivalent of the network layer in the OSI model.

The Sequenced Packet Exchange (SPX) protocol provides connection-oriented transport services, similar to the Transmission Control Protocol in the TCP/IP suite. A NetWare network typically uses the Routing Information Protocol (RIP) to advertise network routes and the Service Advertisement Protocol (SAP) to advertise services. IPX network numbers are 32-bits long, and a host address consists of the network number plus the MAC address in the format *network.host*. Multiple networks are interconnected with an IPX router that uses both RIP and SAP to conform to the NetWare model.

Client-Server Communication

Traditionally in NetWare environments, Novell servers were installed in most LANs so that clients could directly connect to them. The servers also ran the multiprotocol routing (MPR) software and interconnected multiple networks. With the advent of cost-effective and efficient IPX routers, servers were centralized in reliable server farms. IPX client segments became numerous and were handled by IPX routers such as BayRS.

A typical NetWare client follows a well-established connection sequence to communicate with an IPX server. This is illustrated in Figure 6-13, where the client and server are multiple hops away, with BayRS routers interspersed. This is the common procedure followed by a client that does not have a preferred server configured on it—clients with a preferred server have a slightly different connection sequence, but the underlying principle is the same. Understanding the basic client-server communication process is useful in troubleshooting NetWare connection and login problems.

The connection sequence is as follows:

1. The client generates a get nearest server (GNS) request using the SAP protocol. The destination for this request is the broadcast address. Normally, the GNS request is for the file server logon (SAP type 4).
2. Either a local server or a router answers the GNS request directly. The name and network address of an appropriate server from the service (SAP) table is provided to the client. In the case of an IPX router, the service table is constructed using SAP protocol advertisements.
3. The client sends an IPX RIP request for the destination network on which the nearest server resides. The RIP protocol is used for this purpose.
4. A server or a router responds with the next hop address to reach the destination network. BayRS routers listen to RIP requests and maintain their routing tables using periodic RIP advertisements. The router normally provides its own host address as the next hop if it is the only routing device on the specific IPX

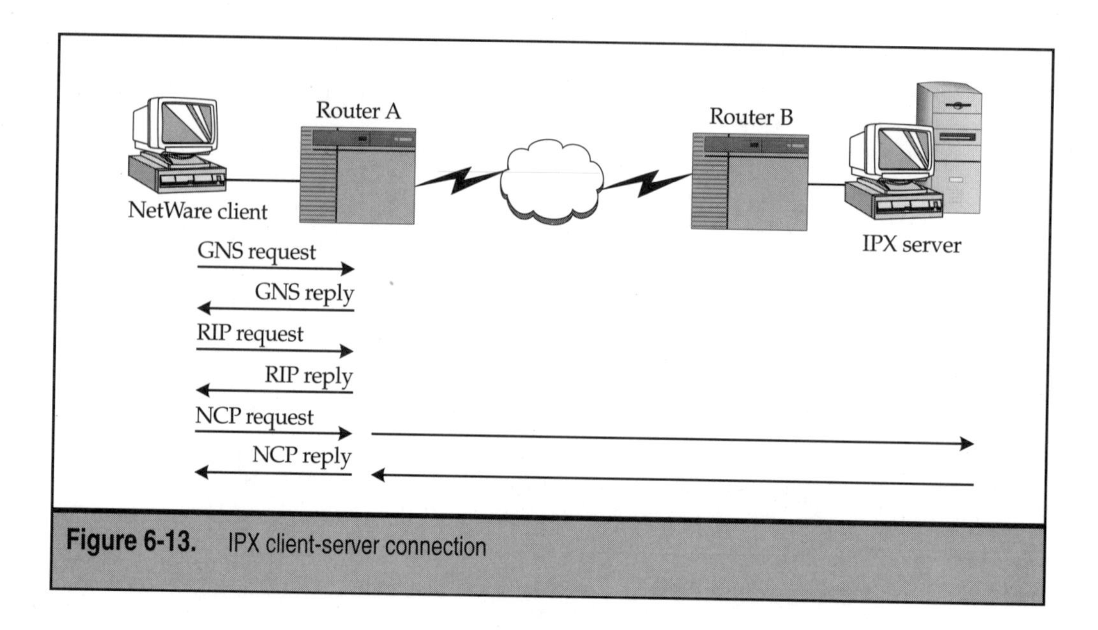

Figure 6-13. IPX client-server connection

network. At this point, the client knows the name of the nearest server, the network the server resides in, and the route to reach the server.

5. The client generates a NetWare Core Protocol (NCP) create connection request directly to the IPX server, using the NCP protocol.
6. Higher-layer negotiations take place between the client and server, primarily the checking of the NetWare version and the maximum packet size to use for the session. The NCP frames are routed directly by the IPX router.

IPX Ping

The IPX ping is a very useful tool for testing IPX connectivity. It is similar in function to the IP ping, but has several operational differences. When an IPX ping is issued, the router sends an IPX configuration request packet to the remote device on socket 0x456. If the device responds, the router considers the connection to be valid. BayRS routers also listen to and respond to pings based on the NetWare Link State Protocol (NLSP); however, they currently do not originate pings based on NSLP.

The syntax of the IPX ping is shown here, followed by an example.

```
ping -ipx network.host [-t timeout] [-r count]
```

```
1m4(24)# ping -ipx 0x100E.0xA2C95E
IPX ping: 0x100E.0xA2C95E is alive
```

NOTE: All NetWare devices, particularly some older clients, may not support IPX pings. If a ping to a known IPX client fails, try pinging a server or another client on the same network to test network connectivity.

The *network* portion is a 4-byte hexadecimal address, and the *host* portion is a 6-byte hexadecimal address, the latter commonly being equal to the MAC address of the destination. When pinging the internal network number of a server, the *host* portion is usually 0x1, with the MAC address used only on the external network. Leading zeros in the *network* and *host* values can be omitted, and the numbers must be prefixed with a 0x because they are hexadecimal values.

NOTE: On Token Ring LANs, the host portion of the destination address must be in byte-swapped (noncanonical) form.

If a route is not available to the destination network where the IPX device resides, then the router will return an unreachable message. In this case, you should review the routes propagated using RIP, and you may need to add a static IPX route from the destination network.

Encapsulation

NetWare networks use multiple encapsulation types, and IPX routers generally support them. BayRS supports all encapsulation types used by NetWare, although the terminology used is slightly different. The encapsulation type on any IPX network must match the value configured on all servers and stations on the same network. Table 6-2 lists the various encapsulation types supported by BayRS, with corresponding NetWare terms.

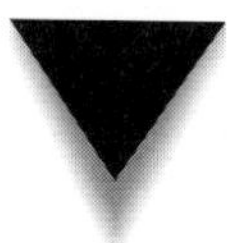

NOTE: NetWare versions earlier than 3.12 by default used the Ethernet_802.3 (Raw) encapsulation on Ethernet networks, called the Novell encapsulation in BayRS. Versions 3.12 and later (including 4.*x*) by default use a frame format of Ethernet 802.2. Some recent client implementations support the ability to autosense the encapsulation type by attempting GNS queries with all supported frame types.

Some common installation problems with IPX arise from encapsulation type mismatches. If a BayRS router is configured with an encapsulation type different than that of the servers on the same segment, then the router will never learn the routes and services

BayRS	NetWare
Ethernet Media	
Ethernet	Ethernet_II
LSAP	Ethernet_802.2 (LLC)
Novell	Ethernet_802.3 (Raw)
SNAP	Ethernet_SNAP
Token Ring Media	
LSAP	Token_Ring
SNAP	Token_Ring_SNAP
FDDI Media	
LSAP	FDDI_802.2
SNAP	FDDI_SNAP

Table 6-2. IPX Encapsulation Types

on the interface. If a newly provisioned IPX interface is operationally up but is not providing routing services, then check the encapsulation type using the **show ipx interfaces** command.

```
show ipx interfaces                                    Apr 27, 2000 06:39:59 [GMT-6]

IPX Interface Configuration Information

Interface Configured Current Network     Host              Encapsulation
Name      State      State   Address     Address           Method
--------- ---------- ------- ----------- ----------------- --------------
E11       enabled    up      00.00.10.0E 00.00.A2.1C.DA.8F ethernet
S21       enabled    up      00.00.40.0F 00.00.A2.D2.C9.52 ppp
E12       enabled    up      00.00.10.0A 00.00.A2.1C.DA.90 novell
... ...
```

Multiple IPX networks can be overlaid on the same physical network. However, each of these networks must have a unique network number and a different encapsulation type. For example, IPX network 0x1000100E with Ethernet encapsulation and network 0x1000100F with Novell encapsulation can be configured on the same physical Ethernet network. The frame format types for the supported encapsulation types are depicted in Figure 6-14. Many IPX clients support the autosensing of the encapsulation type—they sequentially attempt to connect using different encapsulation types until a successful response is returned. This automatic sensing of the encapsulation type may fail if the timeout value on the client is set too low. If the hub or switch providing connectivity to the client is running the spanning tree protocol (STP), then the port may be blocked, initially preventing the forwarding of the first few frames and also preventing connecting to the IPX server. When an IPX client is unable to connect to the server, try manually setting the encapsulation type supported by either the server or the router on the local LAN segment.

Routes and Services

IPX routes are advertised to the RIP destination socket 0x453, and services are advertised to the SAP destination socket 0x452. Like the IP RTM, the IPX RTM maintains IPX routes. Services are likewise maintained by the IPX Service Table Manager (STM). Routes are entered in the routing table when a neighboring router or local server announces remote or internal routes, respectively.

Routing decisions are made on the basis of the tick count, which is an IPX metric measuring the time in ticks (one tick equals 1/18 of a second) required to reach a remote network. Generally, LAN networks use a tick value of 1, and WAN networks use a tick value

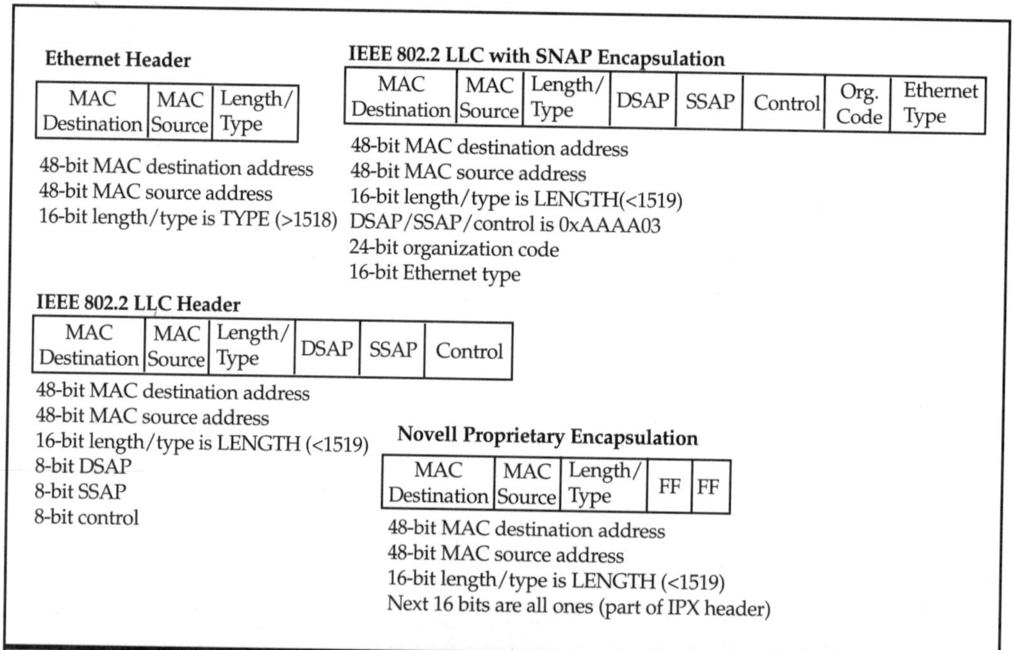

Ethernet Header

MAC Destination	MAC Source	Length/Type

48-bit MAC destination address
48-bit MAC source address
16-bit length/type is TYPE (>1518)

IEEE 802.2 LLC with SNAP Encapsulation

MAC Destination	MAC Source	Length/Type	DSAP	SSAP	Control	Org. Code	Ethernet Type

48-bit MAC destination address
48-bit MAC source address
16-bit length/type is LENGTH(<1519)
DSAP/SSAP/control is 0xAAAA03
24-bit organization code
16-bit Ethernet type

IEEE 802.2 LLC Header

MAC Destination	MAC Source	Length/Type	DSAP	SSAP	Control

48-bit MAC destination address
48-bit MAC source address
16-bit length/type is LENGTH (<1519)
8-bit DSAP
8-bit SSAP
8-bit control

Novell Proprietary Encapsulation

MAC Destination	MAC Source	Length/Type	FF	FF

48-bit MAC destination address
48-bit MAC source address
16-bit length/type is LENGTH (<1519)
Next 16 bits are all ones (part of IPX header)

Figure 6-14. IPX encapsulation frame formats

of 6. The value ultimately depends on estimates provided by multiprotocol routers and servers. By default, BayRS uses the tick count to choose a destination route. If multiple destination routes with the same tick count are available, the route with the lower hop count is used. The hop count is the number of logical network hops that the packet must traverse to reach the destination network.

When multiple equal-metric paths are available, IPX can use multipath routing and perform load sharing across the available routes. Normally, RIP uses only one of the routes, even if the tick value and hop count are the same, with the preferred route usually being the first-learned route. Figure 6-15 shows IPX multipath routing, which can be enabled by configuring the `maximum-path` attribute for the global `ipx` context. Use the **show ipx routes** command to display the IPX routing table.

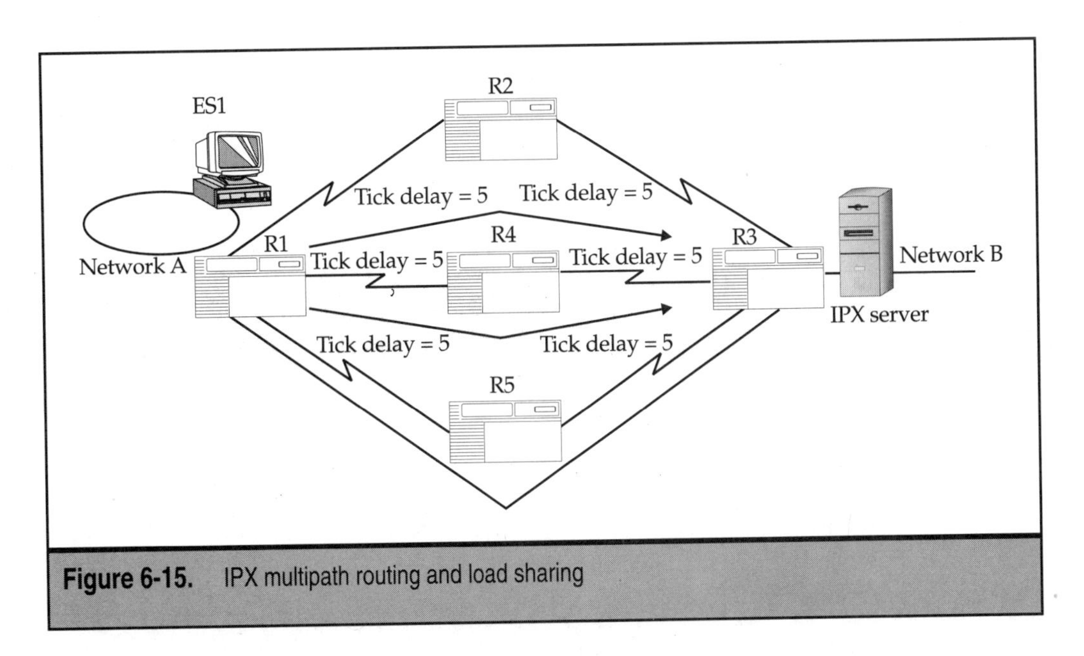

Figure 6-15. IPX multipath routing and load sharing

```
show ipx routes [-dest <arg>] [-net <arg>] [-host <arg>] [-type <arg>]

fxc3(21)# show ipx routes
show ipx routes                                       Apr 04, 2000 14:12:29 [GMT-6]

IPX Routing Table Information

Dest         Interface Nexthop     Nexthop
Network      Name      Network     Host Address      Type   Age    Ticks  Hops
-----------  --------- ----------- ----------------- ------ ------ ------ -----
40.00.10.0E E31        40.00.10.0E 00.00.A2.CB.B5.8C local  2140   1      0
10.AA.4D.8A E31        40.00.10.0E 00.00.A2.1C.A4.2C rip    15     3      2
10.AA.00.14 E31        40.00.10.0E 00.00.A2.1C.A4.2C rip    15     3      2
12.AA.4C.5C E31        40.00.10.0E 00.00.A2.1C.A4.2C static 320    1      0
40.00.25.01 F11        40.00.25.01 00.00.A2.CB.B5.80 local  2120   1      0
10.AA.4D.92 F11        40.00.25.01 00.00.A2.1C.52.91 rip    25     3      2
10.AA.4D.9F F11        40.00.25.01 00.00.A2.1C.52.91 rip    25     3      2
10.AA.4D.9A F11        40.00.25.01 00.00.A2.1C.52.91 rip    25     3      2
10.AA.4C.2B F11        40.00.25.01 00.00.A2.1C.52.91 rip    25     3      2
... ...
```

Servers periodically announce all the IPX services they offer, such as file and print services. They are advertised on the server's external networks, and routers residing on the same network store these services in the SAP table. Routers connected to other LAN and WAN segments rebroadcast these reachable services to other routers.

NOTE: Services are entered in the SAP table when a reachable route exists to the network where the service resides. If a usable route to the service network is not available, BayRS considers the service to be unavailable.

```
#  174: 10/13/1999 00:30:13.796  TRACE    SLOT 10  IPX                    Code:   9
IPX Service no longer available, net 00.aa.5e.9d down:
  Internal Net: 00.00.10.0e Name: MARKETING001 type: 4.
```

The **show ipx services** command displays all usable services in the SAP table maintained by the IPX STM. The **show ipx service-address** command provides both the network number and the host number needed to reach all services. Both commands support optional arguments to filter or search for specific entries. You can ping the service address to test the reachability of the service.

```
show ipx services [-net <arg>] [-socket <arg>] [-type <arg>] [-name <arg>]

fp3(12)# show ipx services
show ipx services                                 Jun 02, 2000 17:43:19 [GMT-6]

IPX Service Table Information

                  Server Name                        type  Network     Age  Hop Sock
-----------------------------------------------      ----- ----------- ---- --- ----
ADMIN-001                                            00.04 00.00.10.4C 15   2   04.51
EMERGENCY-001                                        00.04 00.00.10.4A 15   2   04.51
EMERGENCY-BKP                                        00.04 00.00.10.4E 15   2   04.51
CARDIOLOGY-001                                       00.04 00.00.10.5F 15   2   04.51
RADIOLOGY-001                                        00.04 00.00.10.C2 20   3   04.51
ICARE-001                                            00.04 00.00.10.C5 20   3   04.51
PRINTSRV-001                                         00.47 00.00.10.4D 15   2   80.60
... ...

show ipx service-address [-host <arg>] [-net <arg>] [-name <arg>]

mx2(25)# show ipx service-address
```

```
show ipx service-address                          Jun 02, 2000 17:43:19 [GMT-6]

IPX Service Address Information

                  Server Name                     Network     Host Address
------------------------------------------------ ----------- -----------------
ADMIN-001                                         00.00.10.4C 00.00.00.00.00.01
EMERGENCY-001                                     00.00.10.4A 00.00.00.00.00.01
EMERGENCY-BKP                                     00.00.10.4E 00.00.00.00.00.01
CARDIOLOGY-001                                    00.00.10.5F 00.00.00.00.00.01
RADIOLOGY-001                                     00.00.10.C2 00.00.00.00.00.01
ICARE-001                                         00.00.10.C5 00.00.00.00.00.01
PRINTSRV-001                                      00.00.10.4D 00.06.0D.13.CE.82
... ...
```

We have mentioned before that multiple IPX networks can be provisioned in the same physical LAN, provided that their encapsulation types are unique. Occasionally, servers or routers are misconfigured so that the same encapsulation is used for more than one network address. This causes confusion, and the mismatched network address is reported in the event log. These events are reported for both RIP and SAP packets received on an IPX network interface.

```
#   13: 01/08/1999 21:05:56.307  WARNING  SLOT  5  IPX              Code:  12
IPX received a rip packet on interface 00.00.10.aa.0000a251dc7f with dest
net of 00.00.10.ab.ffffffffffff, src net of 00.00.10.ab.00c40118df4c,
packet discarded.

#   94: 10/17/1999 07:50:45.996  WARNING  SLOT 12  IPX              Code:  13
IPX received a sap packet on interface 00.4c.01.b0.0000a29ce4dd with dest
net of 00.4c.01.dd.ffffffffffff, src net of 00.4c.01.dd.00c4016faa49,
packet discarded.
```

Observe that the RIP and SAP frames are from a valid network and are destined for the broadcast address on the same network. However, the frames were received on a different network interface on the router. The host address displayed in the warning message is useful in locating the device that originated the RIP/SAP frames. Sometimes, when this device is located, you may find that it resides in the proper network segment. In this case, a backdoor bridge or dual-attached server acting as a router may be interconnecting these networks. If you find that the server configuration is correct and that the server resides in the proper network, you will need to isolate and remove the backdoor loop

Triggered Updates

IPX RIP and SAP protocols both can generate immediate updates when a route changes or becomes unreachable. Commonly called triggered updates, these update frames allow a network change to quickly propagate throughout the network. This dramatically decreases convergence time, especially in the case of a route loss. By default, a route update is sent one-half second after a change is detected, and service loss triggers an immediate update without any delay.

By default, triggered updates are enabled globally for the IPX protocol. IPX RIP and SAP interfaces automatically inherit the `triggered-update` values from the global object. An IPX network can become unreachable if a directly attached network goes down or if an externally triggered unreachable route is received. The route is first removed from the routing table by the IPX RTM.

```
#     5: 09/16/1999 10:57:36.899  TRACE    SLOT  5  IPX                Code:   6
IPX Network removed from Table of Networks.
  Net: 00.c4.90.aa NextHopHost: 00.c4.4a.dd.0000a2096dc0
```

Once a route is removed, this change is immediately propagated across all operational RIP interfaces. When a route is removed, all services relying on the route are also removed.

```
#     7: 09/16/1999 10:57:36.923  TRACE    SLOT  5  IPX                Code:   9
IPX Service no longer available, net 00.c4.90.aa down:
  Internal Net: 00.00.40.a1 Name: LOGISTICS01 type: 4.

#     7: 09/16/1999 10:57:36.928  TRACE    SLOT  5  IPX                Code:   9
IPX Service no longer available, net 00.c4.90.aa down:
  Internal Net: 00.00.40.a4 Name: DISPATCH01 type: 4.
```

Triggered updates provide significant benefits while retaining the simplicity of the RIP and SAP protocols. On redundant and meshed topologies, however, triggered updates can cause problems if they are not designed properly. When triggered updates are enabled on redundant interfaces, updates sent by one router are received and resent across other interfaces. Updates received on multiple interfaces simultaneously can cause multiple rebroadcasts and a temporary update loop. If numerous routes and services become unavailable quickly, this can cause a transient stability problem. In redundant configurations, you must disable triggered updates on interfaces where an update loop would otherwise occur. Figure 6-16 illustrates the behavior when triggered updates are configured between two routers and the configuration needed for proper behavior.

Route and Service Filters

IPX networks in large enterprise environments can span numerous locations and may consist of hundreds or thousands of routes and services. Not all servers communicate with each other, and users typically connecting to one group of servers may not connect to others. For example, a group of users may communicate with all marketing servers but never communicate with accounting servers. RIP and SAP are system functions, and processor involvement is required to generate and process these frames. A maximum of 25 routes can be sent in one RIP frame, and a maximum of 7 services can be included in one SAP frame.

Intranets with several hundred routers may not need to maintain extensive RIP and SAP tables for the entire network. SAP in particular consumes significant overhead due to the small number of services that can be included in one packet. On low-speed wide-area links,

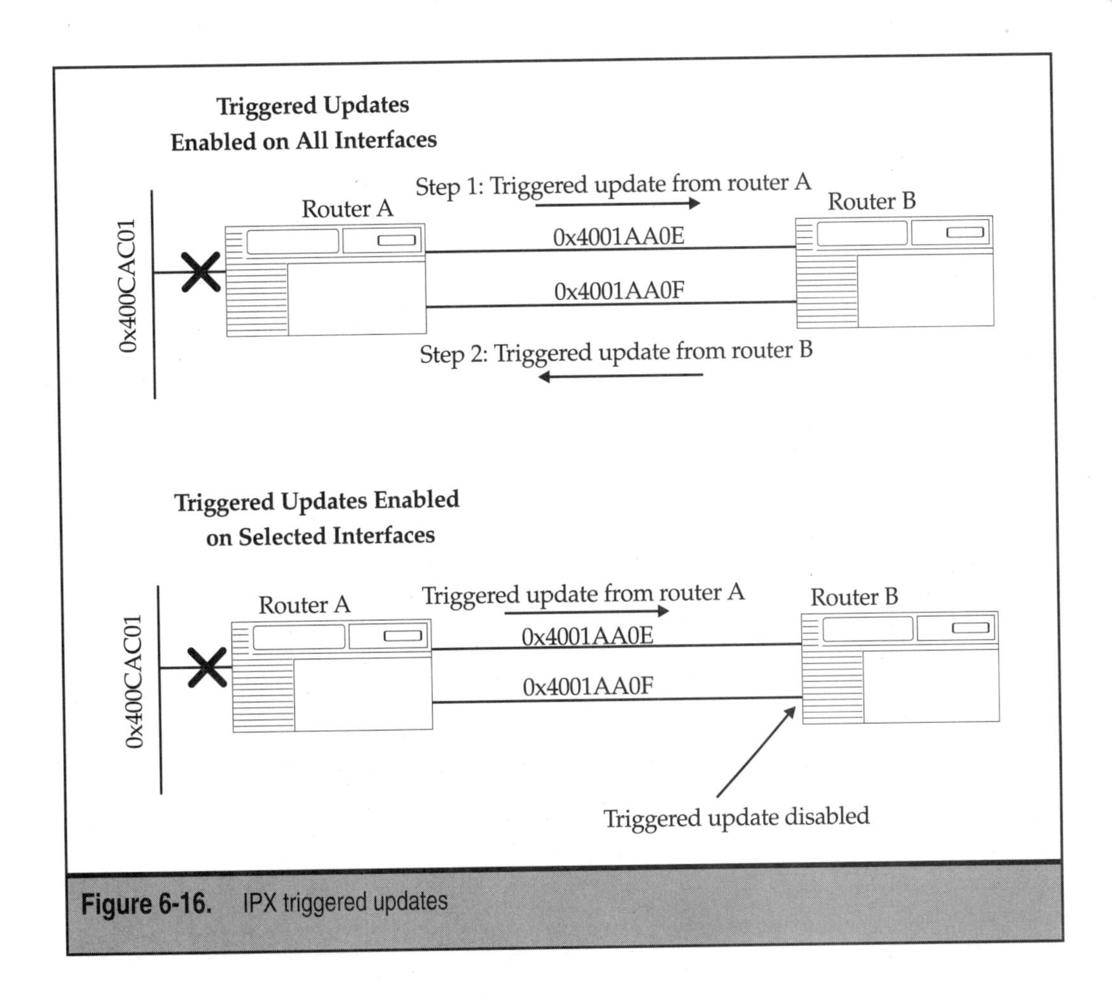

Figure 6-16. IPX triggered updates

RIP and SAP broadcasts can consume significant bandwidth, thus preventing efficient use of the link. RIP and SAP filters can be used in these scenarios to limit route and service entries to a specific set of routers.

Two types of SAP filters are available in BayRS:

- **Service network filter (service-network-filter)** This is used to filter SAPs based on the service type and service network address.
- **Service name filter (service-name-filter)** This is used to filter SAPs based on the service type and service name.

Filters can be applied on inbound or outbound SAP frames, or both. Inbound filters can be used to prevent the local router from learning certain services, and outbound filters can be used to prevent the router from advertising certain services. Filters perform one of three actions (`action`): advertise, accept, or suppress. The suppress action eliminates the service on the inbound or outbound service advertisement; the advertise action applies to outbound SAP frames, and the accept action applies to inbound frames.

Route filters are configured using the `route-filter` object. The filter is applied using a network address and a mask combination and can be applied to inbound or outbound frames. The suppress, advertise, and accept actions can be configured and are mutually exclusive. Networks that become unwieldy due to excessive routing or service table entries must be configured with RIP/SAP filters to optimize performance.

PACKET CAPTURE AND ANALYSIS

Throughout this book, we have used the event log and show commands to understand symptoms and analyze network conditions. On occasion, exact protocol frames may be needed to gain a detailed understanding of a problem. Several types of tools are available for this purpose. Handheld network analyzers can be attached to a network segment to monitor network traffic. More sophisticated products such as sniffers, dynamic and full-fledged protocol analyzers, and service accounting applications can be used to capture frames for comprehensive analysis. The Packet Capture (PCAP) application in BayRS facilitates the capture of frames on an interface and saves this data for offline analysis. A major benefit of the PCAP application is its nonintrusiveness—most other products require external adapters to monitor an interface, and these adapters must be installed during a scheduled downtime. Also, installation of specific adapters and drivers for each media type can be avoided by directly using PCAP.

PCAP Concepts and Lines

In Chapter 2, we discussed the concepts of lines and line numbers. Each physical and logical line has a line driver associated with it. As illustrated in Figure 6-17, the PCAP application resides above the line driver and monitors frames in both directions. When enabled, PCAP performs the following functions:

- Replicate frames in either the inbound or outbound direction, or both.
- Timestamp the replicated frame and store it in a temporary PCAP buffer area.
- Asynchronously pass the frame to the line driver (if it is an outbound frame) or to the data link and network layer (if it is an inbound frame).

When the PCAP application is started, a finite amount of local memory is allocated from the available free memory. After the memory allocation to PCAP, at least 200K of memory must remain free. Otherwise, the GAME operating system will report a warning, and the capture will be canceled. Packet Capture makes an association with a particular data stream by using the line number. Any physical or logical interface having a line number can be easily monitored using the PCAP application.

Most data link and network protocol headers are small and generally occupy the leading 64 bytes. Many network troubleshooting scenarios involve data link and network

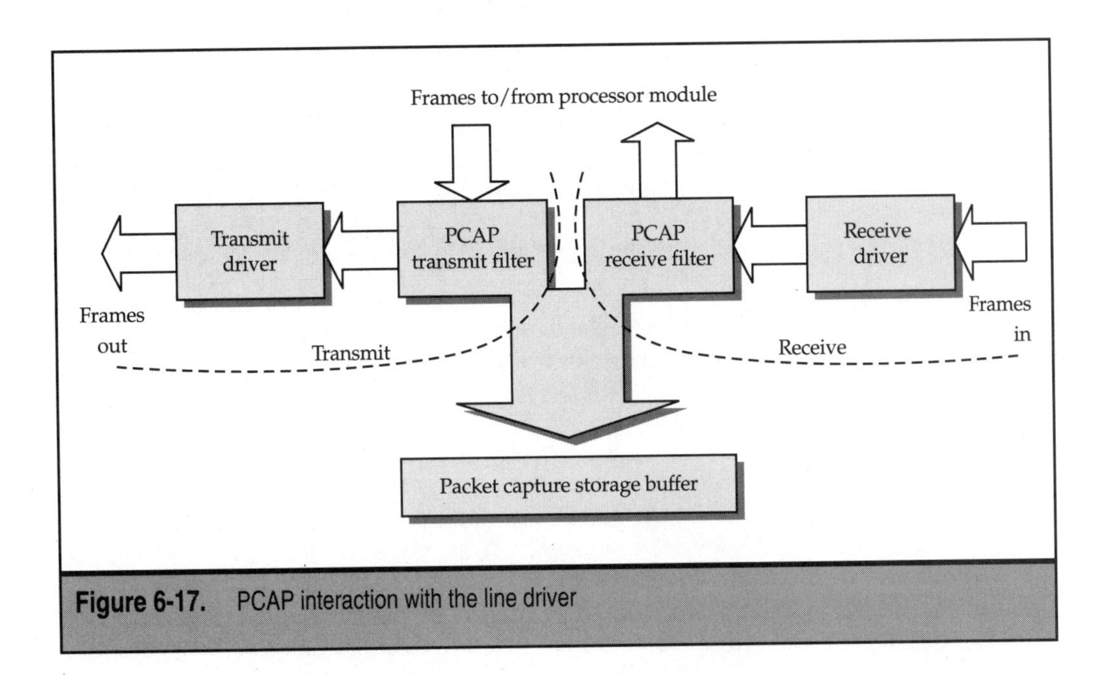

Figure 6-17. PCAP interaction with the line driver

layers and sometimes the higher layers. You usually do not need to capture entire frames because the data contents are rarely relevant in troubleshooting. Since the total memory area available for packet capture is finite, storing the least amount of data necessary from the frame will enable you to capture more frames. For example, with 256K of PCAP memory, 4,096 frames can be stored if the leading 64 bytes of the frames are captured. Alternately, if the captured header data is reduced to the leading 32 bytes, then 8,192 frames can be stored. Both the total PCAP buffer size and the captured packet size play a major role in the ability to capture the frames desired on a line.

Once the total buffer size and the capture packet size are configured, PCAP divides the total allocated local memory by the number of bytes to capture from each frame. The result is a rough estimate of the total number of frames that can be captured with this configuration. PCAP can be configured to terminate in one of three ways:

- **Manual termination** This is the default; PCAP continues to capture packets by overwriting old frames with newer ones.
- **Triggered termination** A particular pattern match on either an inbound or outbound frame can be configured to terminate the PCAP. This approach is desirable when a particular frame indicates the end of the anomalous network behavior.
- **Automatic termination** When the PCAP buffer is full, automatic termination stops the capture. This approach is desirable when a specific set of frame captures is sufficient to identify the problem.

PCAP is configured using the **config packet** command on the TI, which manipulates the `wfPktCaptureEntry` MIB object with an appropriate instance number. PCAP currently cannot be configured using BCC. The instance number of the MIB object is the number of the line on which PCAP will run. Several mandatory attributes must be configured for PCAP to operate successfully. Additional attributes can be configured for advanced filtering and triggers. The PCAP/TAP graphical user interface can also be used to configure, save, and view packet captures using a Windows workstation.

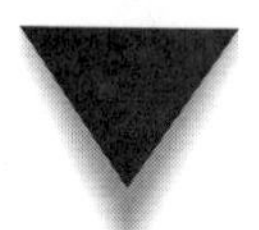

NOTE: PCAP/TAP is a free, downloadable GUI application available from the Router Management Labs web site, at http://www.nortelnetworks.com/servsup/rml. It is available as part of the Optivity network management suite for both UNIX and Windows workstations. The current version of PCAP/TAP is 3.0.

Once PCAP is initialized, it begins to capture frames in the storage buffer. Frames in this storage buffer can be accessed and viewed only after PCAP has been stopped in one of the ways described earlier. After the PCAP is stopped, you can use the **pktdump** command to display the frames in memory as hexadecimal data. Alternately, you can access the storage buffer as a virtual file in memory using TFTP or FTP for retrieval to a PC or UNIX workstation. This file can be saved in PCAP format and then viewed using the TAP application. The saved file can also be converted to the commonly used Network Associates sniffer format using the pktconv application and then viewed on a sniffer.

NOTE: The **pktdump** command for dumping frames offline is available from both the TI and the Site Manager installation directory. The **pktconv** command is available offline from only the Site Manager installation directory.

PCAP Configuration

The first step in provisioning PCAP is to load the Packet Capture application *(pcap)* on the slot with the appropriate line number. Either use the **config packet load** command or set the slotmask for the `wfProtocols.wfPktCaptureLoad` attribute. By default, PCAP is not loaded on any slot automatically. If the **config packet load** command does not specify a slot number, you are prompted for the slot number on which to load the PCAP application.

```
lxc9# config packet load
Enter slot number to load Packet Capture: 3
```

You can then use the **config packet line** command to specify the physical or logical line to capture frames from and a set of capture attributes. The receive and transmit filters are optional and can be used for filtering specific packets for storage or for triggering. The following attributes are required for a successful packet capture:

- **Line number** The line number is necessary for PCAP to attach itself to the appropriate line driver to replicate frames.

- **Capture buffer size** This is the amount of local memory temporarily allocated for the storage of captured frames. By default, no memory is allocated to this capture buffer.
- **Packet capture size** This attribute specifies the number of bytes to capture from each frame. There is no default value for this attribute.
- **Capture direction** By default, PCAP captures only frames in the receive direction. This behavior can be changed so that PCAP captures outbound frames or frames in both directions.

A simple example of a packet capture is shown here. The PCAP application has already been loaded on slot 3 prior to PCAP configuration.

```
lxc9# config packet line


Packet Capture Line Configuration
---------------------------------
Media Types
-----------
1: Ethernet
2: Synchronous or ISDN B Channel
3: T1
4: E1
5: Token Ring
6: FDDI
7: HSSI
9: MCT1
13: ISDN D Channel
15: Bisync.


Enter media type by number: 1

Enter slot number: 3

Enter connector number: 2

The calculated Linenumber is 103102
Please record it for use with other packet commands.


Current available memory in 1Kbyte blocks is 13897
Do not leave less than 200 blocks available unless necessary
```

```
Current value for Capture Buffer size in 1Kbyte blocks is 0
Press return for current value or enter new value: 512

Maximum Packet Save size in 32 Byte blocks is 144

Current value for Packet Save size in 32 Byte blocks is 0
Press return for current value or enter new value: 2

Capture Direction Options
-------------------------
1: Receive
2: Transmit
3: Receive and Transmit

Current value for Capture Direction is 1
Press return for current value or enter new value: 3

Receive Trigger Options
-----------------------
1: Stop Capture when Capture Buffer is Full
2: Stop Capture when Receive Filter One is Matched
3: Stop Capture when Receive Filter Two is Matched
4: No Receive Trigger

Current value for Receive Trigger is 4
Press return for current value or enter new value: 1

Transmit Trigger Options
-----------------------
1: Stop Capture when Capture Buffer is Full
2: Stop Capture when Transmit Filter One is Matched
3: Stop Capture when Transmit Filter Two is Matched
4: No Transmit Trigger

Current value for Transmit Trigger is 4
Press return for current value or enter new value: 1

Packet Capture Receive Capture Filter Configuration
---------------------------------------------------

Receive Filter One Type Options
-------------------------------
```

```
1: Capture
3: Not Used
Current value for Receive Filter One Type is 3
Press return for current value or enter new value:

Packet Capture Transmit Capture Filter Configuration
----------------------------------------------------


Transmit Filter One Type Options
--------------------------------
1: Capture
3: Not Used
Current value for Transmit Filter One Type is 3
Press return for current value or enter new value:

lxc9# set wfPktCaptureEntry.wfPktCaptureControl.103102 1; commit
```

In the preceding configuration, PCAP is set to terminate automatically when the buffer is full. By default, PCAP will continue capturing packets until an explicit termination is issued. Observe that the last command is an MIB set of the `wfPktCaptureControl` attribute, which when, set to 1, starts the packet capture. From the TI, the packet capture can be stopped at any time using the same command, setting it to 2 (stop), as follows:

```
lxc9# set wfPktCaptureEntry.wfPktCaptureControl.103102 2; commit
```

During the packet capture, you can check the current status and the count of packets captured by using the **show packet status** command. With automatic and triggered terminations, the capture state will change dynamically when the buffer is full or when a filter is triggered.

```
lxc9# show packet status

Packet Capture Line Status
--------------------------


Line Number    State       Capture Count
----------- ----------- ------- -----
103102      Up          Started   874
201104      Up          Stopped  1510
```

In the preceding paragraphs, we discussed how to configure a PCAP using the TI. Figures 6-18 and 6-19 show how to configure PCAP the PCAP/TAP application. Figure 6-18 shows all the available interfaces on which PCAP can be configured. Selecting an in-

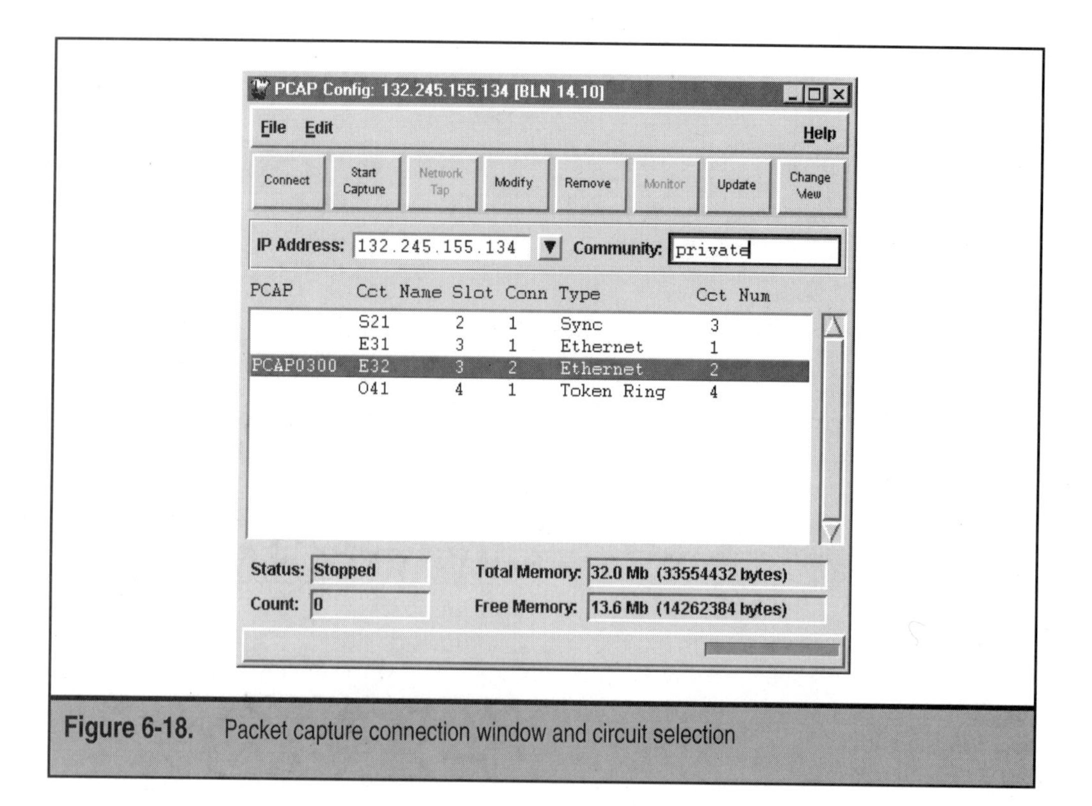

Figure 6-18. Packet capture connection window and circuit selection

terface automatically retrieves the total amount of memory and the amount of free memory available on the slot where PCAP resides. You can use this free memory value as a reference in estimating the maximum amount of memory that can be allocated to PCAP. Clicking the Add/Modify button displays the configuration dialog box shown in Figure 6-19. The three tabs contain the general configuration, receive, and transmit filters. After configuring the desired general parameters, click the Start Capture button in the main dialog box to begin the PCAP operation. During runtime, you can select the interface again to see the active status and the number of frames captured in the lower-left corner of the window.

Saving and Viewing PCAPs

You can access the packet capture buffer in memory after it has been stopped. Locally from the TI, you can use the **pktdump** command to dump the hexadecimal contents of each frame. A general knowledge of frame formats is necessary to decode and under-

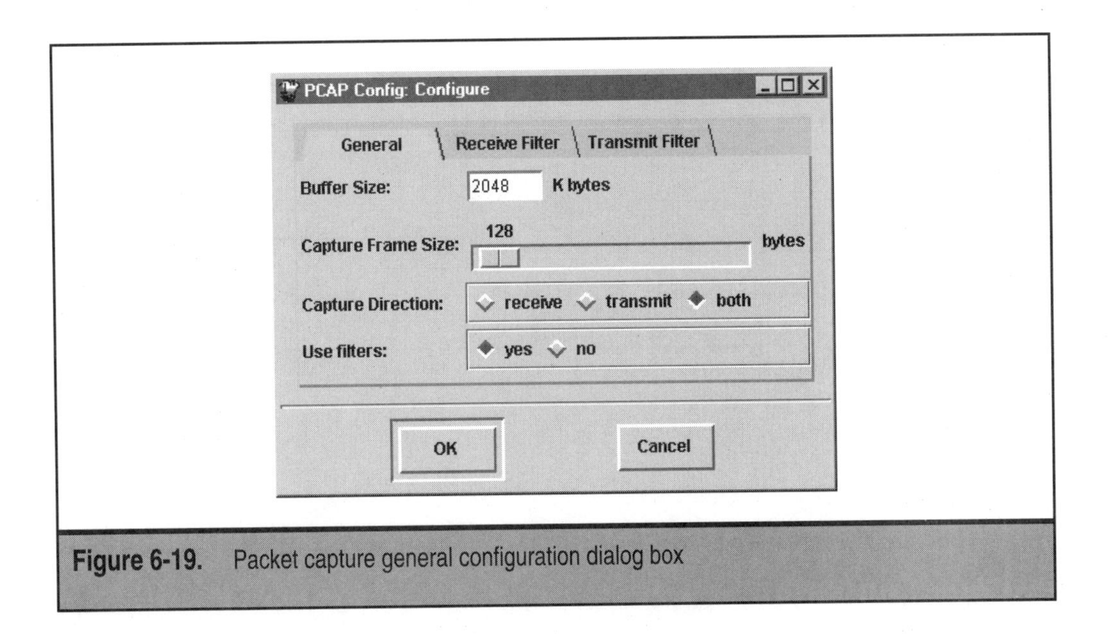

Figure 6-19. Packet capture general configuration dialog box

stand the captured frames. The results of a simple invocation of the **pktdump** command are shown here.

```
lxc9# pktdump 103102

Pkt# 1 04/12/00 10:20:26.430 CSMACD 60 Rx
00000000: 01 80 c2 00 00 00 00 00 a3 00 00 0c 00 26 42 42
00000010: 03 00 00 00 00 00 80 00 00 00 a3 00 00 0c 00 00
00000020: 00 00 80 00 00 00 a3 00 00 0c 80 07 00 00 14 00
00000030: 02 00 0f 00 3f 09 ef df 00 00 01 00 00 00 00 00

Pkt# 2 04/12/00 10:20:28.196 CSMACD 60 Rx
00000000: 00 00 a2 03 c1 66 00 00 a2 03 c1 66 81 02 01 01
00000010: 00 00 00 00 00 1f 80 00 00 00 a3 00 00 0c 00 00
00000020: 00 00 80 00 00 00 a3 00 00 0c 80 07 00 00 14 00
00000030: 02 00 0f 00 51 51 0f a3 00 00 01 00 00 00 00 00
```

It is easy to identify packet 2 because the source and destination addresses are the same, indicating that it is a BOFL packet (see Chapter 5). However, it is usually desirable to have a full-fledged analyzer decode the frames offline for better understanding. The PCAP memory buffer is stored as a virtual file in the file system. The **show packet config** command, when invoked with the appropriate line number, provides the name of the file used to store the buffers in memory.

```
lxc9# show packet config 103102

Packet Capture Configuration for Line 103102
----------------------------------------------
Delete          Created
Disable         Enabled
State           Up
Filename        PCAP0300
Control         Stop
Capture         Stopped
LineNumber      103102
...
```

The file indicated does not physically exist on any flash file system but is a virtual file representing the stored packet buffer. This file can be remotely retrieved for offline storage using FTP or TFTP. Since this is a virtual file in memory, restarting or resetting the slot will erase the contents of the file as if it had never been created. Ideally, once PCAP has been stopped, you should immediately save the file to flash for permanent storage. Using the TFTP client on the router, the virtual file can be saved locally by accessing the router itself as the TFTP server.

```
lxc9# tftp get lxc9 PCAP0300 4:CAP_E31
Transfer with 172.16.25.1 completed successfully.
```

Observe that both the TFTP client and the server are the same router. The virtual file PCAP0300 is saved to the nonvolatile file system on volume 4 as CAP_E31. This saved file can now be transferred to any workstation on the network that has sniffer software or the TAP application for decoding and analysis. The saved file is in the proprietary PCAP format and can be directly viewed using the TAP application. For use with sniffer software, however, it must be converted to the sniffer format using the **pktconv** command from the Site Manager installation directory.

```
C:\WF\SM1400> pktconv cap_e31
PKTCONV Version 1.9
C:\WF\SM1400> dir cap*
cap_e31         cap_e31.enc
```

The extension added by the packet convert application depends on the media interface on which the PCAP operation was performed. For example, Token Ring PCAPs are converted to .trc files, and FDDI PCAPs are converted to .fdc files. Both the PCAP and sniffer-format files can be viewed using the TAP application.

If the PCAP/TAP application is used to perform the capture, then clicking the Monitor button in the PCAP dialog box automatically brings up the TAP application. The application automatically fetches the virtual file and displays it on the screen. When you select a specific frame from the list of frames in the summary pane, a detailed decoding is displayed in the middle pane, with the hexadecimal contents right beneath it. This application is illustrated in Figure 6-20.

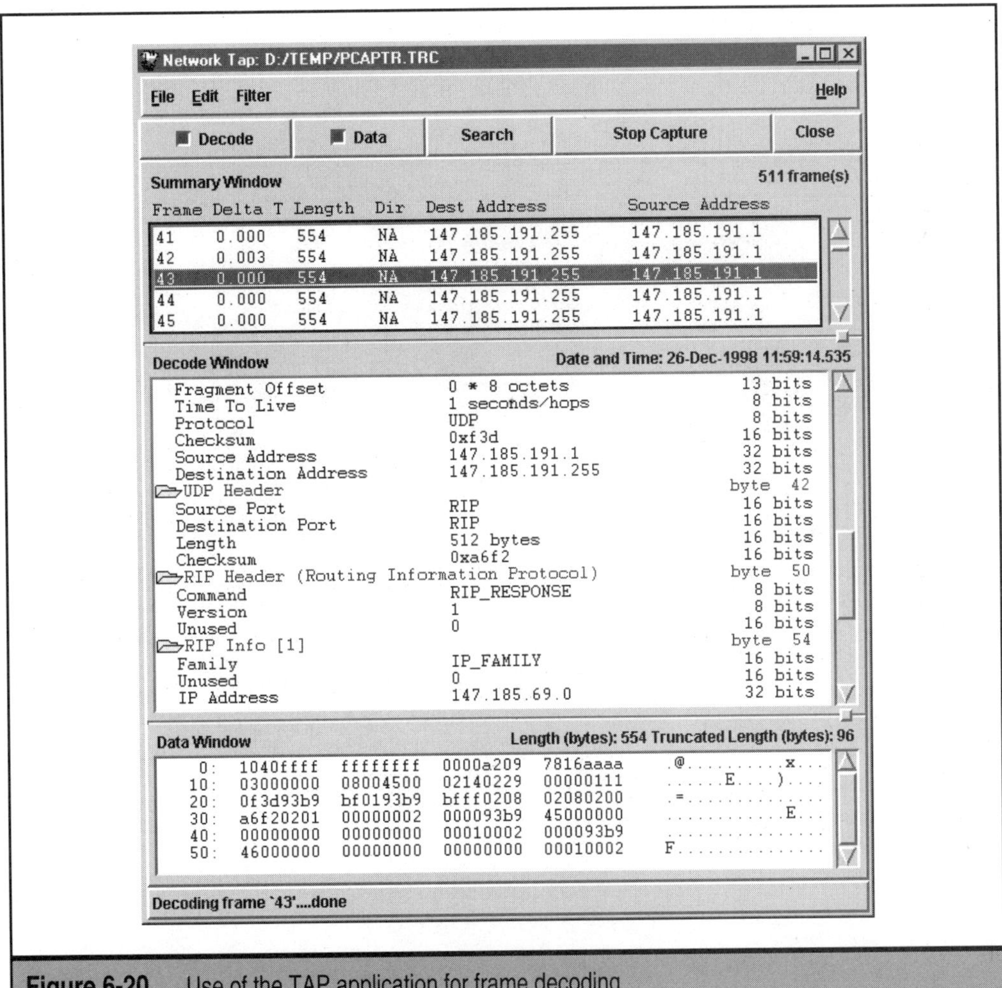

Figure 6-20. Use of the TAP application for frame decoding

Advanced PCAP Techniques

Simple packet captures are used to store all packets transiting a line in either direction or both directions. Since the buffer memory available is usually limited to the maximum amount of free memory, it is sometimes necessary to filter and to save only frames that match one or more criteria. For example, assume that multiple protocols are configured on an Ethernet interface, but a network problem is isolated to the IP protocol. Frames having the IP Ethernet type (0x0800) can be filtered and saved to the PCAP buffer, thereby avoiding traffic for all other protocols. Figure 6-21 shows an example of this advanced filtering. Once the use of filters is enabled on the General tab, the receive and transmit filter configurations become available. On the Receive Filter tab, the MAC (beginning of the frame) reference is used with an offset of 12 bytes (Destination MAC 6 bytes plus Source MAC 6 bytes) and a comparator value of 0800 for IP frames.

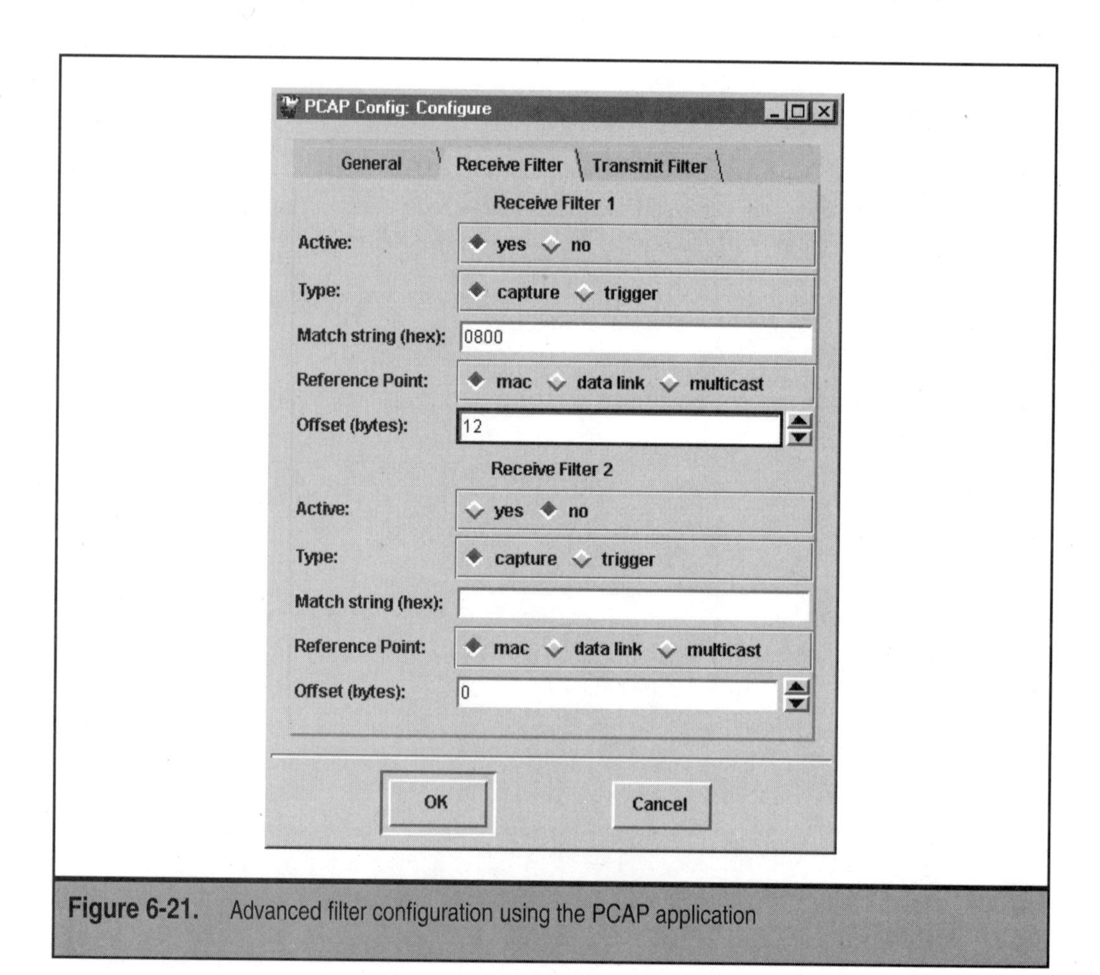

Figure 6-21. Advanced filter configuration using the PCAP application

Filters can also be configured to behave as triggers. If the trigger setting (`wfPktCaptureRxTrigger` or `wfPktCaptureTxTrigger`) is configured, then PCAP will terminate when a frame matching a data pattern is received. This is desirable when certain network issues are difficult to monitor either because they occur at odd hours or it is not possible to predict when they will occur. Setting the trigger will make the PCAP application automatically stop the capture when the desired symptom appears. Two triggers each are available for the receive and transmit directions, which can be applied together (and) or independently (or).

Each filter can match a range of bytes in the frame, using a reference location and an offset. Commonly, the beginning of a frame (MAC) and the end of the MAC addresses (DL) are used as references from which offsets are applied. You thus need to know the frame format for the specific media on which a packet capture is being performed. With proper analysis and preparation, filter matches can be performed well into the network and transport layer headers. For example, data destined for a specific IP multicast address can be copied by looking at the 4-byte destination address in reference to the DL, for which the correct offset must be known. Usually, you should perform a sample run of the packet capture to review the frame format and set the match value, size, reference, and offsets correctly.

SUMMARY

Network troubleshooting is both a science and an art, as the preceding two chapters illustrate. There is no set method for resolving all network issues; however, a systematic approach is always essential for resolving problems quickly. Physical and data link layer issues were discussed in the preceding chapter. In this chapter, we covered the widely used Internet Protocol and techniques for troubleshooting IP connectivity issues. We also discussed differences among the RIP, OSPF, and BGP protocols and described common issues and troubleshooting approaches; OSPF and BGP are more complex in design than RIP and require a detailed analysis of symptoms. We then discussed the troubleshooting of NetWare internetworks that use the equally common IPX protocol. We concluded by discussing sniffers and analyzers, which are also used in troubleshooting, and the Packet Capture (PCAP) application, which is useful for dynamically capturing frames without affecting network connectivity.

REFERENCES

1. *Configuring IP, ARP, RIP, and OSPF Services*, Publication 308627-14.00 Rev 00.
2. *Configuring IP Exterior Gateway Protocols*, Publication 308628-14.00 Rev 00.
3. *Configuring IP Utilities*, Publication 308631-14.00 Rev 00.
4. E. A. Hall and V. G. Cerf, *Internet Core Protocols: the Definitive Guide*, O'Reilly & Associates, June 2000.
5. J. W. Stewart, *BGP4: Inter-Domain Routing in the Internet*, Addison-Wesley, December 1998.
6. J. T. Moy, *OSPF: Anatomy of an Internet Routing Protocol*, Addison-Wesley, February 1998.
7. C. Huitema, *Routing in the Internet*, Prentice-Hall, November 1999.
8. *Configuring IPX Services*, 308633-14.00 Rev 00.
9. L. A. Chappell, *Novell's Guide to LAN/WAN Analysis: IPX/SPX*, IDG Books, March 1998.
10. *IPX Router Specification*, Novell Inc., October 1993.

CHAPTER 7

Network Security and Security Protocols

Nothing is secure but life, transition, the energizing spirit.

—Ralph Waldo Emerson

With the ubiquitousness of the Internet, more consumers and businesses are interconnected, and more easily interconnected, than ever. The increased use of the Internet in almost all aspects of life has brought networking and computer savvy to great numbers of end users. This is very much a welcome aspect of the expansion of the Internet, because it means computer users can be more productive, with fast-paced training and having the know-how needed to deal with basic computer problems.

Enter the dreaded *hacker*, the malicious user whose goal is to create havoc in computer networks. Hackers of various skill levels, including self-anointed gurus, may impair computer and network operation, thereby causing loss of productivity. The results of these attacks range from simple file deletions to large-scale theft of proprietary knowledge. In the past, security concerns typically focused on computers because they were the hubs of information delivery. Today, it is networks that are commonly under attack by hackers because intranets and the Internet are the media that make information delivery possible. Consumer and business traffic constantly flow through the veins of the Internet, and how else better to compromise this data than by compromising the network itself?

Historically, the term *hacker* referred to a skilled computer user or system administrator. The hackers of today share a common trait with those of yesterday—skillful computer and network knowledge. Their typical operations involve gathering confidential data, disrupting computer or network operation (called *denial of service*, or DoS), and compromising data integrity. These activities are of concern to computer system administrators and network administrators alike, and they have increased the burden on the technology to provide improved data, computer, and network security.

Most network and computer installations have a default level of security that is enough to deter the novice hacker. However, the major security threat to networks today comes from within the enterprise itself. A skillful effort to compromise the security of a network or computer is more likely to succeed if the hacker has internal information about the system. For example, a local user usually knows some of the details of a particular server, such as its IP address, its location, and the types of services it provides—information not readily available to a remote hacker.

In addition, in enterprise and service environments, security issues involve not only coordinated attacks by hackers, but also the inadvertent lapses of otherwise knowledgeable, well-intentioned computer users. For example, computer users may unknowingly compromise confidential data because they are unaware of security requirements and its implications for that data.

Overall, security issues arise either from within the enterprise intranet or outside from the public Internet. To provide security, the technology must restrict unauthorized and unwanted traffic, but still allow legitimate internal communications and external business communications without hindrance. In this chapter, we discuss security provisions and fortifications that help protect against various security threats and attacks.

SECURITY POLICY

A uniform, enterprise-wide security policy is important in defining the acceptable use of computer and network resources for personal use and business operations. Many security infringements occur because users do not know what is acceptable use and are unaware of special restrictions placed on computing and networking resources. This policy must not only define security but also cover privacy issues and ethics. Users should be entitled to privacy in the use of their computing systems, and intentional violation of this privacy should also be considered a security breach. Similarly, many network and system administrators consider it unethical to view another user's files, even when they are inadvertently left accessible to other users.

The definition and implementation of a security policy is a significant activity, with network security a key consideration, and the policy must be reviewed periodically. Security policies vary among enterprises, depending on the computing environment. A security aspect that is key in one environment may not even apply in another. Nevertheless, many security features are common. For example, the sharing of user accounts is generally disallowed because access privileges are assigned to users based on their job functions.

The definition of a security policy is an extensive topic by itself and is discussed here only to convey its importance in all enterprise environments. Security measures such as firewalls are more effective when users understand the security policy. Technology that deters security breaches should ideally be designed within the framework of the acceptable use policy, delineating the privileges assigned to internal and external users.

External users are mostly customers and business partners who communicate with a group or an individual within the enterprise. Usually, the policies applied to internal users also apply to external users. In most cases, more stringent policies are applied to business partners such as contractors. With the increased use of the World Wide Web, many business transactions take place over the Internet, and many enterprise customers include not only consumers, but also other businesses. Customers can access many interactive functions on the business web servers of many companies, and a policy statement is needed to define privacy, ethics, and acceptable use for all these external users.

This chapter assumes that a governing security and acceptable use policy applies to most enterprise environments, regardless of the user type, whether internal or external. Application of these security policies will help determine the security configuration of computers and networking devices. In this book, security configuration will revolve around the application of various security levels to BayRS routers.

USER ACCOUNTS

Computing and network devices are usually accessed using a username or user ID, each of which is assigned a privilege level that enables the user to execute certain functions. The higher the privilege level, the more tasks the user can perform on the device. On

BayRS routers, user accounts are primarily used for administration and management. Creation of user accounts and assignment of privilege levels were discussed in Chapter 3.

Three privilege levels can be configured for user accounts in BayRS: Manager, Operator, and User. The User privilege level provides the ability to monitor and view all statistics on the router, but disallows any read-write operations and disruptive changes. The Operator level provides read-write access to the system configuration, but does not allow disruptive administrative tasks. Administrative tasks such as deleting files, running diagnostics, reinitializing a slot, burning the boot and diagnostics PROMs, and transferring files using the **tftp** or **xmodem** command are restricted to the Manager level.

User accounts should be carefully administered to limit the number of Manager- and Operator-level users. Wherever possible, user login and logout should be audited and logged to a syslog server for reference. Two types of user login are reported in the log: one for console access and another for telnet access.

```
#  138: 06/05/1999 03:58:01.832  INFO     SLOT 12  TI                 Code:  60
User pooja logged in from port 1 on slot 12

#   35: 03/26/1999 14:59:46.633  INFO     SLOT  5  TI                 Code:  62
User noc logged in from 178.93.227.159 port 1
```

Each login is reported by the `TI` entity, as is each corresponding logout.

```
#  140: 06/05/1999 04:12:15.947  INFO     SLOT 12  TI                 Code:  61
User pooja logged out from port 1 on slot 12

#   35: 03/26/1999 15:02:32.471  INFO     SLOT  5  TI                 Code:  63
User noc logged out from 178.93.227.159 port 1
```

Of utmost importance are login failures, which must be monitored frequently and can be automatically reported to syslog consoles. Repeated login failures may indicate that an unauthorized user is attempting to gain access to a router.

```
#  165: 09/24/1999 20:05:00.659  WARNING  SLOT  9  TI                 Code:  39
Invalid login attempt by Manager on port 1.

#  166: 09/24/1999 20:05:02.982  WARNING  SLOT  9  TI                 Code:  39
Invalid login attempt by Manager on port 1.
```

As discussed in Chapter 3, the TI login and welcome notices can be used to clearly identify the router, describe the scope of its function, and provide a statement regarding acceptable use. Even users such as network operators with Manager-level privileges may require prior authorization from a network manager before reinitializing a slot or rebooting an entire router. Commands issued at the TI and BCC prompts are also logged to the event log, which can also be sent to a syslog server.

```
#   51: 07/26/1999 16:27:38.049  INFO     SLOT 13  TI                 Code:  64
User agiovino issued command: boot 4:bn.exe 4:staging.cfg
```

Logs should periodically be reviewed to detect intrusion attempts and unauthorized network changes. Some events can be immediately trapped and reported to a network management server, as appropriate.

SecurID

The SecurID authentication mechanism enables the SecurID client on the router, which requires entry of a passcode following the login and password. The passcode is a combination of a user's PIN and a dynamically changing value that must be looked up using a SecurID card and entered with minimal delay. Only an authorized user with the correct login ID, password, and passcode can access the router. SecurID authentication is available only for telnet sessions—console sessions do not require SecurID authentication. A simple SecurID login procedure is illustrated here:

```
Login: ragho
Password:
Passcode:
```

The password and passcode are not echoed on the screen. When SecurID is configured, a BayRS router communicates with a Security Dynamics SecurID server on the IP network. Users must be assigned a PIN and a SecurID card by the SecurID administrator.

CAUTION: If a user enters an incorrect passcode four consecutive times, the SecurID system will automatically remove the user from the active list of valid users. The SecurID administrator must reactivate the user's PIN before the person will be allowed access into any system using the same SecurID authentication.

You configure SecurID by entering the **securelogin** command at the TI console. It cannot be configured using telnet. The IP address of the SecurID server and the TCP port for authentication requests must be obtained beforehand for the configuration. The following example shows SecurID authentication enabled using **securelogin**; user input is highlighted.

```
# securelogin
Do you wish telnet login to require SecurID? (yes/no) yes
What is the IP address of the router being secured? (0.0.0.0) 172.16.193.254
What is the IP address of the SecurID server? (0.0.0.0) 172.16.179.245
What is the TCP port number for SecurID services? (default=755)
You have designated 172.16.179.245 (or 0xac10b3f5) as your SecurID server.
The default port 755 will be used for SecurID services.
Is this information correct? (yes/no) yes
Telnet login now requires SecurID.
Reinitialize Client? (yes/no) yes
Client reinitialized.
```

From this point, users telnetting into the router will be prompted for their passcodes (PINs and card codes) after they enter their passwords.

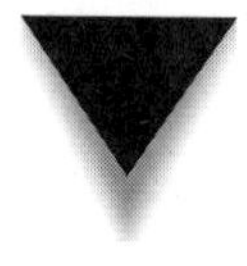

CAUTION: When a processor module is replaced on a multi-slot router like the ASN or BN, both the **password** and **securelogin** commands must be reissued to update the configuration on the new processor. Failure to do so will allow users telnetting into the new slot to bypass the SecurID authentication. If the user is logging in using one of the system logins (Manager or User), then the password authentication will also be bypassed. Always run the **password** and **securelogin** commands immediately after replacing a processor module.

In high-security environments, SecurID provides an additional measure of security for telnet access into a BayRS router. Invalid logins are reported in the event log under the `TI` entity as well as on the SecurID server.

SNMP SECURITY

The Simple Network Management Protocol (SNMP) provides the configuration and management framework for BayRS routers. We discussed the basics of SNMP and the proprietary MIB in Chapter 2. Configuration of SNMP services, such as configuration of managers and communities, was discussed in Chapter 3. SNMP is one of the protocols mentioned earlier that does not inherently provide a robust authentication mechanism.

Communities and Managers

The *community string* is the authentication mechanism used in SNMP. It suffers from some basic security limitations. First, the community string is clear-text password. Any user with a packet sniffer can snoop into data on a broadcast segment and capture the community string with ease. In addition, many SNMP agent implementations generate a reply with a correct community string if the community string supplied by the manager is incorrect. Security breaches are easy to accomplish to gain read-write privileges for a router in this fashion.

When initially configured using the install batch script, BayRS routers have a default configuration using public as the community string and the manager IP address 0.0.0.0 associated with it. The IP address 0.0.0.0 is used to match all source addresses, which essentially allows all SNMP-capable devices to access the router using the public community. When the router is initially provisioned using BCC or a local configuration file created from scratch, SNMP configuration must be performed manually. Although the use of the default public community string from any Site Manager station is desirable during initial configuration, it is not necessary once the circuits have been configured and become operational.

You can do the following to increase SNMP security:

- Eliminate the use of the public and private community strings throughout the network. If public or private strings are required for legacy applications, change the privilege level of these community strings to read-only.
- Create one or more community strings with read-write privileges for specific SNMP managers and applications.

▲ If at all possible, avoid using the wildcard manager IP address, 0.0.0.0. Always configure specific IP addresses for SNMP managers.

Using specific community strings and managers will help the router report invalid SNMP requests, increasing SNMP security. An example of an unauthorized SNMP request is shown here, reported by the SNMP entity.

```
#   25: 05/20/1999 24:29:32.103  TRACE    SLOT  3  SNMP               Code:   8
Agent received unauthorized request from 139.230.44.190 in community private.
```

SNMP authentication failures from unauthorized managers or from managers using the wrong community string can be monitored, and an authentication failure trap can be sent to one or more SNMP management stations. To generate these traps, enable the SNMP global parameter authentication-traps.

```
box->ip->snmp
fp3(12)# authentication-traps enabled
```

Secure SNMP Mode

The steps outlined in the previous section alone may not be sufficient to eliminate all potential SNMP security threats. Skilled hackers may use unauthorized devices to masquerade as authorized managers. To deter potential threats such as these, BayRS supports an optional, proprietary secure SNMP mechanism. An encrypted counter is used to synchronize the SNMP manager and the agent (BayRS). When an SNMP set request is issued, the request includes the encrypted counter, which is compared to the decrypted value by the agent. If the values match, then the request is considered authentic and is accepted. The counter is automatically incremented.

The use of the encrypted counter helps guard against devices masquerading as valid managers, because knowledge of the encryption key and the counter value is required. This method, called SNMP secure mode, also helps prevent message stream modification because the counter values cannot be altered in transit.

SNMP secure mode is configured using the **wfsnmpmode** and **wfsnmpkey** commands. Setting the SNMP mode to 3 enables the proprietary authentication mechanism, and setting it to 1 disables it. The **wfsnmpkey** command sets the key used by the encryption algorithm.

```
# wfsnmpmode 3
# wfsnmpkey up+1m3
```

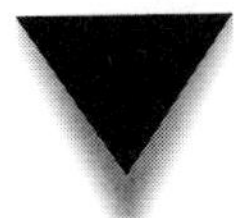

CAUTION: Like the system passwords and SecurID parameters, the secure SNMP key is stored in NVRAM. If you replace a processor module, you must reissue the **wfsnmpkey** command to update the key on the new module.

Ideally, SNMP secure mode should be used in conjunction with the configuration procedures recommended in the previous section. It is still undesirable to use an SNMP wildcard manager (0.0.0.0) or a well-known community string (that is, public or private).

View-Based SNMP Access Control

The SNMP agent on BayRS is otherwise known as the *snmpserver*, and in recent software versions it can provide selective access control using SNMP *views*. An SNMP server view contains a list of MIB objects, attributes, and instances that are specifically included or excluded from the view. Each view is associated with one or more community strings, which affect what SNMP managers can access through the SNMP agent on the router.

The motivation behind view-based access control is to provide multiple sets of MIB views so that users with read-write privileges can modify only MIBs that they are authorized to modify. In multiadministrator environments, such as those following an enterprise merger, and in ISP environments, certain administrators need to have full control over their own interfaces. However, the same administrators (for example, end users in an ISP environment) must not be able to view other interfaces and global protocol-level MIBs that are owned by other users or customers.

SNMP server views are created using the following command:

```
snmpserver view view_name oid_tree {included|excluded|list|delete}
```

The view name uniquely identifies the view subtree, where the OID tree is any representation of a MIB object, MIB object instances, or MIB object attributes. The wildcard attribute or instance (*) can be used, as discussed in Chapter 2. For information on how to set up the OID tree to exclude or include specific objects, instances, or attributes, refer to Chapter 2, which discusses various types of MIB access.

A simple SNMP server view is illustrated here. In this view, attributes of two specific IP interface configurations can be accessed, but the global IP base object cannot be accessed.

```
fp3(3)# snmpserver view customer74 wfIpIntfCfgEntry.*.192.32.254.1.14 included
fp3(4)# snmpserver view customer74 wfIpIntfCfgEntry.*.192.32.272.1.27 included
fp3(5)# snmpserver view customer74 wfIpBase excluded
fp3(6)# snmpserver view customer74 list
   View Subtree                                 View Type    Subtree Index
   wfIpBase                                     excluded     1.3
   wfIpIntfCfgEntry.*.192.32.254.1.14           included     1.1
   wfIpIntfCfgEntry.*.192.32.272.1.27           included     1.2
```

Notice that no instance or attributes were specified for the `wfIpBase` object. The global instance (0) is the only one allowed for the IP base; however, the representation here excludes all attributes for all instances of the `wfIpBase` object. The **list** subcommand is used to view the contents of the MIB subtree. Once the view is created, it is then associated with a community string. If the desired community string has not been configured already, the **snmpserver** command can be used to create a new community. The syntax used with community strings is shown here.

```
snmpserver community community_string {RW|RO|list|delete}
snmpserver community community_string view view_name
```

A new community string ap3c0w is created with read-write privileges and associated with the customer74 view. Managers properly configured to use this community string will be able to view and change parameters on the two IP interfaces, but will not be able to change any global IP parameters.

```
fp3(7)# snmpserver community ap3c0w RW
fp3(8)# snmpserver community ap3c0w view customer74
fp3(9)# snmpserver community ap3c0w list
Community Index     Community Name          Associated View          Access
     2              ap3c0w                  customer74               read-write
```

The **delete** subcommand will remove the configuration when it is used with either a view or a community. Previously created community strings can be directly associated with a view, without having to re-create it first. Managers must then be created for each community string, as described in the router administration tasks (see Chapter 3). Once views have been created and associated with a community string, remember to create appropriate managers that will use the newly configured view. If an SNMP manager attempts to read or write an attribute that has been excluded from the associated view, the following warning event is logged:

```
#  119: 11/02/1999 02:03:58.941  TRACE    SLOT 13  SNMP              Code:   8
Agent received unauthorized request from 204.120.119.56 in community ap3c0w.
```

PROTOCOL SECURITY

BayRS routers support various protocol suites, and some protocols are especially vulnerable to attacks by network hackers. Protocol analyzers and decoders allow network administrators to troubleshoot network problems by checking packets crossing key network devices and broadcast networks for vulnerabilities. However, hackers can use these same network analyzers to determine network vulnerabilities and so find entry points into various systems.

Most protocols, regardless of their operating layer, were designed with the goal of supporting a certain set of features. Security was built into some, primarily as a result of past experiences with other protocols or a previous version of the same protocol. Overall, security is not an inherent feature in many traditional protocols, though it has been improved with recent enhancements such as IP Security (IPSec) and BGP4 with MD5 authentication.

Protocol security on routers must be augmented in at least two areas. In the network layer, data destined for the router and frames in transit over the router require security measures. Higher layers, including transport and session layers, also require similar security measures. BayRS provides various security solutions, including access policies, encryption, secure and authenticated access, traffic filtering, and an integrated firewall.

Switched Networks

Breaches of protocol security in traditional LAN environments are compounded by the shared nature of many topologies. This includes original shared Ethernet, Token Ring, and FDDI networks. Since the local network is shared, an intruder or a curious user can simply use a network analyzer or other promiscuous investigation device to observe data traffic. The knowledge needed to do this is now easily obtained through simple homework; hence, this is a frequent means of intrusion among seasoned hackers.

Switched networks alleviate some of the problem that shared topologies pose. Besides limiting the amount of traffic on many segments, switched networks propagate traffic based on data link addresses. This essentially limits other LAN traffic and makes it difficult for a hacker to sniff traffic destined to stations on other segments. In recent years, switched Ethernet and Token Ring have seen increased deployment to provide more efficient network use and to protect data and are the preferred solution in high-security environments. Ideally, routers should be connected to high-speed switched connections to improve throughput and improve data security within the network.

Traffic Filtering

Traffic filters are rules applied to inbound and outbound frames; when met, they instruct the router to perform one of various special functions, such as blocking, forwarding, logging, and prioritizing. Various types of traffic filters are available, and filter criteria varies among applications—for example, IPX traffic filters use different criteria than IP traffic filters. However, the application of all traffic filters is uniform in each layer of the OSI model. An inbound IP filter, for example, will be applied to all IP frames received on that interface.

Traffic filters are used for two purposes: to control access to resources by restricting data flow and to prioritize data in multiprotocol environments. In the context of security, traffic filters are mostly applied to control data flow by accepting or restricting certain packets. Data frames exhibiting well-known attack patterns and those symptomatic of intrusion attempts can be easily and effectively blocked using traffic filters.

On IP-based intranets and the Internet, traffic filters are configured using the `traffic-filter` object on the appropriate `ip` interface. When the same set of rules and actions apply across multiple interfaces, a `filter-template` object is created in the global `ip` context. This filter template can then be applied to multiple interfaces simultaneously. Multiple traffic filter rules can be configured on an IP interface, with the order of application specified by the `precedence` attribute. When logging detail is enabled in the filter, a properly applied IP traffic filter reports log events like the following:

```
#  153: 01/12/1999 15:21:20.911  INFO     SLOT 11  IP                    Code: 147
IP Traffic Filter - Rule 14, Interface 172.16.99.101, Circuit 57 (Accept packet)
Accepted Pkt - Src: 172.16.57.144, Port: 3421, Dst: 172.16.164.209, Port: 23, Prot: 6
```

```
#   58: 10/05/1999 20:22:57.920  INFO     SLOT 12  IP                 Code: 148
IP Traffic Filter - Rule 8, Interface 10.10.125.1, Circuit 23 (Drop packet)
Dropped Pkt - Src: 10.10.114.87, Port: 9458, Dst: 10.10.174.214, Port: 88, Prot: 6
```

The following example shows a simple IP traffic filter configuration. In this filter, traffic originating from the three subnets (172.16.93.0/24, 172.16.94.0, and 172.16.95.0) is not allowed to reach two specific servers (172.16.52.100 and 172.16.52.101) on the intranet. Endstations on the three networks are otherwise allowed access to devices on the destination network, but their access to these two servers is restricted. This is a very common application, where multiple servers in a server farm have different access levels and groups. Multiple `source-network`, `destination-network`, or `user-defined` criteria can be specified in the same traffic filter configuration.

```
...ip/172.16.48.1/255.255.255.0
mx2(10)# traffic-filter SAP-Restrict
 ...traffic-filter/SAP-Restrict/172.16.48.1
mx2(11)# match
 ...match/filter/SAP-Restrict/172.16.48.1
mx2(12)# source-network range 172.16.93.0-172.16.95.255
 ...source-network/filter/SAP-Restrict/172.16.48.1/172.16.93.0-172.16.95.255
mx2(13)# back
 ...match/filter/SAP-Restrict/172.16.48.1
mx2(14)# destination-network range 172.16.52.100-172.16.52.101
 ...destination-network/filter/SAP-Restrict/172.16.48.1/172.16.52.100-172.16.52.101
mx2(15)# back 2
 ...traffic-filter/SAP-Restrict/172.16.48.1
mx2(16)# actions
 ...actions/filter/SAP-Restrict/172.16.48.1
mx2(17)# action drop
 ...actions/filter/SAP-Restrict/172.16.48.1
mx2(18)# action-log detailed
```

Filters can be applied on both inbound and outbound packets, although this setup must be explicitly configured. Generally speaking, the concept of traffic filters applies to inbound packets. Outbound traffic filters are configured under the protocol priority application. The reason is that inbound traffic filters are naturally applied as part of the appropriate layer protocol function, whereas outbound traffic filtering (protocol prioritization) occurs after all network protocol functions have been performed. Protocol prioritization is discussed in detail in Chapter 8, in the "Traffic Engineering" section.

TCP-Established Criterion

Many IP applications requiring reliable transport use TCP. The TCP protocol, along with UDP, is part of the TCP/IP suite. TCP requires session establishment between the source and the destination. Often, routers residing at the edge of the corporate intranet are configured so that outbound TCP sessions are allowed, but inbound TCP sessions are not. For example, an internal user must be able to browse the Internet freely, but an external user should not be allowed to freely access intranet servers. The criterion used for this purpose is called *TCP established* and is applied to inbound packets. TCP datagrams entering through the public Internet interface are checked to see whether they are part of a session that originated on the internal network or on the external network.

```
 ...filter-template/accept-websurf
mx2(186)# match
 ...match/template/accept-websurf
mx2(187)# tcp-established on
```

To use the TCP-established criterion, enable the `tcp-established` attribute for the `match` object, a child object of each traffic filter. This rule essentially implements a one-sided firewall, protecting the public Internet interfaces from unauthorized users. When traffic filters are used on an Internet-edge router, such as a Backbone node connected to an ISP, the use of the TCP-established criterion helps prevent inbound TCP attacks from the Internet.

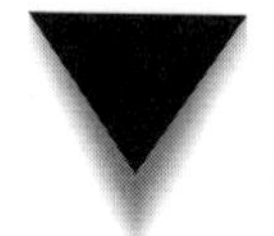

TIP: The functions performed using predefined criteria can also be accomplished using appropriate user-defined criteria. Predefined criteria make the creation of traffic filters more efficient and easier. However, the same TCP-established criterion could be implemented by examining individual TCP flag values in a TCP datagram using a user-defined criterion.

Directed Broadcasts

Network layer protocols are designed to limit the amount of broadcast traffic traveling over the network. Two types of IP broadcasts, directed subnet broadcasts (DSBs) and all-subnet broadcasts (ASBs), are sometimes used in some IP networks. These IP layer broadcasts are set to the subnet broadcast address, or in the case of all-subnet broadcasts, the subnet bits are all set to 1. Subnet broadcasts normally originate within a local subnet and are destined for the same subnet; hence, they are not propagated to other networks. If a direct subnet broadcast is addressed to a destination subnet that is different than the source subnet, then the local gateway will forward the directed broadcast to the destination subnet.

Some applications and devices also generate all-subnet broadcasts in classful environments, where the use of the all-ones subnet is limited to broadcasting. If the global IP parameter `all-subnets` is enabled, then all-subnet broadcasts and directed broadcasts appear the same to a BayRS router. On some parts of traditional IP networks, network devices are still in use that do not fully support subnets. These devices can send frames to the all-subnets or directed broadcast addresses. The network in these cases must facilitate the proper forwarding of these frames to remote subnets. In BayRS, the default IP forwarding configuration is as follows:

- **All-zeros and all-ones subnets (`all-subnets`)** Global IP parameter; disabled. The use of all zeros and all ones in the subnet portion is disabled by default. This allows a router to clearly distinguish between a directed subnet broadcast and an all-subnets broadcast.

- **All subnet broadcasts (`all-subnet-broadcast`)** Interface-specific IP parameter; disabled. When the all-subnets parameter is disabled, this attribute is used to determine whether an ASB is accepted or forwarded for delivery over the interface.
- **Directed broadcasts (`directed-bcast`)** Global IP parameter; enabled. When the all-subnets parameter is enabled, then ASBs and DSBs are treated alike, and this global parameter applies.

Most enterprise networks today use classless routing (CIDR) along with all-zeros and all-ones subnets, so the `directed-bcast` parameter is important in certain networks. In recent years, security threats such as the famed smurf attack have used directed broadcasts to cripple network performance. The directed broadcast remains as is until it reaches the final destination network, where each gateway replicates the frame for all subnets that are intended to receive it. For example, an ASB ping destined for a corporate intranet can originate with an errant user on the Internet and then be forwarded to all the appropriate subnets if all intervening gateways forward it. The numerous destinations on all the subnets will receive this ICMP echo request and generate several hundreds or even thousands of replies to the source—but in most cases in a hacker attack, the source is either a fake address or another ASB destination, thereby causing another cascade of useless data. This affects overall performance of the network, and the availability of the resource is degraded, eventually causing denial of service. These types of directed broadcast attacks have been a key concern of many network administrators in recent years. Directed broadcasts can be summarily disabled from being processed by the router, using the following IP parameter.

```
box->ip
mx2(103)# directed-bcast disabled
```

The default IP forwarding configuration allows only DSBs to be forwarded, which is desirable for proper communication within the intranet. If ASB forwarding is needed, then IP interfaces must be selectively configured to accept and forward these datagrams. On edge routers connected to an ISP, both directed broadcasts and all-subnet broadcasts must be disabled. If `all-subnets` is enabled, simply disabling the `directed-bcast` parameter as shown in the preceding example will be adequate. This will allow the BayRS router to silently drop all incoming and outbound directed broadcasts, which is acceptable because this form of communication is rarely used.

Black Hole Routes

Some networks may be unable to disable directed broadcasts because it is legitimately required by some applications and services. In these cases, the directed broadcasts closest to the ISP router can be disabled, thereby still allowing directed broadcasts to be forwarded within the intranet.

An alternate solution, called the black hole route, can be configured for the directed host-specific route. A black hole route is one whose next-hop gateway address is 255.255.255.255, which causes the BayRS router to drop the frame without forwarding.

Black hole routes are configured as static routes to the directed broadcast address. The next hop is specified as 255.255.255.255, which the router automatically recognizes as a black hole route. In the following black hole route configuration, the destination network 132.245.152.0/24 is protected by configuring a host-specific route to the broadcast address for the subnet: 132.245.152.255/32.

```
box->ip
fp3(14)# static-route 132.245.152.255/32/255.255.255.255
```

Instead of the black hole route, a traffic filter may also be configured to simply drop the frame and log an event.

```
#  166: 06/13/1999 19:06:29.924  INFO     SLOT  2  IP                    Code: 141
IP Traffic Filter - Rule 4, Interface 132.245.155.134, Circuit 19 (Drop packet)
  Dropped Pkt - Src: 209.47.152.192, Dst: 132.245.152.255, Prot: 1
```

For example, a traffic filter applied to the destination address 132.245.152.255/32 can log the event shown in the preceding example and drop the frame. This helps network administrators monitor directed broadcast packets and take appropriate action when they arrive repeatedly. The next example shows the traffic filter used for this purpose. First, a filter template named drop-dsb-net152 is created under the global IP context. Once this template is created, it can be applied to multiple IP interfaces. The filter template thus created is applied to the IP interface 132.245.155.1 on E32.

```
 ...ip
mx2(85)# filter-template drop-dsb-net152
 ...filter-template/drop-dsb-net152
mx2(86)# match
 ...match/template/drop-dsb-net152
mx2(87)# destination-network 132.245.152.255
 ...destination-network/template/drop-dsb-net152/132.245.152.255
mx2(88)# back 2
 ...filter-template/drop-dsb-net152
mx2(89)# actions
 ...actions/template/drop-dsb-net152
mx2(90)# action drop
 ...actions/template/drop-dsb-net152
mx2(91)# action-log on
...actions/template/drop-dsb-net152
mx2(92)# box; ethernet 3/2; ip/132.245.155.1/255.255.255.0
 ...ip/132.245.155.1/255.255.255.0
mx2(93)# traffic-filter drop-dsb-net152
...traffic-filter/drop-dsb-net152/132.245.155.1
mx2(94)# template-name drop-dsb-net152
 ...traffic-filter/drop-dsb-net152/132.245.155.1
mx2(95)# info
  filter-name drop-dsb-net152
  template-name drop-dsb-net152
  precedence 1
  state enabled
```

Multiprotocol Filters

Traffic filters can be configured so that each protocol handles its traffic in a different fashion. BayRS supports predefined traffic filtering criteria for the following protocols:

- Transparent bridge
- Source route bridge (SRB)
- IP
- IPX
- XNS
- OSI
- DECnet Phase IV
- OSI
- Vines
- DLSw
- LLC2 (APPN and LNM)

The predefined criteria are well known and are commonly used for traffic filtering. However, transparent bridging and source route bridging support various encapsulation types, which can be used directly to define a custom set of criteria using bit lengths and values. When a network layer protocol such as IP is configured on an interface, the protocol frames cannot be filtered using bridge filters; the corresponding network layer filters must be configured.

Routing Protocols

Traffic filtering affects protocol traffic uniformly on each interface: that is, all frames received on an interface are checked against each filter configured on it. The predefined and user-defined criteria are useful for easily filtering frames at the network and transport layers. Sometimes it is necessary to have the inbound frame fully processed and then subsequently filter the information selectively. The approach is commonly used with routing protocols, where a whole set of inbound routes are received first and then selectively accepted into the routing table.

Route and Service Filtering

In Chapter 6, we briefly discussed one reason for route and service filtering: security. Routes and services that are remote in reference to a local router usually do not need to be accessible directly. On the other hand, application services such as e-mail and web servers are, in most cases, securely decentralized and so are generally accessible from all locations within the enterprise. For example, IP devices in the logistics department may not need to directly communicate with other IP devices in the corporate division. In this case, the router upstream to the logistics router can filter out the route to the corporate division, thereby not only reducing the size of the routing table, but also preventing unnecessary and unauthorized traffic.

The benefits of route and service filtering may not be readily apparent in some networks, whereas in others, the need may be obvious. For example, suppose IPX servers named MARKETING01 through MARKETING05 contain all confidential marketing requirements and competitive intelligence documents used by a marketing department. Access to these servers is necessary from all devices within the marketing department, but is disallowed from others. User-level security may be applied to control access privileges; however, if a valid route to the server is known, a malicious user may be able to tinker with the destination to gain unauthorized access. Hence, the following set of route and service filters can be applied so that the services are not advertised from the core marketing router and so the route to the servers is suppressed.

```
 ...ipx/100041ac
mx2(16)# server-name-filter "MARKETING.*"
 ...server-name-filter/100041ac/MARKETING.*
mx2(17)# type ffff
 ...server-name-filter/100041ac/MARKETING.*
mx2(18)# action suppress
 ...server-name-filter/100041ac/MARKETING.*
mx2(19)# outbound enabled; inbound disabled
 ...server-name-filter/100041ac/MARKETING.*
mx2(20)# info
  name MARKETING.*
  inbound disabled
  outbound enabled
  type ffff
  action suppress
  protocol any

 ...ipx/100041ac
mx2(22)# route-filter network 0x100042a0 mask 0xfffffff0
...route-filter/100041ac/100042a0/fffffff0
mx2(23)# action suppress
 ...route-filter/100041ac/100042a0/fffffff0
mx2(24)# info
  network 100042a0
  mask fffffff0
  inbound disabled
  outbound enabled
  action suppress
  protocol any
```

The range of marketing servers is represented using the name MARKETING.*, where the trailing .* follows the common regular expression pattern. The service `type` is set to 0xffff so that all services are filtered. The `outbound` parameter indicates that the filter is applied to outbound frames, and the `inbound` parameter applies to the reverse case. The routes to external networks 0x100042a0 through 0x100032af are filtered using the route mask 0xfffffff0 with the `action` parameter set to suppress. Route filters are available for almost all routing protocols. A route import filter affects the acceptance of an incoming route, while an export filter affects the propagation of a route in the local routing table.

NOTE: Route filters and policy filters are unique, although they are both used with routing protocols. Policy filters include accept and announce policies that often directly affect the internal processing of a route by a certain routing protocol.

Authentication and Security

Routing protocols are sometimes vulnerable to attacks because an intruder can supply false routes and redirect traffic maliciously. The routing information protocol (RIP) uses broadcast frames, and a device masquerading as a router can inject RIP frames containing routes to remote destinations with low metrics. This allows a malicious user to easily redirect frames to the masquerading router for monitoring and analysis. The open shortest path first (OSPF) protocol, on the other hand, requires formation of an adjacency, making tampering more involved than with stateless broadcasts, though not impossible for a determined hacker. The border gateway protocol (BGP) is similar to OSPF in that peer formation is necessary for routes to be exchanged.

Routing Information Protocol (RIP) A primary enhancement to RIP1 was made in RIP2 with the introduction of an authentication mechanism. BayRS supports the inclusion of a simple password in RIP2 updates; passwords must match for routers in the same broadcast domain. The authentication mechanism is not enabled in the default configuration of RIP2 interfaces.

Simple password authentication can be enabled on a per-interface basis, which allows a RIP2 router to reject frames that do not have a password configured and those that have an invalid password. BayRS automatically detects misconfiguration of the authentication mechanism between two adjacent routers. A router that has simple password authentication enabled reports the following warning when it receives an update from another router that does not have it enabled:

```
#    2: 10/21/1999 20:42:57.152  WARNING  SLOT 10  IP                   Code:  74
RIP Authentication failed for source 172.16.194.254 on interface 172.16.194.1
Security enabled, received Address Family Identifier 2 Authentication Type 0
```

On a remote source router that recognizes that the simple password security is disabled but is receiving updates with the password, the following warning appears:

```
#   24: 12/27/1999 09:44:43.918  WARNING  SLOT  8  IP                    Code:  75
RIP Authentication failed for source 172.16.194.1 on interface 172.16.194.254
Security disabled, received Address Family Identifier ffff Authentication Type 2
```

Routes are not accepted for entry in the routing table when an authentication type mismatch occurs. This is also true when authentication is enabled but the passwords of the adjacent routers do not match. A password of up to 16 characters can be configured for simple password authentication with RIP2. In the following example, the last character in the locally configured password is mistyped (*chalupu* instead of *chalupa*). Each character is displayed as a hexadecimal number in the warning message.

```
#   72: 08/03/1999 09:32:55.550  WARNING  SLOT  4  IP                    Code:  76
RIP Authentication failed for source 192.168.112.101 on interface 192.168.112.100
Passwords did not match
Received Key   63:68:61:6C:75:70:61:00:00:00:00:00:00:00:00:00
Configured Key 63:68:61:6C:75:70:75:00:00:00:00:00:00:00:00:00
```

To configure simple password authentication, set the `version` attribute of the RIP2 interface to rip2, specify the password in the `authentication` attribute, and set the `authentication-type` attribute. The following example shows how to enable RIP2 on an IP interface with simple password authentication.

```
 ...ip/172.16.147.101/255.255.255.0
fp3(93)# rip
 ...rip/172.16.147.101
fp3(94)# version rip2
 ...rip/172.16.147.101
fp3(95)# authentication-type simple-password
 ...rip/172.16.147.101
fp3(96)# authentication chalupa
 ...rip/172.16.147.101
fp3(97)# info
... ...
version rip2
triggered-updates disabled
authentication-type simple-password
authentication chalupa
... ...
```

Open Shortest Path First (OSPF) Protocol Like RIP2, OSPF provides a minimal level of security on a per-area basis with simple password authentication. Each area can be uniquely configured with a password that must match between routers in the same area. A router receiving a packet without a password or with an incorrect password does not process it. Recall that the formation of adjacent neighbors is essential for OSPF to exchange link state information. Problems with password authentication will be evident soon after neighbor detection.

```
#   20: 04/19/1999 08:00:47.762  WARNING  SLOT  8  OSPF              Code:  51
C1: Packet Rejected: AUTH TYPE(1)
src 172.16.147.101 dst 224.0.0.5 routerid 172.16.19.1
```

If the authentication type is 1, then the neighbor proposed the use of no authentication while the local router proposed the use of simple password authentication (type 2).

When neighboring routers agree on the authentication type, the passwords must match between the two routers for an adjacency to form.

```
#  134: 02/21/1999 11:42:01.516  WARNING  SLOT  8  OSPF              Code:  52
C1: Packet Rejected: AUTH KEY(63:68:61:6C:75:70:61:00)
    src 172.16.157.1 dst 224.0.0.5 routerid 172.16.157.1
```

The configuration of OSPF simple password authentication is similar to that of RIP2, with simple differences. Authentication as a whole is enabled on a per-area basis, and the password is configured on each interface. The following example shows the configuration for the Backbone area and an IP interface 172.16.231.101 residing in it.

```
...area/0.0.0.0
fp3(32)# authentication-type simple-password
 ...area/0.0.0.0
fp3(33)# box; ethernet 12/1; ip 172.16.231.101/24; ospf
 ...ospf/172.16.231.101
fp3(34)# authentication m0nkey9x
```

The vulnerable points in OSPF areas are broadcast networks. Routers periodically send hello multicasts to all other neighbors on a broadcast network. This is also done to discover new neighbors on the network. In enterprise networks where peripheral control of endstations is not strict, an intruder can snoop through the network for hello packets and attempt to form an adjacency. To prevent unauthorized neighbors on a network segment, OSPF interfaces that are usually configured as broadcast interfaces can be set to NBMA so that previously well-known neighbors are statically defined. This static definition of neighbors is complicated; however, it is a definitive solution for preventing unauthorized devices from participating in the OSPF network.

Border Gateway Protocol (BGP) The preceding sections discussed simple authentication mechanisms available for RIP2 and OSPF. In accordance with RFC 2385, BayRS supports the protection of BGP sessions between peers using TCP Message Digest 5 (MD5) signatures. Two peers in a session share a common secret key that is used to compute a digest signature for each TCP packet. The secret key is never sent between the two peers, and the receiving BGP speaker compares the digest with the value computed locally for the same TCP packet. A third-party device masquerading as a peer cannot generate a valid MD5 signature because it does not have the secret key—the MD5 algorithm generates statistically unique digest values for each packet.

TCP MD5 authentication can be configured on a per-peer basis. By default, MD5 authentication for BGP-4 sessions is disabled. The same BGP speaker can have multiple peer sessions, some of which may be selectively authenticated. The TCP stack on each BGP peer follows these steps to determine the authenticity of the message:

- TCP checks the kind value, which must be set to 19. A packet received without this value does not have an MD5 signature, and the following message is reported. This condition indicates that the remote peer is not set to use MD5 authentication.

```
#   85: 01/24/1999 15:37:36.379  WARNING  SLOT  3  TCP                Code:  42
    TCP MD5: No Signature 192.32.104.65,179 - 192.32.104.66,179
```

- If TCP received a datagram with the kind value set to 19, it first checks to ensure that MD5 authentication has been configured and enabled. If a key is not configured on the local router, then the following message is logged. The TCP connection, however, is not closed. This is a configuration issue on the local router, where a key has not been configured.

```
#  164: 12/27/1999 14:08:39.106  WARNING  SLOT  4  BGP                Code: 230
BGP TCP MD5: No Key Configured 172.16.104.33 - 172.16.104.34
```

- If a local key is configured but is encrypted, then the node protection key (NPK) stored locally in the router's NVRAM is used to decrypt the session key. The NPK is configured using the secure shell (ksession) that is accessible only from the TI/BCC console. If an NPK is not available in the NVRAM, then the router reports the following message. When this occurs, enter the secure shell and reenter the NPK so that it is updated on all slots.

```
#  141: 05/16/1999 00:21:36.707  WARNING  SLOT  1  BGP                Code: 228
BGP TCP MD5: No NPK Configured 172.16.104.33 - 172.16.104.34
```

- If a valid key is available, TCP uses it to compute the digest and compare it to the value contained in the received frame. If the signatures match, then the packet is delivered to the BGP process. Otherwise, a warning is logged, and the frame is dropped. This condition indicates that the two peers have mismatched keys.

```
#   33: 05/08/1999 00:21:36.231  WARNING  SLOT  7  TCP                Code:  44
    TCP MD5: Invalid Signature 172.16.104.65,179 - 172.16.104.66,179
```

A simple peer configuration with MD5 authentication enabled is shown in the following example. It is assumed that a NPK has previously been configured through the secure shell. Encryption of the secret key requires the NPK. If the key is stored as clear text in the MIB, then the NPK is not required. If the configuration file is compromised with the clear-text key, then the integrity of all BGP sessions on the router is compromised. In high-security environments, secure storage of the key in encrypted form is recommended.

```
 ...peer/172.16.125.13/172.16.125.14
fxc9(20)# tcp-authentication md5
```

```
 ...peer/172.16.125.13/172.16.125.14
fxc9(21)# tcp-md5-key m0s+ducky1
 ...peer/172.16.125.13/172.16.125.14
fxc9(22)# tcp-md5-key-storage encrypted
```

Firewall-1

BayRS integrates the stateful inspection module and logging capabilities of the Check Point Firewall-1 product. The firewall is a software service in BayRS. It is a sophisticated inspector of data traffic crossing the router and is used to strictly control traffic flow.

Uniform traffic filters usually are configured on a router to accept all traffic and deny certain unacceptable frames. Traffic filters can be applied only on a frame-by-frame basis, and the filtering mechanism does not maintain a history for stateful knowledge of the traffic flow. (The additional information gleaned from maintaining flow and other session-related data is called *stateful knowledge*.)

A firewall, on the other hand, is designed with the intent that all traffic passing through will be dropped unless it is explicitly authorized. The BaySecure Firewall-1, when enabled on supported BayRS platforms, by default disallows all types of traffic, thus behaving like a true firewall. Specific policies must be configured to authorize certain traffic flows through the firewall. This type of implementation is ideal for connections to the Internet, where any and all types of traffic may pass through and would be difficult to categorize. Using a firewall in this scenario normally assumes that all traffic is unauthorized unless previously authorized using a policy.

The firewall in BayRS has three components:

- **Management station** The Check Point Firewall-1 management console software must be installed on a management device, which is used to define security policies and will automatically provide the policies to the BayRS router at startup. Backup management stations can be provisioned for redundancy.
- **Logging station** The firewall service on the router will periodically generate events (different from those in the BayRS event log), which will then be sent to the logging station. In most cases, the logging station is the same as the management station running the Firewall-1 console.
- **Router** The router is the core component of the firewall service. Once the policies are downloaded from the management station, the router uses the stateful inspection module to dynamically filter unauthorized traffic.

The BaySecure Firewall-1 product operates in a more sophisticated fashion than traditional traffic filters and is more efficient in monitoring traffic in a stateful fashion. BayRS does not support other modules supported by the Check Point Firewall-1 product. Routers that have unmanageable traffic filters can be considered for use with Firewall-1. These, however, do not eliminate the need for traffic filters, which may be a better solution in simple implementations.

Firewall Issues

The Firewall-1 software on BayRS significantly changes the behavior of the router. In particular, the firewall forms a shield around the router so that all traffic is dropped. Once the firewall is enabled, the configuration on the router must be saved, and the router must be rebooted for the firewall to become operational. Issues with Firewall-1 commonly arise during initial configuration and startup.

The following checklist is useful in isolating general problems with firewall initialization:

- Check the IP configuration on the router. At least one IP interface must be up and operational for the firewall agent to communicate with the management station.
- The firewall agent uses TCP for reliable communication with the management and logging stations. TCP is a system service that is not enabled by default, but it is automatically enabled when Firewall-1 is configured. Make sure that TCP is loaded and operational on all slots.
- The management station may reside on either the local subnet or a remote network. If the management station is remote, a static route or a default route to reach it must be available on the router.
- Both the management station and the router firewall must have synchronized firewall keys configured using the **fwputkey** command. If the authentication fails, the following message is logged in the router under the `RFWALL` entity. This indicates that either the management station or the router, or both, must have the firewall key reinstalled.

```
# 199: 03/22/1999 16:59:38.909  WARNING  SLOT 11  RFWALL               Code:  26
Authentication of skey failed for automatic LOADME process with server
143.102.99.197
```

- The management station must have at least one security policy configured on it for downloading to the firewall agent on the router. The firewall agent must be represented as a network router object on the management console.
- If the router is being installed for the first time, or if the firewall is being deleted, the configuration must be saved and the router must be rebooted for the changes to take effect. When the router restarts, it will initialize with or without the firewall agent, respectively.

Interaction with NAT and Filters

We mentioned earlier that Firewall-1 behaves like a shield around the router. Data traffic must first be accepted for delivery by the firewall agent and then can be processed by the routing software in the router. There are significant considerations with this behavior, when network address translation (NAT) and traffic filters are implemented along with Firewall-1.

Certain network interfaces currently do not simultaneously support Firewall-1 and NAT. Simultaneous support for Firewall-1 and NAT depends on the BayRS version, so consult the documentation for the version you are using. NAT is commonly used for dynamic translation, but when it is used for static translation, Firewall-1 can easily be configured with a policy to accept data. The following considerations must be applied when configuring the security policy:

- On the local NAT interface side, the firewall must accept inbound frames sourced with the untranslated private address.
- On the global NAT interface side, the firewall must accept inbound frames destined for the translated global address.

In both cases, selected higher-layer protocols can be included or excluded from the policy. The difficulty arises with dynamic translations, where port numbers are dynamically allocated as available, thus making it harder to selectively accept frames on the global interface. The firewall does not directly interact with the NAT software on the router.

The considerations applicable for NAT are also generally applicable to uniform traffic filters. Unlike with Firewall-1 and NAT, traffic filters do not maintain stateful information. Traffic filtering must occur after the data has passed through the firewall. In most cases, the filtering can be performed using a firewall security policy except where the decision must be made within the routing function. Route and service filters are similarly applied only after data has passed through the firewall. If the router configuration requires uniform traffic filters and route or service filters, remember to configure appropriate policies on the firewall agent to pass this traffic to the router for processing.

IP Security

IPSec is a set of emerging standards developed by the IETF for secure IP communication over public networks. The applications for IPSec are broad ranging, from remote access systems to extranet and intranet virtual private networks (VPNs). The security architecture is built around the existing IPv4 Internet protocol and the next-generation protocol, IPv6. In general, the public Internet is an easy target for various threats, such as invasion of privacy, compromise of data integrity, spoofing of identities, and denial of service. Numerous RFCs are applicable to the deployment and use of IPSec, including RFCs 2401 through 2412.

The goal of IPSec is to provide three essential services for IP traffic between security gateways:

- **Confidentiality** This service ensures the privacy of the communication between the sender and the receiver. Confidentiality is achieved by encrypting and decrypting IP frames between security gateways. The Encapsulating Security Payload (ESP) protocol uses the Data Encryption Standard (DES) algorithm to encrypt and decrypt data packets.

- **Integrity** This service helps ensure that the data has not been altered during transit. The ESP protocol uses HMAC MD5 and HMAC SHA-1 algorithms to ensure data integrity between security gateways.
- **Authentication** This service helps ensure that the data transmitted originated at the identified device.

Data between two IPSec gateways is transported using regular IP datagrams. They appear as normal routing packets to intermediate routers. The source and destination gateways are the only devices involved in providing the IPSec services. One of the most desirable aspects of IP security provided by gateways is that it automatically provides data security for applications that do not natively support a security mechanism. IPSec can be applied to any IP traffic passing between two IPSec-capable gateways, without any modification to the existing infrastructure.

Two types of IPSec frame formats are currently defined to provide the services described earlier:

- **Authentication header (AH)** The AH frame format defines the data payload and is used for data integrity and authentication. Either the Message Digest 5 (MD5) algorithm or Secure Hashing Algorithm 1 (SHA-1) is used to ensure the integrity of the data. The unique signature in the payload helps ensure that only the trusted source with a known secret key could have originated the packet.
- **Encapsulating Security Payload (ESP)** The ESP packet also defines a frame format, which contains an optional authentication trailer. The data payload is encrypted using an industry-standard algorithm such as 56-bit DES or even stronger options such as 3-DES. The encrypted data portion and the authentication trailer help ensure confidentiality, integrity, and authenticity.

The two frame formats, AH and ESP, are illustrated in Figures 7-1 and 7-2 respectively.

A router's IP interface is configured to operate as an IPSec security gateway when the IPSec protocol is enabled on the interface. When IPSec is configured, the security gateway can be used to protect data traveling from the trusted local network (for example, the corporate intranet) to an untrusted network (for example, the public Internet). The security gateway uses IPSec security policies to determine how inbound and outbound data should be processed. If data frames matching certain IP criteria require security, the security policies redirect them for processing by the IPSec software; otherwise, the data is sent to the IP routing software for processing as normal frames. The processing of protected data frames is defined using security associations (SAs) in both the inbound and outbound directions. Figure 7-3 shows a simple IPSec environment with trusted and untrusted networks using multiple security gateways.

IPSec can be applied on an end-to-end basis between endstations or between gateways bordering the untrusted network. BayRS routers behave like security gateways bordering the untrusted network, and this type of IPSec configuration is usually called

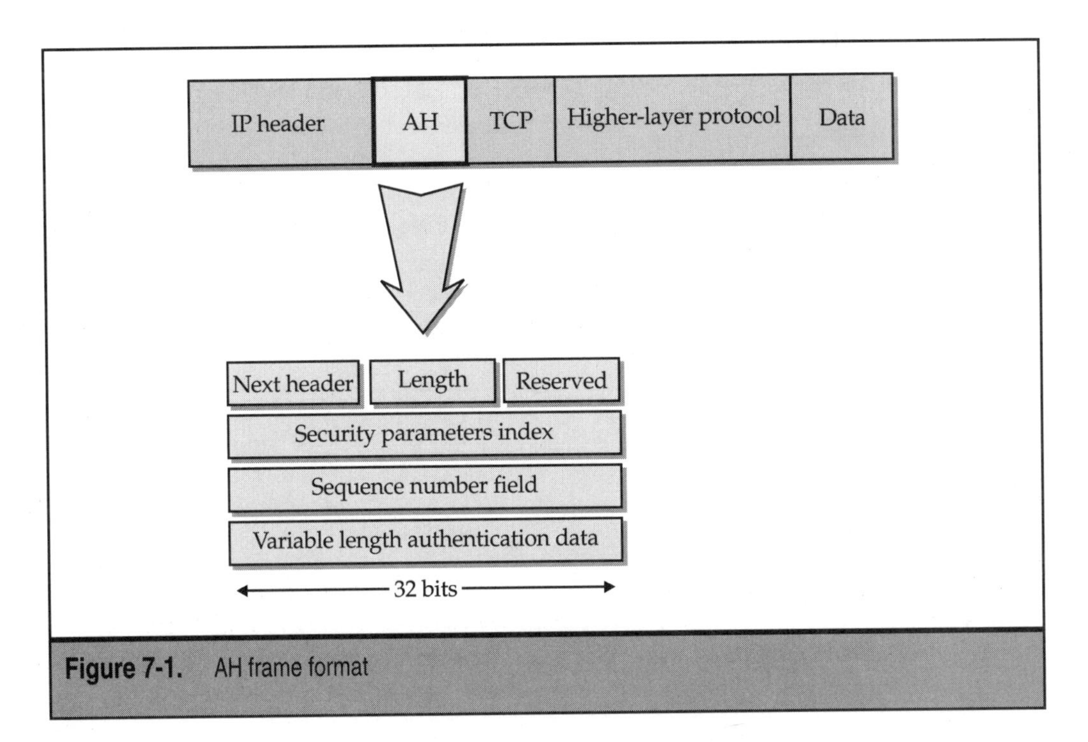

Figure 7-1. AH frame format

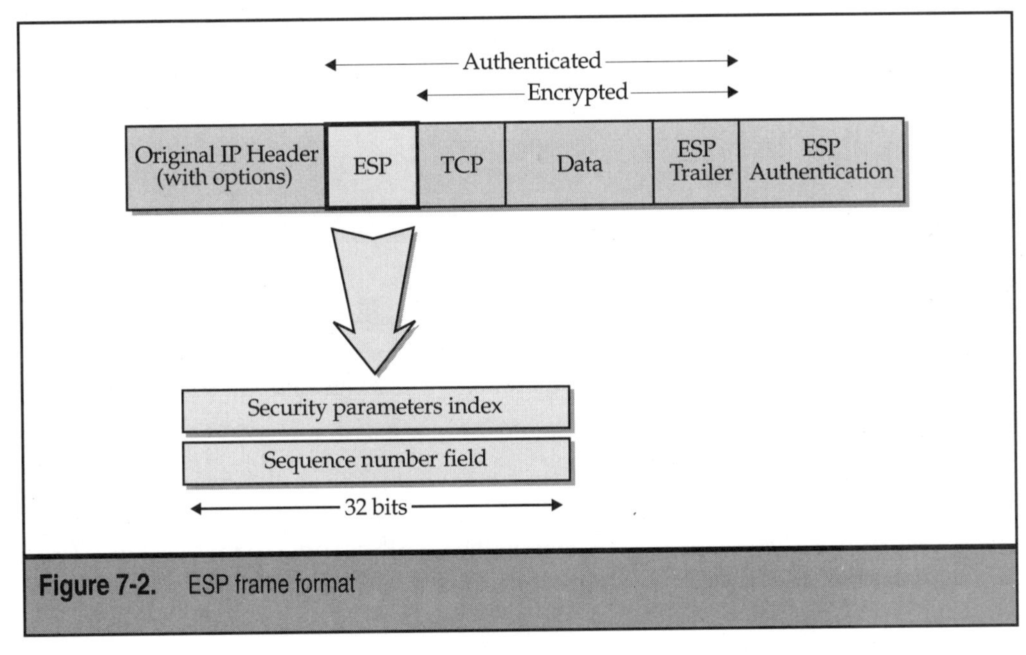

Figure 7-2. ESP frame format

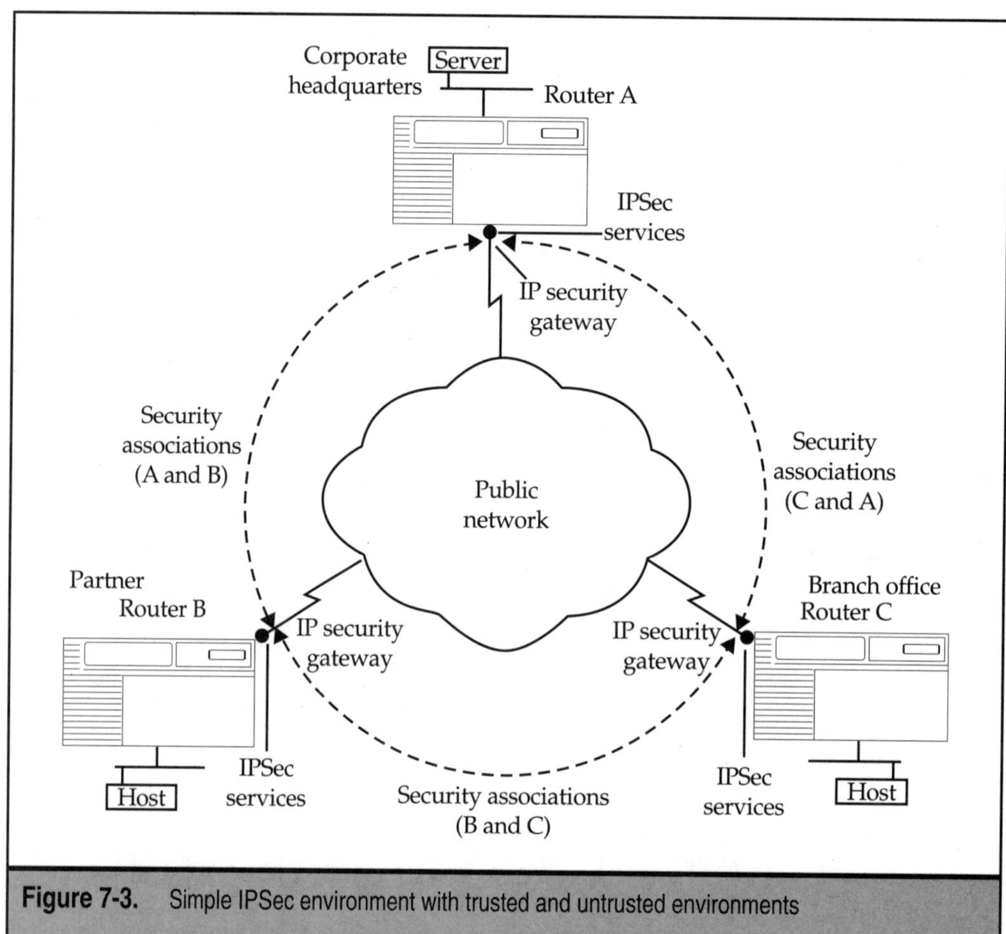

Figure 7-3. Simple IPSec environment with trusted and untrusted environments

tunnel mode, because the secured data is tunneled across the untrusted network without the contents being visible to the intervening routers.

We have already mentioned security gateways. Two other components are crucial for the deployment of IPSec:

- **Security policies** These help define how the security gateway processes data to and from another security gateway.
- **Security associations** These help define the IPSec services for selected datagrams sent between the security gateways.

Once the IPSec protocol has been enabled on an IP interface, the interface begins to behave like a security gateway. The application of security policies and security associations to an IPSec gateway is illustrated in Figure 7-4.

Troubleshooting

The fundamental concepts of IPSec were discussed in the preceding section. To configure and deploy IPSec, consult *Configuring IPSec Services*, listed as a reference at the end of this chapter and available via this book's web site. IPSec requires the interaction of multiple protocols to provide the required services, and this complex interaction makes troubleshooting an involved task. When beginning to troubleshoot an IPSec problem, first make sure that the IP addresses of the security gateways are up and operational. Ping the IP ad-

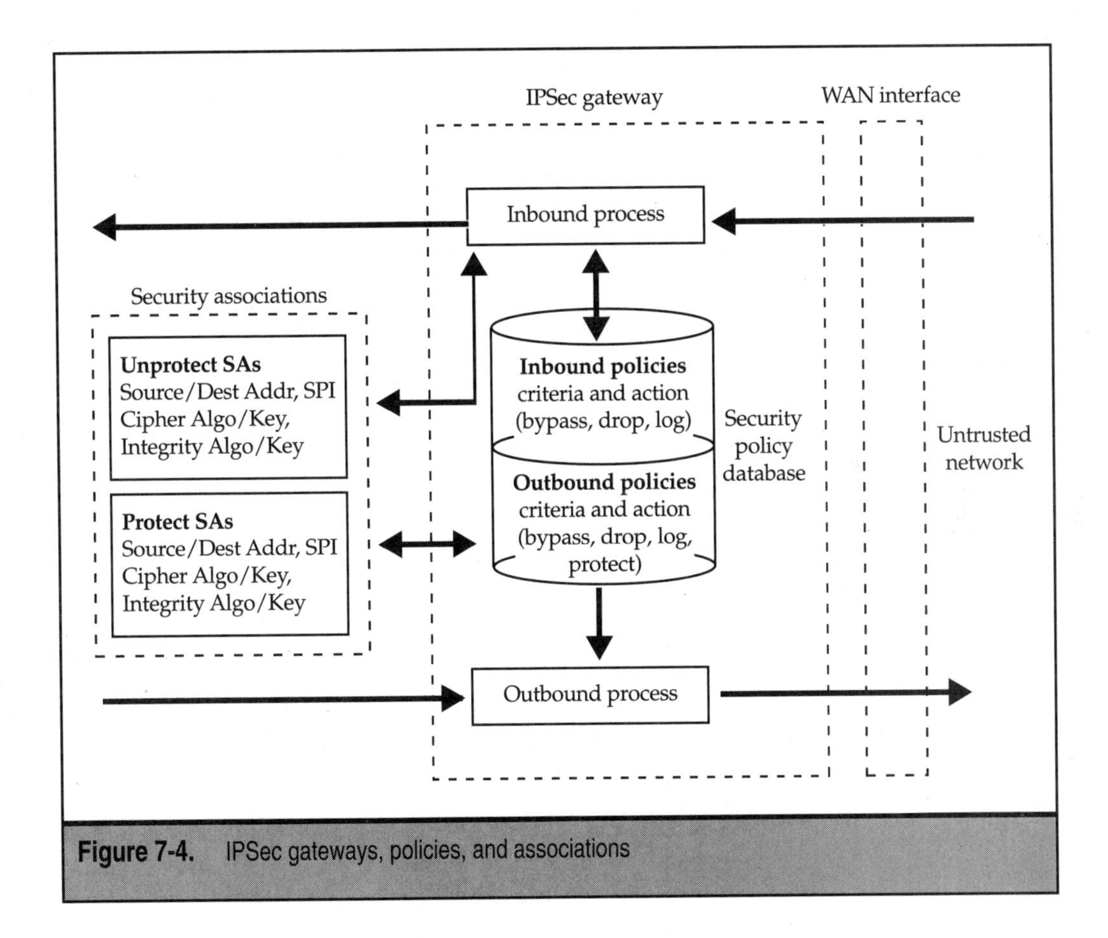

Figure 7-4. IPSec gateways, policies, and associations

dresses of the security gateway interfaces from the local routers to ensure that they are receiving and sending IP datagrams.

The following four log entities are most useful in troubleshooting IPSec problems:

- `IPSEC` The IPSec protocol service on each slot
- `IKE` The Internet Key Exchange protocol, which automates the secure authentication of keys used for security associations
- `CAPI` The Cryptographic Application Programming Interface (CAPI), which provides the encryption, decryption, and digest functions
- `KEYMGR` The Key Manager system service used to verify the validity and integrity of keys and the configuration

Key Manager

The node protection key (NPK) is crucial to the integrity of a router. It is stored in the NVRAM of a router, and in each individual slot in the case of multi-slot routers. A hash value for the NPK is stored in the configuration file to protect the configuration and the router together. The Key Manager is responsible for verifying the integrity of the system keys and, in particular, the NPK. The NPK is set once on a router using the secure shell (ksession), which is accessible only from the TI console.

During initialization of a router, the Key Manager runs through a standard battery of tests. It generates three types of warnings, depending on the inconsistencies found. The following warning indicates that an NPK exists on the routing node, but the configuration does not have an NPK hash. This may indicate a condition where an existing router was reconfigured without security services or where an existing router was booted with an incorrect or a compromised configuration.

```
#   70: 06/05/1999 07:02:57.084  WARNING  SLOT  6  KEYMGR              Code:   4
NPK exists; config missing NPK hash.
```

```
#  167: 03/06/1999 24:23:14.149  WARNING  SLOT 13  IPSEC               Code:  12
Null NPK hash retrieved from config; An NPK has been detected in the NVRAM;
Please update config with kset npk command
```

The preceding warning may indicate a serious condition, where no security services, such as IPSec and WEP, function properly. If the configuration is indeed the right one, reissue the **kset npk** command and save the configuration to flash to update the NPK hash in the MIB.

A more serious problem occurs when the hash value of the NPK stored in NVRAM does not match that in the configuration file. This can occur if the configuration file was corrupted or otherwise compromised, or if the configuration was installed in a router with an existing NPK. In the latter case, a new NPK must be set, or the configuration's hash must be translated to that of the existing NPK.

```
#  128: 12/11/1999 01:20:55.390  WARNING  SLOT  4  KEYMGR           Code:   5
Hash of NPK doesn't match config's NPK hash.

#  131: 03/18/1999 00:03:54.225  WARNING  SLOT  9  IPSEC            Code:  13
Hash of NVRAM NPK does not match config's NPK HASH; Please use ktranslate to
synchronize the config with the router's NPK
```

Occasionally, the configuration file may contain an NPK hash, but the router may not have one stored in the NVRAM. This can occur if processor hardware is replaced and the existing image and configuration files on flash are used.

```
#   59: 03/18/1999 16:03:50.057  WARNING  SLOT  3  KEYMGR           Code:   6
Config has NPK hash; NPK is missing.

#  156: 04/03/1999 14:28:51.442  WARNING  SLOT  8  IPSEC            Code:  14
NPK retrieval from Novram failed;
Please enter a NPK on the router with kset npk
```

CAUTION: When a single-slot router or a processor module in a multi-slot router is replaced, the **kset npk** command must be reissued to update the NPK on all modules. If the NPK is missing from the NVRAM, the new slot may not perform any encryption services if it is configured to do so.

When beginning to troubleshoot an IPSec or encryption problem, always check the KEYMGR entity first to verify that there are no warnings. Key Manager events are reported during initialization only and must be looked at soon after network problems are reported.

Cryptographic API

The Cryptographic API provides all the necessary lower-level functions to the IPSec software. These lower-level functions include encryption and decryption algorithms such as DES and 3-DES, and hashing algorithms such as MD5 and SHA-1. The CAPI is bundled as a separate software application and must be added to the image archive using Image Builder (see Chapter 3). A default CAPI application (capi.exe) is included with each software image, but it is called a stub image and is generally less than 1K in size. The dynamic loader can load this stub image, but it does not provide the required functions. The size of the stub image can be uniquely distinguished in a loadmap.

```
box# loadmap
Loadmap from SLOT 1:
--> ipsec.exe        0x024b28e0  0027896
--> capi.exe         0x025201d0  0000552
--> http.exe         0x024f9990  0156708
--> ppp.exe          0x024763a0  0247084
--> ftp.exe          0x024b95f0  0047628
--> tcp.exe          0x024e4370  0071148
--> snmp.exe         0x02521de0  0031068
--> tn.exe           0x02529750  0044404
```

The Image Builder software must be used to incorporate the full-fledged CAPI application into the archive. If the application is not incorporated correctly, or if it is missing, the IPSEC entity reports the following messages:

```
#    9: 05/12/1999 18:01:28.335  WARNING  SLOT 13  IPSEC                Code:  10
No Cryptographic API (capi.exe) in this image. IPSec cannot proceed.
interface: 132.245.94.11, circuit: 62, code: 15

#    4: 12/11/1999 13:12:19.026  WARNING  SLOT 12  IPSEC                Code:   9
Failed to initialize a session with the Cryptographic API,
interface: 192.32.125.101, circuit: 12, code: 15
```

If IPSec is unable to register with the CAPI application, all required service functions will be unavailable. Data frames requiring protection cannot be encrypted in an ESP or authenticated using AH when this occurs. After the Key Manager checks and before the operation of the IPSec security gateways, the operation of the CAPI application must be verified.

Internet Key Exchange Operation

IKE is the key management technique that dynamically authenticates keys used by security associations such as the ESP and the AH. For BayRS 13.20 and later, IKE automates the key exchange process, not requiring the manual keying methods used for various protocols, including IPSec in BayRS 13.10. During initialization of a session between two security gateways, IKE dynamically authenticates each gateway and the keys used to exchange the security associations.

During initialization, each security gateway must protect the key exchange process because it occurs across a public, untrusted network. The Cryptographic API is used for this purpose and must be available for use by IKE. If the CAPI software is unavailable, IKE reports a warning and cannot proceed further—the capi.exe application must be added to the archive, and the router must be reinitialized.

```
#  110: 02/01/1999 18:42:49.036  WARNING  SLOT 10  IKE                  Code:   4
No Cryptographic API (capi.exe) in this image. IKE cannot proceed, code: 15
```

The Internet Key Exchange implementation in BayRS operates in quick mode (QM) and main mode (MM), and several configuration mismatches can prevent the negotiation and key exchange from occurring. Some of the warning messages are self-explanatory, but they may also indicate other related problems.

```
#   71: 07/10/1999 09:39:26.637  WARNING  SLOT  9. IKE                  Code:   7
IKE SA negotiation could not be started, check if 205.92.145.100 configured
```

The preceding message indicates that the peer security gateway may not have IKE configured; the following message indicates that a peer has been incorrectly associated with the wrong gateway interface.

```
#  173: 01/20/1999 03:25:20.539  WARNING  SLOT  5  IKE            Code:   8
IKE peer 132.245.155.92 not configured for interface 132.245.104.1,
SA negotiation not started
```

The following message indicates that peer IKE parameters are mismatched and must be reverified in the configuration.

```
#  187: 07/02/1999 08:55:58.241  WARNING  SLOT  7  IKE            Code:   9
Cookies did not match properly for IKE peer 205.194.32.11,
interface 206.213.192.50
```

The preceding messages commonly occur in main mode operation.

Preshared secrets between two security gateways must be carefully administered, and care must be taken to ensure that the secret is not compromised. A mismatched preshared key indicates a configuration error, and this condition is reported as a warning by IKE.

```
#  145: 01/28/1999 13:05:52.143  WARNING  SLOT  7  IKE            Code:  14
Malformed encrypted packet received from 191.132.142.101 on
interface 132.245.155.134. Possible cause: misconfigured pre-shared secret
or data corruption.
```

In quick mode (QM) operation, IKE reports the following types of messages when there is a mismatch in the configuration. Generally, correctly matched parameters allow proper negotiation during MM and QM operation.

```
#  131: 07/10/1999 24:11:30.209  WARNING  SLOT  9  IKE            Code:   6
Hash mismatch for the Quick Mode exchange ID 0x9c3

#   64: 04/07/1999 20:24:15.590  WARNING  SLOT  8  IKE            Code:  11
Quick Mode 0x97d6: Error in Key agree object
```

IPSec Operation

The `IPSEC` log entity is useful in isolating and understanding IPSec operation issues. Just as a chain is only as strong as its weakest link, the security policies applied to a security gateway largely determine the integrity of the data. Policies that are too restrictive can render many required applications without a secure path, while a too-lenient set of policies can compromise the security goals.

The TI provides an embedded **ipsec** command, which can be used to set debugging, trace, and audit options during operation. With extended debugging and tracing enabled, IPSec generates additional event messages that can identify several problems. The syntax of the **ipsec set** command is illustrated here; to invoke the command, use **tic** in BCC.

```
ipsec     <sub_command> [<options>]

ipsec set - allows setting of the IPSec extended debug log levels
```

```
Usage:   -d [pol|fwd|inb|tep] [on|off] [<cct name>]: Sets debug log level
         -t [pol|fwd|inb|tep] [on|off] [<cct. Name>]: Sets trace log level
         -a [none|debug|info|warn|fault|trace] [on|off] [<cct name>]:
               Sets the audit logging level
```

Table 7-1 details the debug and trace options; the audit log levels are self-explanatory. By default, minimal audit messages are logged because the auditing for all frames can consume a substantial portion of the event log memory.

Security Policies IP datagrams must correspond to appropriate policies for the desired action to take effect. Normally, BayRS does not generate specific event entries when a policy matches and when a certain action is taken. Using extended tracing and debugging on the policy debug option, it is possible to view policy hits for IP datagrams.

```
# ipsec set -t pol on S34_PPP
# ipsec set -d pol on S34_PPP
```

Note that multiple **set** commands can be issued to specifically turn on or off certain trace and debug messages. Policy matches and processing are typically reported as trace messages.

```
#   37: 05/08/1999 15:37:36.043  TRACE   SLOT  1  IPSEC               Code:  42
spdSearchOutbound: destination address = 142.204.152.11

#   38: 05/08/1999 15:37:36.044  TRACE   SLOT  1  IPSEC               Code:  44
No outbound policy hit ===> DROP
```

Debug Option	Description
Policy (pol)	Trace and debug policy hits are reported in the event log. By default, policy hits for IP datagrams are not logged.
Forwarding (fwd)	Extended events are logged by the forwarding gate, including buffer processing.
Inbound (inb)	IPSec inbound gate messages are logged with this debug or trace bit enabled.
Tunnel Endpoint (tep)	Datagram and other service processing performed by the IPSec Tunnel Endpoint gate report extended debug and trace messages.

Table 7-1. IPSec Extended Debugging and Trace Options

The second trace message states that there was no outbound policy hit for the specified destination packet, indicating that the policy configuration is incorrect or inadequate. The policy illustration applies to the outbound direction on a security gateway; inbound policy events are similar. More detail is available on policy hits when extended logging for IPSEC auditing is turned on. In particular, enabling warning and informational logging can help locate the sources of various problems.

```
# ipsec set -a warn on S148_PPP
# ipsec set -a trace on S148_PPP
# log -ffw -eIPSEC_AUDIT
#  135: 03/10/1999 20:31:34.877  WARNING  SLOT  9  IPSEC_AUDIT        Code:   1
Discarding packet: Interface 134.177.19.120, Circuit 51, Outbound
Cause: No policy match
IP Pkt Info: Src: 134.177.18.240, Dest: 134.177.12.101, Prot: 6, Tot. Len: 210
```

If the local outbound policy is corrected, the IP datagrams can be processed successfully, as indicated in the following event message.

```
#   66: 05/08/1999 15:38:33.464  TRACE    SLOT  1  IPSEC              Code:  47
Interface 142.19.210.1, Circuit 12.
Outbound policy 17 hit...APPLY bit set
```

Application of a local outbound policy must correspond to the application of a remote inbound policy on the peer security gateway. If datagrams are still not being processed and delivered correctly, there may be a problem on the remote gateway. The event log on a remote gateway may report the following message if the security association is misconfigured, which may include incorrect configuration of cipher or encryption keys.

```
#  141: 10/21/1999 05:18:45.850  TRACE    SLOT  8  IPSEC              Code:  36
spdSearchInbound: destination address = 132.245.155.134

#  142: 10/21/1999 05:18:45.852  WARNING  SLOT  8  IPSEC              Code:   6
IPSec Interface 132.245.155.134, Tunnel Endpoint gate failed to process buffer,
IP Pkt Info: Src: 132.245.150.92, Dest: 132.245.155.34, Prot: 50, Tot. Len: 112
```

In these cases, the policies are functioning as configured, and the problem may be the security associations. Aside from looking at the log, check the IPSec policy configuration statistics to determine whether policy hits are actually taking place. The **show ipsec** command supports various subcommands, such as the **policy** subcommand. This subcommand displays a summary of the policies, criteria, and action.

```
# show ipsec policy out
IPSEC Outbound Policy Information
---------------------------------
Circt                    Pol Policy          Low                 High              Act
Num   Interface          Num Name            Range               Range
----- ---------------    --- -------------------------------     --------------- ---
7     142.212.91.1       1   Protect_EngNet  S-142.212.16.0      142.212.19.255  AP
                                             D-142.212.64.0      142.212.67.255
```

```
7      142.212.91.1     2   ICMP_Bypass       P-1                 1                BY
7      142.212.91.1     3   Protect_SvcNet    S-142.212.20.0      142.212.23.255   AP
                                              D-142.212.68.0      142.212.71.255
```

TIP: ICMP datagrams (IP protocol 1) used for testing IP connectivity must be allowed to pass through using a bypass policy, unless they are already protected using another policy with a security association. Although ICMP traffic is a frequent source of network attacks, directed broadcasts can be disabled so that legitimate ICMP datagrams can be allowed. A single protocol-specific policy applied to ICMP can be used instead of multiple policies for specific source and destination addresses.

The application of the policies can be viewed using the **selectors** subcommand. The following example shows the corresponding outbound selectors for the preceding policy list. The Matches column will show increasing values as the policy is applied to more outbound datagrams. A stagnant value in this column indicates that the policy is misconfigured or that IP datagrams matching the criteria have not crossed the router.

```
# show ipsec selectors out
IPSEC Outbound Selector Table Information
-----------------------------------------

                                                          Pol Fragment
Circuit # Interface        Mode     State    Matches Num Number
--------- ---------------  -------- -------  ------- --- --------
7          142.212.91.1    Enabled  Up       8573      1     1
7          142.212.91.1    Enabled  Up       259       2     1
7          142.212.91.1    Enabled  Up       592       3     1
```

Security Associations Policies are linked to security associations, which define specific protection parameters for the matching datagrams. Security associations are unidirectional—a local outbound protect (encryption) SA is matched with an inbound unprotect (decryption) SA on the remote router. This SA pair defines the security services in one direction only; a similar protect/unprotect pair of SAs must be configured for the reverse direction for the same services to be applied, as illustrated in Figure 7-5. BayRS routers operating as IPSec gateways support only tunnel mode SAs and do not use transport mode SAs, the latter normally being used by endstations.

Each security association is uniquely identified by its security parameters index (SPI). The SPI is a 32-bit value used by IPSec in combination with the destination address and the security protocol identifier to generate a number that uniquely identifies the security association. This value ranges from 256 to 65535. The IPSec policies, when configured correctly, help determine the IP datagrams to which each security association applies. SAs can act on an IP datagram by encrypting it or authenticating it, or both.

A frequent problem is the absence of a security association, even though the policy configuration is valid. As the following message illustrates, a corresponding protect SA was not found for outbound datagrams that matched an outbound policy.

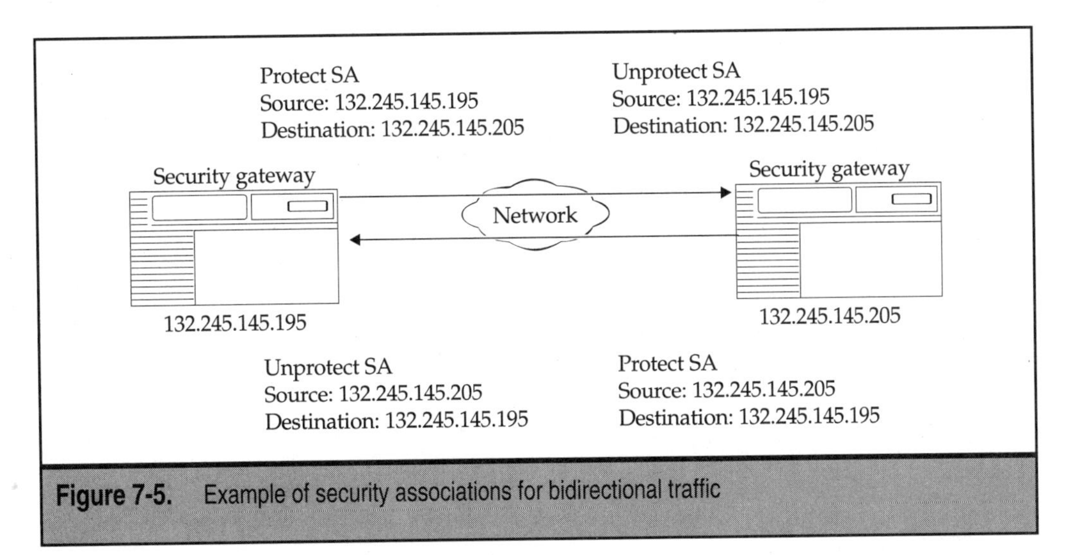

Figure 7-5. Example of security associations for bidirectional traffic

```
#   26: 10/21/1999 00:10:33.512  WARNING  SLOT 14  IPSEC              Code:  11
No Protect SA has been found for an outbound packet which matched an apply policy.
The packet will be dropped.
This can be caused by misconfiguration and needs to be corrected
interface: 132.245.155.137, circuit: 9
```

The absence of an SA for a specific SPI and protocol is also recognized and is reported as an IPSEC_AUDIT warning. Currently, the audit process reports warning and informational messages to indicate protocol operation.

```
#  100: 08/07/1999 03:12:07.466  WARNING  SLOT  4  IPSEC_AUDIT        Code:   2
Discarding packet: Interface 202.124.17.101, Circuit 83, Tunnel Endpoint
Cause: SA search failed
IP Pkt Info: Src:205.192.56.47, Dest: 202.124.17.101, Prot: 50, Tot. Len: 115
IPSec Pkt Info: SPI 272, Sequence Number 928

#  114: 08/07/1999 03:12:12.466  WARNING  SLOT  4  IPSEC_AUDIT        Code:   2
Discarding packet: Interface 202.124.17.101, Circuit 83, Tunnel Endpoint
Cause: No SA found
IP Pkt Info: Src:205.192.56.47, Dest: 202.124.17.101, Prot: 50, Tot. Len: 118
IPSec Pkt Info: SPI 273, Sequence Number 155
```

Several layers of protocol and service interaction must take place for IP datagrams to successfully receive security services. Once the validity of the policy is verified and the correct SA is associated, SA parameters still must be matched between two security gateways. Mismatched SA parameters are another common source of IPSec problems. Cipher and integrity key mismatches, particularly in manual SAs, are easy to identify and resolve using the log. For example, a typographical error may be the cause of the mismatch, since cipher and integrity keys are protected carefully, and Configuration Manager does not redisplay them. Several types of warnings can be seen in the log, as shown here.

```
#  178: 02/25/1999 19:26:21.816  WARNING  SLOT 12  IPSEC                Code:   6
IPSec Interface 142.212.94.101, Tunnel Endpoint gate failed to process buffer,
IP Pkt Info: Src: 142.212.47.241, Dest: 142.212.94.101, Prot: 50, Tot. Len: 80

#   34: 02/01/1999 13:14:17.008  WARNING  SLOT  2  IPSEC                Code:   7
Failed to decrypt the integrity key for a security association,
SA source: 132.245.155.135, SA destination: 132.245.120.101, SPI: 257

#  186: 06/01/1999 24:10:45.384  WARNING  SLOT 10  IPSEC                Code:   8
Failed to decrypt the cipher key for a security association,
SA source: 132.245.155.192, SA destination: 132.245.120.214, SPI: 256, code: 2
```

As always, the **show ipsec** command is useful for checking the summary and statistics of security associations. (Online help on the **show ipsec** command is available.) Check the state of the SA when there are no log events pertaining to an SA.

```
# show ipsec esp sa

IPSEC ESP SA Table Information
------------------------------

                                                   Cipher    Integrity
State   Src              Dest              Spi     Alg       Alg        Manual Key
------- ---------------  ---------------  ------  --------- ---------  -----------
Up      151.47.8.1       151.47.8.2       256     DES       HMAC MD5   0xEB,0xC9...
Up      151.47.8.1       151.47.4.1       257     DES       HMAC MD5   0x40,0x3E...
Up      151.47.8.2       151.47.8.1       256     DES       HMAC MD5   0xEB,0xC9...
Up      151.47.4.1       151.47.8.1       257     DES       HMAC MD5   0x40,0x3E...
```

The State column indicates the operational state of the SA. SAs that are automatically generated using IKE will be dynamically updated, while manual SAs will permanently remain as configured and can be in one of several states. The Down state indicates that an SA is not operational, and the Not Present state indicates that the SA has not been successfully initialized—this can occur when the supporting IP interface is not yet operational. Check the status of the IP interface and ensure that all supporting services, such as IP and UDP, are up. UDP is used by the Internet Key Exchange (IKE) protocol.

To track the use of manual and automatic SAs, use the **show ipsec stats sa** command. In particular, the specific counts of protect and unprotect operations are reported for each SA of the appropriate type.

```
# show ipsec stats sa
IPSEC SA Statistics Information
-------------------------------
```

Src	Dest	SPI	BadAuth Packets	Bad Dec Packets	Bad Pad Packets	Protect Packets	Unprot Packets
151.47.8.1	151.47.8.2	256	1	0	0	540	0
151.47.8.1	151.47.4.1	257	0	0	0	2492	0
151.47.8.2	151.47.8.1	256	2	0	0	0	857
151.47.4.1	151.47.8.1	257	0	0	0	0	3275

Observe that SAs reporting protected packets are usually outbound protect SAs, and those reporting unprotect frames are inbound unprotect SAs. The bad authentication packets indicate frames that could not be correctly authenticated, possibly as a result of incorrect cipher or encryption keys. The bad decryption packets are frames that could not be decrypted successfully, indicating a specific problem with the encryption key. The Bad Pad Packets column shows the frames that had invalid padding detail in the ESP Pad field, a problem likely the result of a network problem or an interoperability issue related to another IPSec device.

IPSec Interaction with Firewall

In the "Firewall-1" section, we discussed considerations when using Firewall-1 with NAT and traffic filters. Similar considerations are applicable when using IPSec. Figure 7-6 shows the order of protocol services. In particular, IP protocols 50 (ESP) and 51 (AH) must be allowed to pass through the firewall for processing by IPSec. Preliminary processing of the IPSec packets can be performed by the firewall—for example, IPSec frames from an unknown security gateway can automatically be dropped by the firewall. It is, however, desirable to let the IPSec software make this determination and report the configuration error or security breach in the log.

WAN Encryption

In the "IP Security" section, we discussed public untrusted networks and the need for encryption services. Whenever business data travels through a network that is not under the exclusive control of the enterprise, security is a major concern. Security is also a significant concern in locally administered networks whose composition is unknown and is rapidly and dynamically changing. Encryption services are often used to protect data by providing confidentiality, integrity, and authenticity. While IPSec services are based on standards and drafts developed by the IETF, they are applicable only to IP and associated protocols. The Nortel Networks WAN Encryption Protocol (WEP), available for PPP and direct mode FR services, can be used to encrypt data at the data link level, regardless of the network layer protocol service. This is useful in networks where multiprotocol data travels over private leased point-to-point and point-to-multipoint lines.

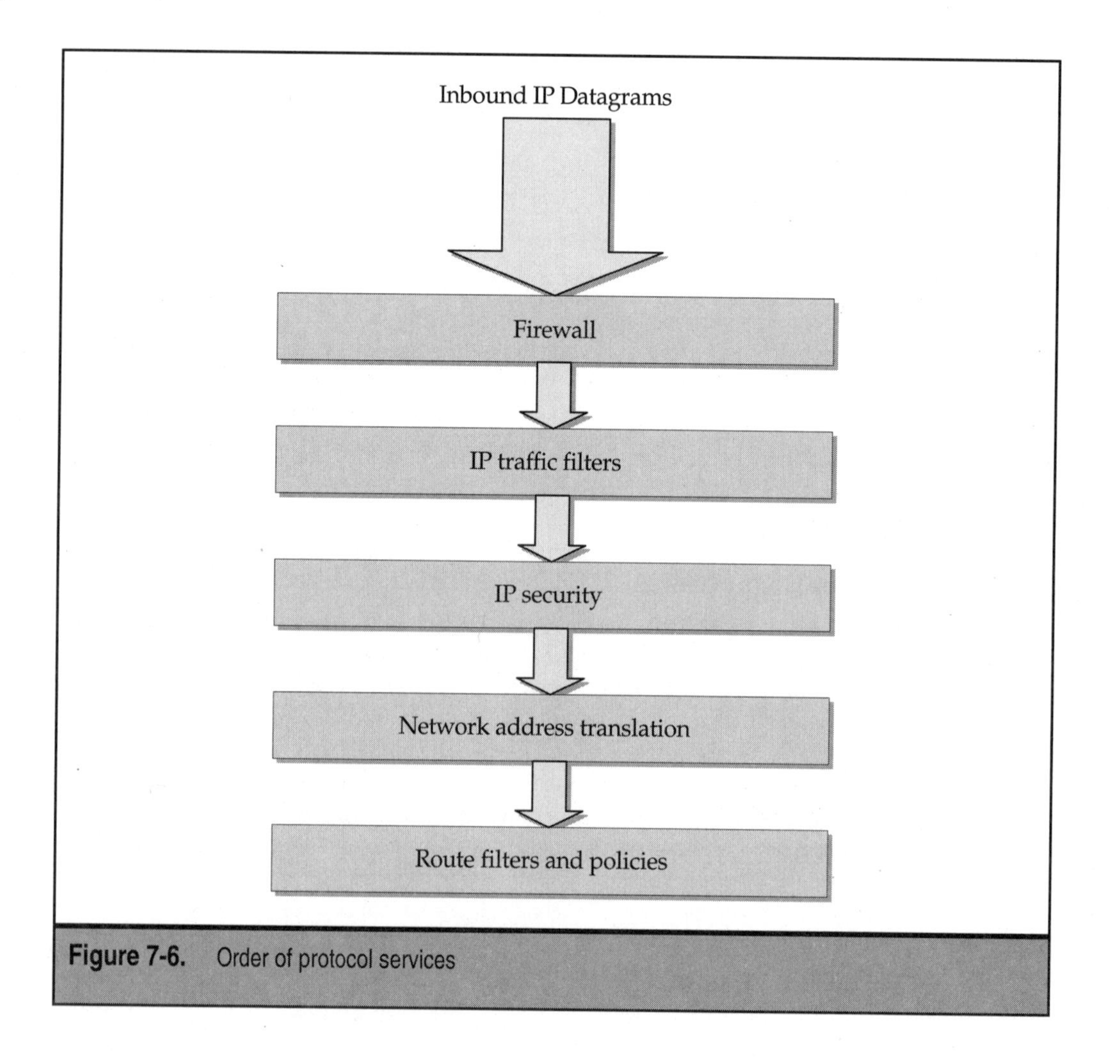

Figure 7-6. Order of protocol services

The WEP protocol is a proprietary implementation that uses a preshared manual key, which is used to automatically manage and generate two levels of dynamic encryption keys. There is a hierarchy of these keys, with the NPK residing at the top of the hierarchy.

- **Node protection key (NPK)** We have already mentioned the NPK, which is used by various services, including WEP, to protect the routing node. It is a 128-bit key that helps protect the integrity of all other keys and systems.
- **Long-term shared secret (LTSS)** The LTSS is uniquely provisioned for each connection to a remote WEP peer. It is used to generate the master encryption key and is preshared between two routers. It must be periodically changed to ensure the integrity of the link.

- **Master encryption key (MEK)** The MEK is automatically generated by an internal calculation based on the LTSS and the current time. It is changed periodically by both routers on each side of a circuit and is used to encrypt the TEK.
- **Traffic encryption key (TEK)** The TEK is also automatically generated using a random number generator on each side of the link. The TEK is unique and is also periodically changed by each router. Each router encrypts its TEK using the MEK and sends this TEK to the peer router for use in decrypting traffic. The TEK changes dynamically and independently on both sides of the circuit.

The hierarchy of the keys is illustrated in Figure 7-7. The configuration and provisioning of higher-layer protocols is performed as usual. When editing PPP and FR protocol services, you must enter the NPK to modify line- and circuit-level parameters. The discussion of NPK in the "IP Security" section equally applies to WEP. The Key Manager performs the same basic checks on the NPK to ensure that the system and the configuration have not been compromised.

The LTSS is encrypted using the NPK and stored in the MIB. The long-term nature of the LTSS does not imply that the key can be left unchanged—you should change the LTSS frequently (for example, every month) to increase the security and integrity of the link. Recall that earlier in the "Key Manager" section we discussed the requirement of keeping the NPK synchronized among all slots on the router as well as in the configuration file. Since the LTSS is encrypted using the NPK, care must be taken to update all slots with the NPK. Also, the NPK hash and the encrypted LTSS must be updated in the configuration

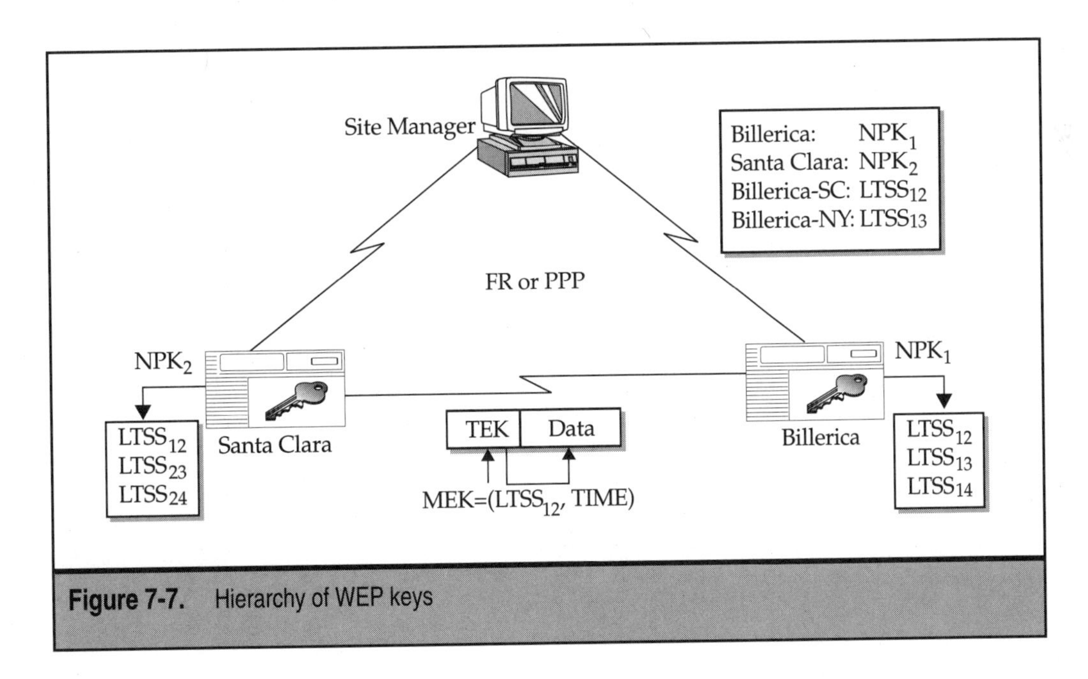

Figure 7-7. Hierarchy of WEP keys

file. Failure to do this causes problems because the LTSS may not be successfully decrypted from the MIB, resulting in the following warnings:

```
#   40: 12/11/1999 00:44:47.662  WARNING  SLOT 10  WEP                  Code:   5
Decryption of Long Term Shared Secret failed for PPP circuit 68.

#   43: 07/06/1999 01:03:58.905  WARNING  SLOT 11  WEP                  Code:   6
Decryption of Long Term Shared Secret failed for FR VC:
DLCI 120, LLIndex 211107, VCID 24.
```

The LTSS is matched between peer routers on a per-circuit basis. If the LTSS is not matched, the generated MEK, and consequently all TEKs, will be invalid. This will result in data loss between the routers because each router will be unable to decrypt the other's frames. The LTSS cannot be directly manipulated in the MIB, yet an unauthorized user or intruder may edit the LTSS in the hope of gaining access to the circuit. This sort of attack will be unfruitful because a similar change must occur on both sides, and there is no deterministic way of making such modifications. If the LTSS is incorrectly modified or corrupted, the WEP protocol reports the following warning:

```
#   65: 05/28/1999 21:13:16.295  WARNING  SLOT  3  WEP                  Code:  10
Error in LTSS authentication.
```

WEP negotiates parameters across a connection, and this negotiation may fail if attributes are mismatched and alternate values cannot be agreed upon. An example is the negotiation of the cipher algorithm. If one router is configured to support 40-bit DES on a link and the other is configured to support 56-bit DES only, then the cipher strength option will never converge. If the remote device is set to 56-bit or 40-bit, it will be able to negotiate down to the 40-bit encryption.

```
#   57: 09/28/1999 03:05:36.223  WARNING  SLOT  1  WEP                  Code:   7
Encryption negotiation failed for PPP circuit 25, reason code = 2

#  128: 04/19/1999 14:20:43.638  WARNING  SLOT 12  WEP                  Code:   8
Encryption negotiation failed for FR VC:
DLCI 412, LLIndex 212101, VCID 52, reason code = 2
```

The master encryption key is periodically computed using the LTSS and the current time. The MEKs are synchronized between two routers on the WEP circuit, and this synchronization requires that the times on both routers be reasonably close. For example, if the times on two WEP peers differ by a few hours, the MEKs generated will be different, and the routers will not be able to communicate.

CAUTION: If either the LTSS is mismatched or the times are significantly different for two peers, then data traffic will not pass between the routers. This will affect remote troubleshooting via telnet, and console access or local telnet access (through the LAN) will be necessary.

Each WEP connection is established as an encrypted virtual circuit over the underlying PPP circuit or direct mode frame relay service record. The **show wep vcs** and **show wep stats** commands are useful for looking at the current state of WEP operation and the data traffic over the line.

```
# show wep vcs

WEP Virtual Circuit Entries
---------------------------

                                                      Actual
   Line       LL                Vc    Connection    Cipher
  Number    Index Cct Name      Id       State       Mode
---------- ----- -------- ---- ------------ ----------
    201101      0 S11          0         Up      56-bit DES

1 WEP virtual circuit(s) configured.

# show wep stats

WEP Performance And Data Statistics
-----------------------------------

   Line       LL                Vc     # Bytes      # Bytes
  Number    Index Circuit      Id    Encrypted    Decrypted
---------- ----- -------- ----- ----------- -----------
    201101      0       S11      0        35200        30754
```

The Connection State column indicates the operational state of the WEP virtual circuit after negotiations have been completed. A nonoperational (Down) state can be caused by convergence failures during negotiation. On PPP circuits, if the physical line is down or LCP negotiations fail, the WEP VC will be down. Likewise, direct mode PVCs must be active for the WEP VC to be up. If the WEP VC is up, but the encrypted and decrypted data count does not increase, there may be mismatched MEKs on both ends; this will occur when the routers' times differ significantly.

TIP: Time synchronization is a key requirement in many enterprise networks, and the Network Time Protocol (NTP) is commonly used on BayRS routers to synchronize the time with a reliable NTP stratum. This not only helps sequence network time and events, but also helps WEP keep synchronized times across multiple links to several routers.

INTRUSION DETECTION

In recent years, the explosive development of the Internet combined with the rapidly increasing community of skilled computer and network users has posed dynamic and new problems. One thing is certain: network managers, engineers, architects, and consultants in various roles need to be agile in detecting security threats and combating them at an early stage. As this chapter was being written, the ILOVEYOU virus was causing various enterprise networks to pay substantial costs in lost productivity. Intrusions like this can occur at almost any layer of the protocol stack, including the application layer. In some cases, as past sendmail attacks and the ILOVEYOU virus have illustrated, the intrusion may occur without the user or proper authorities knowing about the incident.

Using the latest security advisories such as those available from the Computer Emergency Response Team (CERT), network engineers can implement deterrents by using such tools as IPSec, Firewall-1, and Traffic Filters. Active response to these conditions is necessary, and logging security intrusions will help strengthen the security of the network. Whether security breaches are benign or malign; are focused on obtaining valuable, proprietary, or confidential data; or are performed simply to cause denial of service, a proactive approach is required, especially on routers connected to the public Internet. Consider the smurf attacks that plagued many enterprises and ISPs alike but were immediately combated by applying traffic filters or black hole routes or by disabling directed broadcasts.

TIP: CERT security advisories can be obtained from the organization's web site: http://www.cert.org/.

Intrusion detection must not reside at the network layer alone, as we have mentioned before. Again, consider the example of the ILOVEYOU virus. It is easier to detect the application script once the e-mail has been assembled; mail server or anti-virus detection software can be run on all mailboxes to isolate and remove the problem script. Detecting such an intrusion at the data link, network, or transport layers is a substantial task because extensive pattern matches are necessary. On the other hand, issues such as PHF web attacks can still be effectively identified at a higher (HTTP) layer by a router.

SUMMARY

Every systems and network administrator knows that security threats are real, and any enterprise may be under attack for several reasons, or for no reason at all. Early in this chapter, we discussed the need for an enterprisewide security policy that is applied uniformly to users of computer systems and the network. Defining acceptable use of network resources helps steer computing and networking use and provides a framework for clearly delineating unfair or unacceptable practices on the network, especially those that may cause security breaches.

We covered three types of security as applied to BayRS routers: user account security, SNMP security, and protocol security. Users with direct access to BayRS routers, especially those with Operator or Manager privileges, have significant control over network operation and auditing commands, and connection without compromising privacy is essential. Likewise, SNMP security is also important because SNMP managers have full control over the router, and network intrusion can take place at various locations. Protocol security is by far the most important type of security, and we discussed various security applications, including Firewall-1, IPSec, traffic filters, and WEP. The solution implemented depends on the application.

REFERENCES

1. *Configuring Traffic Filters and Protocol Prioritization*, Publication 308645-14.00 Rev 00.
2. *Configuring SNMP, BOOTP and DHCP Services*, Publication 308644-14.00 Rev 00.
3. S. McClure, J. Scambray, and G. Kurtz, *Hacking Exposed: Network Security Secrets and Solutions, Second Edition*, Osborne/McGraw-Hill, 2000.
4. N. Dorasamy and D. Harkins, *IPSec: The New Security Standard for the Internet, Intranets and Virtual Private Networks*, Prentice-Hall, October 1999.
5. *Configuring IPSec Services*, Publication 308630-14.00 Rev 00.
6. *Configuring Data Encryption Services*, Publication 308618-14.00 Rev 00.
7. *Configuring BaySecure FireWall-1*, Publication 308613-14.00 Rev 00.
8. B. Schneier, *Applied Cryptography: Protocols, Algorithms, and Source Code in C, 2nd Edition*, John Wiley & Sons, October 1995.
9. M. Gonclaves and S. Brown, *Check Point Firewall-1: Administration Guide*, McGraw-Hill, September 1999.
10. S. Northcutt, *Network Intrusion Detection: An Analysis Handbook*, New Riders Publishing, July 1999.

CHAPTER 8

Performance Optimization

... (the author) himself is the best judge of his own performance; none has so deeply meditated on the subject; none is so sincerely interested in the event.

—Edward Gibbon, *Memoirs of My Life*

What is performance? Network managers, engineers, and architects all use the word *performance* to describe various outcomes of computing and networking operations. Many assessments of performance center around interface speed and node throughput. Other assessments focus on round-trip delays, real-time delivery, availability, reliability, and service. Protocol service is a key measure of performance, as is real-time delivery of data. Real-time, reliable data transport not only depends on the certain delivery of the packet, but also the timeliness of the delivery. A reliably delivered datagram can become useless if it is delivered beyond the real-time window. Performance in this regard is a weighted combination of several measures and is gauged differently by different network management organizations.

PERFORMANCE MANAGEMENT

Network performance management is unique to each enterprise, and the definition of performance is important in understanding where the network stands and how it is evolving. The concept of the intranet has been around for years; however, until relatively recently, these internal communications systems were referred to as networks or internetworks. Defining performance in these earlier networks, consisting of a few stations connected via a coaxial Ethernet and running a single computing application, was a simple task. Today's intranets, however, are heterogeneous—with their myriad protocols and applications requiring various levels of network connectivity, these networks require redefinition of performance using a new scale.

Defining Performance

Suppose a shipping manager uses a package tracking application twice a day. His definition of performance may be whether the network is constantly available each day from 8 A.M. to 5 P.M. Throughput is not a significant concern to him, although he wouldn't mind an additional two seconds of reduced latency. On the other hand, a CAD engineer may define performance as her ability to back up large blueprint files across the network to multiple servers with ease and minimal delay each day from 4 to 5 P.M.; availability of the network is crucial during this time period, though not at other times. These simple examples illustrate how different users may define performance using a common set of metrics, but different scales.

Knowledge of the specific intranet is the first step in defining performance management. Why can't a network manager simply measure all known metrics and require a high positive value on all scales? In any given network, each particular metric needs to be weighted so it reflects what is important to the business; otherwise, the assessment may not provide a true view of network performance. The first step in defining performance is to know the business needs that drive the intranet. An engineering firm, for instance, may define availability of a core set of services and applications as the key business requirement, with five 9s (99.999%) uptime required. This type of information about business functions is qualitative and is usually obtained directly from users or business managers. Researching network topologies is also useful, to help locate key performance centers such as server farms.

Collecting Statistics

A part of defining performance management for an intranet is data collection. Data collection is the gathering of raw data for use in computing various metrics. Some networks may concentrate heavily on one protocol; for example, a firm that uses IP may have 98 percent IP traffic and 2 percent AppleTalk traffic, the latter used only by the marketing department. A public relations or marketing firm, on the other hand, may have 80 percent AppleTalk traffic and use AppleTalk as the key protocol on the intranet. The process of collecting data that helps gauge the performance of protocols, interfaces, and applications is different for each network.

Raw Data Sets

Some networks may have a high-speed ATM backbone interconnecting several LANs in a campus environment. Measuring the efficiency and use of the ATM backbone is essential in understanding traffic patterns, and a performance metric will help determine whether the backbone is overutilized or underutilized. Once the performance has been defined, specifying a set of raw data for collection is easier. In the example of the firm that uses an IP network, for instance, a CIO looking for aggregate performance statistics may be little concerned with AppleTalk use, and the network manager may not even collect AppleTalk statistics for that reason. Since there are several hundreds and sometimes thousands of attributes that can be polled as raw data for measurements, it is essential to be selective in the definition of data sets.

Collection Periods

After defining a data set for collection, you need to define a collection period. Data collection periods can focus on peak usage times, though many network managers prefer to collect data at various times during the day. Collecting data at peak usage times helps you plan for the worst-case network conditions, because generally, network problems or bottlenecks will appear under peak conditions, when network stress is most likely to occur. However, collecting data throughout the day helps you understand network use during non-peak periods as well. Planning that emphasizes the worst case scenario sometimes results in a costly and unnecessarily overdesigned network. Network Management Systems (NMS) can be configured to poll data sets periodically throughout the day, automating efficient data collection.

Data collection traffic also travels the network in-band, so care must be taken not to excessively poll attributes that generate substantial management traffic. Initial polling of data sets will also help you further refine the data collection period; the period must be small enough so that measurable changes can be recorded.

Collection Tools

Network Management Systems (NMS) applications, based on the SNMP standard, are the most frequently used data collection tools. Most have built-in reporting tools to help categorize and compute performance metrics such as use and deviation. BayRS routers provide both proprietary and standards-based MIB objects for collecting performance data. Some Access platforms support remote monitoring (RMON) using embedded data collection modules (DCM), and others support RMON data collection using embedded software agents. Other types of collection tools, such as promiscuous data collection modules, network probes, and servers, are also useful in collecting performance data.

Recently, more sophisticated devices functioning both as protocol analyzers and service-level accounting software have been deployed in core network interfaces. They promiscuously monitor traffic, providing dynamic detail on traffic at various layers of the OSI model.

Analyzing Performance Data

Raw data collected during periodic polling intervals must be converted into various performance metrics for analysis. One of the important goals of this analysis process is the baselining of performance at certain periods. For example, consider the utilization value of a routed interface at various times during the day: The average utilization rate during lunch hour is computed to be 45 percent. When this value reaches 65 percent on Thursday afternoon, it is easy to characterize the utilization rate as 20 percent above average. This characterization is possible only after a through analysis of the performance metrics over a period of time. It is an outcome of knowing the use and evolution of the intranet.

Several metrics are useful in understanding performance. One of the key metrics is utilization rate, which is the fraction of the total capacity that is being used. An OC3 connection running at 10 percent utilization is transporting 15.5 Mbps of data traffic (that is, one tenth of 155 Mbps), for example. The utilization rate metric is sometimes overused, however; other metrics also play an important role in measuring performance. A WAN link running at 15 percent instantaneous utilization where 50 percent of the frames received are bad packets may be the cause of performance problem. Since a majority of the frames being received are unusable, user connectivity is affected due to delays caused by retransmission. Error statistics normalized using the interface capacity are also useful in such cases as a negative indicator of performance.

In BayRS, the performance statistics can be generally categorized into three areas.

- **Interface statistics** Per-port and per-VC statistics provide physical interface and data link details. This data is usually used to analyze line and driver level performance, including utilization, errors, and availability.
- **Protocol statistics** When multiple protocols are configured on the same interface and on the same slot, granular performance statistics for each protocol

are very useful. Also, you can compute protocol performance on single- and multiple-slot bases for an understanding of overall performance.

- **Kernel statistics** The kernel statistics provide a comprehensive view of the performance of the operating system and the hardware platform.

These three statistical categories have different performance measures. The categories are inter-related, but this division of the performance values helps you understand network use and identify performance bottlenecks that require redesign. An interface statistic may report an undesirable performance value, for example, but the problem may be inadequate buffers caused by network growth.

Interface Statistics

Interface statistics can be gathered at the physical interface and virtual circuit levels. In Chapters 4 and 5, we discussed several interface-level statistics that generally indicate problems such as transmit and receive errors. In performance calculations, the inbound and outbound octet counts are very useful. The MIB-II `ifTable` object has a sequence of `ifEntry` object instances that each represent all media interfaces on the BayRS router. Circuits that have multiple subcircuits each have an individual representation in `ifTable`. Each interface also has corresponding proprietary MIB objects that can be queried for data values. For example, an Ethernet interface will be represented as both an MIB-II–compliant `ifEntry` instance and the proprietary `wfCSMACDEntry` instance.

Table 8-1 summarizes several interface-level MIB objects from the private enterprise tree. Statistics that are otherwise not separately available in the MIB-II object can be gathered directly from the proprietary MIB object. For example, lack-of-resource statistic can be gathered only from the proprietary MIB object, but it is summed with other errors in `ifInErrors` or `ifOutErrors`. When initially baselining performance, you must make sure that errors are not occurring at the interface level and focus on inbound and outbound frames to compute utilization.

Enterprise MIB Object	Description
`wfCSMACDEntry`	Ethernet interfaces, including 10 Mbps, Fast Ethernet, and Gigabit Ethernet
`wfTokenRingEntry`	Token Ring interfaces operating at 4/16 Mbps
`wfFddiEntry`	Fiber Distributed Data Interfaces
`wfSyncEntry`	Serial interfaces, including ports on octal sync modules

Table 8-1. Private Enterprise MIB Objects for Interface Statistics

Enterprise MIB Object	Description
wfBisyncEntry	Binary synchronous communication interfaces
wfHssiEntry	High-speed serial interfaces
wfLogicalLineEntry	Individual DS0s in multichannel T1 and multichannel E1 interfaces; used for both ISDN PRI applications and non-PRI applications
wfDs1E1CurrentEntry wfDs1E1IntervalEntry wfDs1E1TotalEntry	Facilities data link statistics for MCT1 interfaces, for the current 15-minute period, for the preceding 96 intervals, and for the past 24 hours, respectively; can be used only when FDL is available
wfDsuCsuIfEntry	56/64K DSU/CSU interfaces available for the AN/ANH/ARN
wfModemEntry	V.34 modem interfaces
wfIsdnBriInterfaceEntry	ISDN basic rate interfaces
wfFrDlcmiEntry	Frame relay DLCMI interfaces
wfFrVCircuitEntry	Frame relay permanent virtual circuits (PVCs) and switched virtual circuits (SVCs); SVCs are dynamically established and the connection identifiers may change for different connections
wfAtmDxiEntry wfAtmMpeEntry	ATM Data Exchange Interfaces and MPE (multiprotocol encapsulation; RFC 1294 or 1483) interfaces, respectively
wfAtmizerIntfStatsEntry	Asynchronous Transfer Mode interfaces operated by the ATMizer, including OC3, DS3, and E3 interfaces
wfAtmizerVclStatsEntry	ATM virtual circuits operated by ATMizer, including permanent and switched VCs

Table 8-1. Private Enterprise MIB Objects for Interface Statistics *(continued)*

Enterprise MIB Object	Description
wfPppMlStatsEntry	Point-to-point protocol interfaces, including MLPPP interfaces (one per multilink bundle)
wfPppBapStatsEntry	PPP Bandwidth Allocation Protocol (BAP) interfaces
wfLapbPktStatsEntry	Link Access Procedure Balanced (LAPB) interfaces
wfX25VcEntry	X.25 virtual circuits

Table 8-1. Private Enterprise MIB Objects for Interface Statistics *(continued)*

Utilization Lines that are externally clocked such as serial and HSSI lines must have the `external-clock-speed` attribute configured for use by network management applications. Interfaces such as Ethernet and MCT1 lines have a specific data rate, which allows utilization to be calculated easily as a ratio based on the maximum throughput. Since externally clocked lines can operate at various speeds, interfaces are unable to automatically determine the line throughput. The external clock speed is a writable attribute on these lines. Although it is not essential to set the external clock speed for interface operation, we will discuss the merits of having this attribute configured correctly for NMS applications and protocol prioritization.

Interfaces and virtual circuits can generally be classified into two types for the purpose of computing utilization:

- **Half-duplex/shared** Contention media or shared interfaces such as Ethernet and Token Ring fall into this category. These media cannot simultaneously transmit and receive data on the same interface. For example, only one CSMACD station is actively transmitting at any given time on a segment, while all others are receiving the frame. The inability to simultaneously receive data from another station limits the maximum transport capacity on these lines. The utilization value on half-duplex lines is calculated based on the sum of transmit and receive frames.
- **Full-duplex** Serial lines and other bidirectional interfaces are examples of full-duplex interfaces. These lines can transmit and receive simultaneously and in essence make the transmit throughput independent of the receive throughput. Utilization in this case can be computed based on the maximum of these values. Bidirectional point-to-point VCs also fall into this category.

Two utilization formulas are commonly used to compute load at the interface level. The first uses the sum of inbound and outbound bytes and is usually used with half-duplex lines.

$$Utilization = \frac{(\Delta InOctets + \Delta OutOctets) \times 8\ bits\ /\ octet}{Throughput\ (bps) \times (t_2 - t_1)\ seconds} \times 100\%$$

$$where\ \Delta InOctets = InOctets_{t_2} - InOctets_{t_1}\ and\ \Delta OutOctets = OutOctets_{t_2} - OutOctets_{t_1}$$

The various raw attributes used in the utilization calculation are described here.

- **InOctets** The count of the number of octets received on the interface. This attribute is obtained from the physical or logical interface. The MIB-II attribute `ifEntry.ifInOctets` is commonly used by standards-compliant SNMP devices.
- **OutOctets** The count of the number of octets transmitted out of the interface. This attribute is also obtained from the physical or logical interface, and the MIB-II attribute `ifEntry.ifOutOctets` is also commonly used.
- **Throughput** This value is fixed on some lines and variable on others. On Fast Ethernet interfaces, for example, this value is set to 1×10^8, to indicate 100 Mbps. Per-second throughput values are used, because polling period samples are frequently in the range of 5 to 1200 seconds. On virtual circuits, the committed information rate (CIR) or sustained cell rate (SCR) is converted to a bits-per-second value for use as the throughput value.
- **t_x** The t_2 and t_1 values represent the polling times, where $t_2 > t_1$.

The delta values for the InOctets and OutOctets over a measurement period characterize the utilization. This equation is used with half-duplex lines, although you could use the same equation with full-duplex lines. On half-duplex lines, the sum of the inbound and outbound octets will not exceed the line throughput because access is shared. On full-duplex lines, however, since transmission and reception are independent of each other, this equation theoretically can compute more than 100 percent utilization.

Hence, the following equation is specifically geared toward full-duplex lines. The attribute values used are the same as in the preceding equation.

$$Utilization = \frac{MAX\ (\Delta InOctets, \Delta OutOctets) \times 8\ bits\ /\ octet}{Throughput\ (bps) \times (t_2 - t_1)\ seconds} \times 100\%$$

The MAX function is used to select the maximum value of ΔInOctets or ΔOutOctets. This is done so that the peak current utilization rate is defined using whichever direction is receiving the heaviest traffic. For example, if the outbound traffic is only 10 percent of the throughput capacity yet the inbound traffic is consuming 90 percent, then the line can be considered to be heavily utilized because the maximum limit is being reached on the receive side.

TIP: The throughput attribute on externally clocked lines is automatically obtained from the external clock speed attribute by most NMS systems. On BayRS routers, the external clock speed value must be configured manually for accurate utilization statistics. The external clock attribute should be set in the configuration, even when you are using customized applications or scripts, so that it does not need to be maintained elsewhere.

Figure 8-1 shows a simple utilization graph for an Ethernet interface measured throughout the day. Observe that there are peak periods as well as nonpeak periods; peak usage occurs during business hours and late at night when backups are performed over the network. The data has been monotonically smoothed to provide a continuous view of the 24-hour period.

Errors As mentioned before, performance is a measure of several metrics. Error statistics, as ratios based on the line throughput or traffic density, are also a useful metric. Computing the error ratio based on the traffic density is useful in assessing the significance of deviations. For example, 214 receive errors on a DS3 link may appear insignificant overall based on the line speed, but comparing this value to the 8,952 frames received

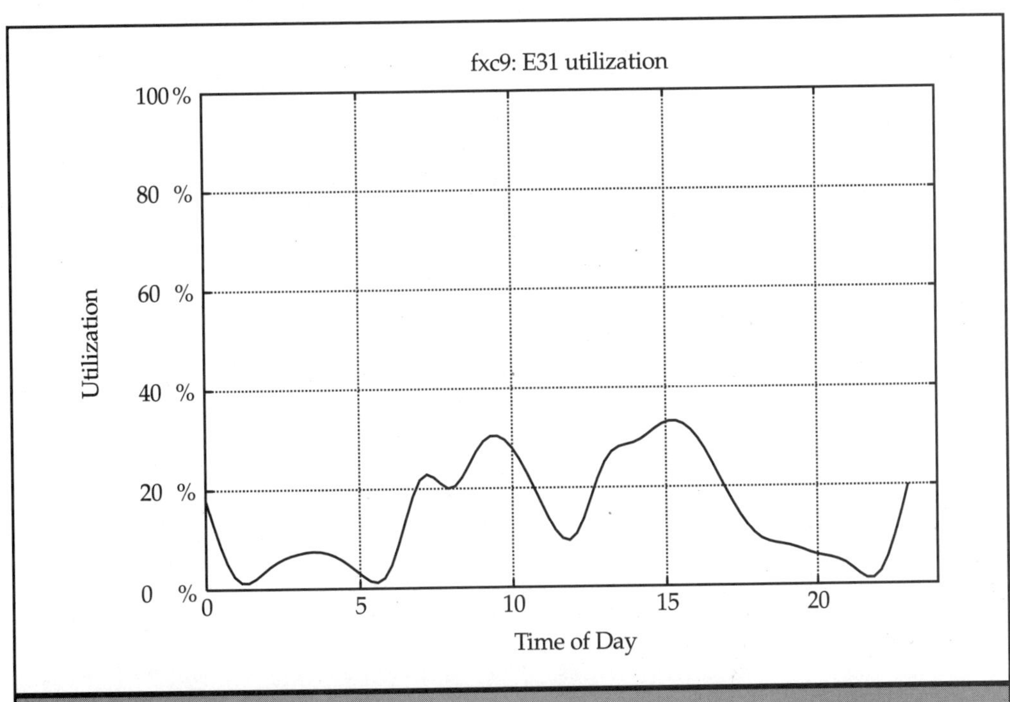

Figure 8-1. Ethernet interface utilization

during the same period portrays a different picture. Comparators and other references should be carefully picked so that the performance computation provides the best application of the metric.

Error statistics can be categorized into three types:

- **Transmit errors** This category includes outbound transmission errors such as underflows, lack of transmission resources, deadlocks, excessive collisions, and many others specific to the media. Internal errors (the third category) on the transmit side are also counted in this category.
- **Receive errors** Similar to transmit errors, this category includes all inbound errors such as overflows, lack of receive resources, bad frames, runts, oversized frames, collisions, and others specific to the media. Internal errors on the receive side are also counted in this category.
- **Internal errors** Internal errors include lack of resources, discards and clips, packet drops, internal MAC errors, memory errors, and others specific to the driver or the link controller. Both receive side and transmit errors are included in this category. The internal error count is primarily used to gauge the number of errors at least partially due to local operation.

Show Commands Instantaneous performance calculations are performed for a brief period immediately preceding the current time. For example, raw attributes are polled currently (t_1) and then polled again after 10 seconds (t_2). Utilization rates are the values most frequently obtained in this fashion, with computations often performed offline. A network management station polls the raw attributes and stores them in a database chronologically for later use. Once the attributes have been polled, they can be used in many types of calculations without communicating with the router.

Delta values and data rates useful in determining traffic density over a specific poll period can be obtained with the **show *media* sample** command. Replace ***media*** with any valid value, such as ethernet, sync, or token-ring; all valid media supported by the underlying platform can be used. The following example illustrates the data rates obtained using **show ethernet sample**, with the syntax shown here.

```
show media sample [-period <arg>] [-circuit <arg>] [-slot <arg>]
```

```
fxc9(4)# show ethernet sample
show ethernet sample                                  May 22, 2000 14:40:10 [GMT-6]


                                                                   Input    Output
Slot/          .--InputRate/<sec.--. .-OutputRate/<sec.--.        Lack of  Lack of
Conn   Circuit (bytes/sec) (Pkts/sec) (bytes/sec) (Pkts/sec) Resource Resource
------ ------- ----------- ---------- ----------- ---------- -------- --------
 3/1   E31            8264         88        6249         52        0        0
 3/2   E32              23          0          25          0        0        0
```

The poll period defaults to 10 seconds, which can be increased to a higher value. The circuit and slot filter options will run samples only on interfaces that match the name or the slot, respectively. Observe that the data rate is displayed in both bytes per second (bps) and packets per second (pps). A zero value in a rate field does not necessarily indicate that the interface is not receiving or transmitting frames; the poll period is usually so small that low traffic density is not observable during the period. For example, notice that the rate of input traffic on E32 in bps is small but a nonzero value, yet the average number of input frames per second during the same period is shown as zero (that is, less than 1).

Protocol Statistics

BayRS routers primarily deal with the lower four layers of the OSI stack: the physical, data link, network, and transport layers. Transport layer functionality is reserved for use by internal and system services on the router, and all forwarding operations are performed at the data link and network layers using bridging and routing services. Since BayRS routers provide multiprotocol functionality at the network layer, we will focus on analyzing and understanding statistics at this layer, including statistics for routing protocols.

Protocol statistics can be categorized into two major types:

- **Global protocol statistics** The global protocol statistics report cumulative protocol operation and service on a slotwide or boxwide basis. These provide information on how the protocol service is interacting with individual protocol interfaces on the slot, and how it is performing in its interaction with higher-layer protocols, including transport and routing services. For example, the global OSPF statistics include the total number of link state entries in the LSDB and the number of times the Dijkstra algorithm has been run on each area.
- **Interface-level protocol statistics** These statistics relate to the operation of a specific protocol interface on a circuit. They include information on the number of packets processed in the transmit and receive directions and errors encountered locally. The types of frames received and frames rejected by the protocol are also classified as interface-level protocol statistics. For example, for OSPF, the counts of hellos and link state updates transmitted and received are available on a per-interface basis.

Baselining Protocol Statistics When analyzing protocol operation, comparators are difficult to find. While media interfaces have a fixed throughput value, protocol traffic has an artificial maximum value equal to the underlying throughput value. Multiple protocols share the same bandwidth on the circuit, and bandwidth consumption varies dynamically. Unlike media traffic, where bytes per second is the measurement unit, protocols use packets per second as the unit of measure. Protocol datagrams vary in size, from the minimum to maximum sizes supported. Assuming that all headers are similar, two protocol datagrams, one twice as large as another, will both consume the same amount of processing time, although the actual transmission time on the media will vary.

These differences make comparison to an absolute value difficult with protocol statistics. The evident solution is *baselining*. Baselining protocol statistics is performed by averaging values over a period of time; subsequent statistics are normalized using this reference baseline average. Every network has unique operational characteristics, and baseline values for a protocol may differ significantly from network to network.

Establishing a unique yet robust baseline is important. For example, when baselining IPX statistics, network engineers must ensure that the protocol is properly functioning, without any errors. If the baseline is established when a key backbone connection is down, then it is an ineffective reference. All network interfaces must be fully operational throughout the baselining period. Say, for example, that an IPX global statistics set yields 150 routes and 380 services on an ASN router, baselined over a week of normal operation. Then suppose that the same set measured several weeks later yields 148 routes with 372 services, which can be characterized as a 97.9 percent service availability and 98.7 percent route availability.

Figure 8-2 illustrates IPX service availability, with the deviation plotted as an impulse from the baseline average of 2,920 services. A positive impulse is considered to be better service availability, and a negative impulse indicates an undesirable unavailability of some services. Notice that the baseline also serves an additional purpose. In large

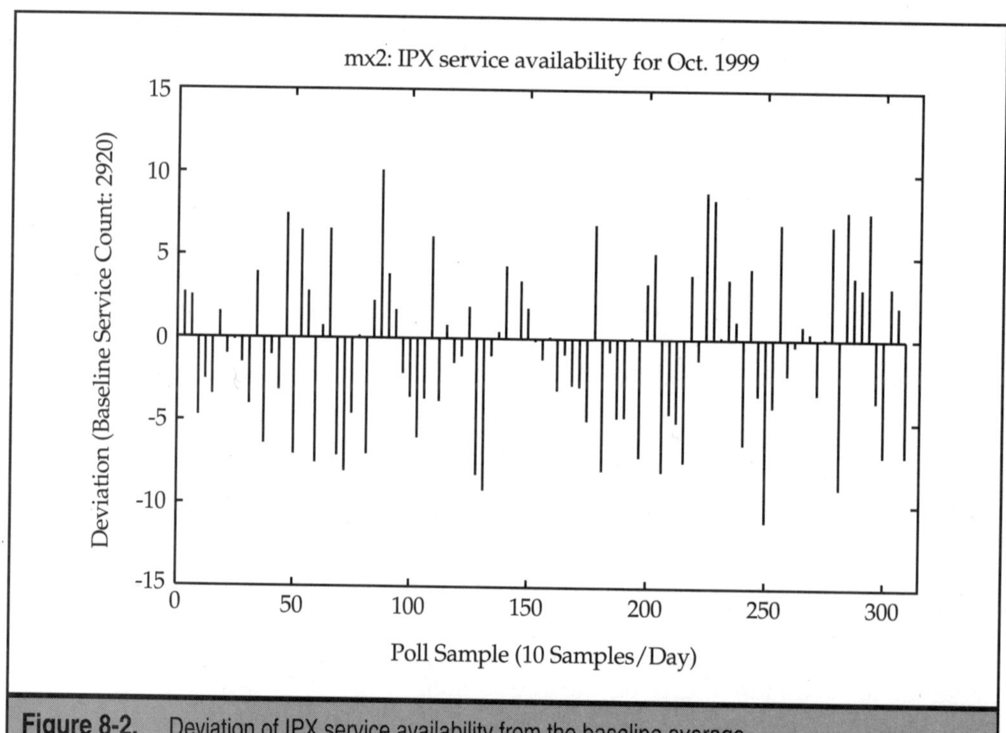

Figure 8-2. Deviation of IPX service availability from the baseline average

enterprise networks, it may be challenging to get an accurate count of services because servers are usually decentralized and managed by multiple divisions. The baseline average of 2,920 services was obtained over a period of 30 days and characterizes typical service availability. Impulse-style charts are useful in gauging better-than-average performance.

Table 8-2 lists global statistics used in performance measurements for common protocols. This is not an exhaustive list, and more MIB objects and attributes can be queried to establish other baselines. Table 8-3 lists several interface-level protocol statistics that are used in performance measurements.

Protocol	MIB	Description and Use
IPv4	`wfIpBase` `ip`	Provides the number of known networks (routes) and local/active IP devices. The MIB-II `ip` object provides the count of receive and transmit frames, errors, discards, unknown protocol frames, unicast and nonunicast frames, reassembled fragments, and reassembly failures.
IPX	`wfIpxBasicSysEntry` `wfIpxAdvSysEntry`	Provides the count of receive and transmit frames, header errors, unknown socket frames, in/out discards, checksum errors, unknown destinations, and encapsulation failures. The advanced object provides the number of circuits, destination networks, reachable services, compressed packets, and filtered frames.
OSPF	`wfOspfGeneralGroup` `wfOspfAreaEntry`	Provides the number of global LSDB entries and the count of SPF calculations on a per-area basis.

Table 8-2. Common Global Protocol Statistics

Protocol	MIB	Description and Use
BGP	wfBgp	Provides the total number of routes currently maintained, routes actively used, and path attributes maintained.
TCP	wfTcp	Provides the number of active and passive connections, connection failures, resets, current sessions, inbound and outbound segments, retransmitted segments, receive errors, and transmit resets.
Transparent Bridging	wfBrTp wfBrTpAggrStats	Provides the total number of entries in the forwarding table, port changes for endstations, packets received and transmitted, and discarded packets.
Source Route Bridging	wfBrSr	Provides the count of IP frames received via endstation support, NetBIOS RIF cache entries, hits and misses, drops, and total number of spanning-tree topology changes.
Network Address Translation	wfNatBase	Provides the number of dynamic mappings, FTP sessions translated, and active N-1 translations.
SNMP	wfSnmp	Provides the number of packets received and sent, bad version frames, invalid community string frames, ASN.1 parse errors, bad PDUs, too big frames, invalid objects and values, and traps sent; also provides the total count of **get**, **getnext**, and **set** requests.

Table 8-2. Common Global Protocol Statistics *(continued)*

Protocol	MIB	Description and Use
IPv4	`wfIpInterfaceEntry` `wfIpIntfStatsEntry`	Provides the count of receive and transmit frames, header and addressing errors, forwarded datagrams, unknown protocol frames, in/out discards, unknown destinations, reassembled frames, reassembly failures, fragmented frames, and fragmentation failures.
IPX	`wfIpxCircEntry`	Provides the count of receive and transmit frames, header errors, unknown socket frames, in/out discards, checksum errors, unknown destinations, and encapsulation failures; also reports the networks, services, and hosts reachable via the interface.
OSPF	`wfOspfIfEntry`	Provides the number of hellos, database descriptions, link state requests, link state updates, and link state acknowledgments received and transmitted.
BGP	`wfBgpPeerEntry`	Provides the count of updates and messages received and transmitted, total routes received from peers, and currently damped routes.
Frame Relay	`wfFrIfDlcmiEntry` `wfFrVCircuitEntry`	Provides the number of VCs configured, polls sent and missed, discards, drops, and faults at the DLCMI. The count of frames sent and received on each VC, FECNs and BECNs received, drops, discards and clips, and traffic-shaped packets is also reported.

Table 8-3. Common Interface-Level Protocol Statistics

Protocol	MIB	Description and Use
PPP	wfPppLineEntry wfPppCircuitEntry wfPppMlStatsEntry	Provides the number of echo requests sent and missed, inbound and outbound link qualities, in/out LQRs, receive and transmit frames, reassembled frames and failures, fragmented frames, priority queued frames, and clips and discards.
Transparent Bridging	wfBrTpInterfaceEntry	Provides the total number of packets received and transmitted on the interface and discarded frames.
Source Route Bridging	wfBrSrInterfaceEntry	Provides the count of frames received and transmitted, invalid ring numbers, SRF drops, translation bridged frames received and transmitted, and STP packets received and transmitted.
Priority Queuing	wfCctOptsEntry wfProtoPriQEntry	Provides the count of clipped packets in various queues, maximum and minimum values encountered in queuing, high-water packets, dropped packets, large frames, packets received, and shaped frames.

Table 8-3. Common Interface-Level Protocol Statistics *(continued)*

Many network administrators and engineers may already have established various baselines for different protocols on the intranet. Metrics at this level should be developed so that they not only gauge current performance, but also can predict future performance and aid in scaling estimates. Protocol performance and scaling are key considerations when assessing intranet evolution and determining future direction. We discuss the requirements and pitfalls of calculating scaling and performance in the section "Performance and Scaling" later in this chapter.

Kernel and Processor Statistics

The third and last set of statistics consists of the kernel and processor performance values. These are among the most frequently overlooked statistics that are useful in spotting internal problems, including design and scaling issues. In Chapters 4 and 5, we discussed the use of local memory and global buffer utilization in troubleshooting. Not all kernel, processor, and driver statistics are used in troubleshooting. In fact, these statistics are just a partial set of the metrics available for assessing overall performance. You should perform normal troubleshooting at the physical, data link, and network layers before delving into statistics dealing with internal operation.

Four sets of kernel and processor statistics can be gathered on BayRS:

- **Operating system statistics** Dynamic statistics pertaining to the GAME operating system are maintained in the `wfKernelEntry` object. These include dynamic values for local free memory, maximum local segment memory free, free buffers, total gates instantiated, active timers, and buffer allocation failures.
- **MIB heap statistics** Dynamic statistics for the MIB heap manager indicate the number of free segments and pages and the largest and smallest memory segments allocated. The MIB heaps are usually not a significant source of internal statistics; however, they can be very useful on routers where routing, service, and other tables are maintained dynamically in the MIB. If MIB support is enabled for certain tables (for example, `wfIpBaseMibTables`), memory is consumed in maintaining them.
- **Backbone (PPX/SPEX) statistics** Useful on stacked ASNs and Backbone node platforms, error statistics specifically pertaining to PPX or SPEX traffic can be obtained from the backbone MIB, `wfBackboneEntry`. Attributes such as overflows, framing errors, CRC errors, and arbitration errors can be obtained from this object. Most routers in normal operation will have no errors or have insignificant errors. However, observing these errors when and if they occur is useful in understanding Backbone subscription issues.
- **Application resources** BayRS also maintains per-application resource statistics that include CPU use, local memory consumption, and global buffer use. Concurrent resource use values are available in the `wfResourceUseEntry` object, and cumulative peak and current statistics for each slot are maintained in the `wfResourceTotalEntry` object.

Operating system attributes from `wfKernelEntry`, particularly those dealing with the use of local and global memory, were discussed in Chapter 4 and so are not addressed here. When baselines are created on a per-slot basis, the free local memory and free buffer values should be used. Figures 8-3 and 8-4 show examples of baseline kernel statistics, local memory and global memory (buffers) utilization measured as ratios of the maximum values. Calculating kernel and processor statistics relative to the maximum values is helpful in evaluating issues relating to scaling and future growth.

Performance and Scaling

The term *performance* generally refers to the current operation of the interface, protocol, or kernel, quantified using known values. *Scaling* relates to the amount of future growth

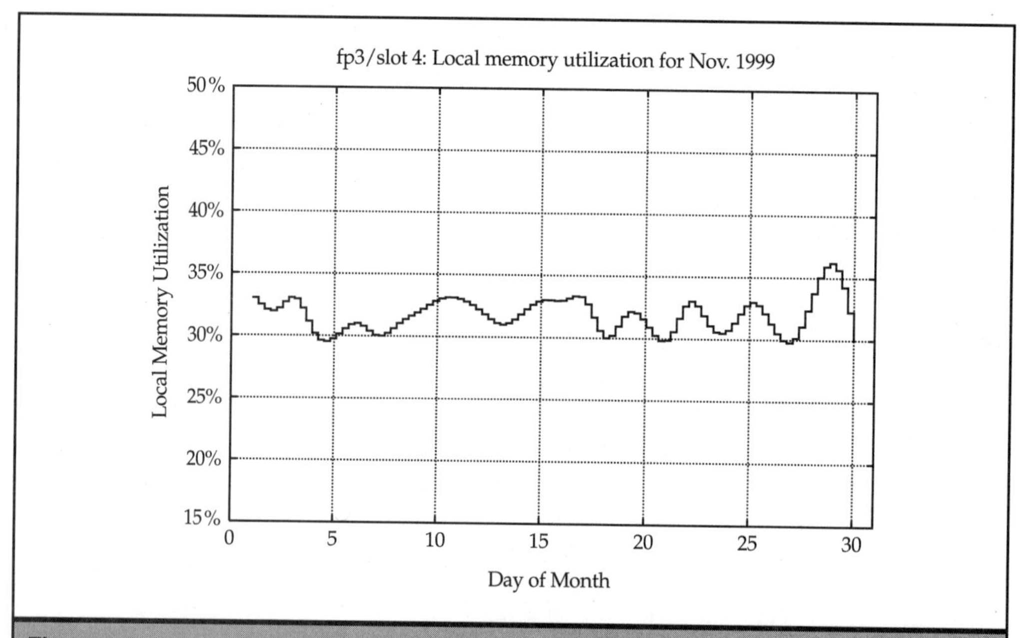

Figure 8-3. Baselining local memory use on a slot

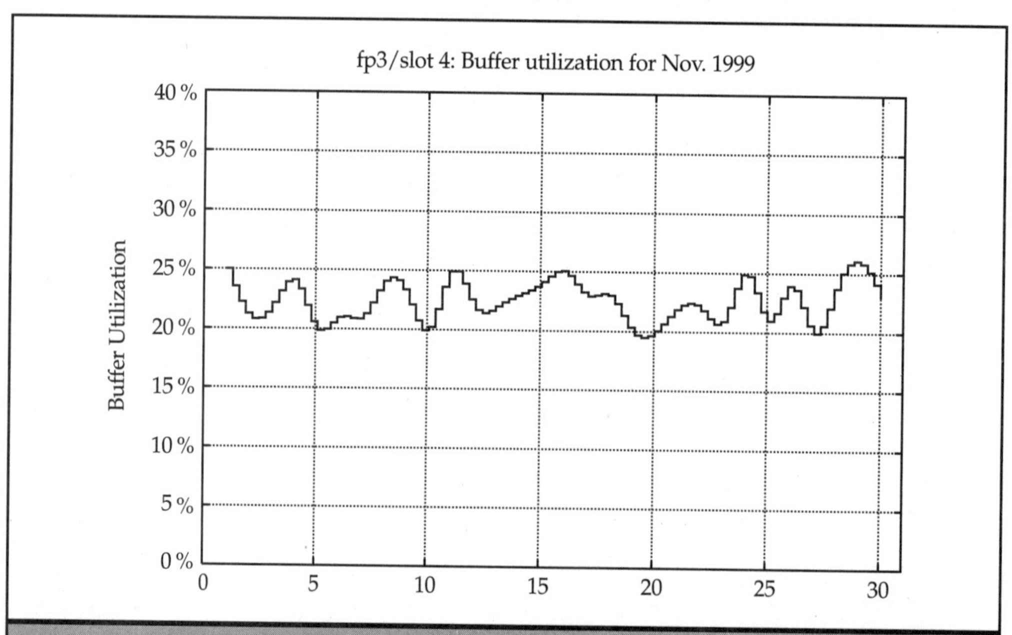

Figure 8-4. Baselining global buffer use on a slot

that a software or hardware component can sustain without a significant degradation in performance. Scaling arises as a result of normal network growth, which is a side effect of business growth.

Performance and scaling are treated differently, and differ significantly from each other in many ways. The scalability of a network is an important consideration during initial deployment, as well as in the future, when the network grows. Although many general guidelines are applicable during initial design, network growth often comes in spurts and its form is difficult to predict with any level of accuracy. Understanding the network's capacity for growth, potential for engineering topology and load changes, and future needs relies on the computation of performance metrics.

Since each network is distinctively designed, the performance metrics at the interface, protocol, kernel, and processor levels will be unique and characteristic of that network only. Baselining, as described in the previous section, is useful in establishing a comparator for almost all metrics. As network changes are implemented in phases, performance metrics are usually recomputed to understand the effect of the changes. The differential in most metrics is used to calculate how much the individual software or hardware component can be scaled to accommodate future growth. As a rule of thumb, you can apply the principle of linear scaling: that is, the performance will theoretically change (increase or decrease) by the same amount if the change in load is the same, in comparison to a previous change. For example, if the performance increases by 40 percent when the load is decreased by 20 percent, another 10 percent reduction in the load will theoretically increase the performance by 20 percent. Although this is not always the case, it is a reasonable assumption when a previously defined peak performance value is not available.

$$k \cong \frac{\Delta Utilization}{\Delta Reference} \qquad Units$$

where k is the linear scaling constant

$$and\ \Delta Utilization = Utilization_{t_2} - Utilization_{t_1}\ and\ \Delta Reference = Reference_{t_2} - Reference_{t_1}$$

The application of linear scaling is not always straightforward; scaling can be affected by external factors, including other network changes in addition to the one being analyzed. For example, computing OSPF scaling values while new IPX routes are being injected will affect CPU use calculations. It is also difficult to individually measure the impact of each change, because multiple network changes are often scheduled during a short downtime period. Real-time round-trip calculations should never be estimated using linear scaling; this method is appropriate only for load calculations such as interface, memory, buffer, and CPU use. The scaling constant is an approximation and must be used with care or recomputed when the resulting value deviates significantly from the expected value.

Consider an example where the forwarding cache on an IP interface is being increased on a heavily used Ethernet circuit. The `wfIpIntfStatsCacheMisses` attribute has reportedly been increasing, and so the `cache-size` value will be changed from 128 to accommodate 256 entries. Once the cache size for the IP interface has been modified, you notice that the IP application's local memory consumption (in `wfResourceUseEntry`) has increased by about 7,680 bytes. CPU use has dropped a little, but insignificantly, with the average CPU use reported at 4 percent on the slot. Using local memory consumption as the utilization metric and the cache size as the reference metric, you can compute a value of 60 bytes per cache entry for the forwarding cache. This value can be used in estimating future memory consumption when the forwarding cache size is increased on other interfaces. Again, observe that this value is a linear approximation—as the cache size is increased, the scaling constant may change a bit, but it should remain comparable to the approximation. In this example, increasing the forwarding cache to 1,024 entries reduced the cache overhead, resulting in 58 bytes per cache entry. Whenever there is reason to doubt the usability of a scaling value, perform the computation again in a controlled setting to obtain a new approximation.

TRAFFIC ENGINEERING

Traffic engineering (TE) has become an increasingly important application in almost all enterprise data centers. Traffic engineering was, until recently, a minor application in ISP environments, but it is now being used frequently to provide and guarantee service-level agreements (SLAs) for various service tiers. The use of numerous applications for various business purposes means that different levels of service and connectivity requirements need to be imposed throughout the network. All data traffic typically passes through several key data centers or backbone devices, where a majority of the traffic engineering is accomplished.

Prior to the explosive development of the Internet, private intranets and campus networks accomplished the task of traffic engineering merely by having a specific set of applications and protocols operating on the network. In today's intranets, characterization of traffic is increasingly difficult. Traffic engineering must deal not only with prioritizing certain protocols and flows of traffic, but also efficiently and dynamically using bandwidth. Applications such as voice and video over IP require bandwidth guarantees for a predefined quality of service (QoS), which must be achieved on multiprotocol interfaces that also have to share bandwidth with other traffic. In this section, we discuss several traffic engineering techniques used with BayRS routers to achieve SLAs both within the intranet and at the service edge.

Protocol Prioritization

The most commonly used TE technique is protocol priority queuing (PPQ), sometimes simply called priority queuing (PQ). Protocol prioritization is an outbound processing mechanism that sorts and forwards data in specific queues. Data traffic is sorted into prioritized delivery queues, called priority queues, using predefined criteria. The sorting process sends network data frames into one of the priority queues—high, normal, and low. The data dequeuing process for the interface uses these priority queues to help resequence frames during outbound transmission. Protocol prioritization affects only traffic leaving the router and does not affect the sequence of traffic in the receive direction.

Priority traffic filters, similar in concept to uniform traffic filters, are configured on an interface to sort frames into their appropriate queues. They are also sometimes called outbound traffic filters because all other filtering mechanisms are applied in the incoming direction only. Once protocol prioritization is enabled on a circuit, the high, normal, and low queues are automatically created. Data traffic that is not sorted explicitly is sent to the normal queue.

Three categories of prioritization action can be taken on outbound traffic for traffic engineering purposes:

- **Prioritization** Frames matching specific filter criteria can be prioritized into the high queue for expedited processing. This allows critical and time-sensitive frames to exit the interface quickly, with minimal latency. Bulk data transfer applications and noncritical application protocols can be prioritized into the low queue, which is processed after other traffic, including normal data frames. Application response times and round-trip delays can be reduced for high-priority applications using this action.
- **Redirection** Selected frames can be redirected or routed to specific next hops or to the first available routing hop from a list of adjacent routers. This not only affects queuing for outbound transmission, but also alters the automatic routing table lookup and forwarding process. For example, IP frames from a specific server can be forwarded to a list of high-priority next hops to control forwarding direction.
- **Accept/Drop** Some networks assign an acceptable value to certain applications, even if they are not critical to business operation. Other applications that have no direct business value yet affect network operation significantly are silently discarded. An example is 10,000 instantiations of a news/ticker screen saver application used in a large enterprise. While traffic for this application may be acceptable to a media firm, it may add little business value to a manufacturing firm. Accepting or dropping traffic at TE nodes helps achieve better service for crucial data.

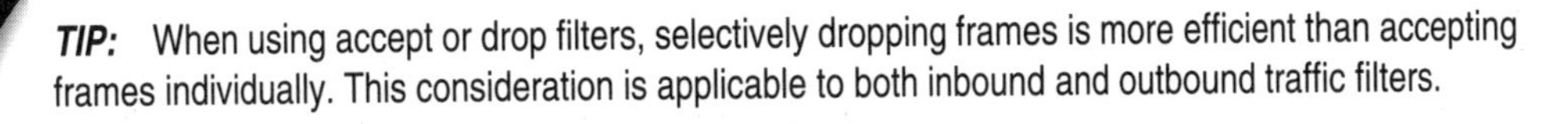

TIP: When using accept or drop filters, selectively dropping frames is more efficient than accepting frames individually. This consideration is applicable to both inbound and outbound traffic filters.

Packets are classified similar to the way they are with uniform traffic filters, where a set of criteria, when matched, allows a specific action to be performed. Once frames are sorted into their respective queues, the dequeuing process takes place. This is the mechanism by which the line or VC driver empties the priority queues by sending traffic to the outbound transmit queue, as illustrated in Figure 8-5.

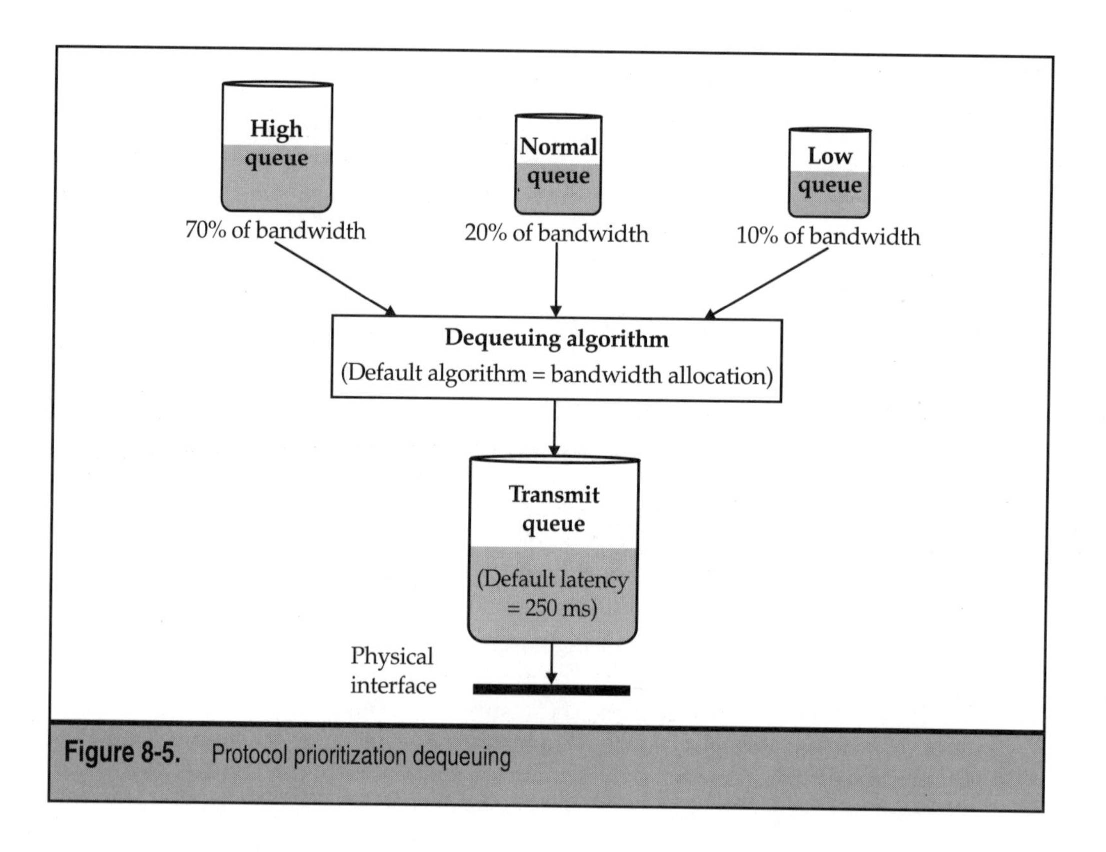

Figure 8-5. Protocol prioritization dequeuing

Two types of dequeuing algorithms are used for protocol prioritization:

- **Bandwidth allocation** This is the default dequeuing algorithm. Each priority queue is assigned a percentage of the outbound bandwidth, which is configurable. The default allocation is 70 percent bandwidth to the high queue, 20 percent to the normal queue, and 10 percent to the low queue. When the amount of data transmitted from a queue reaches the allocated bandwidth percentage, the next queue is serviced, and so forth. This dequeuing process is depicted in Figure 8-6.
- **Strict dequeuing** The strict dequeuing algorithm forwards frames to the transmit queue in order, high queue first, then normal, and then low. Each time a queue is serviced, the high queue is rescanned to ensure that a higher priority queue is always serviced prior to a lower priority queue. This queuing method is desirable for constant bit rate traffic that is being sorted into the high queue. This dequeuing process is illustrated in Figure 8-7.

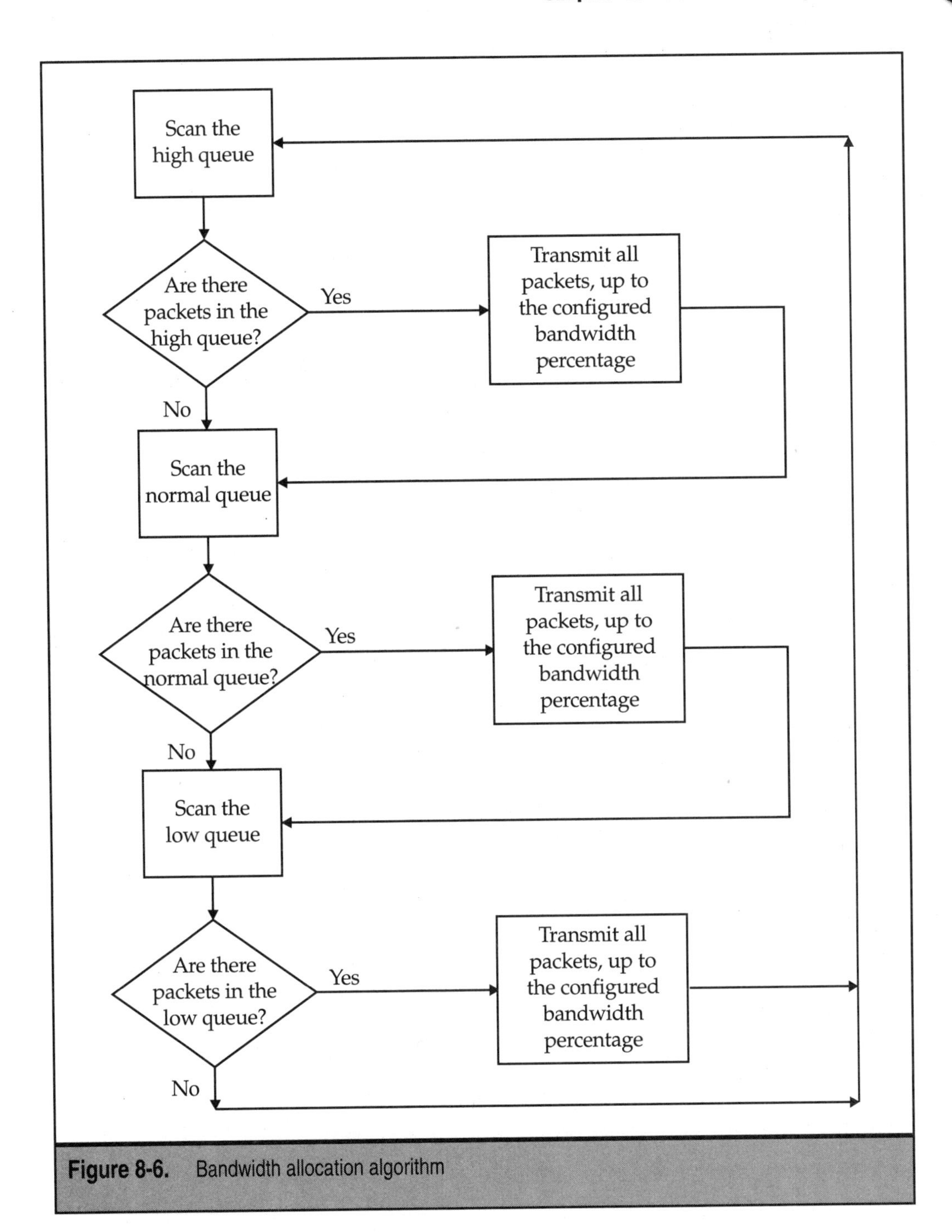

Figure 8-6. Bandwidth allocation algorithm

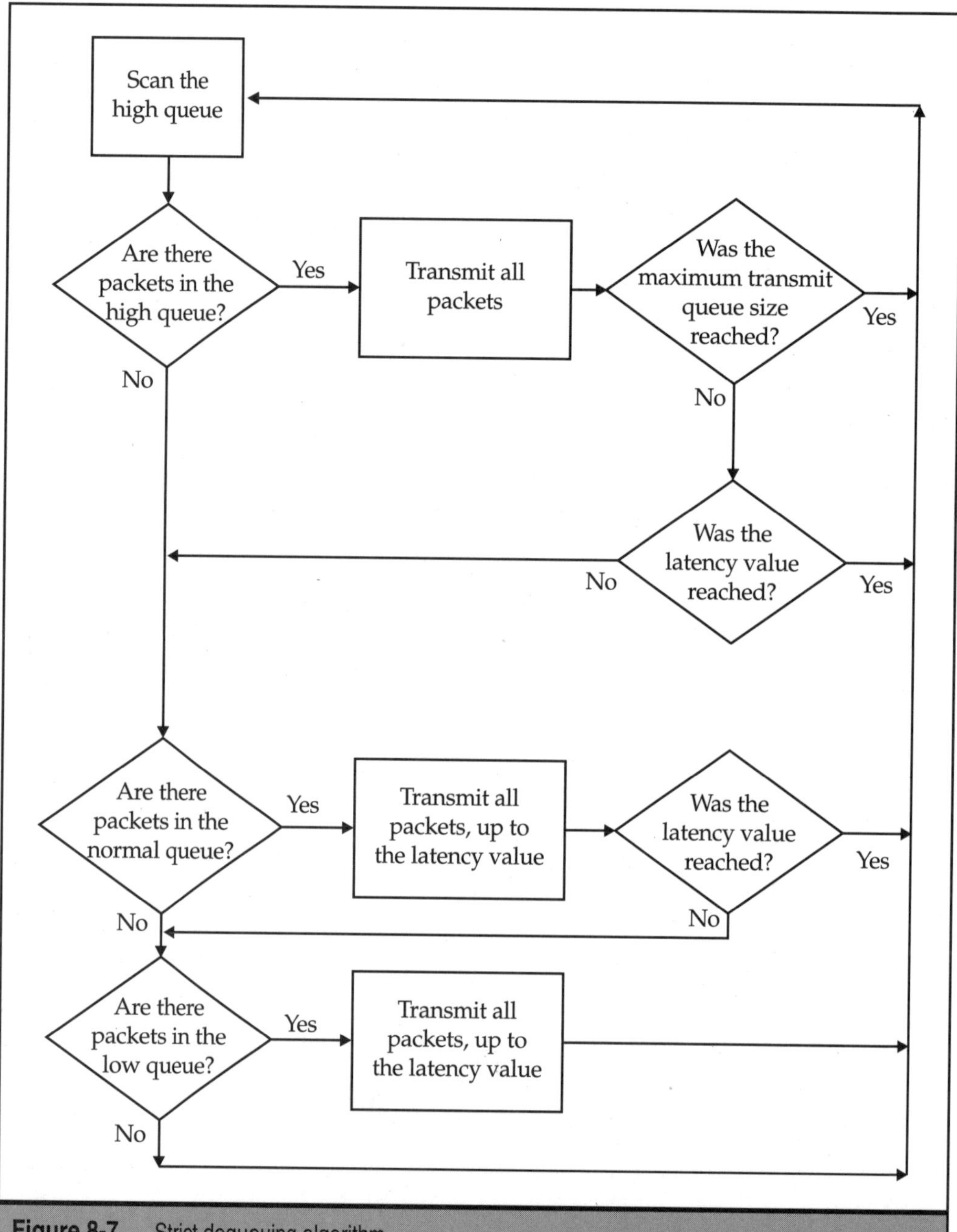

Figure 8-7. Strict dequeuing algorithm

Tuning Protocol Prioritization

Initially, protocol prioritization is configured with default queuing parameters to help you understand the impact of the traffic engineering. The default configuration parameters are generally acceptable for use in simple and average-size networks. When you clearly understand the effects of protocol prioritization, you can tune the parameters to help achieve optimal performance in a BayRS router.

Protocol prioritization statistics are maintained in the `wfCctOptsEntry` MIB object. These statistics indicate the number of frames processed by each queue, the maximum mark ever reached by a queue, and the number of clipped packets. In particular, the high-water mark indicates the peak usage of the queue. If the high queue size is set to 25 frames but the high-water mark reports 4, the other buffers in the queue are unused. The clipped packet count for each queue helps pinpoint oversubscribed queues. If, at a given moment, a priority queue is already filled up to the queue size, newly sorted packets are clipped from the queue. This indicates that the priority queue is receiving an excessive amount of traffic, an issue that can be resolved by increasing the queue size.

```
box# mget wfCctOptsEntry.12.*
wfCctOptsEntry.wfCctOptsHiClippedPkts.3 0
wfCctOptsEntry.wfCctOptsHiClippedPkts.4 7841
wfCctOptsEntry.wfCctOptsHiClippedPkts.7 6
```

The tuning of protocol priority largely depends on the type of dequeuing algorithm used. With bandwidth allocation, the following parameters can be tuned:

- Bandwidth percentage for each queue
- Queue size

With strict dequeuing, the bandwidth percentage allocation is not used. The maximum latency time applicable to the interrupt and high queues are applied instead. These are the tunable parameters for strict dequeuing:

- Queue size
- Latency

Bandwidth Consider an example where users of a certain application are requesting better response times. This SNA application has already been prioritized into the high queue, which has been allocated 20 percent of the bandwidth in the dequeuing algorithm. Observing the circuit statistics, you can see that the high queue is not reporting any clipped frames, but the high-water mark of 18 is close to the queue limit of 20 buffers. The normal queue, with 70 percent bandwidth allocation, reports a high-water mark of 3, with a queue limit of 20. This indicates that neither queue is reaching the queue limits, but the normal queue is being serviced more frequently due to the bandwidth allocation. The high-water mark of 3 in the normal queue indicates that the packets are being dequeued into the transmit queue soon after they are sorted. However, the same is not the case for the high queue, where packets remain in the queue longer before dequeuing takes place. To remedy this, the bandwidth allocation for the high queue is increased from 20 percent to 30 percent, and that of the normal queue is decreased from 70 percent to 60 percent. The sum of all allocated bandwidth percentages must equal 100 percent when using the bandwidth allocation algorithm. An error in the bandwidth allocation will result in the warning shown here, and the interface will default to using 70 percent, 20 percent, and 10 percent, respectively, for the high, normal, and low queues.

```
#   63: 11/26/1999 02:31:42.077  WARNING  SLOT 13  PPRI              Code:  15
S131: Percent for Priority Queuing should total 100, defaulting
```

TIP: The external line speed must be configured correctly on the appropriate media interface. Priority queuing performs calculations based on the line speed, and this is not known if the interface is externally clocked. If PPQ appears to have been configured correctly but the performance is unacceptable, check the line speed first and ensure that it is set correctly.

Queue Size Now consider an example where each of the three priority queues has a size, or queue depth, of 20 buffers. The high queue is reporting clipped frames, as illustrated in Figure 8-8. There are no clipped packets in the normal and low queues. The two key items you observe are the high-water mark in the low queue and the clipped frames in the high queue. The low queue has a high-water mark of 6, which indicates that it is sparingly used. The high queue, however, has had a significant number of frames clipped because the queue limit was reached.

The solution is to reconfigure the queue sizes of the three priority queues. For example, decrease the low queue size by 10 and increase the high queue size by the same amount.

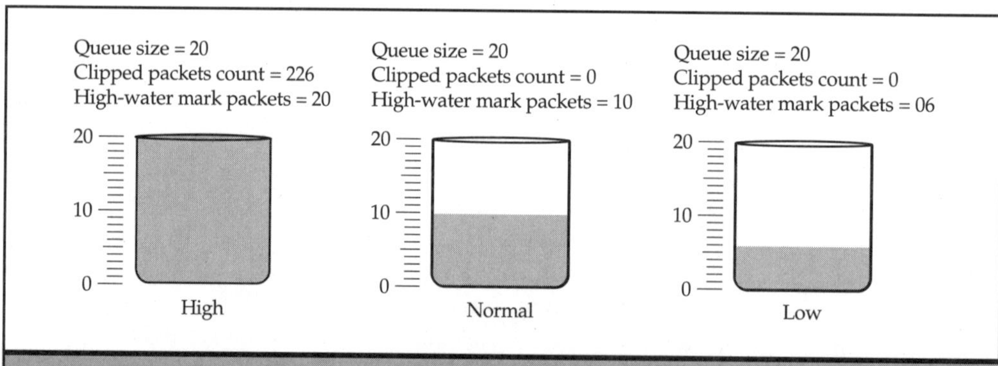

Figure 8-8. Queuing statistics before tuning the queue size

This reallocation keeps the total number of priority queue buffers at 60. Figure 8-9 illustrates the queue depths and statistics after the reallocation, showing that frames are no longer being clipped. The high-water marks are at reasonable values for each queue, and the queue sizes can be reconfigured when the network demands it.

TIP: Increasing the queue size is not always a solution to clipped packets. If a priority queue (especially the high queue) is being used unfairly, then the benefit of traffic shaping is reduced. For example, the high queue should be reserved for packets that are truly in need of high-priority servicing. If most traffic is classified as high priority, then performance of the prioritized interface may actually be degraded due to excessive clipping.

Transient clipping is acceptable as long as the behavior does not recur periodically and predictably. For example, a network manager may find it appropriate to let an outbound queue clip frames when excessive traffic is being queued. During periods of peak traffic, for example, you may let frames to the normal queue be clipped if the best-effort delivery process determines that the queue limit is being reached. Before making queue depth changes, always monitor the clipping behavior over a period of several days, particularly during peak usage hours.

Latency The latency attribute is used by the strict dequeuing algorithm to dictate the amount of normal- and low-priority traffic that can be delivered to the transmit queue at any given time. In essence, the latency is the maximum amount of time that a high-priority frame can be subjected to delay. By ensuring that the high-priority queue is serviced with minimal latency, you can reduce application response times.

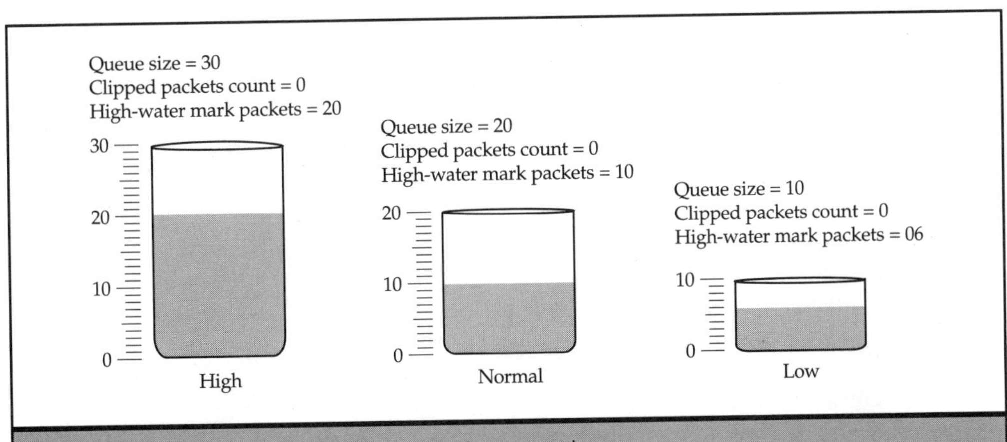

Figure 8-9. Queuing statistics after tuning the queue size

Latency depends on the line speed because the faster the physical media is able to transmit, the lower the latency. The following equation is used to compute the latency for a specific interface whose data rate is known.

$$Latency \cong \frac{Queued\ bytes \times 8bits\ /\ byte}{Line\ speed(bps)}$$

By default, BayRS uses a latency of 250 ms, which is acceptable for most traffic between interactive applications. Reducing the latency directly limits the throughput of queued data but increases response time. For example, halving the latency will halve the amount of bytes that can be queued. On the other hand, increasing latency will increase the throughput of queued data by increasing the queue service time, though it will also increase the response time. There is an optimal value for latency applicable to each interface, and this can be determined using empirical data on the router.

Traffic Shaping

Traffic shaping is another technique used on wide-area lines to remove bandwidth bottlenecks and effectively use the available bandwidth. Data traffic is normally very bursty, and traffic shaping smoothes this data flow into an optimal form so that the traffic has a quality of service (QoS) that is inherently lacking at the data link layer. Bandwidth on wide-area links is significantly more expensive than that on the local area network, and efficient use of this bandwidth is crucial in many enterprise networks. Large enterprises use frame relay, ATM, or PPP links in the wide area to interconnect hub data centers. BayRS supports traffic shaping on frame relay and ATM interfaces, as covered in this section. Our initial discussion will focus on frame relay traffic shaping (FRTS).

BayRS traffic shaping generally involves three factors: reliable transport, rate enforcement, and QoS. Frame relay and ATM can be considered to be inherently unreliable, with the loss of a frame or cell in the media detected only by a higher-layer protocol. With frame relay, the discard eligibility (DE) bit in each frame indicates whether the network may preferentially drop the frame when there is congestion. Different techniques are employed to provide a basic level of reliable transport. For example, the WAN Compression Protocol (WCP) maintains compression histories that are used for retransmission of frames.

Rate enforcement is another component of traffic shaping, where hub sites typically have access lines with high data rates, and remote sites have lines with lower data rates. Lack of rate enforcement can lead to the "big pipe, little pipe" problem, where the hub site sends more data to a remote site than the remote site can receive. For example, consider a central hub router with a 1.544 Mbps T1 link into a frame relay cloud with several remote sites that have an access line of 64 Kbps each. The central hub router can easily saturate the link to a remote router by sending excessive amounts of data. Rate enforcement is the solution. The committed information rate (CIR), for example, of each remote site can be configured on the corresponding VC at the central site, allowing the hub router to limit the amount of traffic sent. Rate enforcement is also necessary for protocol prioritization, which was discussed in the previous section. Protocol prioritization is difficult to achieve on variable-speed lines without rate enforcement. With the combination of protocol prioritization, rate enforcement, and filtering, a basic level of QoS can be achieved on wide-area lines. The Circuit Resource Manager (CRM) also provides QoS functionality by providing resource management and reservations for committed flows.

The use of both protocol prioritization and traffic shaping creates two levels of queues. Traffic shaping at the VC level is performed first, and then protocol prioritization has its own queues at the driver level. This setup is illustrated in Figure 8-10. Notice that the driver that performs the queuing and filtering directly handles unshaped VCs. Observe the presence of a very high-priority *shaping* queue at the driver level, on which frames are not reprioritized or requeued but are sent directly at the CIR. This is because the frames have already been prioritized by the VC, and the driver will attempt to transmit the frames as quickly as possible. When using reserved traffic flows (such as with RSVP), traffic shaping is a necessity on a VC because it is the only mechanism that can provide and guarantee the reservation.

Committed Burst

You are likely already familiar with the concept of the committed information rate (CIR), which is discussed in detail later. The committed burst (B_C) is the number of committed bits that the router will transmit over a measurement period (T_C). On average, the router will attempt to send at least B_C bits over the period T_C. If this value is set to zero along with the throughput (CIR), then traffic shaping is disabled. When this value is increased or decreased, the measurement period is indirectly affected, thereby allowing fine control over the traffic shaping on an individual VC. The router periodically measures the bits transmitted to actively shape traffic on each VC. The committed burst is related to the CIR in the following form:

$$B_C = CIR / T_C$$

The minimum value of B_C is arrived at using the maximum frame size. Committed bursts should be measured so that a maximum-sized frame can be transmitted fully during the measurement period. For example, if the maximum frame size is 1,600 bytes on a serial interface running frame relay, then the B_C must have a minimum value of 1,600 × 8 bits/byte, or 12,800 bits.

Excess Burst

Similar to the committed burst, the excess burst (B_E) is also a configurable parameter that indicates the amount of excess data bits that can be transmitted above the B_C when there is no congestion. On an interface with a high line rate such as a T1, it may be desirable to burst above the CIR temporarily to transmit large frames and bursty flow. This is one of the beneficial aspects of frame relay, and B_E can be set to any value so that the sum of B_C and B_E does not exceed the line rate. On VCs where the CIR must be strictly enforced without any excess burst, the B_E parameter must be set to zero.

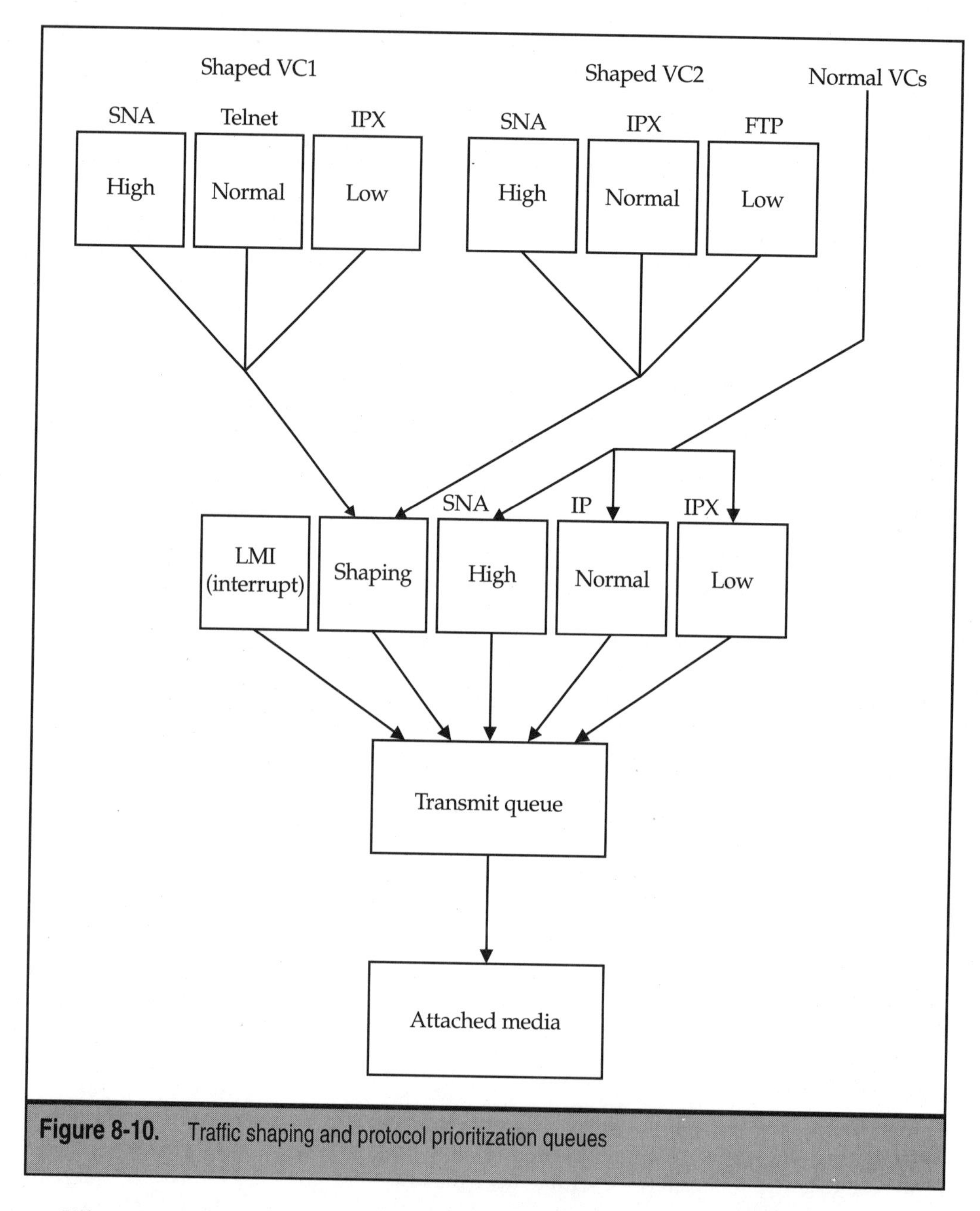

Figure 8-10. Traffic shaping and protocol prioritization queues

When the B_E value is nonzero, each VC will attempt to use the excess burst available as long as there are no congestion notifications (CNs). The traffic is automatically throttled back to the committed burst when congestion is observed, if the `congestion-method` attribute is set to throttle. This is illustrated in Figure 8-11, where the router automatically throttles back to B_C. Both B_C and B_E are measured over the interval T_C.

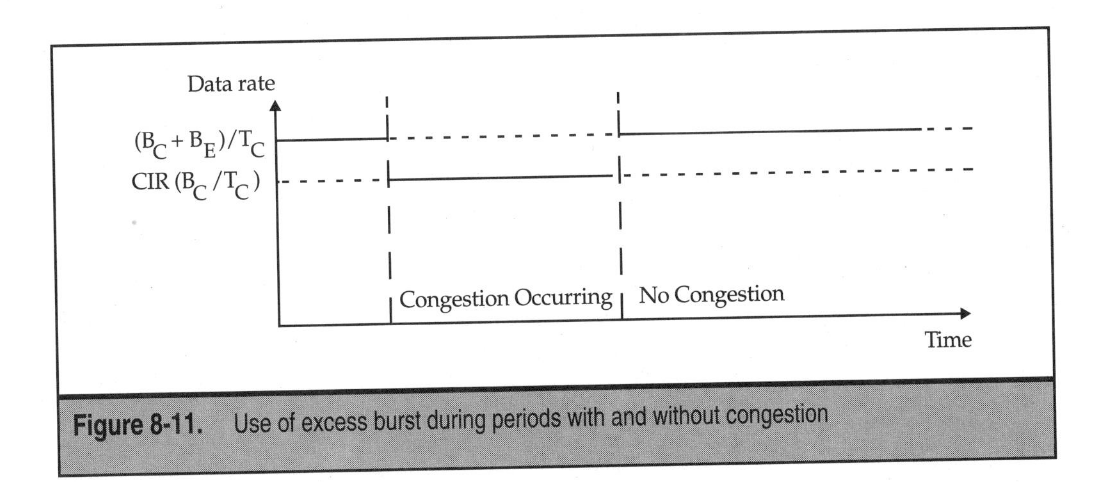

Figure 8-11. Use of excess burst during periods with and without congestion

Committed Information Rate

The CIR is the rate at which each VC can transmit data to the network on a sustained basis. Data traffic up to the CIR is *committed* to be available as throughput by the provider. The CIR is a contract agreed to by the carrier, provided in bits per second. This value is configured on a per-VC basis, and traffic up to this rate is guaranteed to be delivered to the destination. Data traffic can be burst above this committed information rate, but it may be subject to discard during periods of network congestion.

The user-configurable throughput value on frame relay interfaces is called `cir`. Measured in bits per second, this attribute defines the amount of traffic that the VC will attempt to transmit on average. If this value is set to zero, then traffic shaping is considered to be disabled. Since many VCs can exist on the same interface, the summation of all configured throughputs may potentially exceed the line rate. This condition, called *oversubscription*, is allowed because in normal use, not all VCs may simultaneously require the committed throughput.

We have already discussed B_C, B_E, and T_C. The measurement interval T_C is defined using the following equations.

$$T_C = B_C / CIR \quad if \quad B_C \neq 0 \text{ and } CIR \neq 0$$

$$T_C = B_E / \text{Line rate} \quad if \quad B_C = CIR = 0$$

Various CIR, B_C, B_E, and T_C combinations can be specified in three major configurations, listed in Table 8-4. The B_C and CIR values can be fine-tuned within the same category to achieve exacting control over traffic shaping.

A zero value for the throughput (CIR) generally creates a best-effort service. The carrier has not committed to a specific throughput, which allows best-effort service with minimal cost. When set to zero, the throughput value must be complemented with a setting of zero for the committed burst (B_C) value. The excess burst (B_E) may be set to a value that adds up to the maximum line rate. For traffic shaping to have the desired effect, B_C must also be set

Configuration	CIR	B_C	B_E	T_C
Zero CIR	= 0	= 0	> 0	B_E/Line rate
CIR with no excess burst	> 0	> 0	= 0	B_C/CIR
CIR with excess burst	> 0	> 0	> 0	B_C/CIR

Table 8-4. Traffic Shaping Configurations

to zero, and B_E must be set to a nonzero value, for use in the measurement period calculation. With a nonzero CIR, strict CIR enforcement or the use of excess burst can be configured. In both cases, the measurement period must be carefully configured using B_C and CIR. For example, a configuration of 8K for B_C and 64 Kbps for the CIR will result in a T_C of 125 ms. As a general guideline, begin traffic shaping design using 250 ms as the measurement interval: that is, use a committed burst equal to one-fourth of the CIR.

The exact value of the CIR provided by the carrier need not be configured as the throughput value on the router. The committed burst, excess burst, and CIR can be adjusted to the desired level and condition of traffic shaping. You should, however, begin traffic shaping tests and tuning with the exact value of the CIR supplied by the carrier. Errors in the B_C and CIR configuration will be reported as events in the log, and traffic shaping will behave as if it were disabled.

```
#   55: 11/14/1999 13:15:42.181   INFO     SLOT  7  FR                   Code: 169
VC (207104.0.191): Traffic Shaping config is invalid; CIR: 0 Bc: 16000.
```

Figure 8-12 illustrates the use and behavior of traffic shaping. The various attributes we have discussed are shown as rates and absolute values. The CIR and line rate are measured in bps, shown as a linear line at an angle; the amount of bits proportionally increases with time, shown linearly on the *x*-axis. B_C and B_E, measured using absolute bits, are represented horizontally on the *y*-axis. The bold lines depicted at an angle represent frames that are being transmitted over time. These frames are sloped at the same angle as the line rate, indicating that outbound frames are transmitted at the line rate. The first three frames are within the CIR. The succeeding frame exceeds the CIR because it goes beyond the B_C, though this frame still has not exceeded the burst B_E; both the fifth and sixth frames exceed the B_C and B_E. Frames that exceed the B_C are subject to discard, but this can be selectively controlled using PPQ, where certain frames can be preferentially marked DE (discard eligible).

Frame relay traffic shaping statistics are useful in determining the actual performance and operation of traffic shaping. In particular, traffic shaping statistics pinpoint large frames and frames dropped. When these values increment frequently, many frames are too large to be efficiently transmitted with the configured committed burst. Generally,

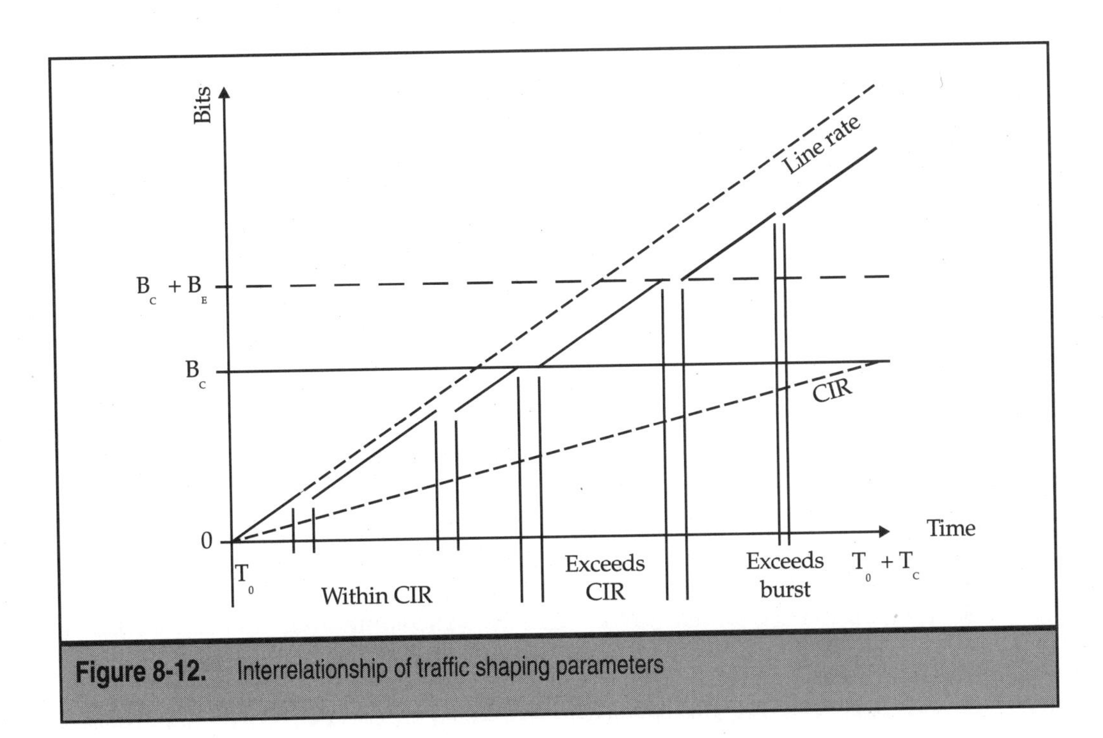

Figure 8-12. Interrelationship of traffic shaping parameters

you do not adjust the CIR when this occurs, but you increase the B_C in small increments. The average frame size on each network varies; if this value is previously known, it can be used to compute the minimum value of the B_C as a starting point. Otherwise, start with the recommended one-fourth CIR value to fine-tune the B_C for efficient traffic shaping.

Congestion Control Methods

The preceding paragraphs discussed four key parameters in traffic shaping: B_C, B_E, CIR, and T_C. Traffic shaping is incomplete without the use of the congestion control method attribute, which dictates how a VC should behave when congestion notifications (FECNs and BECNs) are received. We discussed congestion behavior and troubleshooting in Chapter 5, in the "Frame Relay" section. Several congestion control methods are available:

- **Throttle** This behavior is illustrated in Figure 8-11. Traffic is automatically throttled back to the CIR when congestion is observed. Data continues to be transmitted at the CIR.
- **Throttle, then shut down** Initially, the VC throttles back to the CIR, but if congestion persists, then the VC is shut down. Some noncritical VCs can be configured in this fashion to allow other VCs on the same physical interface to continue to operate, depending on business needs.

- **Shutdown** The least desirable option is to shut down the VC when congestion notifications occur, but this method is beneficial when multiple VCs exist on the same physical interface.
- **Disabled** The application of a congestion method can be disabled; the receipt of congestion notifications will be ignored, and traffic will continue to be shaped as if there is no congestion.

Throttling and shutdown behaviors are discussed in detail in Chapter 5. In the traffic-shaping context, throttling back to the CIR is the ideal solution because the provider has committed to the throughput. The VCs continue to be shaped as required by limiting the outbound rate to the CIR. Protocol priority queuing will prioritize time-sensitive frames and send them out first. Frames prioritized into the low queue by default are marked discard eligible so that the network can discard the lower-priority frames during periods of congestion. This allows the router a reasonable amount of control over what can be dropped in the frame relay cloud.

Switched Virtual Circuits

Traffic shaping on SVCs is similar to that on PVCs. The committed burst, excess burst, and throughput values are derived from the network during call establishment. The congestion method and the actual use of traffic shaping on the SVC are locally administered. If an SVC need not have traffic shaping, it can be disabled on the VC. The network provides the link layer (LL) core values, essentially defining the throughput and burst values.

Three traffic shaping configurations can be set on a service record using SVCs:

- **Network-supplied LL core values** SVC traffic shaping is enabled, and the network-supplied LL core values are used in traffic shaping computations. The only other attribute that can be modified is the congestion method.
- **Locally administered LL core values** SVC traffic shaping is enabled, but the LL core values are configured manually on the VC. Both inbound and outbound values are configured for the throughput, committed burst, and excess burst. The actual value used by the SVC may be different because both routers and the network negotiate throughput and burst rates. The QoS provided by traffic shaping depends on the locally configured LL core values, and the actual values may need to be periodically checked to ensure that the desired QoS is being provided.
- **Locally administered LL core values without TS** Traffic shaping is disabled, and the LL core values are manually configured. CIR enforcement is not performed when traffic shaping is disabled; however, there is a potential risk of oversubscribing the interface. To ensure that the interface is not oversubscribed, periodically check the VC and driver statistics to see if frames are being clipped or otherwise dropped.

Guidelines and Queue Limits

Traffic shaping is best suited for application at a central hub router where several remote spoke routers converge. Using a consolidated view of the wide area network, start working on the traffic shaping design from the hub router. Initial configurations of B_C should start with one-fourth of the CIR, which can later be tuned to accommodate larger-sized frames; if the average frame size is relatively small in a network, the B_C can be reduced to improve the effectiveness of traffic shaping.

Before making traffic shaping changes, always observe the traffic shaping and protocol priority statistics for an extended period of time. A transient period with a significant number of large packets in a bursty fashion may indicate that there were large packets, but traffic shaping should be designed on the basis of the *average* frame size, not necessarily the largest. External probes can be used to obtain an understanding of frame size and density, to help you arrive at an optimal value for B_C.

Data Compression Data compression is another optimizer than can be used along with traffic shaping. It increases the throughput of a connection while also improving reliability. Compression maintains a history of transmitted packets, and selected frames can be retransmitted reliably. It increases the throughput cost, however, because it is resource intensive, but you can compensate for this cost by considering the characteristics of both data compression and traffic shaping. Compression tuning is a consideration on its own and is discussed later in this chapter. The considerations discussed here, when applied to traffic shaping parameters, help tune the behavior of traffic shaping.

The average compression ratio should be measured over an extended period of time. If B_E is set equal to B_C, then the VC will attempt to transmit twice the amount of traffic committed. With compression enabled, the compressor will attempt to compress the data at the rate it receives it. In general, a 2:1 compression ratio will allow the line driver to empty traffic onto the line at the committed rate without loss of data. Note, however, that compression ratios are dynamic and depend on the traffic pattern. Throttling is the preferred congestion control method in this case, to allow the router to queue traffic without loss of data when congestion occurs. Line and VC errors may significantly affect shaping and compression performance because they may result in out-of-sync compression histories, causing retransmissions on the line.

When one end of the link is configured with compression, the other end must also be configured the same way. However, traffic shaping need not be configured on both ends. As mentioned before, it is beneficial to traffic shape at the central hub router. If the CIR and the line rate are equal at the remote spoke router, configuring traffic shaping provides no additional benefit. Using compression, however, increases the throughput of the end-to-end connection. Observe that compression occurs at the driver level and traffic shaping at the VC level. Since shaping occurs before compression, the effective compression ratio is reduced because only the preshaped traffic is compressed. Note that the compressor can compress data only at the rate it receives it, and the transmit queue should optimally empty at the throughput rate, preferably including the excess burst available. What you need to understand is that the data traffic pattern may result in lower compression ratios in some cases without the use of shaping or priority queuing. In some networks, shaped traffic may actually result in improved compression because the compressor acts on a series of frames from the same protocol.

Queue Depth When traffic shaping is configured on a VC, the router allocates a set of buffers for it. By default, 200 buffers per interface are allocated for traffic shaping, an amount further divided by the number of shaped VCs for uniform suballocation. When using protocol priority queuing in conjunction with traffic shaping, defaults of 30, 200, and 30 buffers are allocated per interface for the high-, normal-, and low-priority queues, respectively. On frame relay interfaces, reconfiguration or redistribution of the queue depths may be essential when there are numerous shaped VCs on the same interface. The following attributes are used to modify the queue depths:

- **high-queue-limit** The number of buffers allocated to the high-priority queue
- **normal-queue-limit** The number of buffers allocated to the normal-priority queue
- **low-queue-limit** The number of buffers allocated to the low-priority queue

The queue depths should be changed only after extended monitoring of the discards and clips at the VC or interface level. The attributes are configurable both at the interface level and the VC level, whichever is appropriate. Increasing the queue depths may sometimes be detrimental to router operation because the allocated buffers are not available for use by other subsystems and may remain unused. It can also result in excessive queuing during peak periods of traffic that may be better handled using selective discards.

Differentiated Services

In the past, the Internet and intranets provided best-effort delivery of IP datagrams throughout the network. The Internet Protocol by itself does not provide special service classes, except the type of service (TOS) bits that have been designed into the protocol header. Differentiated services (DiffServ) is an architecture build using the TOS bits, renamed as the DS (DiffServ) field. Unlike PPQ, which is used mostly within intranets and has a predefined priority structure, DiffServ defines an environment for service providers and enterprise networks to provide various tiers of service depending on the classification of data traffic. Both PPQ and DiffServ have unique applications, and DiffServ is a preferred choice at the network edge because the architecture is scalable. The classification, when performed by an edge router such as BayRS, allows the network core to be actively involved in providing the required class of service.

The edge router is tasked with the job of filtering frames and marking appropriate data packets for the desired class of service. Two methods are employed for configuring the filters on the edge routers:

- **Dynamic** An external server called the *bandwidth broker* is configured with broad-ranging filters that are dynamically downloaded to the router using the Common Open Policy Service (COPS) protocol. The filters and appropriate actions are automatically installed for dynamic operation. The advantage of dynamic filters is the support of the COPS protocol, providing scalability on several routers throughout the network.
- **Manual** Similar to uniform traffic filters and priority outbound filters, manual filters instruct the router to mark specific packets with the desired class of service on the network. Manual configuration of filters does not require an external bandwidth broker and enable local administration of the filters. However, management of manual filters may be cumbersome in large networks.

The benefits of both dynamic and manual configurations are easily seen. Dynamic configuration requires the provisioning and maintenance of a bandwidth broker but provides scalability. The manual mode of operation provides simplicity in classification maintenance and is adequate in many small networks; however, maintenance becomes a complication when numerous routers are involved, as is the case in most large enterprise networks.

A DiffServ network includes several components:

- **Bandwidth broker** The bandwidth broker functions as a resource manager, providing details on all reserved data flows and filter configurations for managing bandwidth allocation. The broker is responsible for installing trigger and flow filters on edge routers so that they can classify incoming packets. The trigger filter identifies the beginning of a traffic flow, and the bandwidth broker is notified. Then the flow filter becomes effective and identifies traffic belonging to the specific data flow. Also, packets in the flow are marked with the specific classification indicated by the flow filter. The COPS protocol is used by the bandwidth broker to communicate with the edge router. Bandwidth brokers are necessary in dynamic filter environments and are not needed if filters are manually configured.
- **Policy server** The policy server is responsible for policy admission control, which is essential for managing network services and user authentication. Data flows through the router are monitored, and the policy server is aware of flows assigned to specific users and applications; it may also dictate limits on traffic flows reserved for certain users or applications. It polices data flows to ensure that high-priority data traffic is limited. Excess high-priority traffic can be detrimental to the network because network resources may become oversubscribed. Often the bandwidth broker (also known as the COPS server) and policy server are administered on the same server for ease of administration. The policy server is not a mandatory component in the DiffServ architecture.
- **Edge router** The edge router resides at the periphery of the network. Generally, the network is an endless fabric, and identification of the "edge"

router may be difficult in some environments. The edge router in a DiffServ architecture is considered to the router farthest from the domain of control; also, it borders the next network, which is usually a service provider network. This router is responsible for classifying data flows and marking the DS bits in the IP header. The DS bit marking essentially defines the class of service that the flow requires through the network.

- **Core network device** This device is usually a high-performance router that is very efficient in the rapid forwarding of data. Based on the DS bit marking and the class of services it provides, it routes data flows through the network efficiently, ensuring that the required Class of Service (CoS) is met. Since this is a high-performance device, it is almost never tasked with the job of classifying flows; that job is left solely to the edge router.

The key components in a DiffServ network are illustrated in Figure 8-13. The BLN and ASN behave as edge routers, and the bandwidth broker and policy servers reside in the ISP/enterprise core network. Notice the presence of a host and server manager in the figure. The server manager is not a mandatory component of a DiffServ network; it is used as an intermediary between the host (endstation) and the bandwidth broker to communicate reservation requests.

Figure 8-14 shows a simple operation of a DiffServ trigger and flow filter. In the illustration, an external bandwidth broker provides the trigger and flow filter for flow identification and processing. When the data traffic begins to flow, the trigger filter is matched, and the bandwidth broker is notified. Then a flow filter is installed on the edge router, and the DS bits are marked and sent to the appropriate queue. Notice that PPQ is an integral feature, with the high-, normal-, and low-priority queues used to ensure that the desired tier of service is provided.

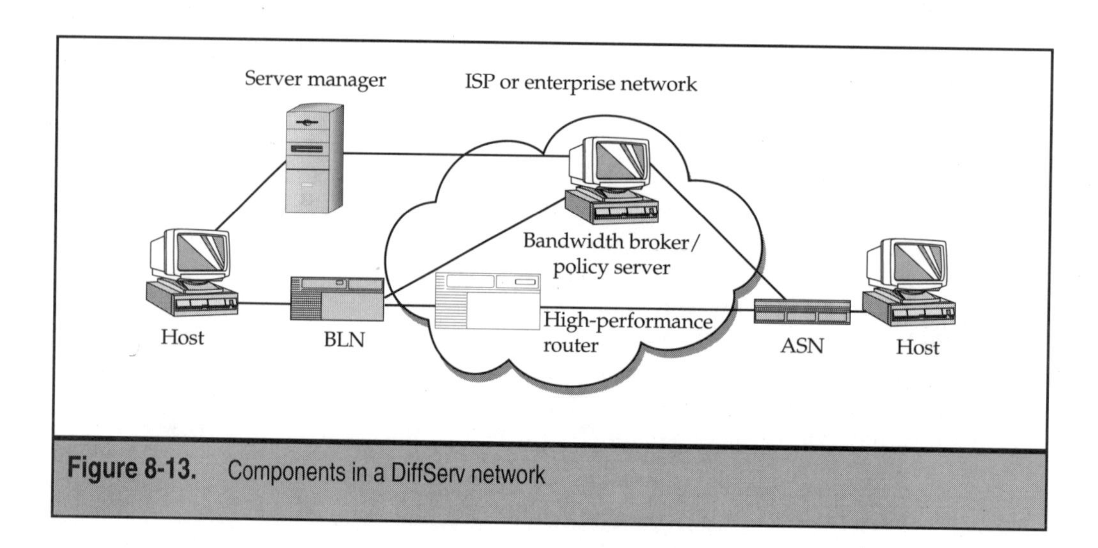

Figure 8-13. Components in a DiffServ network

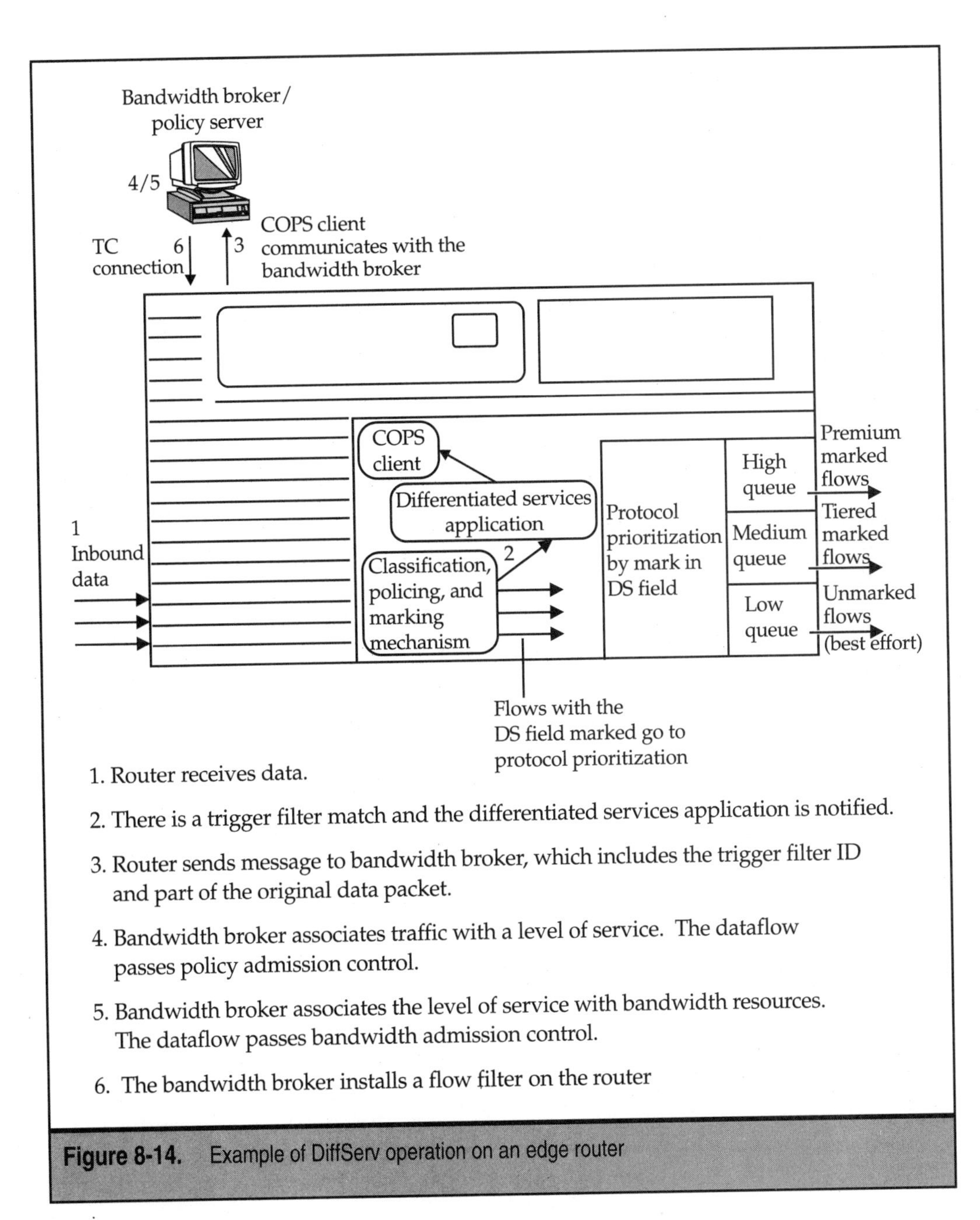
Bandwidth broker/
policy server
4/5
TC
connection
6
3
COPS client
communicates with the
bandwidth broker
COPS
client
Differentiated services
application
1
Inbound
data
2
Classification,
policing, and
marking
mechanism
Protocol
prioritization
by mark in
DS field
High
queue
Medium
queue
Low
queue
Premium
marked
flows
Tiered
marked
flows
Unmarked
flows
(best effort)
Flows with the
DS field marked go to
protocol prioritization
1. Router receives data.
2. There is a trigger filter match and the differentiated services application is notified.
3. Router sends message to bandwidth broker, which includes the trigger filter ID
and part of the original data packet.
4. Bandwidth broker associates traffic with a level of service. The dataflow
passes policy admission control.
5. Bandwidth broker associates the level of service with bandwidth resources.
The dataflow passes bandwidth admission control.
6. The bandwidth broker installs a flow filter on the router

Figure 8-14. Example of DiffServ operation on an edge router

Although DiffServ is still evolving to comprehensively provide various tiers of service to meet different SLAs, it is increasingly being used in many enterprise intranets to provide various classes of service. DiffServ poses several challenges because end-to-end SLA guarantees are difficult to achieve in today's service environments. As the evolving

DiffServ architecture settles into a more solid form, more networking devices will feature the true ability to guarantee the desired class of service. Within enterprise intranets, a combination of DiffServ and PPQ on BayRS routers will help achieve a dependable CoS with QoS agents. When connecting to ISP routers that provide differentiated services reliably using BayRS or other devices, the same configuration can be used with simple administrative changes.

This chapter does not present a detailed discussion of differentiated services because of the ongoing IETF effort in the DiffServ working group, where the specific details of behavior aggregates still are being defined. Readers are encouraged to consult the reference section for further information on DiffServ. The book's web site also includes references to DiffServ documentation for BayRS versions.

Load Balancing

In the not-too-distant past, reliability and availability concerns commonly were met by adding redundancy to the network. Additional LAN interfaces were added to bridged networks using the IEEE 802.1d Spanning Tree Protocol to ensure availability in the event that an interface was lost. The addition of redundant links is not without cost, however, and bandwidth needs and bottlenecks have led to the load-balancing concept. Redundant links are usually set up as passive connections, becoming active only when the primary link goes down. Load balancing goes one step further, actively using all links simultaneously, and thereby using all the available bandwidth efficiently. If one of the links in the active bundle goes down, the other links automatically assume the load, compensating for the lost connection. This benefit is an important consideration in providing RAS on committed SLAs and improving performance.

Multiline and Multilink

Two commonly used load balancing techniques are multiline and multilink. Multilink is standards based and is used exclusively with PPP links. Multiline connections can be configured between two BayRS routers running PPP, frame relay, Bay Standard (HDLC), or ATM DXI. On PPP links, multiline is simple to configure, though multilink is more efficient in managing multiple links in a bundle that have disparate speeds. PPP multilink also provides sequencing that helps prevent retransmission of frames arriving out of order; frames are automatically resequenced at the remote router.

Multiline configuration is a proprietary implementation for load balancing and redundancy. Figure 8-15 shows a frame relay multiline configuration using PVCs, and Figure 8-16 shows a PPP multilink configuration. Frame relay multiline requires the use of identical DLCIs on the load-sharing interfaces. The number of lines in a multiline bundle depends on the software version, but in general, up to four lines can be bundled together.

During the initial configuration of a circuit, multiple lines can be selected for addition into the bundle. Also, during dynamic operation, additional lines can be added to an existing circuit to make it a multiline or multilink circuit. PPP multilink automatically performs load balancing and line selection; however, line selection is an important consideration with

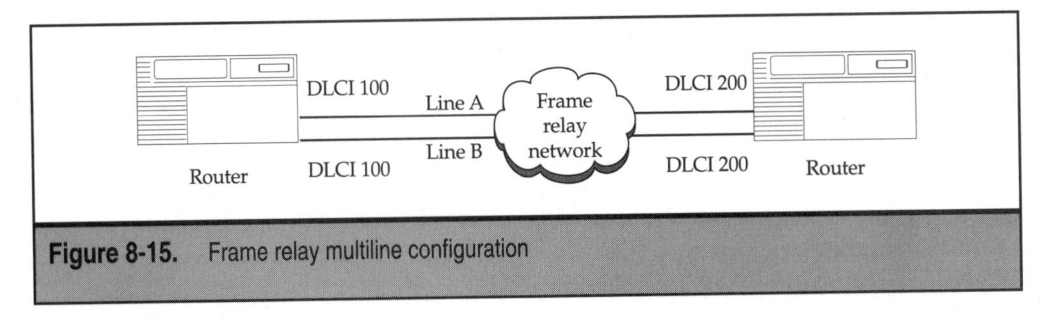

Figure 8-15. Frame relay multiline configuration

a multiline configuration. If the network uses higher-layer protocols that automatically resequence frames arriving out of order, the *random* chooser method can be employed. If frames for a certain destination should always use the same line, thereby ensuring orderly packet delivery, the *address-based* method can be used. Both methods have their advantages and disadvantages. The random chooser method uses the lines fairly but can cause frames to arrive out of sequence. The address-based method ensures orderly frame delivery, but traffic patterns may cause certain lines to be used more heavily than other lines.

Route and Traffic Balancing

Multiline and multilink provide redundancy and load balancing at the physical and data link layers. However, this is not always adequate, and load sharing service is often necessary at the network layer. In IP networks, the equal-cost multiple paths (ECMP) technique is used to maintain multiple equal-cost paths to the same destination in the routing table. When redundant links are available, often there are multiple routes to one or more destinations with the same metric. These paths are generally considered equivalent, and ECMP routing allows fair use of these routes to share the load across multiple next-hop gateways.

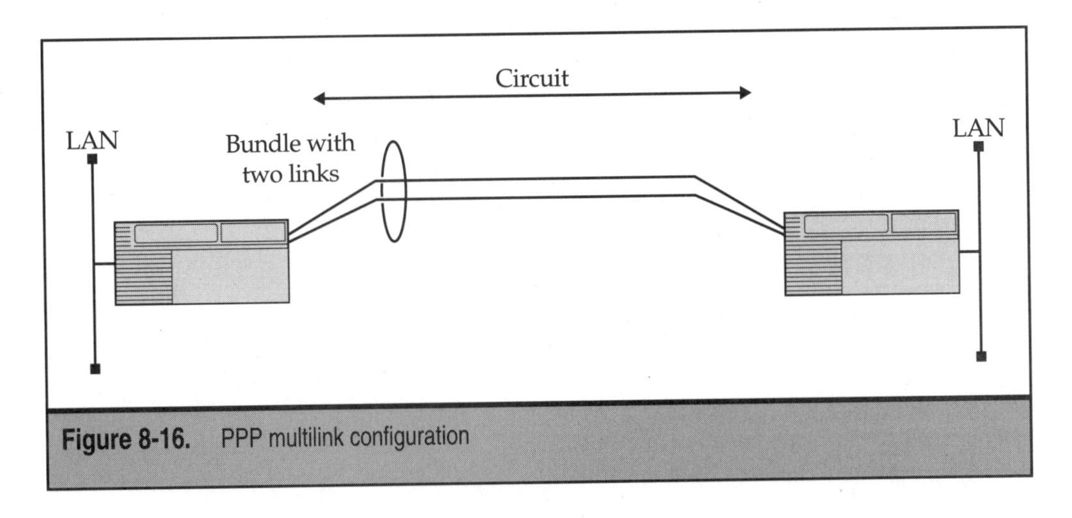

Figure 8-16. PPP multilink configuration

By default, the IP Routing Table Manager, as discussed in Chapter 5, installs only one best route to a destination network. Best routes are submitted by each routing source, such as directly attached networks, RIP, and OSPF. The precedence level and weight of a protocol determine the best-route source; however, when multiple routes are submitted either by the same protocol or by multiple protocols, there may be multiple equal-cost paths, in which case, the first one submitted is the one used. BayRS can be configured to accept and actively use more than one of these equal-cost routes. The following example illustrates the default selection of one active route (marked with an asterisk) to the destination network 172.16.23.0/24, but another equal-cost route is available.

```
fxc9(104)# show ip routes -A 172.16.23.0/24
show ip routes -A 172.16.23.0/24                        May 27, 2000 20:58:30 [GMT-6]

 Network/Mask         Proto       Age Sl      Cost NextHop Address     AS  Weight
 ------------------ ------ -------- -- -------- ---------------- ----- --------
*172.16.23.0/24       RIP          15  3         2 172.16.155.129          7b9e0002
 172.16.23.0/24       RIP          15  3         2 172.16.155.130          7b9e0002

   Total Networks on Slot 3 = 15
   Total Hosts on Slot 3 = 214
```

Two global IP attributes, `rip-max-paths` and `ospf-max-paths`, specify the number of maximum equal-cost routes that can be accepted from RIP and OSPF routing protocols, respectively. When entered with equal costs and precedence levels, static routes are automatically submitted when the next hop is available. A maximum of five routes can be actively used for ECMP load balancing. While these attributes specify the count of ECMP routes, they do not specify how traffic will be balanced across all available paths. For this purpose, the `ecmp-method` attribute, another parameter of the global `ip` object, is configured with one of the following values:

- **Round robin (`round-robin`)** All available ECMP routes are used one after another, thereby making fair and efficient use of the paths. This provides fair route use; it also provides fair data density across all paths. However, this choice is less well suited for higher-layer protocols that do not handle out-of-order delivery of frames. Since frames between the same data session may travel multiple links at different speeds, this option is best suited for protocols that can resequence frames.
- **Source and destination hash (`src-dest-hash`)** A simple hash value is computed using both the source and destination IP addresses. Using this hash value, a modulo is calculated with *n* number of routes available. The modulo *m* indicates the route number that will be used out of the available *n* paths. For any given source and destination pair, the path indicated by *m* will remain constant, ensuring frame delivery in order. This option is useful for sequenced frame delivery in networks where data traffic occurs between a wide range of host addresses.

- **Destination hash (`dest-hash`)** This value is similar to the source and destination hash, but the hash is computed with only the destination address. All frames for a specific destination take the same path. This option is useful in networks where orderly frame delivery is important to the destination, but where the range of destination addresses is disparate enough to efficiently use all available links. Use of both hash configurations can result in some paths being overused and others being underused, depending on traffic patterns between source and destination IP addresses.

The following example shows an ECMP configuration with round-robin path use, allowing the submission of up to four paths by static sources, RIP, and OSPF. Later, when we review the routes to the 172.16.23.0/24 network again, we observe that all available routes are marked active.

```
box
fxc9(111)# ip

box->ip
fxc9(112)# ecmp-method round-robin

box->ip
fxc9(113)# rip-max-paths 4

box->ip
fxc9(114)# ospf-max-paths 4

fxc9(115)# show ip routes -A 172.16.23.0/24
show ip routes -A 172.16.23.0/24                   May 27, 2000 20:58:30 [GMT-6]

 Network/Mask        Proto        Age Sl      Cost NextHop Address      AS  Weight
 ------------------- ------ -------- -- -------- ---------------- ----- --------
*172.16.23.0/24      RIP           15  3         2 172.16.155.129          7b9e0002
*172.16.23.0/24      RIP           15  3         2 172.16.155.130          7b9e0002

   Total Networks on Slot 3 = 15
   Total Hosts on Slot 3 = 214
```

The Border Gateway Protocol performs true route and traffic balancing with both IBGP and EBGP configurations. Since BGP interacts with many other routing protocol sources, especially in the case of multi-hop connections to other speakers, it not only load balances traffic but also balances routes to other BGP speakers. In a *route balancing* configuration, BGP submits one route to the IP Routing Table Manager for each BGP next hop and in turn uses all available ECMP routes in round-robin fashion. In this setup, the routes are fairly distributed across all known BGP destinations. The *traffic balancing* configuration, on the other hand, allows BGP to submit all available routes natively and directly to the IP RTM for all BGP next hops. This allows traffic balancing across all BGP next hops for each destination network. Table 8-5 summarizes the available BGP route and traffic balancing configurations.

Method	Configuration	Description
IBGP route balancing	`ibgp-ecmp-method route-balance`	IBGP submits one route to the IP RTM for each BGP next hop. It uses all available IGP ECMP routes to the destination in turn in a round-robin fashion.
IBGP traffic balancing	`ibgp-ecmp-method traffic-balance`	IBGP submits multiple routes to the IP RTM for the destination, using all active IGP ECMP next-hop gateways.
EBGP route balancing	`ebgp-ecmp-method route-balance`	EBGP submits one route to the IP RTM for each BGP next hop. It uses all available ECMP routes to the destination in turn in a round-robin fashion.
EBGP traffic balancing	`ebgp-ecmp-method traffic-balance`	EBGP submits multiple routes to the IP RTM for the destination, using all active IGP ECMP next-hop gateways.

Table 8-5. Route and Traffic Balancing for IBGP and EBGP

NOTE: The `ibgp-ecmp-method` attribute is configured in the global `bgp` context, whereas the `ebgp-ecmp-method` parameter is specified for the `peer` object. The EBGP ECMP method can be configured differently for each EBGP peer.

Each configuration has its own benefits, depending on the network topology. The route balancing configuration provides efficient maintenance of the routing table because the table size is minimized. However, a specific destination network always uses the same BGP peer as the next hop unless a network change occurs. For example, if destination network A receives more traffic and is routed via BGP speaker X, and if destination network B receives low traffic and is routed via BGP speaker Y, then the next hop X is used more heavily. Traffic balancing alleviates this utilization problem by ensuring fair use of all routes with all available next hops, but the routing table is burdened with numerous routes. In the same example of two ECMP peers X and Y, a local BGP speaker will have to submit twice as many routes to all known destinations. Both the IBGP and EBGP ECMP methods are independent of each other; IBGP may use traffic balancing internally in the local AS, and EBGP may use route balancing via its EBGP peers.

Compression

A natural solution to the bandwidth requirement is the use of compression on wide-area links. Even when bandwidth requirements are adequately met, compression provides an added level of throughput that any enterprise network would desire. We earlier briefly discussed compression in the context of traffic shaping and broached the topic of reliable transmission. In addition, although it is a secondary result of the compression function, most compression mechanisms use history buffers, and these aid in the reliable transmission and retransmission of lost packets. Most data link layer protocols do not provide this level of functionality. Recovering from a lost frame within a single connection span reduces response times significantly and maintains the integrity of established sessions. Compression is an excellent bandwidth enhancer, whether it is performed on an entire frame or limited to certain headers.

BayRS provides two compression solutions in two types of applications. The WAN Compression Protocol (WCP) and Hi/fn LZS are two compression solutions based on different Lev-Zimpel (LZ) algorithms. Both solutions are available for application in hardware and software, where the hardware application is typically geared toward high-speed lines, and the software application is geared toward slower-speed connections. The Hi/fn LZS, also called STAC, is based on the well-known Stacker application and is used exclusively with PPP lines. It also provides interoperability over PPP connections to third-party router products. WCP is a proprietary solution and is commonly used when both routers on a link are running BayRS.

Software compression is generally used with slow-speed lines and on Access platforms. For example, a frame relay PVC operating at 64 Kbps on an ARN is best suited to use software compression. On a BN router, a PPP link over a serial interface will be appropriate for use with hardware compression. Hardware compression can be enabled either using a compression daughtercard or using the integrated compression engine on the FRE2-060E. The FRE2-060E, described in Chapter 1, is a processor module that contains an integral compression coprocessor. To enable compression, both sides of a link must be configured to use compression. On PPP lines, the Compression Control Protocol (CCP) automatically negotiates the use of parameters on both ends of a link. WCP also performs the necessary negotiations on both sides of a compression link.

Compression can be enabled on frame relay, X.25, and PPP (leased- and switched-line) connections. Under the appropriate configuration context, the `wcp` and `hifn` child objects can be created to enable the desired configuration mechanism. On frame relay lines, the `wcp-control` attribute for a `pvc` or `svc-options` object can be selectively enabled or disabled to control compression on a specific VC. By default, creating the `wcp` object for a service record will automatically enable WCP on all VCs present in the service record.

Quality of Service

The demand for quality of service (QoS) comes in various forms. Increasing bandwidth alone does not solve all of today's problems, though it helps accommodate many high-density and real-time applications such as videoconferencing and Internet telephony. The goal of QoS is to provide a predictable quality of service, with minimal delay (in real time), low variance in delay (jitter), low packet loss, and a guaranteed amount of bandwidth. Many QoS solutions and protocols are in development today for data networks, and the traffic engineering mechanisms discussed earlier help to establish a desirable level of QoS end to end in BayRS environments.

In general terms, QoS dictates the requirement that certain network elements, such as applications, workstations, and routers, provide an assured level of consistent data delivery. QoS requirements between applications vary. QoS is provided by two categories of services. The first type, integrated services, also referred to as resource reservation, is based on the idea that network resources will be allocated for a QoS request from an application. This resource reservation is subject to administrative and bandwidth management policies. The second type is prioritization, also called differentiated services. With prioritization, network traffic is classified and allocated network resources based on administrative and bandwidth management policies. Network devices provide preferential access and processing to traffic classified as requiring higher-priority service.

QoS services as described here can be applied to a specific traffic flow (such as the flow belonging to a specific application session) or to an aggregate flow pattern (such as a common protocol). Whether QoS is applied to a specific flow or service depends on the application's requirements and the ability of the network to support QoS needs. In the integrated services category, the Reservation Protocol (RSVP) is used to guarantee network resource reservation for specific flows and, occasionally, aggregate flows. With differentiated services, prioritized services are provided to aggregate traffic flows. In preceding sections, we discussed protocol prioritization and DiffServ services that help provide QoS at a higher layer for aggregate flows with simple policy definitions. BayRS supports both RSVP and DiffServ, which, when combined with other traffic engineering techniques, provide QoS for many defined flows and aggregate flows in enterprise networks.

Other QoS protocols and solutions do exist and are evolving. BayRS also supports a preliminary version of the Multiprotocol Label Switching (MPLS) protocol, which provides bandwidth management for aggregate traffic flows. An integrated QoS solution must be provisioned on an end-to-end basis; enabling QoS services at core ingress and egress points alone is inadequate for providing true QoS.

PERFORMANCE TUNING

Performance management and traffic engineering concepts applied to BayRS were discussed earlier in this chapter. We covered the topics of baselining and analysis of performance statistics and the use of TE techniques to improve performance. Tuning configuration parameters to optimum levels can also enhance performance on certain nodes and the network as a whole. Many protocols and drivers automatically adjust resource use for optimal performance. However, not all resources are controlled in this fashion, and manual configuration of certain parameters provides the necessary fine-tuning. In this section, we discuss some tuning concepts for the services that benefit most from resource allocation changes.

Protocols

Network and routing protocols consume memory dynamically for various tables, such as routing, forwarding, and service tables. Some of these tables are seeded with an initial table size that is increased when necessary. While the extent of resource use depends on the network size, these initial table sizes can be administratively specified. This specification allows the protocol to allocate enough memory during initialization so that the need for future allocation is reduced, and it also facilitates faster learning of routes and services. Tables, caches, and MIB objects are the components that most frequently require resources for individual protocols.

This chapter has focused mostly on the Internet Protocol and related protocols. With IP, the routing and host tables maintained by the RTM grow dynamically. Initially, the routing table contains only the local routes, but as routing protocols submit their individual best routes, the active routing table gains entries. This also is true for the ARP cache (host table), which grows as more traffic from hosts crosses the router. The `wfIpBaseNetworks` and `wfIpBaseHosts` attributes indicate the number of active networks and hosts in the routing and host tables, respectively. The `wfIpBaseEstimatedNetworks` and `wfIpBaseEstimatedHosts` attributes control the amount of memory that is preallocated to these tables during initialization. A default of 500 entries is allocated; in large enterprise networks, additional resources must be dynamically allocated. Using the baseline active network and host count, the estimated network and host values can be configured so that incremental allocations do not have to be made during initialization. This can help speed up the initialization of the IP protocol. Subsequent steady-state operation is not affected by these attributes.

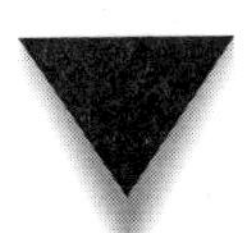

TIP: Overallocation of resources, particularly memory and buffers, wastes resources. Always use a baseline average value for resource allocation and increase it when appropriate. Allowing the router to dynamically allocate resources is preferable to overallocation of resources.

In a way similar to the use of routing, forwarding, and service tables, many other protocols use data caches. The DNS proxy server, for example, caches resolved hostnames, and the embedded HTTP server caches recently accessed URL data. NetBIOS similarly can cache server names to reduce the number of broadcast frames in the network. Cache sizes for these protocols can be increased when the corresponding cache miss attribute increases significantly. Occasional and transient failures of cache lookups are acceptable, especially during periods when entries are being learned. Increasing cache sizes for optimization must also be performed carefully because resources can be overallocated. Some cache timeout periods can also be increased so that frequently reused entries remain in the table longer than the default value specifies. Increasing cache timeout values has a downside, however: some cached details are more likely to become stale faster than others. For example, a cached DNS name resolution is unlikely to change dynamically (except in the case of DHCP hosts) or frequently. However, a cached log page from the embedded web server can become significantly outdated if loaded after a minute. The acceptable timeout values depend on the protocol and the type of information cached.

Buffers and Local Memory

In Chapter 4, in the context of troubleshooting local memory and global memory use, we discussed how buffers and local memory are allocated and shared between the kernel and all subsystems. Local memory can be carved so that an optimal amount is allocated for all the protocols and dynamic tables. Care should be employed to ensure that free memory is available for both dynamic network growth (that is, for tables and caches) and future provisioning (that is, for new protocols and interfaces). In general, you should leave at least 10 to 20 percent of local memory free for network evolution.

Global buffers are used both deterministically and nondeterministically. For example, a line driver on an interface is allocated a predefined number of buffers for transmit and receive queues. This is a deterministic use, because this minimum amount is always necessary for the requisitioning subsystem. Dynamic buffer use can vary dramatically, and we discussed ways of identifying buffer starvation in Chapter 4. There are several methods for computing buffer use over extended periods of time to understand traffic patterns. Data traffic is the most common source of dynamic buffer use. Peak buffer use should ideally never reach the total buffer count, and an average of at least 10 to 20 percent of the total buffers must be available. Both local and global memory can be carved appropriately to improve performance on platforms that support memory carving.

SUMMARY

Performance enhancement is an ongoing process in many enterprise networks. An ideal network provides both unlimited bandwidth and guaranteed QoS. In the absence of this ideal network, optimal use of available resources is paramount for meeting the requirements of the business. Even simple enhancements can improve overall network performance dramatically. To enhance the performance, understanding the current network performance is essential, to determine where improvements should be made. Quality-of-service improvements do not increase the bandwidth, but in fact help utilize the existing bandwidth more efficiently and appropriately to meet business needs. This chapter also discussed some analytical techniques and traffic engineering processes to help you get better mileage from average and not-so-average networks.

There is no simple method for achieving the perfect network; however, a performance-optimized network is an achievable goal.

REFERENCES

1. G. Huston, et al, *Internet Performance Survival Guide: QoS Strategies for Multi-Service Networks*, John Wiley & Sons, February 2000.
2. D. Verma, *Supporting Service Level Agreements in IP Networks*, Macmillan Technical Publishing, July 1999.
3. K. Kilkki, *Differentiated Services for the Internet*, Macmillan Technical Publishing, June 1999.
4. U. D. Black, *QoS in Wide Area Networks*, Prentice-Hall, December 1999.

APPENDIX A

Software Suites, Feature Availability, and Release Policy

This appendix provides a listing of various software suites, their availability on all BayRS platforms, and the feature sets supported by each suite. The current BayRS release policy describing the various type of software releases is also included for reference. Table A-1 summarizes the software suites and platform availability.

Each software suite supports a standard feature set that uniquely distinguishes it from others. For example, support for the IPX protocol is not available in the IP Access suite but is available with the Corporate suite. The support for various feature sets in the software suites are detailed in Table A-2.

Suite	AN	ARN	ASN	BN	System 5000
LAN			√	√	√
WAN			√	√	√
System			√	√	√
IP Access	√	√			
Remote Office	√	√			
Corporate	√	√	√	√	√
ATM				√	
VNR				√	√

Table A-1. Software Suites and Platform Availability

Feature	System	LAN	WAN	IP Access	Remote Office	Corporate	ATM & VNR*
Network Protocols							
IP with RIP, OSPF, BGP/EGP	√	√	√	√	√	√	√
Novell IPX with RIP/SAP		√			√	√	√
AppleTalk Phase 2		√			√	√	√
OSI		√				√	√
DECnet Phase IV		√				√	√
Banyan VINES		√				√	√

Table A-2. Feature Availability

Feature	System	LAN	WAN	IP Access	Remote Office	Corporate	ATM & VNR*
Network Protocols *(continued)*							
Xerox XNS		√				√	√
ST-II	√	√	√	√	√	√	√
IBM Integration							
Source Route Bridging	√	√	√	√	√	√	√
LAN Network Manager (LNM) Agent		√			√	√	√
Data Link Switching (DLSw) for Ethernet and Token Ring	√	√	√		√	√	√
DLSw for SDLC		√			√	√	√
Transparent Sync Pass-Through	√	√	√	√	√	√	√
Advanced Peer-to-Peer Networking (APPN)						√	√
Bridging							
Transparent (Ethernet and Token Ring)	√	√	√	√	√	√	√
Translation Bridge (Ethernet–Token Ring)	√	√	√	√	√	√	√
Translation Bridge (Ethernet–FDDI)	√	√	√	√	√	√	√
Translation Bridge (Token Ring–FDDI)	√	√	√	√	√	√	√
Native Mode LAN (NML)		√		√	√	√	√
Wide Area Networking							
HDLC Encapsulation	√	√	√	√	√	√	√
Point-to-Point Protocol (PPP)	√	√	√	√	√	√	√
Frame Relay			√	√	√	√	√
ISDN BRI	√	√	√	√	√	√	√

Table A-2. Feature Availability *(continued)*

Feature	System	LAN	WAN	IP Access	Remote Office	Corporate	ATM & VNR*
Wide Area Networking *(continued)*							
ISDN PRI	√	√	√	√	√	√	√
SMDS			√	√	√	√	√
X.25			√	√	√	√	√
ATM DXI	√	√	√	√	√	√	√
Dial on Demand	√	√	√	√	√	√	√
Bandwidth on Demand	√	√	√	√	√	√	√
Dial Backup	√	√	√	√	√	√	√
Traffic Management							
Data Compression (PPP)	√	√	√	√	√	√	√
Data Compression (X.25 and Frame Relay)			√	√	√	√	√
Traffic Prioritization	√	√	√	√	√	√	√
Uniform Traffic Filters	√	√	√	√	√	√	√
Multiline Circuits	√	√	√	√	√	√	√
Node Management							
EZ Install/EZ Update (HDLC)	√	√	√	√	√	√	√
EZ Install/EZ Update (Frame Relay)			√	√	√	√	√
SecurID	√	√	√	√	√	√	√
Dynamic Loader	√	√	√	√	√	√	√
RMON	√	√	√	√	√	√	√
Availability							
Software Fault Isolation and Recovery	√	√	√	√	√	√	√
Online Operational Servicing	√	√	√	√	√	√	√
Flash Card Partitioning	√	√	√	√	√	√	√

Table A-2. Feature Availability *(continued)*

Feature	System	LAN	WAN	IP Access	Remote Office	Corporate	ATM & VNR*
Availability *(continued)*							
Interface Redundancy	√	√	√	√	√	√	√
Router Redundancy	√	√	√	√	√	√	√
Dynamic Reconfiguration	√	√	√	√	√	√	√
Redundant Power	√	√	√	√	√	√	√

* ATM and VNR suites support the same features as the Corporate suite. The ATM suite includes support for ATM modules, using PVCs only. The VNR suite adds support for SVCs over ATM interfaces in addition to the features provided by the ATM suite.

Table A-2. Feature Availability *(continued)*

RELEASE POLICY

The following is excerpted from Bay Networks (Nortel Networks) publication 301653-A Rev 00 and is reused with permission. References to Bay Networks should be read as Nortel Networks, the successor.

With the release of BayRS Version 12.10, Bay Networks is introducing a new methodology for releasing Bay Networks routing software.

The goals of the new BayRS Release Policy are to:

- Enable you to maintain a stable and mature network while having access to new software capabilities at your discretion.
- Deliver new versions of BayRS on a consistent and predictable schedule.
- Improve access to BayRS to enable more effective planning for major network software upgrades.
- Make it easy for you to determine which software releases contain new functionality and which consolidate software fixes.

The BayRS development process is now channeled into two separate, but parallel, development streams after a new Major Release:

- One development path will ensure that stability is maintained with the then current Major Release.
- A separate development path will provide new functionality beyond that found in the then current Major Release.

Figure A-1 provides a conceptual view of these two development paths. There are now four distinct types of BayRS releases:

- **A Major Release** (12.00, 13.00, etc.) Designated by a change in the number to the left of the decimal point. Provides a roll-up of all new features, functionality, and fixes or significant software architecture changes and enhancements introduced through the combination of Maintenance, Feature, and Revision Releases made since the most recent Major Release. No new features or functionality are introduced.
- **A Maintenance Release** (12.01, 12.02, etc.) Designated by a change in the second number to the right of the decimal point. Provides a collection of software bug fixes for the then current Major Release. No new software or hardware features or functionality for either software or hardware is introduced beyond that already contained in the then current Major Release.
- **A Feature Release** (12.10, 12.20, etc.) Designated by a change in the first number to the right of the decimal point. Provides new features and functions for software and/or hardware, and software bug fixes from corresponding Maintenance Releases.
- **A Revision Release** (12.01 (Rev. 4), etc.) Designated by a second decimal point followed by a number. A software bug fix or collection of software bug fixes integrated with any then currently supported Major Release, Feature Release, or Maintenance Release.

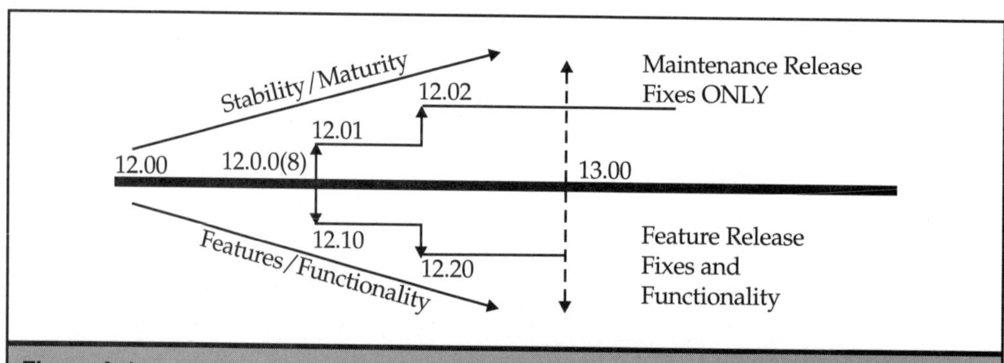

Figure A-1. BayRS parallel development process

Major Releases

A Major Release of BayRS will occur approximately every 12 months and provide a stable and mature suite of features and functionality introduced since the prior Major Release, which you can integrate into your existing network with little or no disruption to your network operations.

A Major Release brings the two separate development streams into a single software baseline. It integrates the features and functionality found in the prior Major Release and the Feature Releases introduced since the then current Major Release, as well as any software bug fixes made by Maintenance and Revision Releases subsequent to the introduction of the then current Major Release. Therefore, a new Major Release will be equivalent, with respect to features, functions, and bug fixes, to the most recent Feature and Maintenance Releases.

To provide for a continuous and predictable stream of new software features and functionality, a new Feature Release will occur approximately at the time of a new Major Release. This Feature Release will introduce incremental features and functionality beyond that included in its then corresponding Major Release.

Under this new BayRS Release Policy, each Major Release will remain on the Bay Networks Price List for two years. Whenever a new Major Release occurs, the older of the two Major Releases then on the Bay Networks Price List will be removed.

Whenever a Major Release is removed from the Price List, Bay Networks will continue to support the removed Major Release for an additional one-year period by providing Revision Releases. Telephone support will be available for an 18-month period after the Major Release is removed from the Price List.

Maintenance Releases

A Maintenance Release will occur approximately each three to four months after a Major Release occurs. A Maintenance Release replaces the then current Major Release (or the prior Maintenance Release) on the Price List and will become the current shipping version of BayRS. Each Maintenance Release will be a consolidation of the prior bug fixes made through Revision Releases, and will roll up all such software bug fixes provided by the recent Revision Releases.

Feature Releases

A Feature Release is a new type of software release that gives you quicker access to new hardware and software features and functionality.

A Feature Release will occur approximately every four months. This release schedule enables you to more easily schedule migrations to new releases.

You should be deliberate in your decision to implement a Feature Release. The decision to migrate to a new Feature Release should be driven by whether or not you need the features or functionality introduced in that release. Major Releases will give you access to new products and features on a matured software base.

Revision Releases

A Revision Release will occur as soon as is practical to resolve a particular reported customer problem.

All Revision Releases will be available only through the Bay Networks web site. This is meant to improve your access to specific software bug fixes and to provide a quicker and more efficient means of informing you of, and allowing you access to, software bug fixes that have been made for a specific Major Release.

ROUTER ELEMENT MANAGEMENT RELEASE POLICY

With each release of BayRS (i.e., Major, Maintenance, Feature, and Revision), device management will be included that supports the newly introduced features and functionality. Beginning with BayRS Feature Release 12.20, new device-specific management functionality will be provided using either the Bay Command Console (BCC) or Site Manager.

Through calendar year 1999, both Site Manager and BCC support will be provided for all BayRS Releases. Once a base level of BCC functionality has been established, then only the BCC will be used for management of new Feature Release features and functionality. During this transition period, Site Manager may also be used to configure features and functionality carried forward from a Major Release.

BayRS PROBLEM ESCALATION AND RESOLUTION

The problem escalation guidelines for all BayRS software releases will not change from the process currently followed. You will continue to contact Customer Support to obtain support for all BayRS software releases. The only significant difference will be in how you will obtain a software fix made for a Feature Release.

Under this BayRS Release Policy, Bay Networks Sustaining Engineering will be responsible for addressing any problems reported in Major Releases and Maintenance Releases. This provides you with the same response that you currently experience with BayRS releases.

Because Feature Releases focus on maintaining a consistent and reasonably predictable release schedule, software fixes for Feature Releases will be provided at Bay Networks discretion. Any software fix generated for a Major Release or Maintenance Release will also be provided for the then current and future associated Feature Releases. However, if a software fix is required for a new feature found only in a Feature Release, Bay

Networks reserves the right to instead direct you to the next Feature Release or Major Release. Additionally, if you request a software fix for a prior Feature Release, you will instead be directed to upgrade to the then current Feature Release or Major Release.

To ensure consistency in software bug fix integration among Major, Maintenance, and Feature Releases:

- All Revision Releases generated for a Major Release will be integrated into the next Feature Release.
- All software bug fixes generated for a Feature Release will be integrated into the next Major Release.
- Priority will be given to fixing reported problems for Major and Maintenance Releases over those reported for a Feature Release.

Once the next Major Release occurs, you need to upgrade to it to ensure timely response from Bay Networks for support of your mission-critical production network.

BayRS DOCUMENTATION

- **Major Releases** (12.00, 13.00, etc.) A documentation CD and hard-copy release notes will be produced for all Major Releases. The documentation, including release notes, will also be available on the Bay Networks web site. You may print the documentation from the web site or the documentation CD, or you may order manuals from the Bay Networks Price List.
- **Feature Releases** (12.10, 12.20, etc.) Documentation for new features and functionality will be provided on the Bay Networks web site and on a documentation CD for a Feature Release. You may obtain printed manuals from the Bay Networks Price List only.
- **Maintenance Releases** (12.01, 12.02, etc.) The then current documentation CD (12.00, for example) and hard-copy release notes documenting the software bug fixes provided in the Maintenance Release will be produced and shipped. Customers can also obtain the release notes for a Maintenance Release from the Bay Networks web site.

SOFTWARE DISTRIBUTION

BayRS licenses are available for purchase from the Bay Networks Price List. Customers who have a software maintenance agreement in place with Bay Networks are provided with new Major, Maintenance, and Feature Releases of BayRS through software CD and related documentation.

Under this BayRS Release Policy, Major, Feature, and Maintenance Releases will be available for purchase through the Price List for users without a software maintenance agreement, and through the Bay Networks web site for customers under a current software maintenance agreement.

This software distribution strategy will:

- Provide a better mechanism for Bay Networks customers and channel partners to obtain new Releases of BayRS.
- Provide timely access to Revision Releases prior to installation of a new Major Release.

If you want continued and immediate access to all available Major, Maintenance, and Feature Releases, you need to have a Bay Networks software maintenance agreement to gain controlled access to the BayRS location on the Bay Networks Customer Service web site.

APPENDIX B

Router and Processor LEDs

All routers and modules have LEDs indicating the status of various components. Front-panel status indicators are useful in quickly identifying hardware failures. Sometimes you may also need to look at back-panel, adapter, and expansion module LEDs for interface-specific status indications. On Backbone routers, you may need to remove the front bezel and the EMI shield to look at the processor LEDs to determine the status of the processor and link modules.

ACCESS NODE AND ACCESS NODE HUB

Tables B-1 and B-2 summarize the LED status indicators on the front and back panels of AN and ANH-12 routers, illustrated in Figures B-1 and B-2. The back panel of AN/ANH-12 routers will look different depending on the unit configuration and whether an expansion interface is installed.

The ANH-8 router differs slightly from the ANH-12 router in that its expansion capacity is limited, and its front panel provides more comprehensive status indicators. LED indicators on the ANH-8 are summarized in Table B-3 and illustrated in Figure B-3.

ADVANCED REMOTE NODE

The ARN has several status LEDs. The base module LEDs provide comprehensive information on all components, including adapter and expansion module status. Expansion modules and adapter modules each have a unique set of status indicators. Base module LED indicators are illustrated in Figure B-4 and summarized in Table B-4. Adapter module

LED	Condition	Meaning
Power	On	Unit is plugged into a power source.
Run	Flashing	Diagnostics are currently executing (testing lasts one to three minutes).
	On	Unit is fully operational in a steady state.
Boot	On	Unit is booting to the runtime software.
Diag	On	One of the diagnostic tests failed.

Table B-1. AN and ANH-12 Front-Panel LEDs

LED	Condition	Meaning
XCVR/UTP Ethernet Ports		
TX	On	The LED lights briefly when the interface transmits a frame.
RX	On	The LED lights briefly when the interface receives a frame.
CL	On	The LED lights briefly when a collision occurs on the segment.
Repeater Ports		
MDI	On	The media-dependent interface setting (crossover switch) on port 12 is in the straight-through setting.
	Off	Port 12 is in the crossover setting.
LINK	On	There is an operational connection to a downstream CSMA/CD transceiver.
PART	On	The specific port (1 through 12) has been autopartitioned by the repeater, indicating a defective transmitter downstream.
Token Ring (MAU) Ports		
RCVR	On	The interface is receiving data from the ring.
NSRT	On	The interface has successfully inserted (NSRT) into the ring.
WFAIL	On	The hardware detected a wire failure (WFAIL).
Serial Ports		
RLSD1	On	The receive line signal detection (RLSD) circuitry detected a signal on port 1 (COM1).
RLSD2	On	The receive line signal detection (RLSD) circuitry detected a signal on port 1 (COM1).

Table B-2. AN/ANH-12 Back-Panel LEDs

LED	Condition	Meaning
ISDN/BRI Ports		
ISDN BRI	On	The port is active and operating.
56/64K DSU/CSU Ports		
Tx	On	The LED lights briefly when the interface transmits a frame.
Rx	On	The LED lights briefly when the interface receives a frame.
CD	On	A Carrier Detect (CD) signal is present, indicating a valid connection to the network.
Test	On	The interface is performing loopback testing.
FT1/T1 DSU/CSU Ports		
Red Alm	On	There is a red alarm condition (see Chapter 4).
Yellow Alm	On	A yellow alarm condition was received from the network (see Chapter 4).
Loop	On	The port is in loopback testing mode.
Sync	On	The port is receiving an operational T1 carrier signal from the network and is synchronized.

Table B-2. AN/ANH-12 Back-Panel LEDs *(continued)*

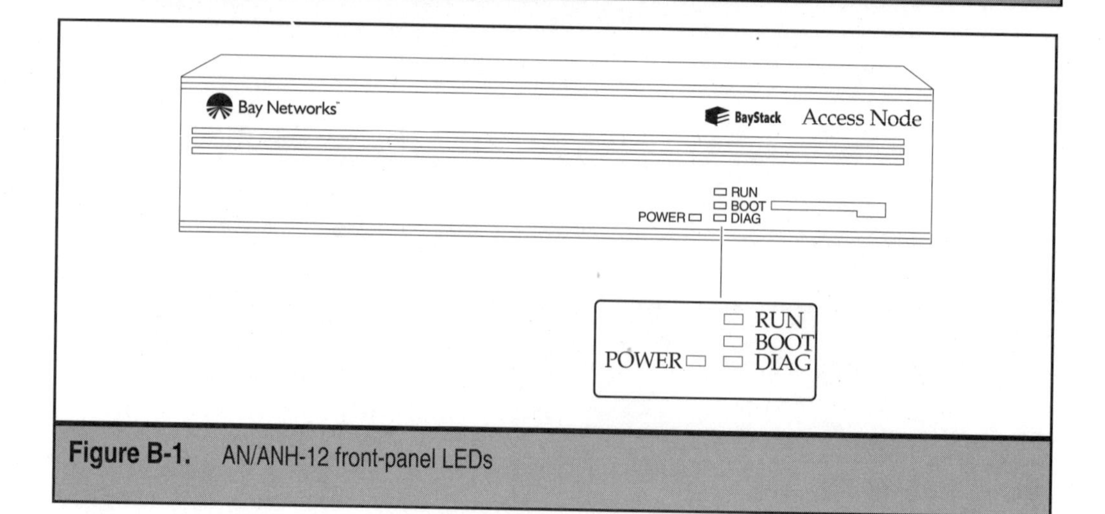

Figure B-1. AN/ANH-12 front-panel LEDs

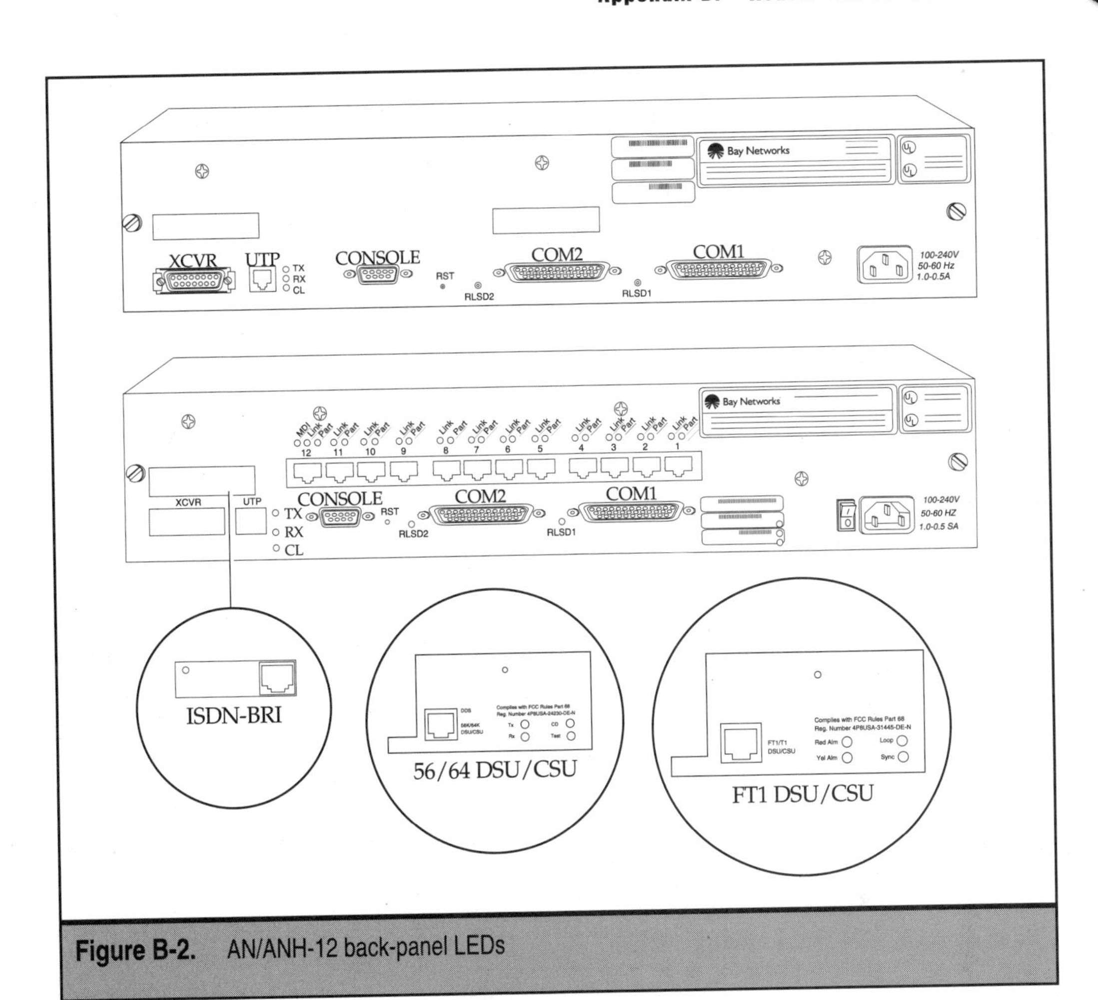

Figure B-2. AN/ANH-12 back-panel LEDs

LED	Condition	Meaning
Power	On	Unit is plugged into a power source.
Fault	On (amber)	One of the diagnostics tests failed. If the Run light is flashing, the unit is currently executing a self-test.
Boot	On	Unit is booting to the runtime software.
Run	Flashing	Diagnostics are currently executing (testing lasts one to three minutes).

Table B-3. ANH-8 Front-Panel LEDs

LED	Condition	Meaning
	On	Unit is fully operational in a steady state.
DCM	On	The RMON data collection module (DCM) is installed and operating.
AUI Part	On (amber)	The AUI port has been partitioned from the repeater unit. Autopartitioning occurs when there are an excessive number of consecutive collisions or an excessively long collision/jam signal.
DCD1	On	Serial port 1 (COM1) is operationally active (a Data Carrier Detected signal was received).
DCD2	On	Serial port 2 (COM2) is operationally active.
LAN	On	The LED turns on briefly every time the repeater successfully transmits a frame.
Col	On (amber)	The LED turns on briefly during a collision in the AUI repeater.
Partition	On	The specific repeater port (1–8) has been autopartitioned by the repeater unit, indicating a defect in the transmitter downstream.
Link	On	There is an operational connection to a downstream CSMA/CD transceiver.

Table B-3. ANH-8 Front-Panel LEDs *(continued)*

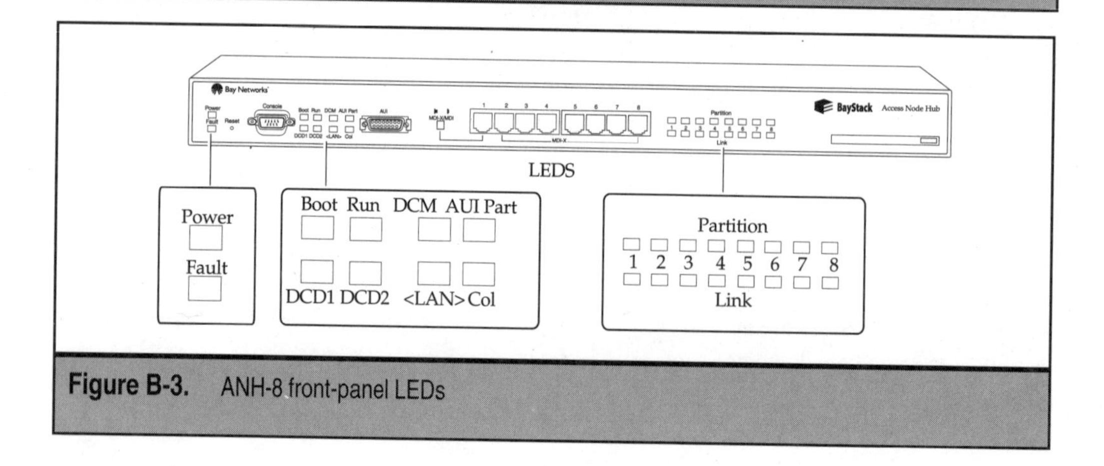

Figure B-3. ANH-8 front-panel LEDs

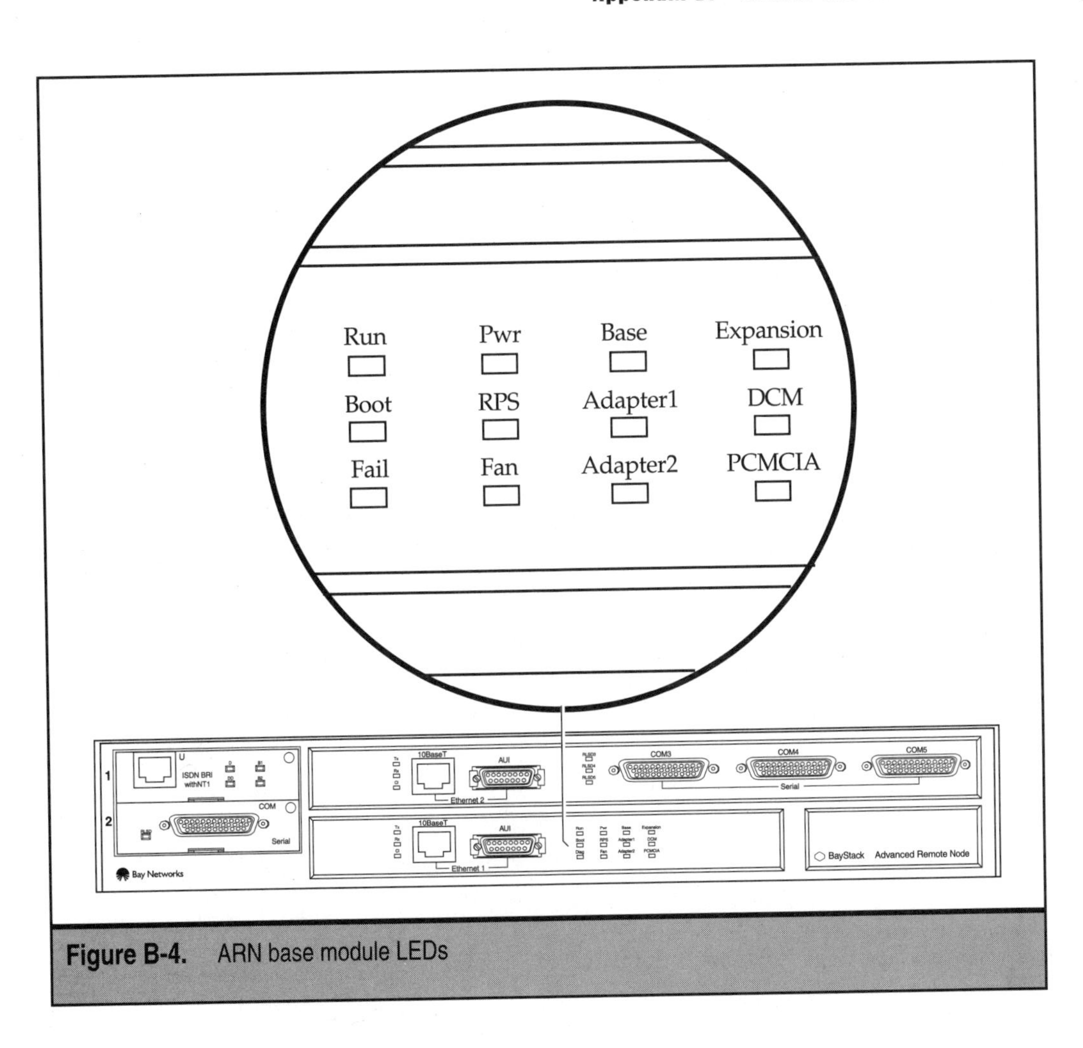

Figure B-4. ARN base module LEDs

LED	Condition	Meaning
Run	Flashing	Diagnostics are currently executing (testing lasts one to three minutes).
	On	Unit is fully operational in a steady state.
Boot	On	Unit is booting to the runtime software.
Pwr	On	Unit is plugged into a power source.
Fail	On	One of the diagnostic tests failed.

Table B-4. ARN Base Module LEDs

LED	Condition	Meaning
RPS	On	The optional redundant power supply unit (RPSU) is operating.
Fan	On	One of the cooling fans has stopped operating.
Base	On	Base module is operating properly.
	Flashing (slow)	Base module is running diagnostics.
	Flashing (rapid)	One of the base module diagnostic tests failed.
Adapter1	On	Adapter module 1 is operating properly.
	Flashing (slow)	Adapter module 1 is running diagnostics.
	Flashing (rapid)	One of the adapter module 1 diagnostic tests failed.
Adapter2	On	Adapter module 2 is operating properly.
	Flashing (slow)	Adapter module 2 is running diagnostics.
	Flashing (rapid)	One of the adapter module 2 diagnostic tests failed.
Expansion	On	Expansion module is operating properly.
	Flashing (slow)	Expansion module is running diagnostics.
	Flashing (rapid)	One of the expansion module diagnostic tests failed.
DCM	On	RMON data collection module (DCM) is installed and operating.
	Flashing (slow)	DCM is running diagnostics.
	Flashing (rapid)	One of the DCM diagnostic tests failed.
PCMCIA	On	PCMCIA flash is installed and has been detected.

Table B-4. ARN Base Module LEDs *(continued)*

LED indicators are illustrated in Figure B-5 and summarized in Table B-5. Expansion module LEDs are similar to the corresponding interface indicators on the base and adapter modules.

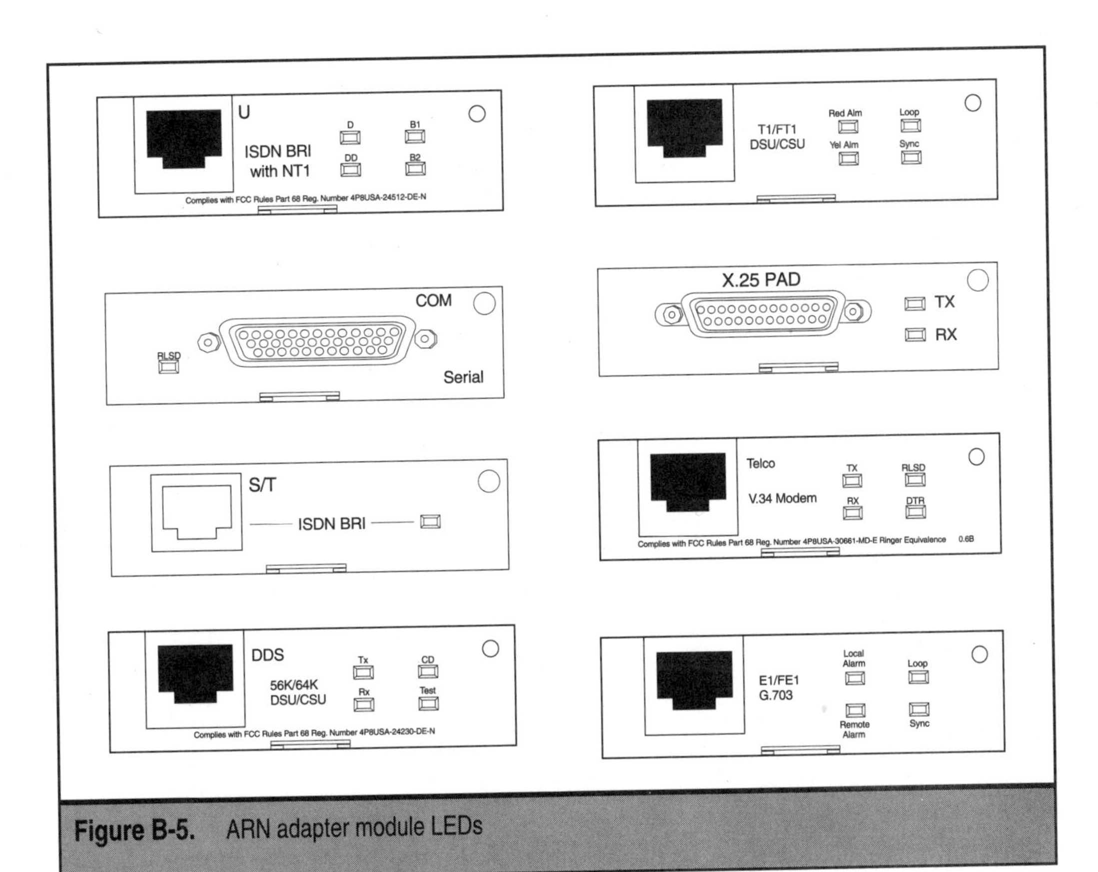

Figure B-5. ARN adapter module LEDs

LED	Condition	Meaning
Serial Ports		
RLSDx	On	The interface received a line signal detection (RLSD). On expansion modules, the numeric suffix (for example, RLSD2) indicates the number of the serial interface (COM2) that received the line signal detection.
ISDN BRI S/T/U Ports		
D	On	On some S/T interfaces, this LED is unlabeled and the only LED present. This LED indicates that the router is using the D-channel for signaling.

Table B-5. ARN Adapter Module LEDs

LED	Condition	Meaning
DD	On	A Data Detect (DD) signal was received. The interface has an operational connection to the external device, usually the ISDN switch.
B1	On	The first bearer channel (B1) is currently in use.
B2	On	The second bearer channel (B2) is currently in use.
DSU/CSU Ports		
TxD	On	The interface is transmitting an outgoing frame to the network.
RxD	On	The interface is receiving an incoming frame from the network.
CD	On	A Carrier Detect (CD) signal was received. The interface detected a carrier signal on the line, thereby validating the presence of a connection to another device.
Test	On	The interface is in test/loopback mode.
FT1/FE1 DSU/CSU Ports		
Red Alm/Local Alarm	On	The interface has lost either the signal or the framing pattern (see "Media State" in Chapter 4).
Yellow Alm/Remote Alarm	On	An RAI was received from the remote end, and the port has entered the yellow alarm state (see "Media State" in Chapter 4).
Loop	On	The port is currently in loopback mode.
Synch	On	The port has a synchronized signal with the T1 network.
X.25 PAD Ports		
TxD	On	The packet assembler/disassembler (PAD) is transmitting data to the network.
RxD	On	The PAD is receiving data from the network.

Table B-5. ARN Adapter Module LEDs *(continued)*

LED	Condition	Meaning
V.34 Modem Ports		
TX	On	Briefly lights when the interface transmits data to the network.
RX	On	Briefly lights when the interface receives data from the network.
RLSD	On	A Received Line Signal Detection signal was received. The router has activated the modem and it is ready for use.
DTR	On	A Data Terminal Ready signal was received. The modem is ready to transmit data.

Table B-5. ARN Adapter Module LEDs *(continued)*

ACCESS STACK NODE

The ASN1 and ASN2 both feature the same indicators on the front panel. Table B-6 summarizes the front-panel LED indicators. The front-panel LEDs are illustrated in Figure B-6. The boot sequence on ASN platforms is unique and follows these steps:

1. Initial power-on.
2. Diagnostics running (Run LED flashes while Boot and Diag are off).
3. Bootstrapping (Boot LED on).
4. Steady-state operation (Run LED on).

LED	Condition	Meaning
Power	On	Unit is plugged into a power source.
Remote	On	The optional RPSU is operating and delivering power. This LED is only present on ASN units equipped to support a redundant power supply.

Table B-6. ASN Front-Panel LEDs

LED	Condition	Meaning
Run	Flashing	Diagnostics are currently executing. If the Diag LED is simultaneously on, then one of the diagnostics tests has failed (includes SPEX failures).
	On	Unit is fully operational in a steady state.
Boot	On	Unit is booting to the runtime software.
Diag	On (runtime)	One of the connectors failed a diagnostic test.
	Flashing	If alternately flashing with Run, SPEX has failed a diagnostic test.

Table B-6. ASN Front-Panel LEDs *(continued)*

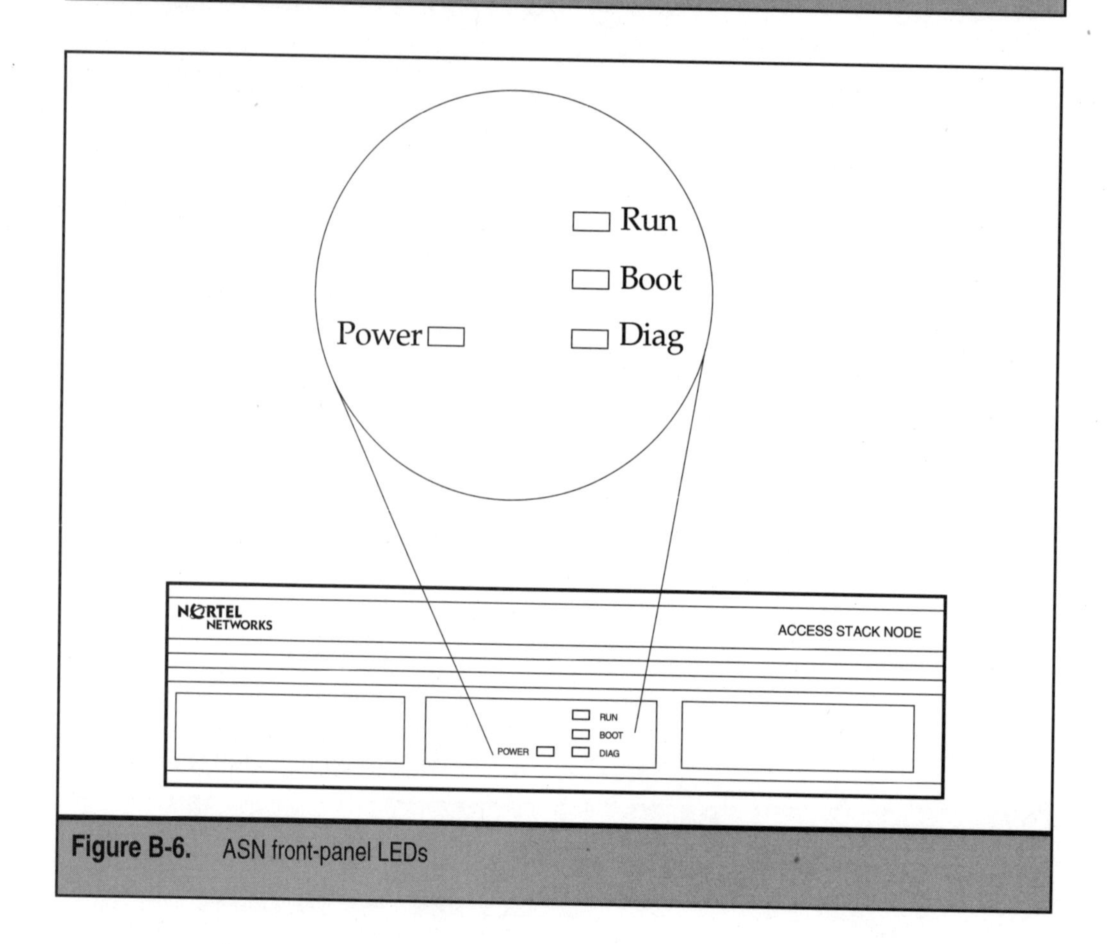

Figure B-6. ASN front-panel LEDs

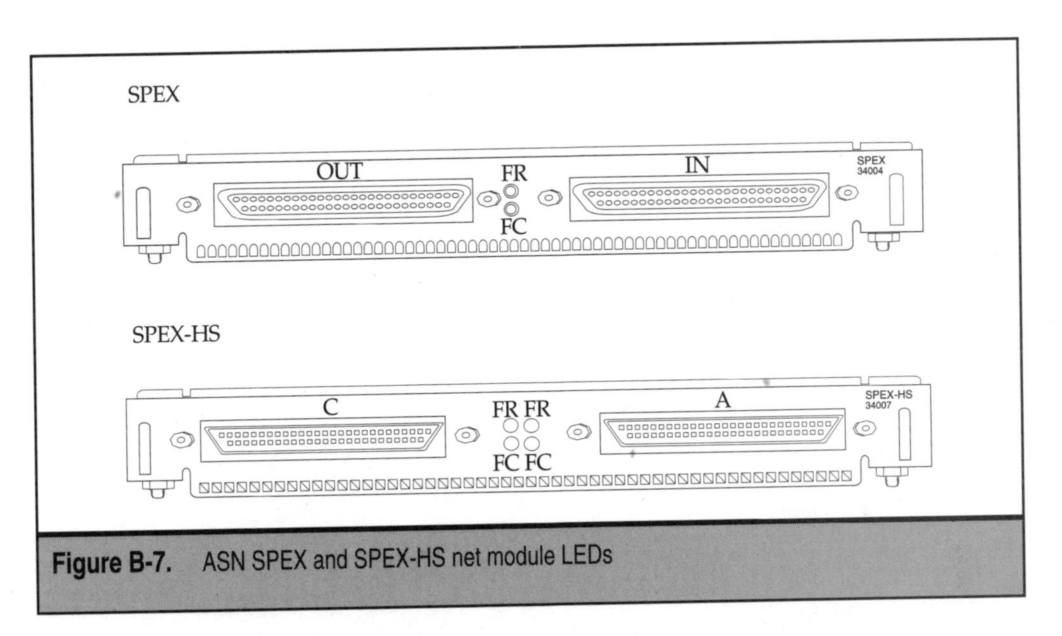

Figure B-7. ASN SPEX and SPEX-HS net module LEDs

The SPEX and SPEX-HS modules were discussed in the platform overview in Chapter 1. They are used to connect multiple ASN units together in a stack. Figure B-7 shows these two modules, and Table B-7 summarizes their LED indicators.

LED	Condition	Meaning
SPEX		
FR	On	The LED lights whenever the net module transmits a frame on the interconnect to another unit.
FC	On	The net module uses flow control (FC) to slow the data transmission whenever the receiving unit indicates congestion.
SPEX-HS		
FR	On	The LED lights whenever the net module transmits a frame on the interconnect to another unit. Two FR LEDs are present: one each for the A and C connectors.
FC	On	The net module uses flow control (FC) to slow the data transmission whenever the receiving unit indicates congestion. Two FC LEDs are present: one each for the A and C connectors.

Table B-7. ASN SPEX and SPEX-HS Net Module LEDs

The ASN is used for a wide range of applications, from mid-size Access routers to large-size regional Backbone routers. A fully stacked ASN router in many respects can be considered equivalent to a BLN router. Because it can support a wide range of applications and high-density of ports, the ASN supports various media interfaces. Tables B-8 and B-9 summarize the status indicators for most net modules used with the ASN platform. The

LED	Condition	Meaning
Dual Sync Net Module		
DCD1	On	Interface COM1 has a Data Carrier Detect (DCD) signal.
DCD2	On	Interface COM2 has a DCD signal.
HSSI Net Module		
LC	On	The local DTE interface is in line loopback operation.
TM	On	The downstream DCE device is in test mode (TM).
CA	On	The DCE (CA; communications adapter) is operational and is communicating with the DTE interface.
TA	On	The local DTE (TA; terminal adapter) is operational and is communicating with the DCE.
FDDI Net Modules (MM and SM)		
SD	On	A Signal Detect (SD) signal was received. There is an incoming signal on the interface.
TX	On	The corresponding interface (A or B) is transmitting a frame.

Table B-8. ASN Net Module LEDs (Part 1)

LED	Condition	Meaning
P or S	On	The LED lights only when a dual bypass switch is present. During normal operation, both the Primary (P) and Secondary (S) LEDs remain lighted. When the A and B connectors are bypassed by the bypass switch, then the LEDs turn off.
Dual Sync/ISDN BRI Net Module		
DCD1	On	Interface COM1 has a Data Carrier Detect (DCD) signal.
DCD2	On	Interface COM2 has a DCD signal.
BRI	On	The interface is currently active and operational.

Table B-8. ASN Net Module LEDs (Part 1) *(continued)*

LED	Condition	Meaning
Quad BRI Net Module		
1, 2, 3, 4	On	The corresponding BRI interface is currently active and operational.
10Base-T Dual-Ethernet Net Module		
XMT and RCV	On	The corresponding port is transmitting (XMT) and receiving (RCV) data during normal operation.

Table B-9. ASN Net Module LEDs (Part 2)

LED	Condition	Meaning
100Base-T Ethernet Net Module		
LNK	On	The port has an established link with another device (usually a hub or switch), either in 10-Mbps or 100-Mbps mode.
COL	On	The LED lights briefly whenever an Ethernet collision is detected on the segment. This occurs only in half-duplex mode, and this LED will normally remain off in full-duplex operation. When full-duplex operation with congestion control is enabled, then this LED indicates that the interface is congested and is asserting flow control.
TX	On	The interface is transmitting a frame.
RX	On	The interface is receiving a frame.
SPD	On	The interface is in 100-Mbps operation.
	Off	The interface is in 10-Mbps operation.
DPX	On	The interface is in full-duplex mode at the corresponding operational speed.
Dual Token Ring Net Module		
RCVR	On	The Token Ring is receiving data on the ring.
NSRT	On	The Token Ring port has successfully inserted into the ring.
WFAIL	On	A wire fault condition, such as the insertion of an improper cable, occurred.

Table B-9. ASN Net Module LEDs (Part 2) *(continued)*

LED	Condition	Meaning
Dual MCT1 Net Module		
RED	On	There is a red alarm condition on the corresponding interface, where either an LOS or OOF has occurred (see Chapter 4).
YEL	On	There is a yellow alarm condition, otherwise known as the RAI (see also Chapter 4).
LOOP	On	The interface has been put into a loop using one of the known mechanisms.
SYNC	On	The interface is receiving a valid synchronized signal and framing from the network.
MCE1 Net Module		
LCL ALM	On	The interface is in a local alarm state. See Chapter 4 for a discussion of media states and alarms.
RMT ALM	On	A remote alarm indication was received by the local interface.

Table B-9. ASN Net Module LEDs (Part 2) *(continued)*

corresponding net modules are illustrated in Figures B-8 and B-9. Similar to the Backbone link modules, the ASN net modules have a FAIL (F) LED. Diagnostics failure on a connector or the entire net module will turn on the F LED. A warning message is generated in the log, identifying the diagnostic failure.

Some net modules such as the quad-sync do not have individual status indicators due to the small amount of space available for the high-density interface connectors. The fail (F) LED is common across all units and flashes during diagnostic tests at startup. During

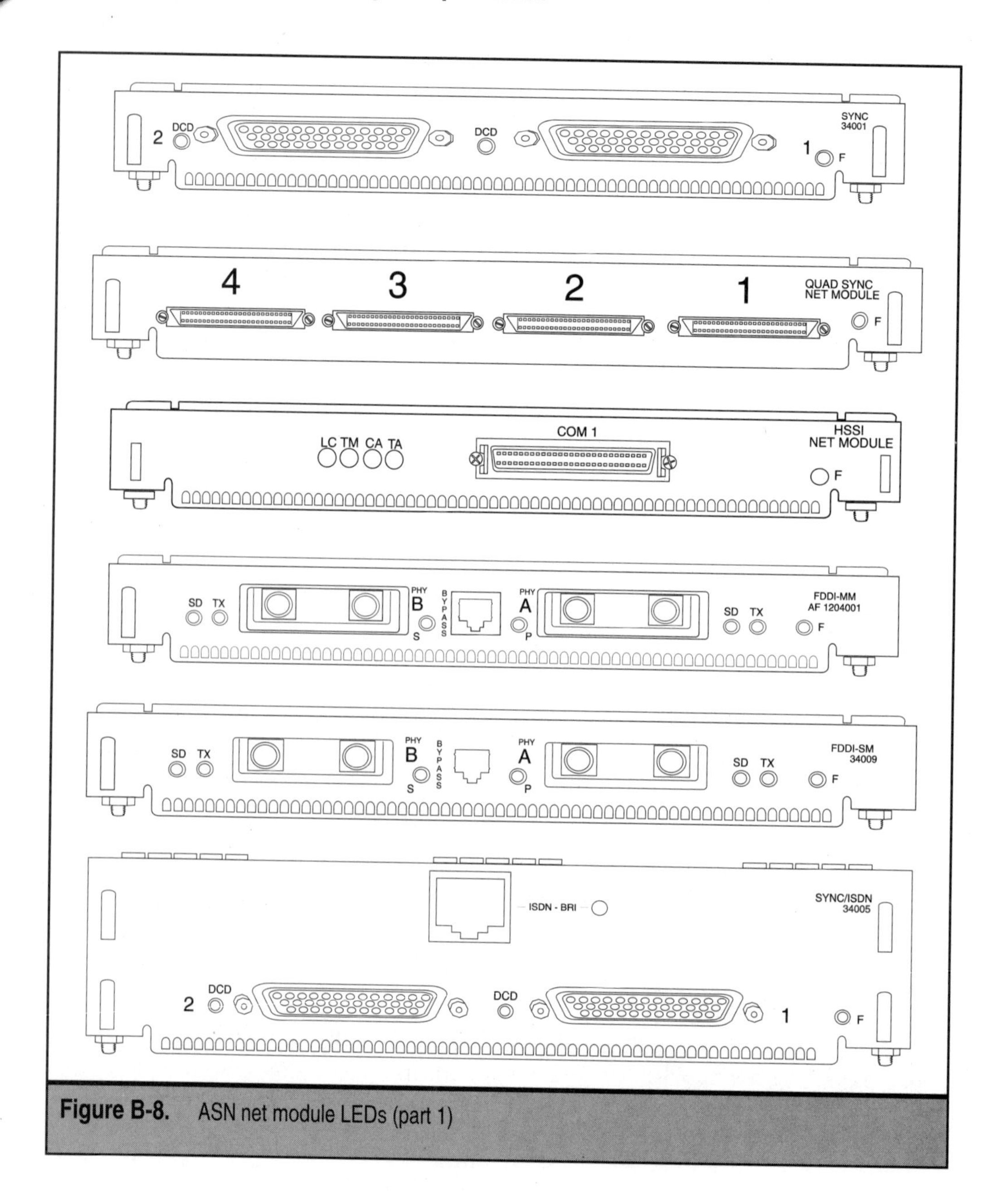

Figure B-8. ASN net module LEDs (part 1)

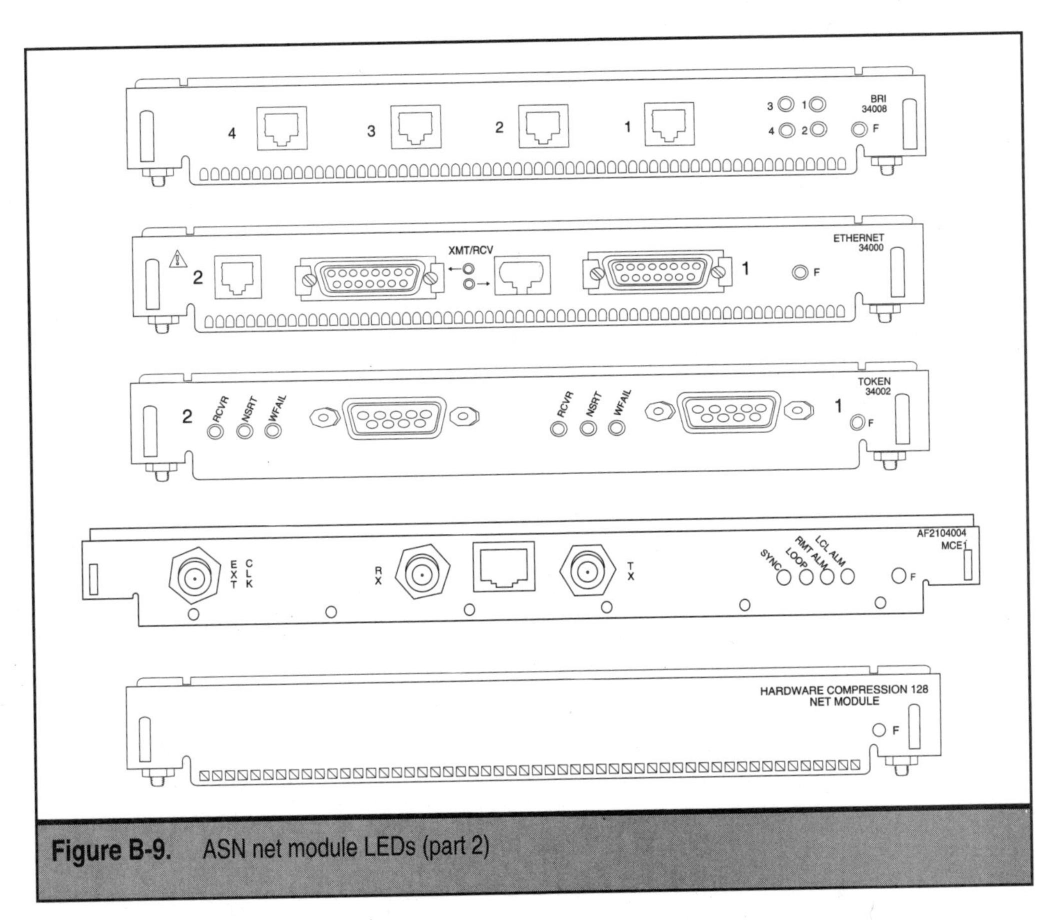

Figure B-9. ASN net module LEDs (part 2)

steady-state operation, the F LED will remain on if either the net module or one of the connectors fails diagnostics testing. The processor module will automatically also turn on the Diag LED in the front panel when one of the F LEDs is lighted.

SYSTEM 5000 ROUTER

A System 5000 router is generally considered equivalent to a single ASN router inserted into the System 5000 chassis. Each router supports its own net modules, except the ATM routers, which are considered equivalent to ARE processors. Each router module has a unique annunciator panel at the top, and many LEDs, as shown in Figure B-10. Table B-10 summarizes the LED indictors for these modules.

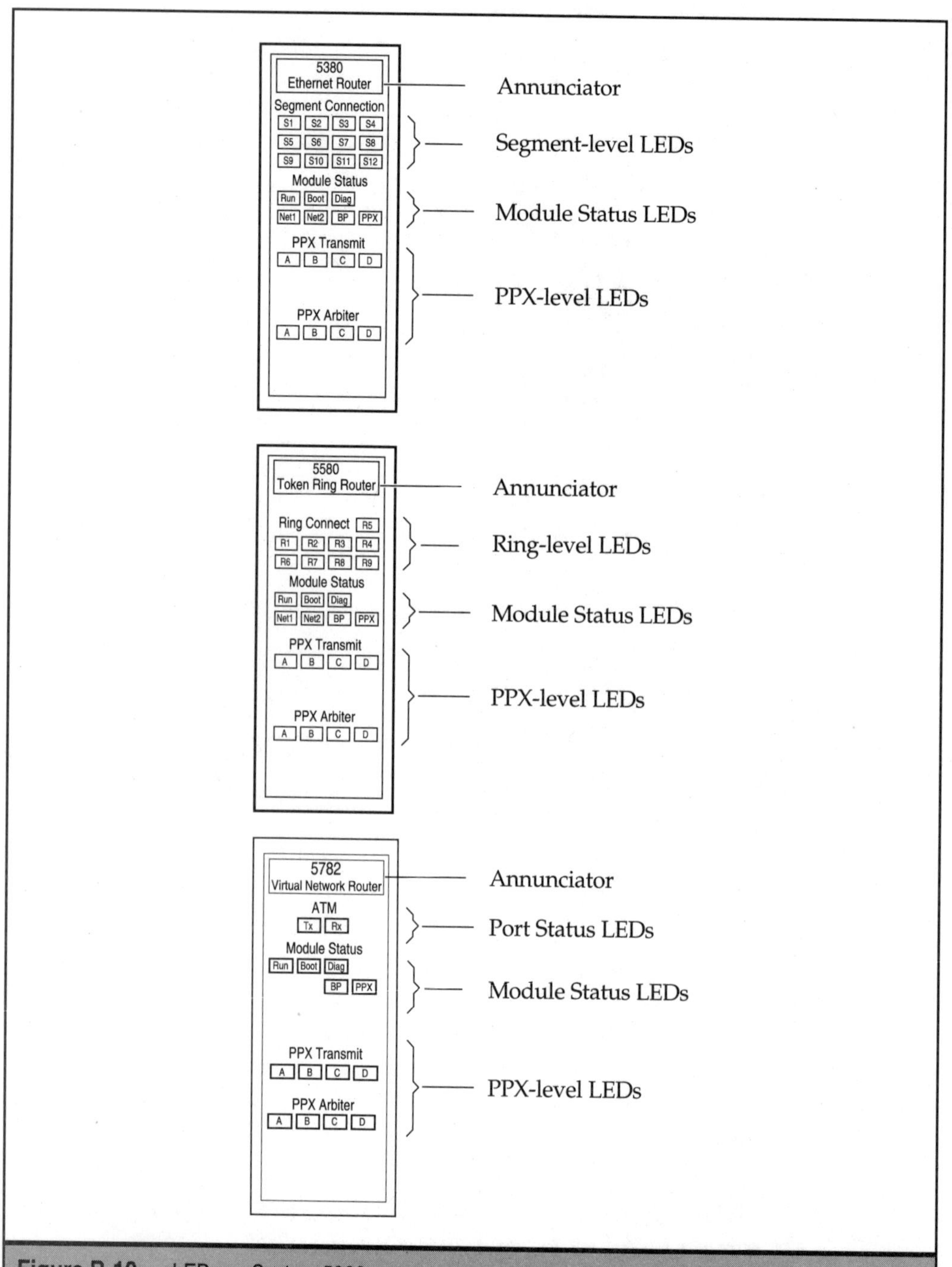

Figure B-10. LEDs on System 5000 routers

LED	Condition	Meaning
Annunciator	On (green)	The annunciator is the bicolor indicator at the top. When green, this LED indicates that the module is operational. This light remains off when the module is not receiving power.
	On (amber)	The module is booting, resetting, or running diagnostics. If it does not turn green after a few minutes, a diagnostics or boot failure occurred.
Module Status		
Run	On (green)	The router is active and fully operational.
	On (flashing)	The router is running diagnostic tests.
	Off	The router is booting or has failed bootstrap.
Boot	On	The router is booting to the runtime image.
Diag	On (amber)	One or more of the diagnostic tests failed. The annunciator may also glow amber when a test fails.
	On (flashing amber)	A fatal error occurred during diagnostics testing, and the router module cannot proceed. Critical errors occurred, such as failures in PPX interface tests.
Net 1	On (green)	The net module in position 1 has been detected and is fully operational.
	On (amber)	The net module in position 1 failed diagnostic testing.
	Off	No net module is installed in this position.
Net 2	On (green)	The net module in position 2 has been detected and is fully operational.

Table B-10. LEDs on System 5000 Routers

LED	Condition	Meaning
	On (amber)	The net module in position 2 failed diagnostic testing.
	Off	No net module is installed in this position.
BP	On (green)	The backplane interfaces on the router to the appropriate (System 5000) backplane are operational.
	On (amber)	One or more of the backplane interfaces on the router module failed.
PPX	On (green)	The router has detected the PPX backplane in the system, and the interface is operational.
	On (amber)	One or more of the PPX backplane interconnects has failed.
PPX Status		
A, B, C, D (Transmit)	On (green)	The router is currently transmitting a frame on the indicated PPX rail.
	Off	The specified rail is not currently used.
A, B, C, D (Arbiter)	On (green)	The local router module is performing arbitration functions for the specified PPX rail.
	Off	The local router module is not the arbiter of the specific PPX rail, and another router module may be arbitrating the rail.
Ethernet (5380) Router		
S1–S12	On (green)	The 5380 router is attached to the specific LAN segment on the backplane.
	Off	The router does not have an attachment to the LAN segment on the backplane.

Table B-10. LEDs on System 5000 Routers *(continued)*

LED	Condition	Meaning
Token Ring (5580) Router		
R1–R9	On (green)	The 5580 router is connected to the specific ring on the backplane.
	On (amber)	The router is wrapped on the specific ring on the backplane.
	Off	The router is not connected to the specific ring on the backplane.
ATM (578x) Router		
Tx	On	The module is transmitting ATM cells on the backplane connection.
Rx	On	The module is receiving ATM cells on the backplane connection.

Table B-10. LEDs on System 5000 Routers *(continued)*

The following table shows the PPX backplane rail LEDs and their corresponding chassis slot assignments. System 5000 routers can be installed as stand-alone routers without the use of the PPX interconnect to other routers in the same chassis.

Chassis Slot Number	Corresponding Rail LED
6	D
7	A
8	B
9	C
10	D
11	A
12	B
13	C
14	D

Net modules on the Ethernet and Token Ring 5000 routers are identical in design and operation to the net modules on ASNs, although their form factor is different. For net module–specific status indicators, consult the appropriate tables (B-8 and B-9) in the "Access Stack Node" section.

BACKBONE NODE

The Backbone Node family consists of the BLN1, BLN2, and BCN routers. All Backbone Nodes share an identical front panel, which provides comprehensive status information on the entire router. More detailed and granular status indicators are discussed later in this section. Figure B-11 shows the front-panel indicators; Table B-11 summarizes the LED indicators.

In some cases, the front-panel indicators alone may be insufficient to determine the source of a problem. When the Diag LED is on for an extended period of time after bootup or lights during steady-state operation, first check the F LED on all the link modules (or the corresponding runtime indicator for the appropriate processor) after removing the front bezel and the EMC shield. This will provide a clear view of all the processor modules. Table B-12 details all the status indicators for the various processor modules, which are illustrated in Figure B-12. In particular, on all processor modules, you should pay attention to an amber or red LED indicating diagnostics failure. The position of the Diag LED depends on the processor module. Processor modules that do not have any active LEDs or that have LEDs that appear frozen may have failed.

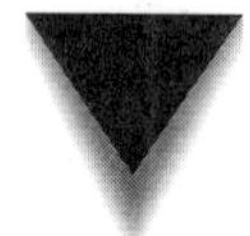

CAUTION: Consult the appropriate system service manual before attempting to remove or service any processor and link modules. Always use an anti-static wrist strap to prevent damage to system components, even if you are removing a module that you suspect has failed. Almost all processor and link modules support hot-swap operation, and care should be employed to avoid affecting other processor or link modules in the system. Consult the hardware compatibility matrix referenced in the book's web site (http://www.mahalingam.com) for details on hot swap support.

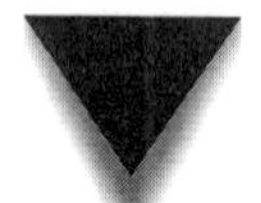

NOTE: In the following tables, the reference to the F LED does not refer to the FAIL LED; instead, it refers to the F LED on the processor itself, which is described in the same table.

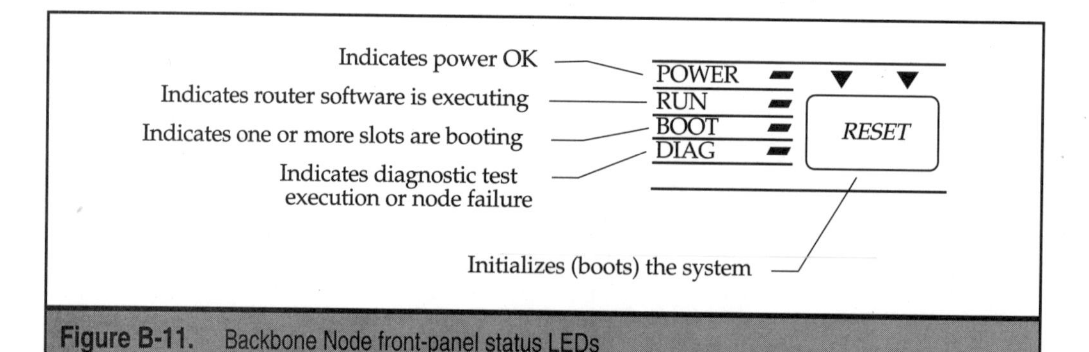

Figure B-11. Backbone Node front-panel status LEDs

LED	Condition	Meaning
Power	On	The system is receiving power, and the source is within normal operating limits for both AC- and DC-powered systems.
Run	On	One or more processor modules have initialized to the runtime image and are in steady-state operation. Both the power and run LEDs remain on during normal operation.
Boot	On	One or more processor modules are bootstrapping to the runtime image. This is a transient state, and all processors must eventually transition to the run state.
Diag	On	Either diagnostic tests are executing on one or more processor modules, or one or more system components failed during diagnostics testing or normal operation. Fan failures are also considered diagnostic failures.
Power (PS)	On	Each 620W power supply module has its own status indicator in the rear. Note that this LED is not part of the front panel. A green light indicates that the module is receiving power from a source and is fully operational.

Table B-11. Backbone Node Front-Panel Status LEDs

LED	Condition	Meaning
System Resource Module		
A	On	The SRM-F is receiving 5 volts of power from the chassis. If the power switch is set to the off position, the SRM-F will not be used.

Table B-12. Backbone Node Processor Module LEDs

LED	Condition	Meaning
B	On	The SRM-F is held in reset due to a hardware failure.
FRE-1 Daughterboard during Diagnostics		
1–7	On	The LEDs indicate the number of the diagnostic test that is currently executing. Each LED indicates the corresponding bit position—when on, the LED represents a 1, and when off, the LED represents a 0. The currently executing diagnostic test can also be seen using the harpoon diagnostics console monitor.
8	On	Diagnostic tests are currently executing.
9	On	The CPU has received a reset instruction.
FRE-1 Daughterboard during Runtime (F On)		
1–6	On	This set of LEDs represents a sequential counter that increments every second. The least significant digits of the seconds counter are represented in binary form; these LEDs are sometimes called the *racing* LEDs.
7	On	The processor module in this slot is currently running the TI soloist process. Resetting the slot will reset the TI and will cause another slot to assume the TI soloist function.
8	On	The processor module is booting or resetting, or PROM programming protection is disabled. PROM programming protection is disabled when the **prom –w** command is issued during PROM upgrades (see Chapter 3).

Table B-12. Backbone Node Processor Module LEDs *(continued)*

LED	Condition	Meaning
9	On	The CPU received a reset instruction.
FRE-1 Motherboard		
A	On	The PPX backbone is held in reset due to either a hardware or software failure.
B	On	The PPX direct memory access (DMA) hardware logic is accessing the DRAM.
C, D	On	The current diagnostic test set is being executed (see Table B-13).
E	On	Diagnostic tests are currently executing.
F	On	The GAME operating system is active and operational.
G	On	The CPU is accessing the DRAM.
H	On	The hardware is resetting.
I	On	The processor module is transmitting on PPX rail A (backbone 0).
J	On	The processor module is transmitting on PPX rail B (backbone 1).
K	On	The processor module is transmitting on PPX rail C (backbone 2).
L	On	The processor module is transmitting on PPX rail D (backbone 3).
M	On	The processor module is performing flow control on PPX rail A (backbone 0).
N	On	The processor module is performing flow control on PPX rail B (backbone 1).
O	On	The processor module is performing flow control on PPX rail C (backbone 2).
P	On	The processor module is performing flow control on PPX rail D (backbone 3).

Table B-12. Backbone Node Processor Module LEDs *(continued)*

LED	Condition	Meaning
FRE-2 Motherboard during Runtime		
Q	On	The processor module is booting or resetting, or the PROM programming protection is disabled. PROM programming protection is disabled when the **prom –w** command is issued during PROM upgrades (see Chapter 3).
R	On	The processor module in this slot is currently running the TI soloist process. Resetting the slot will reset the TI and will cause another slot to assume the TI soloist function.
S–X	On	This set of LEDs represents a sequential counter that increments every second. The least significant digits of the seconds counter are represented in binary form; these LEDs are sometimes called the *racing* LEDs.
FRE-2 Motherboard during Diagnostics		
Q–X	On	The LEDs indicate the number of the diagnostic test that is currently executing. Each LED indicates the corresponding bit position—when the LED is on, it represents a 1, and when the LED is off, it represents a 0. The currently executing diagnostic test can also be seen using the harpoon diagnostics console monitor.
FRE-2 Motherboard		
1	On	The local processor module is running the Technician Interface.

Table B-12. Backbone Node Processor Module LEDs *(continued)*

LED	Condition	Meaning
2, 3, 4	On	The link module interface is requesting access to the DRAM.
5	On	The link module interface is currently accessing the DRAM.
6, 7	On	These LEDs indicate internal state information for the link module interface.
A	On	The PPX backbone is held in reset due to either a hardware or software failure.
B	On	The PPX direct memory access (DMA) hardware logic is accessing the DRAM.
C, D	On	These LEDs indicate the current diagnostic test set being executed (see Table B-13).
E	On	Diagnostic tests are currently executing.
F	On	The GAME operating system is active and operational.
G	On	The CPU is accessing the DRAM.
H	On	The hardware is resetting.
I	On	The processor module is transmitting on PPX rail A (backbone 0).
J	On	The processor module is transmitting on PPX rail B (backbone 1).
K	On	The processor module is transmitting on PPX rail C (backbone 2).
L	On	The processor module is transmitting on PPX rail D (backbone 3).
M	On	The processor module is performing flow control on PPX rail A (backbone 0).
N	On	The processor module is performing flow control on PPX rail B (backbone 1).
O	On	The processor module is performing flow control on PPX rail C (backbone 2).

Table B-12. Backbone Node Processor Module LEDs *(continued)*

LED	Condition	Meaning
P	On	The processor module is performing flow control on PPX rail D (backbone 3).
FRE-2-060E Motherboard during Runtime		
Q	On	The processor module is booting or resetting, or PROM programming protection is disabled. PROM programming protection is disabled when the **prom –w** command is issued during PROM upgrades (see Chapter 3).
R	On	The processor module in this slot is currently running the TI soloist process. Resetting the slot will reset the TI and will cause another slot to assume the TI soloist function.
S–X	On	This set of LEDs represent a sequential counter that increments every second. The least significant digits of the seconds counter are represented in binary form; these LEDs are sometimes called the *racing* LEDs.
FRE-2-060E Motherboard during Diagnostics		
Q–X	On	The LEDs indicate the number of the diagnostic test that is currently executing. Each LED indicates the corresponding bit position—when the LED is on, it represents a 1, and when the LED is off, it represents a 0. The currently executing diagnostic test can also be seen using the harpoon diagnostics console monitor.
FRE-2-060E Motherboard		
1	On	The local processor module is running the Technician Interface.

Table B-12. Backbone Node Processor Module LEDs *(continued)*

LED	Condition	Meaning
2, 3, 4	On	The link module interface is requesting access to the DRAM.
5	On	The link module interface is currently accessing the DRAM.
6, 7	On	These LEDs indicate internal state information for the link module interface.
A	On	The PPX backbone is held in reset due to either a hardware or software failure.
B	On	The PPX direct memory access (DMA) hardware logic is accessing the DRAM.
C, D	On	These LEDs indicate the current diagnostic test set being executed (see Table B-13).
E	On	Diagnostic tests are currently executing.
F	On	The GAME operating system is active and operational.
G	On	The CPU is accessing the DRAM.
H	On	The hardware is resetting.
I	On	The processor module is transmitting on PPX rail A (backbone 0).
J	On	The processor module is transmitting on PPX rail B (backbone 1).
K	On	The processor module is transmitting on PPX rail C (backbone 2).
L	On	The processor module is transmitting on PPX rail D (backbone 3).
M	On	The processor module is performing flow control on PPX rail A (backbone 0).
N	On	The processor module is performing flow control on PPX rail B (backbone 1).
O	On	The processor module is performing flow control on PPX rail C (backbone 2).
P	On	The processor module is performing flow control on PPX rail D (backbone 3).

Table B-12. Backbone Node Processor Module LEDs *(continued)*

LED	Condition	Meaning
ARE Motherboard, Group 1		
1	On	The processor module is transmitting on PPX rail A (backbone 0).
2	On	The processor module is transmitting on PPX rail B (backbone 1).
3	On	The processor module is transmitting on PPX rail C (backbone 2).
4	On	The processor module is transmitting on PPX rail D (backbone 3).
5	On	The processor module is performing flow control on PPX rail A (backbone 0).
6	On	The processor module is performing flow control on PPX rail B (backbone 1).
7	On	The processor module is performing flow control on PPX rail C (backbone 2).
8	On	The processor module is performing flow control on PPX rail D (backbone 3).
ARE Motherboard, Group 2		
1	On	The backbone is requesting access to the virtual buffer memory (VBM).
2	On (amber)	The PPX backbone is resetting.
3	On (amber)	The transmit ATMizer is resetting. The ATMizer is responsible for performing segmentation and reassembly (SAR) or, on the transmit side, segmentation.
4	On (amber)	The receive ATMizer is resetting. The receive ATMizer performs the reassembly functions.

Table B-12. Backbone Node Processor Module LEDs *(continued)*

LED	Condition	Meaning
ARE Motherboard, Group 3 (Diag, Boot, and Runtime)		
1	On (amber)	The link module is resetting.
2	On	The operating system has been successfully loaded and is fully operational.
3	On (red)	Diagnostic tests are currently executing.
4	On (amber)	The ARE processor is currently booting to the runtime image.
5	On (amber)	The ARE processor is resetting.
6–13	On	When diagnostic tests are running (LED 3 is on), LEDs 7 through 13 indicate the current diagnostic test in hexadecimal notation. Each LED, when on, indicates a 1 in the corresponding bit position. When the processor is in steady-state operation (LED 2 is on), the LEDs represent a sequential counter that increments every second. The least significant digits of the seconds counter are represented in binary form; these LEDs are sometimes called the *racing* LEDs.
14, 15	On	When LED 3 is on (diagnostic tests are executing), these LEDs indicate the diagnostic test set that is currently executing (see Table B-14). If the processor module has stopped diagnostics, the frozen state of these LEDs indicate the last set of tests that failed.
16	On (amber)	The Technician Interface soloist process is running on this slot.

Table B-12. Backbone Node Processor Module LEDs *(continued)*

LED	Condition	Meaning
17–25	On	The transmit and receive ATMizers are active and operational. These LEDs indicate bus activity.
26	On (amber)	Memory coherency operations are being performed between the two PowerPC microprocessors. Both PowerPC processors share memory between their data caches when necessary, and this data must be kept synchronized to prevent stale data.
27–31	On	The processor bus is active.
32	On	The ARE processor module is receiving 3V power.
FRE4 Motherboard, Group 1		
1–4	Unused	Unused.
5	On	Address bus is busy.
6	On	Address retry is underway.
7	On	Data bus is busy.
8	On	The TI soloist process is executing on this slot.
9–10	Unused	Unused.
11–18	On	When diagnostics are running (LED 11 is on), LEDs 12 through 18 indicate the current test in hexadecimal notation. Each LED, when on, represents a 1 value in the corresponding bit position. When the processor is in steady-state operation (LED 22 is on), the LEDs represent a sequential counter that increments every second. The least significant digits of the seconds counter are represented in binary form; these LEDs are sometimes called the *racing* LEDs.

Table B-12. Backbone Node Processor Module LEDs *(continued)*

LED	Condition	Meaning
19	On	The FRE4 processor module is resetting.
20	On (amber)	The FRE4 processor module is booting.
21	On (red)	The FRE4 processor module has failed.
22	On	The FRE4 processor module has loaded the runtime software and is operational.
FRE4 Motherboard, Group 2		
23	On	The link module is resetting.
24	On	The I2C memory controller is resetting.
25	On	The PPX is held in reset due to a software or hardware problem.
26	On	The direct memory access (DMA) controller is resetting.
27	On	The DMA logic is accessing the buffer memory.
28	On	The FRE4 processor module is performing flow control on PPX rail D (backbone 3).
29	On	The FRE4 processor module is performing flow control on PPX rail C (backbone 2).
30	On	The FRE4 processor module is performing flow control on PPX rail B (backbone 1).
31	On	The FRE4 processor module is performing flow control on PPX rail A (backbone 0).
32	On	The FRE4 processor module is transmitting on PPX rail D (backbone 3).
33	On	The FRE4 processor module is transmitting on PPX rail C (backbone 2).
34	On	The FRE4 processor module is transmitting on PPX rail B (backbone 1).
35	On	The FRE4 processor module is transmitting on PPX rail A (backbone 0).

Table B-12. Backbone Node Processor Module LEDs *(continued)*

LED	Condition	Meaning
FRE4 Motherboard, Group 3		
36	On	The FRE4 processor module is receiving 2.6V power.
37	On	The FRE4 processor module is receiving 3.3V power.

Table B-12. Backbone Node Processor Module LEDs *(continued)*

C LED	D LED	E LED	Diag Test Set
On	On	On	CPU
Off	On	On	PPX Backbone
On	Off	On	Link Module
Off	Off	On	HDCM

Table B-13. FRE Diag Test Indicators

LED 14	LED 15	LED 3	Diag Test Set
Off	Off	On	CPU
On	Off	On	PPX Backbone
Off	On	On	Link Module
On	On	On	ATM

Table B-14. ARE Diag Test Indicators

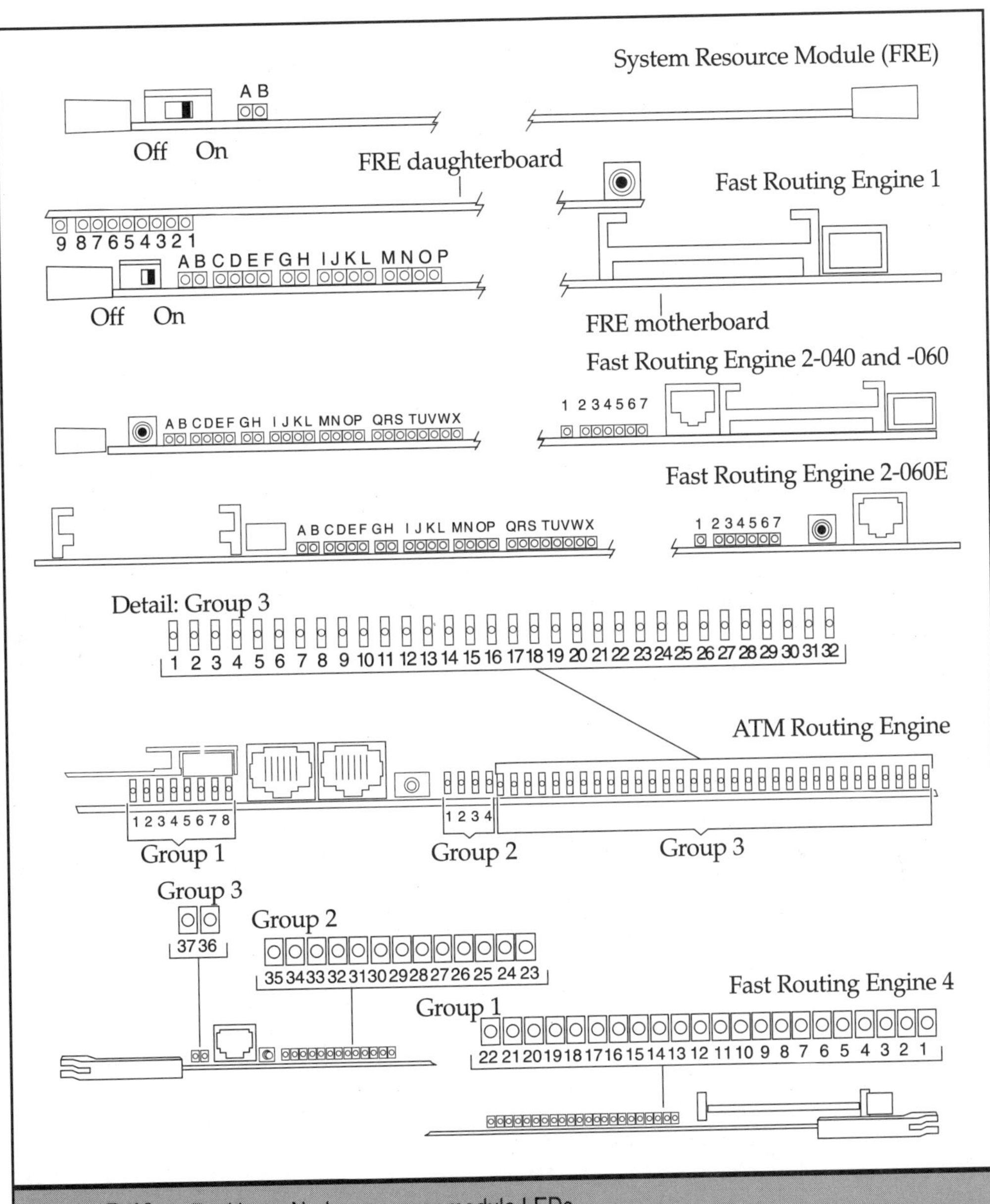

Figure B-12. Backbone Node processor module LEDs

The Backbone Node family supports a comprehensive array of media interfaces with various link modules. Figure B-13 shows the available Ethernet link modules for the

Figure B-13. Ethernet and Dual 100Base-T Ethernet link modules

FRE1/FRE2 processors, and Figure B-14 shows all the available Ethernet/sync modules. Status indicators for these modules are summarized in Table B-15.

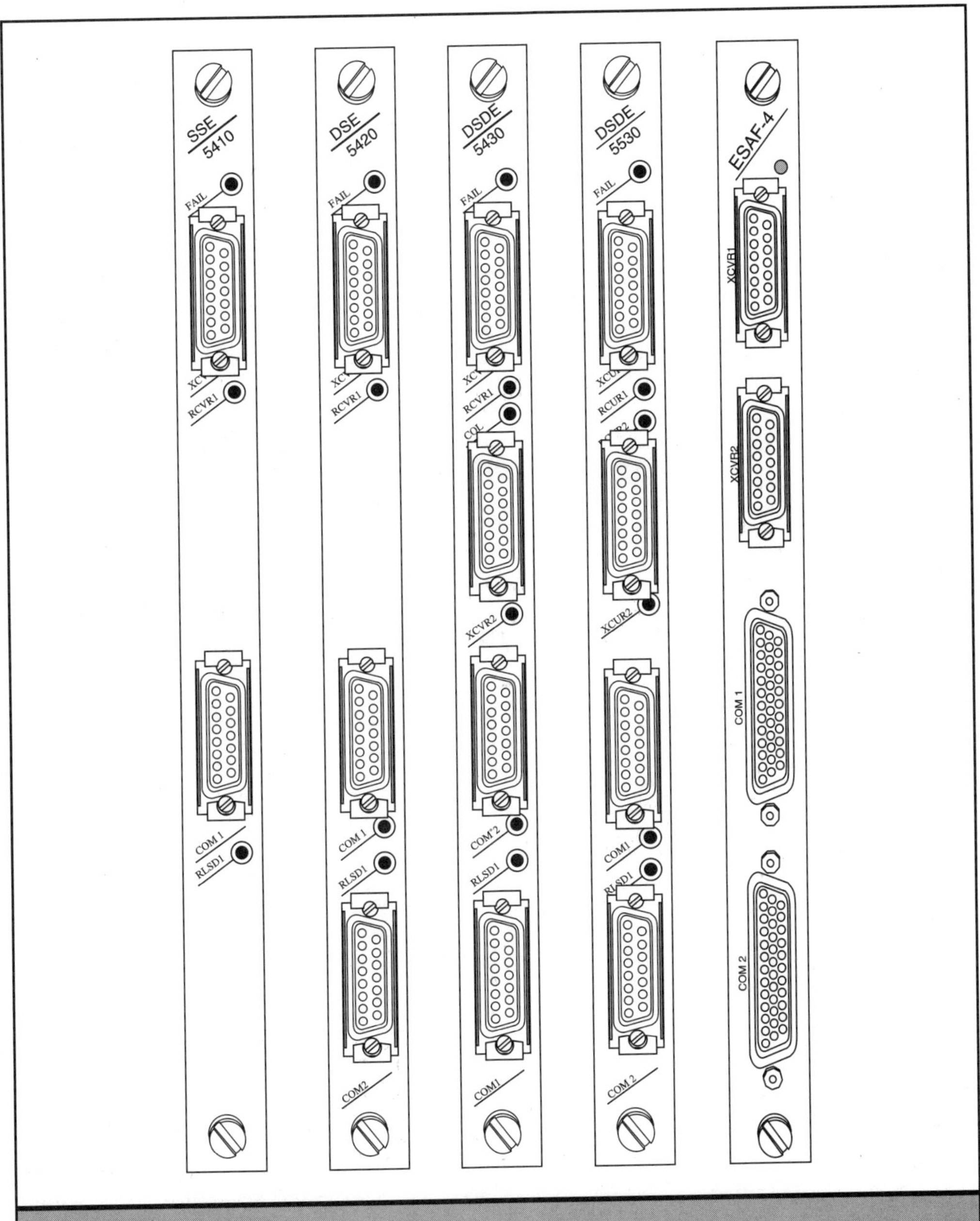

Figure B-14. Ethernet/sync link modules

LED	Condition	Meaning
Ethernet Link Modules		
COL	On	The interface detected a collision.
RCVR RCVR1 RCVR2	On	The corresponding interface is active and receiving data.
XMIT	On	The corresponding interface is transmitting a frame.
XCVR1 XCVR2	On	Ethernet traffic (either transmitting or receiving) is present on the corresponding port.
FAIL	On	Diagnostic testing is in progress, or power-up diagnostic tests failed, or the link module detected a hardware problem and the module is waiting for an automatic attempt to reinitialize the diagnostic tests.
Dual 100Base-T Link Module		
COL	On	In half-duplex mode, this status indicates that the corresponding port detected a collision on the wire. In full-duplex mode, this status indicates transmit congestion.
LNK	On	The port has established a valid link with another Ethernet device (hub or switch) on the corresponding port.
TX and RX	On	The corresponding port is either transmitting or receiving a frame.
SPD	On	The port is in 100-Mbps operation.
	Off	The port is in 10-Mbps operation.
DPX	On	The corresponding port is in full-duplex operation.
FAIL	On	This LED flashes during diagnostic testing. It glows steadily if one of the connectors or the link module itself fails a diagnostic test.
RLSD1 RLSD2	On	The corresponding serial port has the Carrier Detect (CD) signal.

Table B-15. Ethernet, Ethernet/Sync, and Dual 100-BaseT Link Module LEDs

Token Ring and Token/sync modules are illustrated in Figure B-15. The status indicators for these modules are summarized in Table B-16.

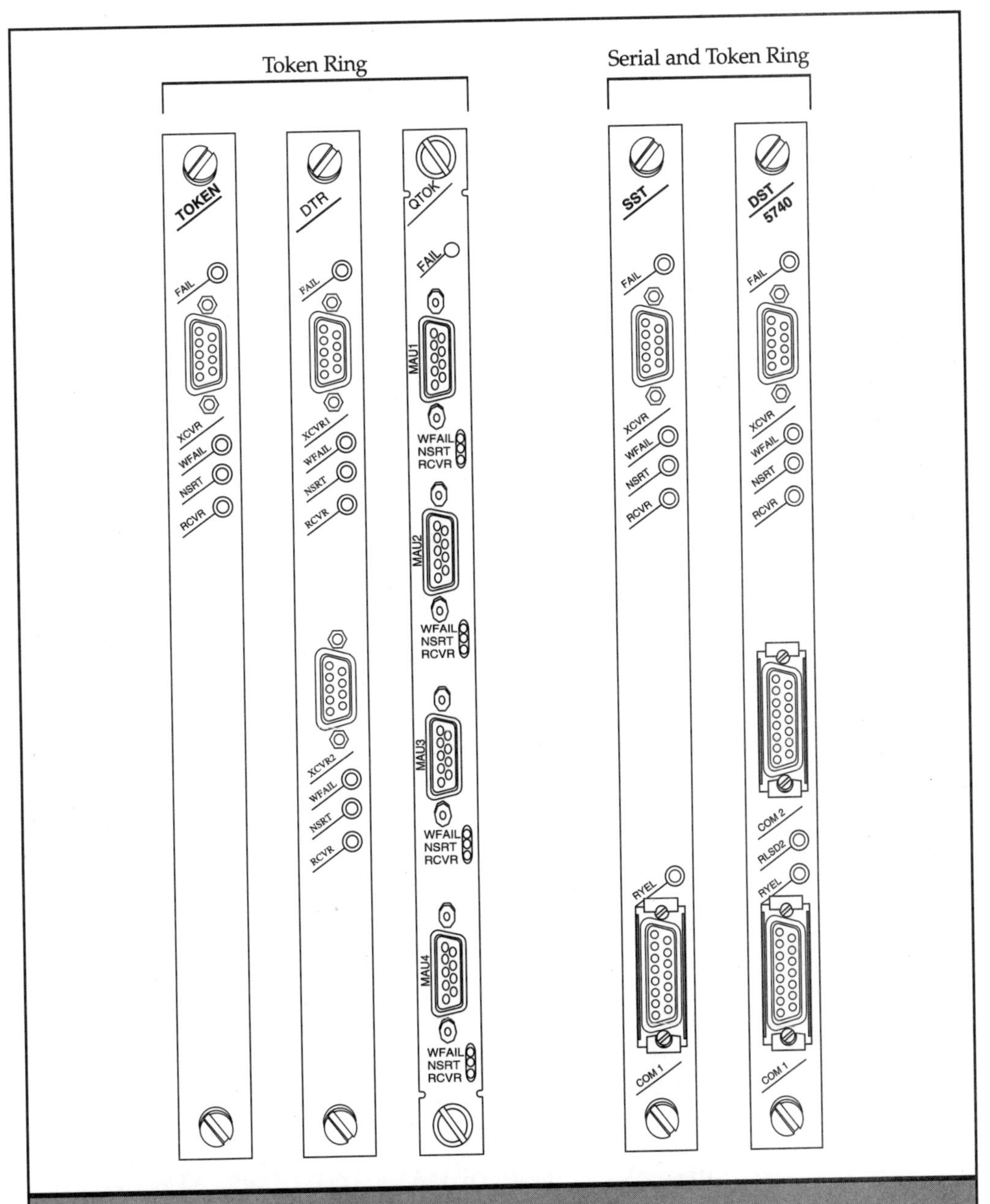

Figure B-15. Token Ring and Token/sync link modules

LED	Condition	Meaning
Token Ring Link Modules		
RCVR	On	Data is being received across the Token Ring interface.
NSRT	On	The router's interface is inserted into the active ring.
WFAIL	On	A wire fault occurred either in the transmit or receive direction.
FAIL	On	Diagnostic testing is in progress, or power-up diagnostic tests failed, or the link module detected a hardware problem and the module is waiting for an automatic attempt to reinitialize the diagnostic tests.
Token/Sync Link Modules		
RLSD1 and RLSD2	On	The corresponding port has the Carrier Detect (CD) signal.
RCVR	On	Data is being received across the Token Ring interface.
NSRT	On	The router's interface is inserted into the active ring.
WFAIL	On	A wire fault occurred either in the transmit or receive direction.
FAIL	On	Diagnostic testing is in progress, or power-up diagnostic tests failed, or the link module detected a hardware problem and the module is waiting for an automatic attempt to reinitialize the diagnostic tests.

Table B-16. Token Ring and Token/Sync Link Module Status LEDs

Figure B-16 shows three link modules: FDDI, ATM DS3/E3, and ATM OC3. Each has a unique set of status indicators, which are described in Table B-17. DS3 and E3 link modules have identical LEDs. Similarly, SONET/SDH MMF and SMF modules have identical LEDs.

Figure B-16. FDDI, ATM DS3, and ATM OC3 link modules

LED	Condition	Meaning
FDDI Link Module		
SD	On	An incoming signal was detected.
TX	On	The FDDI interface is transmitting a frame.
L1, L2, and L3	Variable	The LED indicates the line state (see Table B-18).
S and P	On	The LEE indicates whether the router is absent or present in the FDDI ring. The S (Secondary) and P (Primary) LEDs are lighted only when the optional dual-switch module is connected to the bypass connector. The S and P LEDs turn off when FDDI dual-switch module bypasses the physical connectors.
FAIL	On	The link module is running diagnostic tests or has failed diagnostic tests.
ATM DS3, E3 and OC3 Link Modules		
MODULE RESET	On	The module is resetting and is inactive. SONET/SDH interfaces continue to generate idle frames to maintain framing.
ATMC RESET	On	The ATM coprocessor is resetting. During initialization, the ARE downloads microcode to this controller on the link module, during which time the LED stays in this state.
SW LOAD	On	The driver application has been successfully loaded into the link module memory and is ready to forward traffic. If the LED is flashing, a software fault occurred.
TEST	On	The port is either disabled or has been administratively looped back.

Table B-17. FDDI and ATM Link Module LEDs

LED	Condition	Meaning
RDI	On	A Remote Defect Indication (RDI) signal was received. This LED lights when the framer receives an indication from the far-end device that a receive failure occurred and the framer is thus unable to maintain frame synchronization. This status generally applies only when the interface is operational (that is, SW LOAD is on).
LINK	On	The interface has signal and framing synchronization.
	On (flashing)	The interface detected a signal but is unable to synchronize with the frame.
	Off	Neither a signal nor framing is detected. The LINK LED is applicable only when the interface is operational (that is, SW LOAD is on).
FAIL	On	Diagnostic tests are executing, a diagnostic test failed, or diagnostic tests have not been run.

Table B-17. FDDI and ATM Link Module LEDs *(continued)*

L1	L2	L3	Line State (LS)
On	On	On	Idle 16 (ILS16)
On	On	Off	Halt (HLS)
On	Off	On	Master (MLS)
On	Off	Off	Quiet (QLS)
Off	On	On	Idle 4 (ILS4)
Off	On	Off	Reserved
Off	Off	On	Active (ALS)
Off	Off	Off	Noise

Table B-18. FDDI Interface Line States

The recently introduced FRE-4 processor supports several high-density link modules, illustrated in Figure B-17. These link modules operate only with FRE-4 processor mod-

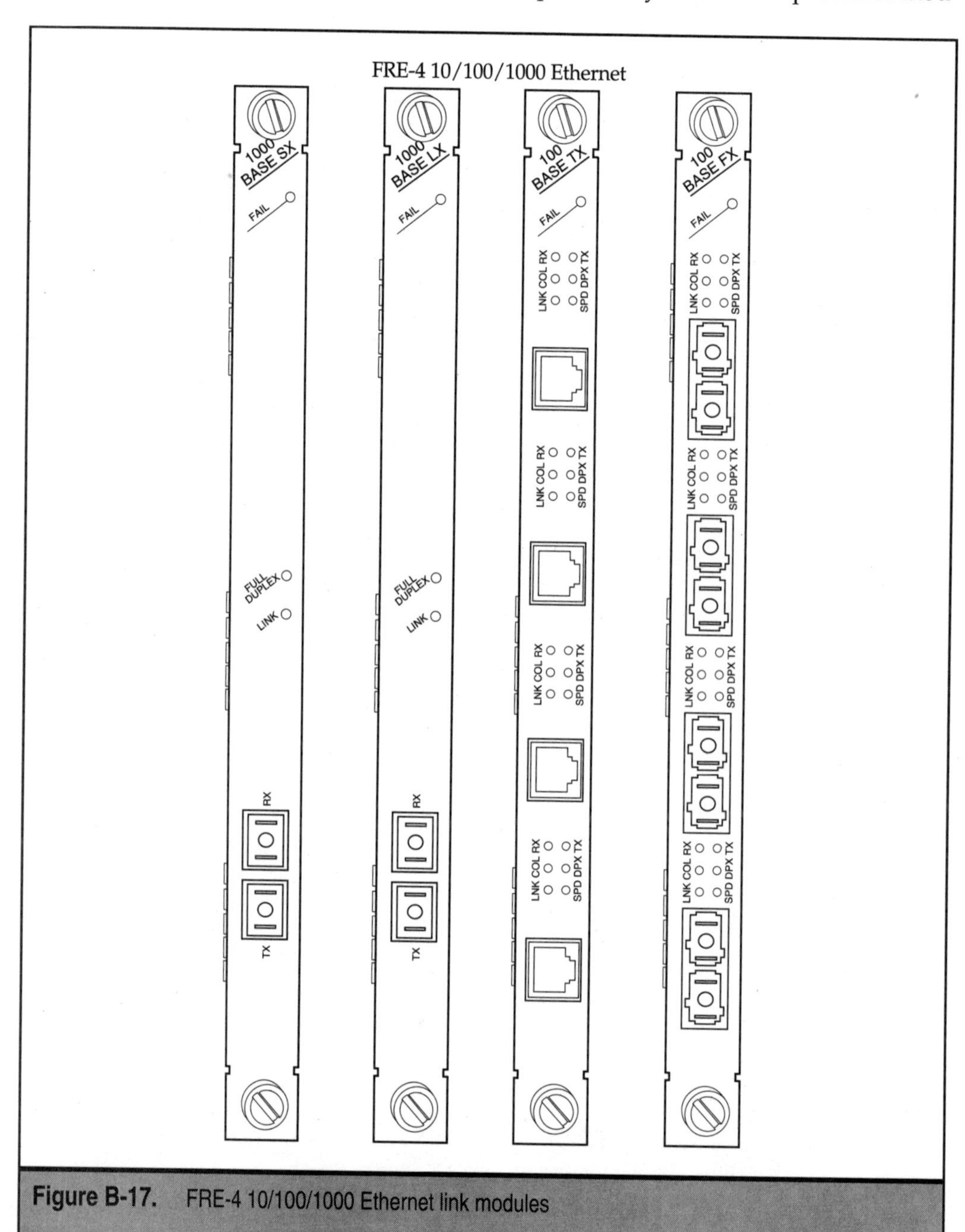

Figure B-17. FRE-4 10/100/1000 Ethernet link modules

ules—they cannot be installed along with FRE1, FRE2, or ARE processor modules. The status indicators on these modules are described in Table B-19.

LED	Condition	Meaning
1000 Base-LX and -SX Link Modules		
LINK	On	The interface has established a valid and operational link with another Gigabit Ethernet device.
FULL DUPLEX	On	The port is in full-duplex mode and is sending and receiving data.
FAIL	On	Diagnostic testing is in progress, a diagnostic test failed, or the module is waiting for an automatic attempt to reinitialize diagnostics.
10/100 Base-TX Link Modules		
LNK	On	The interface has a valid link to another Ethernet device such as a hub or repeater.
COL	On (amber)	In half-duplex mode, this LED briefly lights when the interface detects a collision on the wire. In full-duplex mode, this LED indicates transmit congestion.
TX and RX	On	The interface is transmitting and receiving data.
SPD	On	The interface is operating at 100 Mbps.
	Off	The interface is operating at 10 Mbps.
DPX	On	The interface is operating in full-duplex mode.
	Off	The interface is operating in half-duplex mode.

Table B-19. FRE-4 10/100/1000 Link Module LEDs

LED	Condition	Meaning
FAIL	On	Diagnostic testing is in progress, a diagnostic test failed, or the module is waiting for an automatic attempt to reinitialize diagnostics.
100 Base-FX Link Modules		
LNK	On	The interface has a valid link to another Ethernet device such as a hub or repeater.
COL	On (amber)	In half-duplex mode, this LED briefly lights when the interface detects a collision on the wire. In full-duplex mode, this LED indicates transmit congestion.
TX and RX	On	The interface is transmitting and receiving data.
SPD	On	The interface is operating at 100 Mbps.
	Off	The interface is operating at 10 Mbps.
DPX	On	The interface is operating in full-duplex mode.
	Off	The interface is operating in half-duplex mode.
FAIL	On	Diagnostic testing is in progress, a diagnostic test failed, or the module is waiting for an automatic attempt to reinitialize diagnostics.

Table B-19. FRE-4 10/100/1000 Link Module LEDs *(continued)*

The quad- and octal-sync link modules are used purely for wide-area connectivity. The single port high-speed serial interface (HSSI) link module is also used for WAN connectivity. These link modules, along with the SRM-L, are shown in Figure B-18. Table B-20 summarizes the LED indicators.

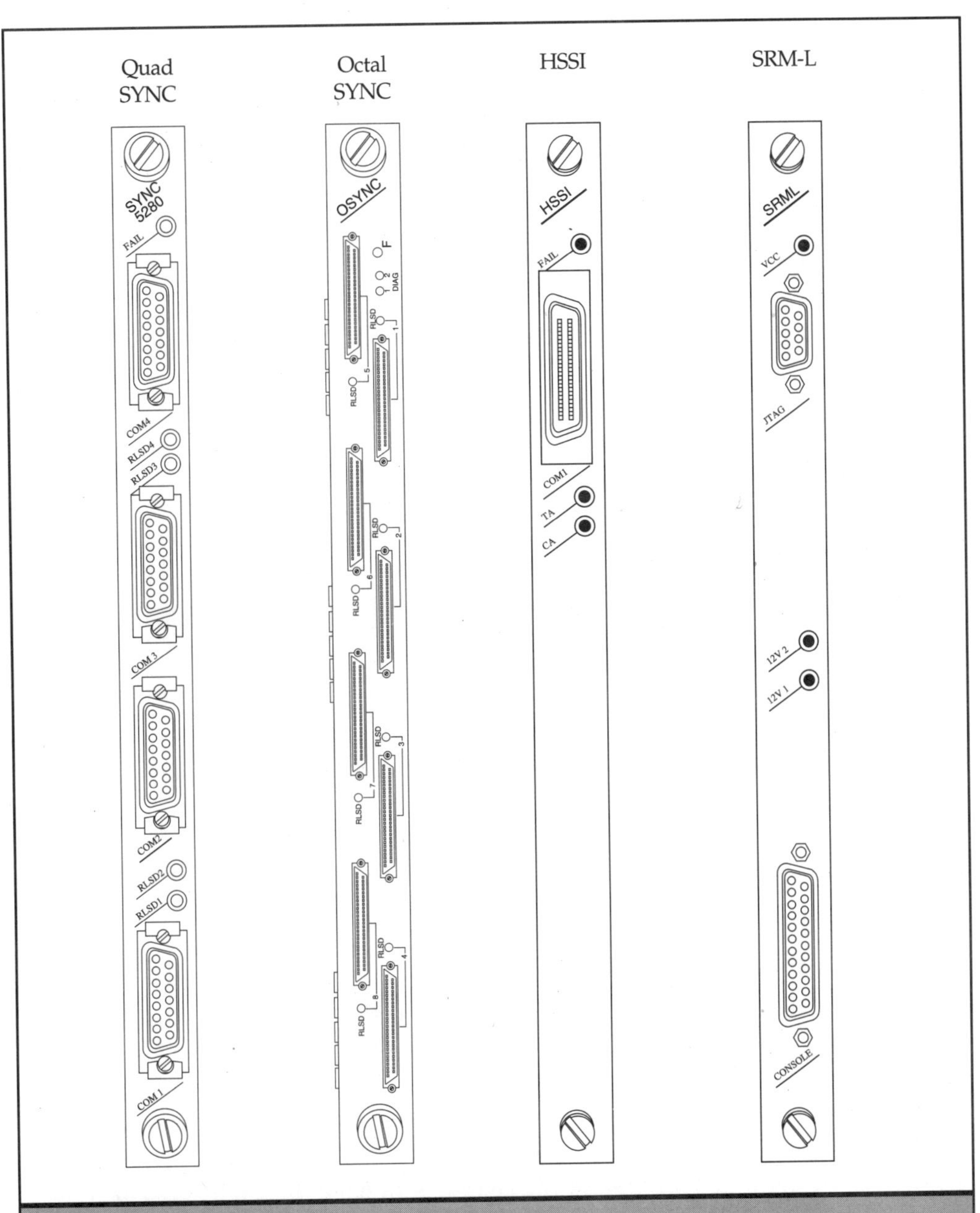

Figure B-18. Quad-sync, octal-sync, HSSI, and SRM-L link modules

LED	Condition	Meaning
QSYNC and OSYNC Link Modules		
RLSD1 through RLSD8	On	The corresponding interface has the Carrier Detect (CD) signal.
DIAG1 and DIAG2	Unused	Unused.
FAIL	On	Diagnostic testing is in progress, a diagnostic test failed, or the module is waiting for an automatic attempt to reinitialize diagnostics.
HSSI Link Modules		
CA	On	The DCE (CA; communications adapter) is active and operational and is ready to receive and transmit data.
TA	On	The DTE (TA; terminal adapter) interface on the router is ready to receive and transmit data.
FAIL	On	Diagnostic testing is in progress, a diagnostic test failed, or the module is waiting for an automatic attempt to reinitialize diagnostics.
SRM-L Link Modules		
VCC	On	Power supply voltages are within the normal operating range.
12V1	On	The first power supply bus is operating properly.
	Off	The F7 fuse is blown (if the system is a BLN or BLN-2), or the F9 fuse is blown (if the system is a BCN).
12V2	On	The second power supply bus is operating properly.
	Off	The F6 fuse is blown (if the system is a BLN or BLN-2), or the F12 fuse is blown (if the system is a BCN).

Table B-20. QSYNC, OSYNC, HSSI, and SRM-L Link Module LEDs

Figures B-19 and B-20 show another common set of modules used for wide-area connectivity. Figure B-19 shows the T1 and MCT1 modules, and Figure B-20 shows

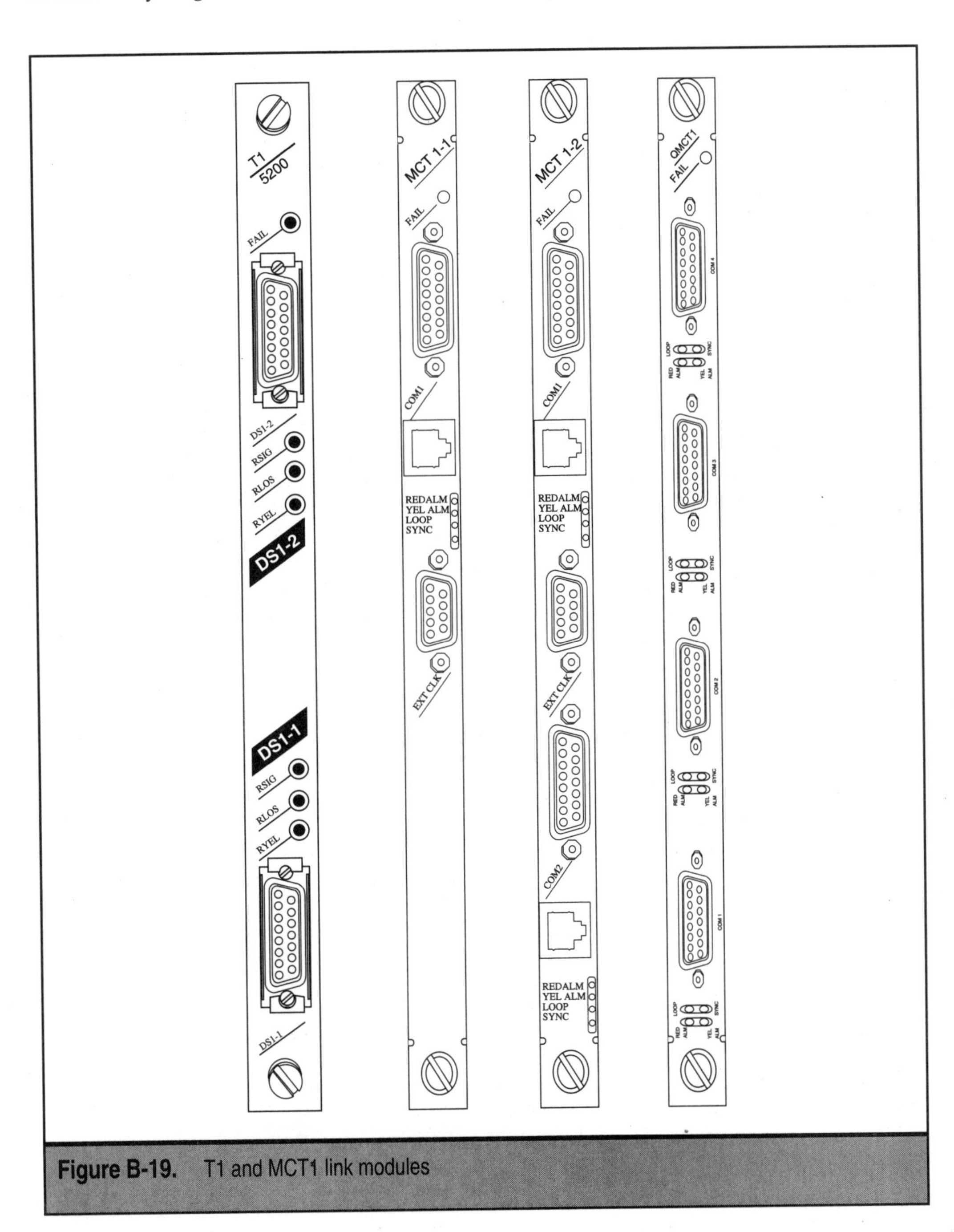

Figure B-19. T1 and MCT1 link modules

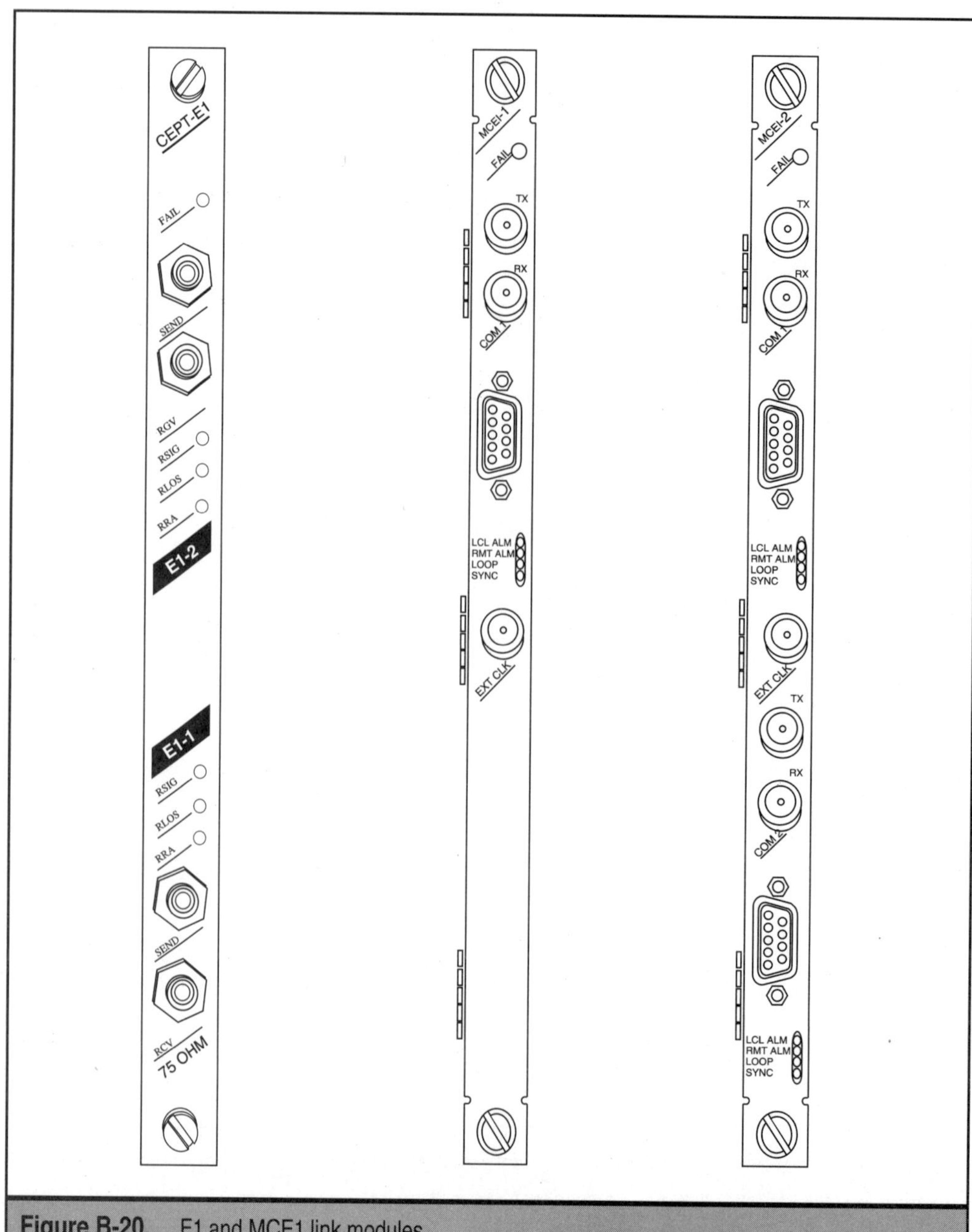

Figure B-20. E1 and MCE1 link modules

the E1 and MCE1 modules. The status indicators for all of these modules are summarized in Table B-21.

LED	Condition	Meaning
T1 and MCT1 Link Modules		
RLOS	On	A Receiver Loss (RLOS) signal was received. The interface received a signal with at least two framing errors in the past four frames.
RSIG	On	A Receiver Signal Loss (RSIG) signal was received. The interface received a signal of less than 0.5V or detected more than 32 consecutive zeros.
RYEL	On	A Received Yellow Alarm (RYEL) signal was received. Either RLOS is on at the remote end or RSIG/RLOS is on at the local end.
FAIL	On	Diagnostic testing is in progress, a diagnostic test failed, or the module is waiting for an automatic attempt to reinitialize diagnostics.
RED ALM	On	A loss of frame (LOF) or loss of signal (LOS) condition occurred.
YEL ALM	On	A yellow alarm was received, indicating that the remote end is in a red alarm state.
LOOP	On	The port is in line or payload loopback mode, or the port is isolated from the network due to an engaged relay on the port.
SYNC	On	The port is configured and is synchronized with the T1 network.
FAIL	On	Diagnostic testing is in progress, a diagnostic test failed, or the module is waiting for an automatic attempt to reinitialize diagnostics.

Table B-21. T1/MCT1 and E1/MCE1 Link Module LEDs

LED	Condition	Meaning
E1 and MCE1 Link Modules		
RLOS	On	A Receiver Loss (RLOS) signal was received. The interface received a signal with at least two framing errors in the past four frames.
RSIG	On	A Receiver Signal Loss (RSIG) signal was received. The interface received a signal of less than 0.5V or detected more than 32 consecutive zeros.
RRA	On	A Received Remote Alarm (RRA) signal was received. Either RLOS is on at the remote end or RSIG/RLOS is on at the local end.
FAIL	On	Diagnostic testing is in progress, a diagnostic test failed, or the module is waiting for an automatic attempt to reinitialize diagnostics.
LOC ALM	On	A loss of frame (LOF) or loss of signal (LOS) condition occurred.
RMT ALM	On	A remote local (LOC) alarm was received.
LOOP	On	The port is in loopback mode.
SYNC	On	The port is configured and is synchronized with the E1 network.

Table B-21. T1/MCT1 and E1/MCE1 Link Module LEDs *(continued)*

APPENDIX C

Console/Modem Cables and Modem Parameters for Dial Access

Out-of-band access is the reliable method of managing BayRS routers, especially for remote troubleshooting using analog modem lines. Table C-1 lists the console and modem cables used with various BayRS platforms and their connector types. The cables listed can be built manually, using the pinout and connector information in the Cable Guide, which is available on the book's web site.

The serial console port parameter, `modem-enabled` (`wfSerialPortModemEnable`), must be set to enabled when a modem is connected to the router's console interface. No additional configuration is required. If a regular console terminal is to be attached, the `modem-enabled` parameter for the `console` object must be set to disabled, and the appropriate console cable must be used. Table C-2 summarizes the configuration parameters for modems when connecting them to the console port for dialup access.

Platform	Connector Type	Console Cable	Modem Cable
BLN1, BLN2, BCN	25-pin D-sub female	7525	7850
AN1/ANH, ASN1	9-pin D-sub female	7526	7825
AN2, ARN, ASN2	9-pin D-sub male	7527 with null modem crossover	7527

Table C-1. Console and Modem Cables

Parameter	Value
CTS (Clear to Send)	Enabled. In combination with DCD, this allows both the router and modem to send traffic without requiring the RTS signal as was used in earlier half-duplex lines.
DTR (Data Terminal Ready)	Enabled. When the router's console port is operational, it asserts the DTR signal. If the DTR requirement is enabled on the modem, then it will answer calls when the router asserts DTR.
DCD (Data Carrier Detect)	On when a data carrier is detected. Transition of this signal to an off state indicates a carrier disconnect, which is sensed by the router.
DSR (Data Set Ready)	Enabled. This enables the modem to assert the DSR signal, which the router detects to provide console service.
RTS (Ready to Send)	Ignored. RTS was widely used in historic half-duplex protocols and is not used in the router (see CTS).
Auto Answer	Enabled to answer in one or two rings with DTR active.
Local Character Echo	Disabled.
Supervisory Functions	Disabled.
Baud Rate	9600 bps.
Data Bits	8.
Stop Bits	1.
Parity	None.
Synchronous/Asynchronous	Synchronous.

Table C-2. Modem Configuration Parameters for Dialup Access

APPENDIX D

Technician Interface Command List

TI COMMAND SUMMARY

The following are valid Technician Interface commands:

alias	arrayenv	atmarp	bcc	bconfig	boot
cd	clearlog	commit	compact	copy	cutenv
date	delete	diags	dinfo	dir	disable
echo	enable	enumenv	exec	export	firewall
format	fwputkey	get	getcfg	getenv	gosub
goto	help	history	if	ifconfig	instenv
ip	ip6	isdb	kexit	kget	kpassword
kseed	ksession	kset	ktranslate	let	list
loadmap	log	logout	mibget	more	mrinfo
mtrace	octetfmt	on	osi	osidata	partition
password	ping	pktdump	printf	prom	readexe
record	reset	restart	return	run	save
set	setenv	show	source	sprintf	stamp
stop	securelogin	snmpserver	string	system	telnet
tftp	type	unalias	unmount	unsetenv	verbose
wfsnmpkey	wfsnmpmode	wfsnmpseed	xmodem		

TI COMMAND SYNTAX

<table>
<tr><th>Command</th><th>Syntax</th></tr>
<tr><td>!</td><td>[<repeat count>]</td></tr>
<tr><td>alias</td><td>[<name> [["]<alias_value>["]]]</td></tr>
<tr><td>arrayenv</td><td>[-a] <variable name> "<string1>" ["<string2>" ...]</td></tr>
<tr><td>atmarp</td><td>table [<options>] <IP address></td></tr>
<tr><td>bcc</td><td></td></tr>
<tr><td>bconfig</td><td><image | config> <local | network> [<IP address> <pathname>]</td></tr>
<tr><td>bconfig</td><td>-d <image | config></td></tr>
<tr><td>boot</td><td>[<vol>:<image_name>|- <vol>:<config_name>|-]</td></tr>
<tr><td>cd</td><td>[<vol>:] [<directory>]</td></tr>
<tr><td>clear</td><td><sub_commands> <flags></td></tr>
</table>

```
Command     Syntax
clearlog    [<slot ID>]
commit
compact     <volume>:
copy        <vol>:<filename1> <vol>:<filename2>
cutenv      -s -d<delimiter> [-f<list>|-c<list>] <variable>
            "<text string>"
date        [<mm/dd/yy>] [<hh:mm:ss>] [<+|-><hh:mm>]
delete      <vol>:<filename>
diags       [<slot ID>]
dinfo
dir         <vol:>
disable     <entity> <option>
echo        [["]<string>["]]
enable      <entity> <option>
enumenv     <start #> [+<incr.> <variable name> [<variable
            name> ...]
exec        [-load|-unload] <command name>
export      {<variable name> ...}
firewall    <sub_command>
format      <volume>:
fwputkey    [<key_string> <ip_address>] | [clearkey]
get         {<obj_name>|<obj_id>}.{<attr_name>|<attr_id>|*}
            [.{<inst_id>|*}]
getcfg
getenv      [<variable name>]
gosub       :<label name>:
goto        :<label name>:
help        [-all|<command>]
history     [n]
if          "<string1>" [<=>|<!=>] "<string2>" [then];
            command(s) ;
ifconfig    [-s] [-d|-enable|-disable] <xcvr>|[-r4|-r16]
            <mau> [<IP addr> <mask> [<Next Hop>]]
```

<table>
<tr><th>Command</th><th>Syntax</th></tr>
<tr><td>ifconfig</td><td>[-s] [-fr [-annexd|-lmi|-annexa]] | [-int_clk] | [-d|-enable|-disable] <com> [<IP addr> <mask> [<Next Hop>]]</td></tr>
<tr><td>instenv</td><td><variable prefix> <mib-object name> [<mib-instance-pattern>]</td></tr>
<tr><td>ip</td><td><sub_command> <flags></td></tr>
<tr><td>ip6</td><td><sub_command> [<options>]</td></tr>
<tr><td>ipsec</td><td><sub_command> [<options>]</td></tr>
<tr><td>isdb</td><td><sub_command> [-s<slot>] [-c<connector>] [-p<port>] [<vol>:<filename>]</td></tr>
<tr><td>kexit</td><td></td></tr>
<tr><td>kget</td><td><sub_command></td></tr>
<tr><td>kpassword</td><td></td></tr>
<tr><td>kseed</td><td></td></tr>
<tr><td>ksession</td><td></td></tr>
<tr><td>kset</td><td><sub_command> [<flags>]</td></tr>
<tr><td>ktranslate</td><td><old_npk></td></tr>
<tr><td>let</td><td><var. name> = <expression></td></tr>
<tr><td>list</td><td>[[<instances> [<obj_name>]]]</td></tr>
<tr><td>loadmap</td><td>[<slot list> | all] [<filepath>]</td></tr>
<tr><td>log</td><td>[<vol>:<logfile>] [-d<date>] [-t<time>] [-e"<entity>"] [-f<severity>] [-s<slot ID>] [-p[<rate]] [-c<code #>]</td></tr>
<tr><td>log</td><td>[-x|-i] [-e"<entity>"] [-f<severity>] [-s<slot ID>]</td></tr>
<tr><td>log</td><td>-z [-s<slot ID>]</td></tr>
<tr><td>logout</td><td></td></tr>
<tr><td>mibget</td><td>[-n] [-p <pattern>] <object> <attribute var. array> <inst. id> <value var. array> <next_inst var.></td></tr>
<tr><td>more</td><td>[on | off] [# of lines per screen]</td></tr>
<tr><td>mrinfo</td><td>[-r retry_count] [-t timeout_count] multicast_router</td></tr>
</table>

<table>
<tr><th>Command</th><th>Syntax</th></tr>
<tr><td>mtrace</td><td>[-M] [-O] [-U] [-s] [-w wait] [-m max_hops] [-q nqueries] [-g gateway] [-e extrahops] [-S statint] [-t ttl] [-r resp_dest] [-i if_addr] source [receiver] [group]</td></tr>
<tr><td>octetfmt</td><td><variable name> <format option> <MIB object></td></tr>
<tr><td>on</td><td>ERROR :<label name>:</td></tr>
<tr><td>osi</td><td><subcommand> [<options>]</td></tr>
<tr><td>osidata</td><td>-s <SLOT> -t <lsp_l1 | lsp_L2 | path_L1 | path_L2 | adj_L1 | adj_L2 | adj_ES> -i <ID></td></tr>
<tr><td>partition</td><td>create|delete [<vol>:]</td></tr>
<tr><td>password</td><td>[<login-id>]</td></tr>
<tr><td>pause</td><td><seconds></td></tr>
<tr><td>permit</td><td>[-file [<vol>:]<filename>] | [<command> [<attribute>]] | [<mib object>]</td></tr>
<tr><td>ping</td><td><-IP| -IPV6| -IPX|-OSI|-VINES|-AT|-APPN> <hostname|address> [-t<timeout>] [-r<repeat count>] [-s<size>] [-p] [-a<address>] [-m<mode_name>] [-iifindex] [-v] [-n]</td></tr>
<tr><td>pktdump</td><td><linenumber> [-s<start>] [-c<count>]</td></tr>
<tr><td>printf</td><td><format string> <p1> <p2> . . . <pN></td></tr>
<tr><td>prom</td><td>[-v|-w] <vol>:<ROM Update File> <slot ID> [<slot ID> . . .]</td></tr>
<tr><td>readexe</td><td><vol>:<filename></td></tr>
<tr><td>record</td><td>open [-fileonly] [-pause] <vol>:<filename>
pause [on|off]
close</td></tr>
<tr><td>reset</td><td>[<slot ID>]</td></tr>
<tr><td>restart</td><td>[<slot ID>]</td></tr>
<tr><td>return</td><td>:<label name>:</td></tr>
<tr><td>revoke</td><td><command> [<attribute>]</td></tr>
<tr><td>rsvp</td><td><sub_command></td></tr>
<tr><td>run</td><td><vol>:<filename> [<p1> [. . . <p9>]]</td></tr>
<tr><td>save</td><td>{config|aliases|perm} <vol>:<filename></td></tr>
</table>

<table>
<tr><th>Command</th><th>Syntax</th></tr>
<tr><td>save</td><td>log [<vol>:<logfile>] [-d<date>] [-t<time>] [-e"<entity>"] [-f<severity>] [-s<slot ID>]</td></tr>
<tr><td>securelogin</td><td></td></tr>
<tr><td>set</td><td>{<obj_name>|<obj_id>}.{<attr_name>|<attr_id>}.<inst_id> <value></td></tr>
<tr><td>setenv</td><td><variable name> "<text string>"</td></tr>
<tr><td>show</td><td><entity> <option></td></tr>
<tr><td>snmpserver</td><td>view [view-name] [oid-tree] [included | excluded | list | delete] community [community-name] view [view-name] [RO | RW | list | delete]</td></tr>
<tr><td>source</td><td>{aliases|env|perm} <vol>:<filename></td></tr>
<tr><td>sprintf</td><td><variable name> <format string> <p1> <p2> ... <pN></td></tr>
<tr><td>stamp</td><td></td></tr>
<tr><td>stop</td><td><slot ID></td></tr>
<tr><td>string</td><td>load|unload</td></tr>
<tr><td>system</td><td></td></tr>
<tr><td>tarp</td><td><sub_command> <flags></td></tr>
<tr><td>telnet</td><td>[-d] [-e escape_char] [hostname|address [port]]</td></tr>
<tr><td>tftp</td><td>{get|put} <name|address> <vol>:<file_spec> [<vol>:<file_spec>]</td></tr>
<tr><td>type</td><td>[-x] <vol>:<filename></td></tr>
<tr><td>unalias</td><td>{<alias name>|*}</td></tr>
<tr><td>unmount</td><td><volume>:</td></tr>
<tr><td>unsetenv</td><td>[<variable name> ...| [-l] [-g] *]</td></tr>
<tr><td>verbose</td><td>[on | off]</td></tr>
<tr><td>xmodem</td><td>rb|sb [ylwpn] filename</td></tr>
<tr><td>wfsnmpkey</td><td><key_string> [encryption_alg_id]</td></tr>
<tr><td>wfsnmpmode</td><td><proprietary(3) | trivial(1)></td></tr>
<tr><td>wfsnmpseed</td><td><community> <manager> [-|<val1>] [-|<val2>] [-|<val3>] [-|<val4>] [-|<val5>]</td></tr>
</table>

APPENDIX E

Event Log Entities

Chapter 2 covered the essentials of the event log. The event log is commonly used for observing router operation and analyzing symptoms during troubleshooting. Table E-1 lists the various event log entities, their meaning, and the corresponding entity code. The entity code is usually used in combination with the event code on NMS systems to understand events sent using SNMP traps.

Entity	Description	Code
ACE	Advanced Communications Engine Board	10
AHB	ATM Half Bridge	150
AOT	Polled Asynchronous over TCP/IP	138
APPLETALK	AppleTalk Protocol	36
APPN	Advanced Peer-to-Peer Networking Protocol	86
ARP	Address Resolution Protocol	19
ASR	Application Sensitive Routing	146
ASYNC	Asynchronous Protocol	87
ATM	Asynchronous Transfer Mode	78
ATMDXI	Asynchronous Transfer Mode/Data Exchange Interface Protocol	49
ATM_LE	Asynchronous Transfer Mode LAN Emulation Service	100
ATM_SIG	Asynchronous Transfer Mode Signaling Service	95
ATMINTF	Asynchronous Transfer Mode Interface Service	76
BGP	Border Gateway Protocol	52
BGP3	Border Gateway Protocol Version 3	53
BGP4	Border Gateway Protocol Version 4	72
BISYNC	Bisynchronous over TCP/IP (BOT) Protocol	105
BOD	Bandwidth on Demand: Dial Backup and Dial-on-Demand Services	54
BOOT	Router Boot Operation	22
BOOTP	Bootstrap Protocol	59

Table E-1. Logging Entities

Entity	Description	Code
BOT	Binary Synchronous Communication Transport Service	104
CDM	Configuration Data Manager	103
CONVSTR	Conversation Steering	134
COPSC	Common Open Policy Service Client	168
CRM	Circuit Resource Manager Service	91
CSMACD	Carrier Sense Multiple Access/Carrier Detect (Ethernet) Driver	9
DCMMW	Data Collection Module Middleware	96
DECNET	DECnet IV Protocol	4
DHCP	Dynamic Host Configuration Protocol Service	161
DIFFSERV	Differentiated Services	167
DLS	Data Link Switching (DLSw) Protocol	50
DMAP	Direct Memory Access Processor	39
DNS	Domain Name Service	117
DOS	Disk Operating System	28
DP	Data Path Service	6
DSUCSU	Digital Service Unit/Channel Service Unit	111
DS1E1	Multichannel T1/E1	63
DVMRP	Distance Vector Multicast Routing Protocol	88
DVS	Dial VPN Service	159
E1	E1 Driver	35
EGP	Exterior Gateway Protocol	46
ENTITY	Entity MIB Subsystem	135
FDDI	Fiber Distributed Data Interface Driver	8
FLOP	Floppy Disk Controller	32
FNTS_ATM	Fujitsu Network Transmission Systems ATM	116
FR	Frame Relay Protocol	25
FRPT	Frame Relay PVC Pass-through	115
FRSW	Frame Relay Switch	65
FR_MLINK	Frame Relay Multilink	169

Table E-1. Logging Entities *(continued)*

Entity	Description	Code
FR_SVC	Frame Relay Switched Virtual Circuits	137
FR_SVC_API	FR SVC Application Programming Interface	147
FS	File System	64
FTP	File Transfer Protocol	88
GAME	Gate Access Management Entity	5
GFS	Generic File System	33
GNS	Generic Number Service	127
GRE	Generic Route Encapsulation	114
HELLO	Proprietary Hello Discovery	129
HSSI	High-Speed Serial Interface Driver	27
HTTP	Hypertext Transfer Protocol	145
HUB	Access Node Repeater	71
HWCOMP	Hardware Compression Driver	99
HWF	Hardware Filter	37
IFTAB	Interface Service Subsystem	128
IGMP	IGMP Service	83
IKE	Internet Key Exchange	170
IP	Internet Protocol	2
IP6	Internet Protocol Version 6	100
IP6_POLICY	Ipv6 Policy Services	141
IPEX	IP Encapsulation on X.25	93
IPSEC	IP Security	163
IPSEC_AUDIT	IPSec Auditing	164
IPX	Internet Packet Exchange Protocol	30
ISDB	Intelligent Serial Daughterboard	152
ISDN	ISDN Service	79
ISDN_BRI	ISDN BRI Driver Service	80
KEYMGR	Key Manager	118
L2TP	Layer 2 Tunneling Protocol	151

Table E-1. Logging Entities *(continued)*

Entity	Description	Code
LAPB	Link Access Procedure Balanced	73
LAPF	Link Access Procedure for Frame Relay	108
LB	Learning Bridge Protocol	1
LLC	Logical Link Control Protocol	48
LNM	LAN Network Manager Protocol	51
LOADER	Dynamic Loader	55
MCT1	Multichannel T1	67
MIB	Management Information Base	13
MIP	Mobile IP (for DVS)	113
MODEM	Modem Service	110
MODEMIF	Modem Interface	57
MODULE	Module Driver	21
MOSPF	Multicast OSPF	139
MPC	Multiprotocol Client (for MPLS)	158
MPLS_LDP	Multiprotocol Label Switching, Label Distribution Protocol	165
MPLS_MLM	MPLS Multiprotocol Label Manager	166
MPS	Multiprotocol Server	156
NAT	Network Address Translation	141
NBASE	NBASE	75
NBIP	NetBIOS over IP Service	77
NHRP	Next-Hop Resolution Protocol	148
NLSP	Network Link State Protocol	97
NML	Native-Mode LAN	81
NOV_SYNC	Nonvolatile Memory Synchronization	61
NTP	Network Time Protocol	109
NVFS	Nonvolatile File System	11
OSI	Open Systems Interconnection Protocol	38
OSPF	Open Shortest Path First	12
PCAP	Packet Capture	62

Table E-1. Logging Entities *(continued)*

Entity	Description	Code
PIM	Protocol Independent Multicast	106
PING	PING MIB	85
PORTMTX	Port Matrix	133
PPP	Point-to-Point Protocol	44
PPRI	Protocol Priority Service	90
PTY	Pseudo TTY Driver	41
QLLC	Qualified Link Level Control	122
RADIUS	Radius Service	115
RAESA	RMON Alarms and Events Subagent	112
RARP	Reverse Address Resolution Protocol	45
RCMDS	Remote Commands	116
RFWALL	Firewall Service	119
RIP6	Routing Information Protocol Version 6	120
RMONSTAT	Remote Monitoring Statistics	155
RREDUND	Router Redundancy Service	98
RSVP	Resource Reservation Protocol	139
RUIBOOT	Remote User Interface Boot	102
S5_CHASSIS	System 5000 Chassis	136
SDLC	Synchronous Data Link Control	74
SMDS	Switched Multimegabit Data Service	24
SMDS_SW	SMDS Switch	68
SNMP	Simple Network Management Protocol	3
SPAN	Spanning Tree Protocol	16
SR	Source Routing Bridge Protocol	29
ST2	ST2 Service	94
STA	Statistical Thresholds and Alarms	56
STAC	STAC (Hi/Fn) Compression	143
STATS	Statistical Data Collection	92
SVCSEC	ASR/NHRP Protocol Behavior for SVCs	149
SWSERV	Switched Access Service	58

Table E-1. Logging Entities *(continued)*

Entity	Description	Code
SYNC	Synchronous Driver	20
SYS	System Record	60
SYSLOG	System Log	89
T1	T1 Driver	34
TAG1Q	IEEE 802.1Q Tagging	157
TBL	Table Subsystem	14
TCP	Transmission Control Protocol	47
TELNET	Telnet Server Protocol	40
TF	Traffic Filter	15
TFTP	Trivial File Transfer Protocol	7
TI	Bay Networks Technician Interface	0
TI_RUI	Technician Interface/Remote User Interface (Site Manager)	18
TNC	Telnet Client Protocol	70
TOKEN_RING	Token Ring Driver	26
TOPOLOGY	Topology Discovery	130
TTY	Teletypewriter	17
VCCT	DLSw/APPN Boundary Function	154
VINES	Virtual Networking System Protocol	23
VLAN	Virtual LAN	126
VRRP	Virtual Router Redundancy Protocol	160
VST	Virtual Spanning Tree for VLANs	131
WEP	WAN Encryption Protocol	117
WCP	Data Compression	84
X25	X.25 Protocol	43
X25PAD	X.25 Packet Assembler and Disassembler	153
XCTL	Switch Control Driver	124
XMODEM	Xmodem/Ymodem Protocol	42
XNS	Xerox Networking Systems Protocol	31

Table E-1. Logging Entities *(continued)*

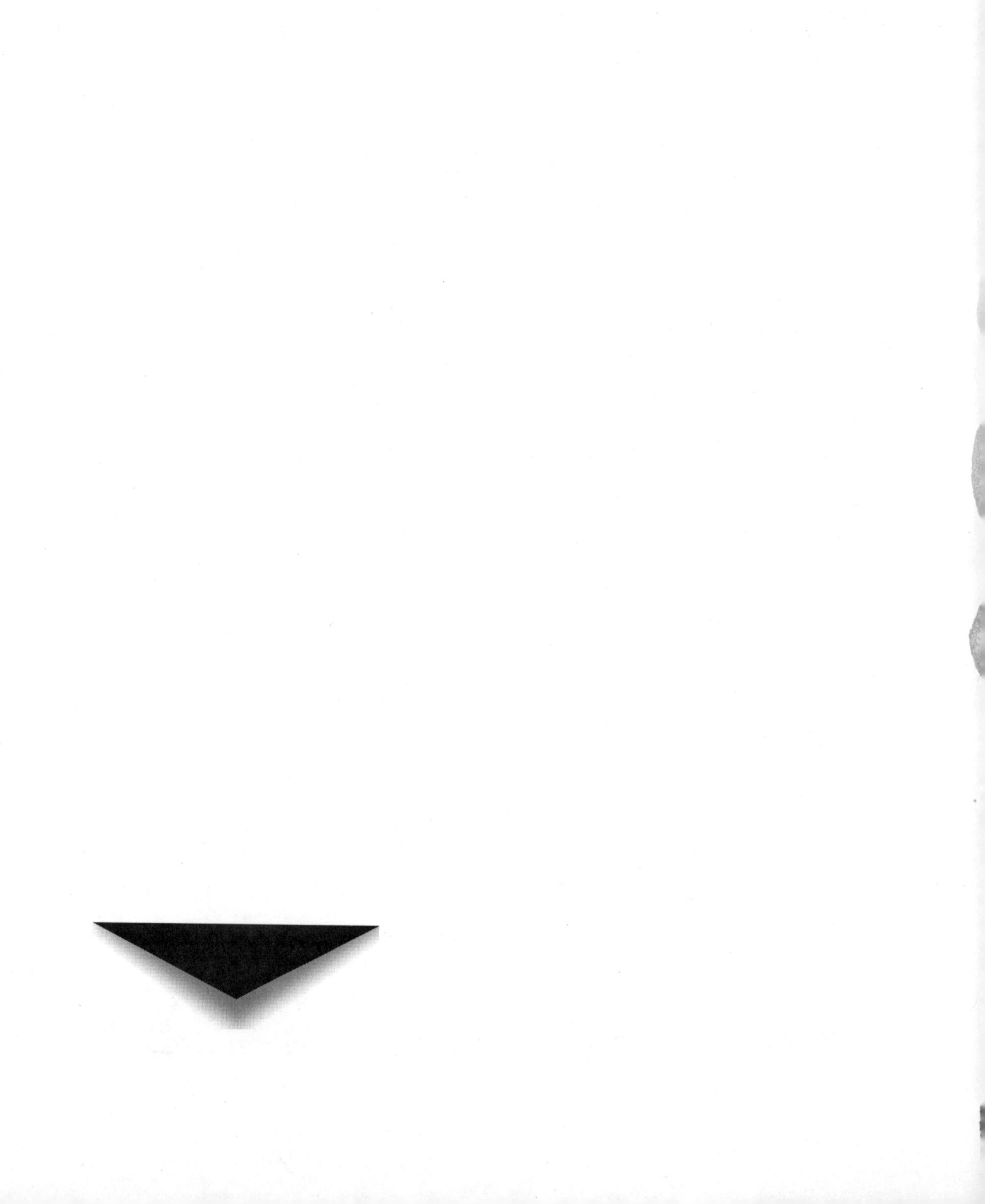

APPENDIX F

TI and BCC show Commands

BAY COMMAND CONSOLE show COMMANDS

The **show** command in BCC currently supports the following commands. Each command has various arguments and options, which can be seen by pressing the help (?) key or simply pressing ENTER after typing `show` *<subcommand>*.

```
access
atm
bgp
bri
bridge
ces
classical-ip
config
console
dial
dlsw
dns
dsucsu
dsx3
dsync
dvmrp
ethernet
fddi
firewall
frame-relay
ft1e1
ftp
gre
hardware
hifn
hssi
http
hwcomp
igmp
instance
interface
ip
ipx
isdn-switch
lane
llc2
mct1e1
modem
mpoa
mtm
nat
nhrp
ntp
ospf
ppp
process
radius
rarp-server
sdlc
serial
snmp
sonet
span
srb
switched-h110
syslog
system
tcp
tdm
telnet
tftp
token-ring
translation-bridge
voice
vrrp
wcp
```

TECHNICIAN INTERFACE show COMMANDS

The TI **show** commands are supported using individual batch scripts. Batch scripts are unique to each software version because they use the corresponding MIB objects. The scripts are distributed with the software suite and Site Manager, or can be separately downloaded from the Assurance Online web site. A link to the Assurance Online web site is available on the book web site, http://www.mahalingam.com/.

The syntax for the TI **show** commands are identical to those in BCC; using a question mark (?) or pressing ENTER without any subcommands will display all available syntax options. If the **show** command generates an error, it usually indicates that the batch script version is incompatible with the software version. Use the **show *command* version** (**show gre version**, for example) and the **stamp** command to compare the batch script version and the software version respectively. If necessary, update the script files to the appropriate version corresponding to the BayRS software.

ahb	circuits	gre	mpoa	rip6	sws
aot	console	hardware	mpr	rptr	sync
appn	csmacd	hifn	nat	rredund	syslog
asr	date	hssi	nbip	rsc	system
at	dcmmw	http	nhrp	rsvp	t1
atm	decnet	hwcomp	nml	sdlc	tag1q
atmarp	dls	igmp	ntp	smds	tcp
atmdxi	drivers	ip	osi	snmp	telnet
atmsig	ds1e1	ip6	ospf	sonet	tftp
aurp	dsx3	ipx	packet	span	token_ring
autoneg	dvmrp	iredund	ping	sr	vines
bgp	e1	isdn	ppp	srspan	wcp
bisync	egp	l2tp	process	srsvp	wep
bootp	fddi	lane	protocols	st2	x25
bot	fr	lnm	protopri	sta	xb
bridge	frsw	memory	radius	state	xns
buffers	ftp	mospf	rarp		

APPENDIX G

BayRS Applications and Drivers: Function and Dependencies

This appendix lists applications and their functions and common dependencies. A dependency is an application that is required for another application to function. Although most common dependencies are listed in Table G-1, a listing of all potential dependencies is beyond the scope of this book.

When using Image Builder to edit images, remove drivers only as a last resort. The dependencies of applications are straightforward—for example, ATM LAN Emulation (LANE) cannot function without ATM Signaling software. However, the dependencies for chipset and link module drivers are more complex and are too numerous to list easily in this appendix. The absence of coprocessor images, link modules, and chipset drivers is usually identified by the Not Present state in the corresponding media interface.

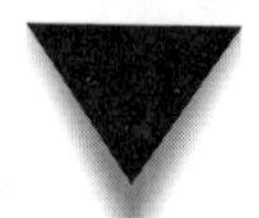

NOTE: The exe, ppc, and ppb extensions have been omitted. Certain platforms that do not support a particular application or driver will not contain the application in the corporate suite.

Application	Function and Common Dependencies
ahb	ATM half bridge (depends on atm)
aot	Asynchronous transport over TCP/IP (depends on tcp, ip)
appn_cp	Advanced Peer-to-Peer Networking (APPN) protocol control point
appn_ls	APPN link station
arp	Address Resolution Protocol (depends on ip)
asn_mod	ASN module driver (mandatory)
at	AppleTalk Phase 2
atm	Asynchronous Transfer Mode (ATM)
atm_dxi	ATM Data Exchange Interface
atm_le	ATM LANE emulation (depends on atm, atmsig)
atmalc	ATM link module line driver
atmalcop	ATM link module coprocessor driver (FRE2)
atmc	ATM link module coprocessor driver (ARE)
amc5000	ATM link module coprocessor driver (578x)

Table G-1. BayRS Applications and Drivers

Application	Function and Common Dependencies
atmc_mod	PowerPC-based ATM link module driver
atmsig	ATM signaling (depends on atm)
atmz_are	ATMizer driver
atmz_drv	ATMizer microcode
bcc	Bay Command Console
bgp	Border Gateway Protocol (depends on tcp, ip)
bot	Bisynchronous transport over TCP/IP (depends on tcp, ip)
capi	Cryptographic Application Programming Interface
copsc	Common Open Policy Services (depends on ip)
crm	Circuit Resources Manager
dcmmw	Data collection module middleware
de100	Dual-Ethernet 100 link module driver
debug	GAME kernel debugging application (mandatory for debugging purposes)
dhcp	Dynamic Host Discovery Protocol server (depends on ip)
diffserv	Differentiated services (depends on copsc, if necessary)
dls	Data link switching (depends on tcp, ip)
dns	Domain Name Service (depends on ip)
drs	DECnet routing services
ds2180	MK2180 Dual Sync interface driver
ds2181	MK2181 Dual Sync interface driver
dsde2	Dual-sync, dual-Ethernet link module driver
dst	Dual-sync, token link module driver
dtok	Dual-token link module driver
dvmrp	Distance Vector Multicast Routing Protocol (depends on ip, igmp)
dvs	Dial-VPN service
e1	E1 link module driver
egp	Exterior Gateway Protocol (depends on ip)
enet2	Ethernet2 link module driver

Table G-1. BayRS Applications and Drivers *(continued)*

Application	Function and Common Dependencies
fddi	Fiber Distributed Data Interface link module driver (depends on fsi)
fmpb	Quad Token Ring link module driver
fr	Frame relay (FR)
fr_svc	FR switched virtual circuits (depends on fr)
frpt	FR PVC pass-through (depends on fr)
fsi	FDDI station management (SMT) and chipset driver
ftp	File Transfer Protocol server (depends on ip, tcp)
gig_enet	Gigabit Ethernet link module driver
hdlc	High-level data link control
hdwancop	High-density SYNC/BRI I/O module coprocessor driver
hdwanlm	Octal sync link module driver
hfsi	HSSI SMT and chipset driver
hifn	Hi/Fn compression service
hilance	LANCE chipset driver
hssi	High-Speed Serial Interface (HSSI) link module driver (depends on hfsi)
http	Hypertext Transfer Protocol (HTTP) server (depends on tcp, ip)
hwcomp	Hardware compression software
hwf	Hardware filter driver
igmp	Internet Group Membership Protocol (depends on ip)
ike	Internet Key Exchange protocol
ilacc	ILACC chipset driver
ip	Internet Protocol (IP), version 4
ip6	Internet Protocol, version 6
ipex	IP encapsulation of X.25 (depends on ip, x25)
ipsec	IP security (depends on ip, capi)
ipx	Internetwork Packet Exchange
isac	ISAC, ISDN D-channel driver

Table G-1. BayRS Applications and Drivers *(continued)*

Application	Function and Common Dependencies
isdb	Intelligent serial daughterboard
isdn	Integrated Services Digital Network (depends on isac)
krnl_an	Access Node kernel
krnl_arn	Advanced Remote Node kernel
krnl_asn	Access Stack Node kernel
krnl_bn.exe	FRE1/FRE2 kernel
krnl_bn.ppc	ARE kernel
krnl_fre4bn	FRE4 kernel
krnl_s5000.exe	5380/5580 kernel
krnl_s5000.ppc	5780/5782 kernel
l2tp	Layer 2 Tunneling Protocol
lance	LANCE chipset driver
lapb	Link Access Procedure Balanced
llc	Logical link control
lm	Line Manager driver
lnm	IBM LAN Network Manager
mct1cop	MCT1 link module coprocessor driver
mct1e1	DS1E1 driver for the MCT1/MCE1 link modules
mospf	Multicast OSPF (depends on ip)
mpls_ldp	Multiprotocol Label Switching, Label Distribution Protocol (depends on atm)
mpls_mlm	MPLS Multiprotocol Label Manager (depends on atm)
mps	Multiprotocol server (for MPLS) (depends on atm)
munich	MCT1 link module driver
nat	Network address translation (depends on ip)
nbase	NBASE
nbip	NETBIOS over IP (depends on ip)
nept7wan	ARN expansion module driver
netboot	Network boot service
nhrp	Next-Hop Resolution Protocol

Table G-1. BayRS Applications and Drivers *(continued)*

Application	Function and Common Dependencies
nsc_100m	NSC 100M Ethernet chipset driver
ntp	Network Time Protocol (depends on ip)
osi	Open Systems Interconnect
ospf2	Open Shortest Path First, version 2
pcap	Packet capture
pim	Protocol-independent multicast
ping	Ping service and MIB
ppp	Point-to-Point Protocol
qenet	Quad Ethernet link module driver
qsync	Quad Sync link module driver
quicsync	Sync interface driver on QUICC platforms
radius	Remote Authentication Dial-In User Service (depends on ip)
raesa	RMON alarms and events subagent
rarp	Reverse Address Resolution Protocol
rip6	Routing Information Protocol, version 6
rmonstat	RMON statistics
rredund	Router Redundancy Protocol
rsvp	RSVP (IntServ) protocol
run	Embedded and batch script execution application
sdlc	Synchronous data link control
seeq_100m	SEEQ 100M Ethernet chipset driver
sgig	Gigabit Ethernet chipset driver
sh_csmac	Show csmacd embedded script
sh_fr	Show frame relay embedded script
sh_ip	Show IP embedded script
sh_snmp	Show SNMP embedded script
sh_sync	Show sync embedded script
sh_tcp	Show TCP embedded script
sh_tftp	Show TFTP embedded script

Table G-1. BayRS Applications and Drivers *(continued)*

Application	Function and Common Dependencies
snmp	Simple Network Management Protocol (depends on ip)
sqe100	Signal quality error tester for 100 Mbps Ethernet
sysl	Syslog service (depends on ip)
t1	T1 link module driver
tag1q	802.1Q tag protocol
tcp	Transmission Control Protocol (depends on ip)
tftp	Trivial File Transfer Protocol (depends on ip)
tms380	TMS380 Token Ring chipset driver
tn	Telnet server (depends on ip)
tnc	Telnet client (depends on ip)
treectrl	Web server applet service (depends on http, ip)
vcct	DLSw/APPN boundary function application
vines	Banyan Vines protocol
vrrp	Virtual Router Redundancy Protocol (depends on ip, ipx)
wan	Common wide area service interface
wcp	WAN Compression Protocol
wep	WAN Encryption Protocol
x25	X.25 protocol (depends on lapb)
xm	Xmodem file transfer protocol
xns	Xerox networking service

Table G-1. BayRS Applications and Drivers *(continued)*

APPENDIX H

Glossary Reference

A glossary is essential for every technical book, and I initially set out to include a comprehensive one. However, during my initial work on a glossary, I realized that there must be many available references to networking terms already. So instead I have attempted to introduce each concept with a corresponding definition within the main text. Acronyms are expanded fully where they initially occur, and acronyms and product terms specific to BayRS and Nortel Networks implementations are defined in the appropriate context.

For additional information on networking terminology, I strongly recommend the following web sites. This is not an exhaustive list, however; if you already have a favorite glossary site or reference book, consult that as well.

- **Bay Networks Guide to Networking Terms**
  http://support.baynetworks.com/library/tpubs/terms/
  (Also available as a printed publication: 345-1108B.)
- **Webopedia: Online Computer Dictionary for Internet Terms and Technical Support**
  http://www.webopedia.com/
- **CNET Glossary**
  http://coverage.cnet.com/Resources/Info/Glossary/

Index

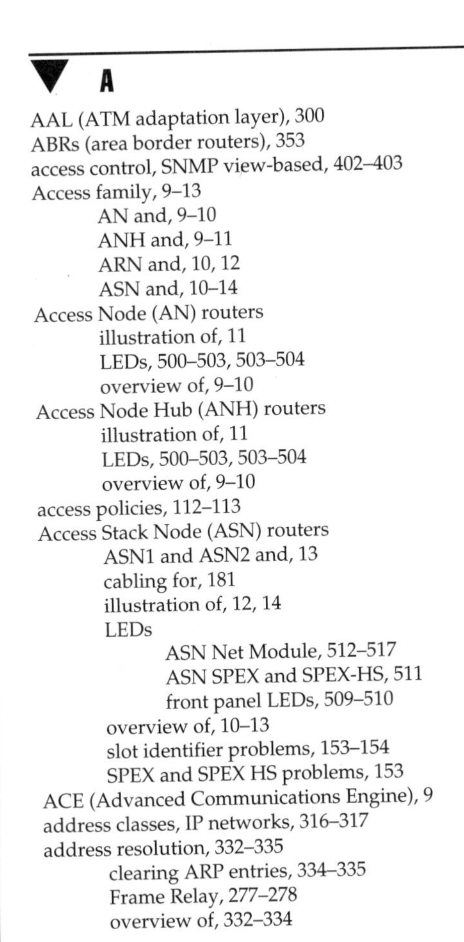

A

B

C

D

E

F

K

L

M

N

O

P

Q

R

T

U

V

W

X

Y